IMMOBILIZED ENZYMES
PREPARATION AND ENGINEERING

IMMOBILIZED ENZYMES

PREPARATION AND ENGINEERING

Recent Advances

J.C. Johnson

NOYES DATA CORPORATION
Park Ridge, New Jersey, U.S.A.
1979

Library of Congress Catalog Card Number: 79-13616
ISBN: 0-8155-0760-7
Printed in the United States

Published in the United States of America by
Noyes Data Corporation
Noyes Building, Park Ridge, New Jersey 07656

Library of Congress Cataloging in Publication Data

Johnson, Jeanne Colbert, 1920-
Immobilized enzymes, preparation and engineering.

(Chemical technology review ; no. 133)
Includes index.
1. Immobilized enzymes--Patents. I. Title.
II. Series.
TP248.E5148 661'.1 79-13616
ISBN 0-8155-0760-7

FOREWORD

The detailed, descriptive information in this book is based on U.S. patents, issued since July 1974, that deal with immobilized enzymes. This title contains new developments since our previous title *Immobilized Enzymes, Preparation and Engineering Techniques,* published in 1974.

This book serves a double purpose in that it supplies detailed technical information and can be used as a guide to the U.S. patent literature in this field. By indicating all the information that is significant, and eliminating legal jargon and juristic phraseology, this book presents an advanced, commercially oriented review of recent developments in immobilized enzymes.

The U.S. patent literature is the largest and most comprehensive collection of technical information in the world. There is more practical, commercial, timely process information assembled here than is available from any other source. The technical information obtained from a patent is extremely reliable and comprehensive; sufficient information must be included to avoid rejection for "insufficient disclosure." These patents include practically all of those issued on the subject in the United States during the period under review; there has been no bias in the selection of patents for inclusion.

The patent literature covers a substantial amount of information not available in the journal literature. The patent literature is a prime source of basic commercially useful information. This information is overlooked by those who rely primarily on the periodical journal literature. It is realized that there is a lag between a patent application on a new process development and the granting of a patent, but it is felt that this may roughly parallel or even anticipate the lag in putting that development into commercial practice.

Many of these patents are being utilized commercially. Whether used or not, they offer opportunities for technological transfer. Also, a major purpose of this book is to describe the number of technical possibilities available, which may open up profitable areas of research and development. The information contained in this book will allow you to establish a sound background before launching into research in this field.

Advanced composition and production methods developed by Noyes Data are employed to bring these durably bound books to you in a minimum of time. Special techniques are used to close the gap between "manuscript" and "completed book." Industrial technology is progressing so rapidly that time-honored, conventional typesetting, binding and shipping methods are no longer suitable. We have by-passed the delays in the conventional book publishing cycle and provide the user with an effective and convenient means of reviewing up-to-date information in depth.

The table of contents is organized in such a way as to serve as a subject index. Other indexes by company, inventor and patent number help in providing easy access to the information contained in this book.

15 Reasons Why the U.S. Patent Office Literature Is Important to You –

1. The U.S. patent literature is the largest and most comprehensive collection of technical information in the world. There is more practical commercial process information assembled here than is available from any other source.

2. The technical information obtained from the patent literature is extremely comprehensive; sufficient information must be included to avoid rejection for "insufficient disclosure."

3. The patent literature is a prime source of basic commercially utilizable information. This information is overlooked by those who rely primarily on the periodical journal literature.

4. An important feature of the patent literature is that it can serve to avoid duplication of research and development.

5. Patents, unlike periodical literature, are bound by definition to contain new information, data and ideas.

6. It can serve as a source of new ideas in a different but related field, and may be outside the patent protection offered the original invention.

7. Since claims are narrowly defined, much valuable information is included that may be outside the legal protection afforded by the claims.

8. Patents discuss the difficulties associated with previous research, development or production techniques, and offer a specific method of overcoming problems. This gives clues to current process information that has not been published in periodicals or books.

9. Can aid in process design by providing a selection of alternate techniques. A powerful research and engineering tool.

10. Obtain licenses — many U.S. chemical patents have not been developed commercially.

11. Patents provide an excellent starting point for the next investigator.

12. Frequently, innovations derived from research are first disclosed in the patent literature, prior to coverage in the periodical literature.

13. Patents offer a most valuable method of keeping abreast of latest technologies, serving an individual's own "current awareness" program.

14. Copies of U.S. patents are easily obtained from the U.S. Patent Office at 50¢ a copy.

15. It is a creative source of ideas for those with imagination.

CONTENTS AND SUBJECT INDEX

INTRODUCTION

Because of the catalytic specificity of enzymes, considerable attention has been directed toward finding methods of using them in both laboratory and industrial applications. Enzymes are commonly water-soluble, and for that reason, many enzymes are uneconomical to use in large-scale batch-type operations since the enzymes can generally be used only one time in the absence of rather costly enzyme recovery and purification steps. In recent years, however, techniques have been devised to fix active enzymes on mostly water-insoluble materials that can be readily removed from a reaction, thus permitting reuse of the insolubilized or immobilized enzyme.

Prior methods for immobilizing enzymes have usually been classified as: (1) physical adsorption; (2) ionic bonding as on an ion exchange resin; (3) physical entrapment such as inclusion in a microporous gel or fiber or by microencapsulation; (4) crosslinking of the enzymes; and (5) covalent bonding of the enzyme to a support. With increasing sophistication, the technology now combines one or more of these techniques to produce the immobilized enzymes, and the ligands or coupling groups used to covalently bind the enzyme to a support are designed to place the enzyme at optimum distance from the support. Thus, the immobilization methods used as chapter headings in this review must not be regarded as rigid classifications.

This review covers approximately 200 processes disclosed in 211 patents issued since July 1974. Covalent bonding of the enzyme to a support is the means of immobilization in the majority of the processes. Emphasis is also placed on the production of enzyme products that can be used in continuous processes; that is, they permit continuous flow of substrate over the immobilized enzyme in a column or reactor without disintegration of the support with plugging of the column.

Industrial applications of the immobilized enzymes are found in the production of sugars from starch and in the production and analysis of various pharmaceutical and drug-related products. Increasing attention is being shown to the use of coupled enzymes in affinity chromatography and immunoassay procedures.

Definitions of abbreviations:

DE Dextrose equivalent refers to the total reducing sugar content of the dissolved solids in starch hydrolysates expressed as percent dextrose.

GU Glucose activity unit is the amount of enzyme which catalyzed the production of one gram of dextrose per hour at 60°C at pH 4.5.

IU The amount of pullulanase which catalyzed the liberation of 1 μmol of maltotriose per minute from a 0.5% of pullulan at pH 5.0 at 45°C.

IGIU International glucose isomerase unit is that amount of glucose isomerase needed to convert 1 μmol of glucose to fructose per minute at 60°C at pH 6.85.

The abbreviations used to designate enzyme depositories and their catalog numbers are:

ATCC American Type Culture Collection, Rockville, Maryland.

NRRL Northern Regional Research Laboratory, U.S. Department of Agriculture, Peoria, Illinois.

IFO Institute of Fermentation, Osaka, Japan.

IMMOBILIZATION BY ADSORPTION

Adsorption of enzymes to water-insoluble supports, whether organic or inorganic, has been the simplest insolubilization technique. It has been attractive because it requires merely exposing the enzyme in solution to the support material. The ease of adsorption, however, is offset by the corresponding ease of desorption.

USE OF POROUS ALUMINA-MAGNESIUM OXIDE SUPPORT MATERIALS

Porous Alumina with Specific Amounts of Magnesium Oxide for Supporting Isomerase

D.L. Eaton and R.A. Messing; U.S. Patents 3,982,997; September 28, 1976 and 3,992,329; November 16, 1976; both assigned to Corning Glass Works have found that a very high enzyme loading per gram of carrier for an immobilized glucose isomerase composite can be achieved by incorporating a critical amount of MgO in a porous Al_2O_3 enzyme support material. Specifically, it has been found that a very efficient composite can be prepared by adsorbing glucose isomerase enzymes to the internal surfaces of a high surface area (at least 5 m^2/g), porous, inorganic carrier having an average pore diameter of 100 to 1000 A and comprising by weight, between 0.84 and 3.80% MgO and Al_2O_3. Preferably, the porous MgO-Al_2O_3 enzyme carrier is in particulate form having an average particle size (U.S. Standard Sieve), preferably between 30 and 45 mesh with the average pore diameter being between 150 and 250 A.

The general method for preparing the MgO-Al_2O_3 porous carriers involves starting with alumina particles having an average particle size of 300±200 A. These particles are then mixed with a solution consisting of varying amounts of magnesium ions to form a slurry which is mixed well. The magnesium ions can be added from a variety of available sources such as $MgCl_2 \cdot 6H_2O$ or $Mg(OH)_2$. The slurry is then gently dried to remove water. This drying step tends to shrink the individual particles together such that the ultimate dried product is porous and has an average pore size approximating the average particle size of the starting materials. The gentle drying can be accomplished via a number of methods such as simple air drying, drying with gentle heat (~100°C), spray drying the slurry, and like methods. The main requirement in the drying step is that it be gentle enough to preserve the skeletal pore structure formed as the particles shrink together.

After drying, the porous body is strengthened by firing it to a temperature below the sintering point; e.g., fired to 400° to 600°C for 1 to 16 hours. The resulting product can then be comminuted, if necessary, and the individual porous particles sorted according to desired mesh size range which is preferably between 30 and 45 mesh, U.S. Standard Sieve. Alternatively, the slurry can be spray dried to the desired particle size range prior to firing.

After the porous particles are prepared, they can be used to immobilize the glucose isomerase molecules by adsorption to internal surfaces of the porous bodies. By using porous particles having an average pore diameter of less than 1000 A, in particle sizes of 30 to 45 mesh, a very high surface area per gram (e.g., greater than about 5 m^2/g) is assured for maximum enzyme loading. It was found that enzyme loading is increased significantly if, prior to adsorption of the enzymes, the carriers are reacted with an aqueous citrate solution (e.g., 0.1 M citric acid or sodium citrate solution, pH 7.0).

Examples 1 through 10: Ten sample carriers were made consisting of alumina and from 0 to 28.6% MgO (by weight). By using the amounts of each ingredient as shown in Table 1, the sample carriers were made by first adding to the distilled or deionized water sufficient glacial acetic acid to bring the solution to 0.1 M. Then a slurry was formed by adding the alumina with vigorous stirring. The stirring was continued until a smooth, creamy mixture was obtained, approximately 15 to 30 minutes. The pH was then adjusted to 2.0 to 3.0 and the magnesium compound added to the slurry either as a liquid, or as a solid. This mixture was then blended at a high speed for an additional 15 to 30 minutes.

The resulting blend was then formed into particles or spheres by either slip casting or spray drying. The slip cast material was then broken and sorted according to particle size by conventional means. Both the slip cast and spray-dried material were fired at 600°C for 16 hours. Prior to adsorption of the enzymes, the effect of MgO additions on the pH of the carriers was determined by mixing 1 gram of each carrier with 9 grams of distilled water for 15 minutes to achieve an equilibrium, and then measuring the pH of the mixture with a conventional pH meter.

Table 1

	Ingredients				Final Product		
Example	H_2O	Al_2O_3	$MgCl_2 \cdot 6H_2O$	$Mg(OH_2)$	Carrier Shape (30-45 mesh)	pH	MgO (% by wt)
1*	489	400	0	0	spheres	4.4	0
2**	100	100	0	0	particles	4.4	0
3**	100	99.16	4.3	0	particles	7.0	0.84
4*	244	200	14.2	0	spheres	7.5	1.4
5*	244	200	22.3	0	spheres	8.2	2.2
6*	244	200	29.4	0	spheres	8.1	2.9
7*	244	200	38.6	0	spheres	8.4	3.8
8**	100	100	0	10.3	particles	8.8	6.65

(continued)

Table 1: (continued)

	. . . . Ingredients				. . . . . . Final Product.		
Example	H_2O	Al_2O_3	$MgCl_2 \cdot 6H_2O$	$Mg(OH_2)$	Carrier Shape (30-45 mesh)	pH	MgO (% by wt)
9**	100	100	69.2	0	particles	8.9	12.0
10**	100	100	0	58.1	particles	9.3	28.6

*Pounds **Grams

Note: All porous bodies had an average pore diameter within range of 150-250 A.

Each of the above carrier samples was used to immobilize glucose isomerase by reacting the enzyme preparation with each carrier to adsorb the enzyme onto the internal surfaces of the pores. About 15 ml of the enzyme preparation was used for each 15 grams of carrier. Prior to the actual adsorption step, each carrier sample was initially washed with distilled water by fluidizing the carrier sample in a column. The washed carriers were then reacted with a 0.1 M citrate solution in a shaking bath for 1 hour. Then, the enzyme preparation was added and the adsorption was allowed to proceed for 24 hours with shaking to facilitate the adsorption process.

The final product was then rinsed with distilled water and the individual samples were assayed with the following results where E_0 represents the enzymatic activity per gram and $E_{0\ (equiv)}$ represents a normalized value associated with an increased loading observed using the irregularly shaped particles (cf spheres).

Table 2

Ex. No.	MgO (%)	pH*	Shape	E_0	E_0 (equiv)
1	0	4.4	Spheres	203	203
2	0	4.4	Particles	387	(200)
3	0.84	7.0	Particles	805	(600)
4	1.4	7.5	Spheres	650	(650)
5	2.2	8.2	Spheres	898	898
6	2.9	8.1	Spheres	899	899
7	3.8	8.4	Spheres	909	909
8	6.65	8.8	Particles	916	(720)
9	12.0	8.9	Particles	768	(600)
10	28.6	9.3	Particles	200	(50)

*Of carrier.

To be commercially feasible, the immobilized glucose isomerase should have an enzymatic loading of at least 500 units of activity per gram of carrier under a continuous isomerization (flow-through) process. As can be seen from Table 2, this loading level is obtained when the percent by weight MgO in the $MgO\text{-}Al_2O_3$ porous carrier is 0.84 to 12.0% MgO with best results obtained when the carrier consists of 0.84 to 3.8% MgO. Although the exact mechanism(s) whereby the MgO content results in improved loading is not fully understood, it can be appreciated from the data in Table 2 that the carrier pH may play a role in determining loading amount since the desired minimum loading of at least 500 activity per gram occurs on carriers having a pH of 7.0 to 8.9. Hence, it is thought that the addition of MgO may not only serve to satisfy a portion of the enzymes Mg^{++} needs but also set carrier pH parameters which limit both higher and lower loadings.

Regenerating Isomerase Supports by Pyrolysis plus Citrate Washing

Highly porous $MgO-Al_2O_3$ support materials useful for the immobilization of glucose isomerase can be regenerated for reuse in the process disclosed by *L.R. Bialousz, E.R. Herritt, D.J. Lartigue and W.H. Pitcher, Jr.; U.S. Patent 3,965,035; June 22, 1976; assigned to Corning Glass Works.* The regeneration comprises pyrolysis under conditions sufficient to remove substantially all carbonaceous matter, followed by treatment with a neutralized citrate solution.

Preferred carrier materials consist of highly porous particles consisting of $MgO-Al_2O_3$ having incorporated between 0.84 and 12.0% by weight MgO, the particles having an average pore diameter, preferably between 150 and 250 A and an average particle size, preferably of 30 to 45 mesh, U.S. Standard Sieve. The preferred regeneration steps involve subjecting such particles to a temperature of 500° to 900°C in the presence of oxygen for a period of time sufficient to remove substantially all carbonaceous matter from the carrier.

Thereafter, the particles are allowed to cool and are then exposed to a neutralized aqueous solution of citrate ions long enough to remove contaminants such as metal ions which would minimize economical reuse of the carrier. Preferably the citrate solution consists of 0.1 molar citrate solution at a pH between 6.0 and 10, very preferably a sodium citrate solution at a pH of 7.0, and the pyrolyzed carrier is incubated with the citrate solution for at least 15 minutes, preferably at room temperature. The glucose isomerase solution used was an aqueous solution with a glucose isomerase activity of 2,700 IGIU/ml. The enzyme was derived from a *Streptomyces* sp. organism.

The composites were prepared by reacting 10 ml of the enzyme solution with 10 grams of porous carrier as follows: The carriers are initially washed with distilled water in a fluidizing column. The carrier is then placed in a flask to which 10 ml/g carrier of 0.05 M magnesium acetate is added and the flask is placed in a shaker bath for 1 hour.

The solution is then decanted and the enzyme is added and this mixture is allowed to react in the shaker bath for 24 hours to facilitate enzyme adsorption. The product is then rinsed with distilled water and the immobilized enzyme composite can be stored in water or as a wet cake until used.

In preparing the composites of the examples, approximately 10-gram (wet weight) quantities of each composite were prepared by the above methods. Nine separate composites were prepared with new carrier (unregenerated). Each new carrier was then spent by placing it in a plug flow-through column through which the glucose solution was flowed continuously under assay conditions for at least 30 days. Then, the indicated number of samples was regenerated by pyrolysis, and pyrolysis followed by citrate treatment, as indicated. The pyrolysis step involved heating the carriers to a temperature of 500° to 600°C for 1 hour in the presence of an oxygen source. The citrate solution treatment involved pumping 0.1 M citric acid solution neutralized to pH 7.0 (with NaOH) through a packed bed (plugged flow-through column) of the pyrolyzed carrier for about 1 hour, the amount of citrate solution being about 3 ml/g of pyrolyzed carrier treated.

Examples 1 and 2: Regeneration of MgO-Al_2O_3 Porous Carriers — The table below shows the initial activities of adsorbed glucose isomerase preparations in laboratory column operation using 50% glucose (Cerelose) feed containing 0.005 molar $MgCl_2$ at pH 8.4 and 60°C. The benefits of citrate treatment are evident.

Carrier Description (examples)	Number of Samples	Average Initial Activity (units per gram)
New carrier	6	854
Pyrolysis and citrate (1)	4	895
Pyrolysis (no citrate)	1	781
New carrier	3	688
Pyrolysis and citrate (2)	2	736
Pyrolysis (no citrate)	1	616

In Situ Regeneration of Supported Isomerase Using Sodium Hypochlorite Solution

J.L. Gregory and W.H. Pitcher, Jr.; U.S. Patent 4,002,576; January 11, 1977; assigned to Corning Glass Works regenerate the highly porous and particulate MgO-Al_2O_3 carrier materials useful for the immobilization of glucose isomerase by circulating a sodium hypochlorite solution through the carrier particles under conditions sufficient to remove substantially all carbonaceous material and contaminants.

In preferred embodiments, the sodium hypochlorite solution consists of at least 5% by weight sodium hypochlorite in water and the regeneration is carried out in a fluidized bed containing the spent composite. Very preferably, the amount of solution used is at least 5 ml/g of spent composite and that solution is circulated through a bed of the composite particles for at least 15 minutes.

Details of this process are shown in the following examples where the MgO-Al_2O_3 carrier consisted of 30 to 45 mesh porous particles having an average pore diameter of 190 to 210 A and consisting of 2.2% by weight MgO. A one-time use of the immobilized glucose isomerase using such carriers consisted of placing in columns 15-g quantities of the composite consisting of the glucose isomerase adsorbed to the MgO-Al_2O_3 particles in accordance with U.S. Patent 3,982,997 and then continuously passing a glucose-containing solution through the column at a flow rate of 3 to 4 ml/min. The glucose solutions contained 0.005 M $MgCl_2$ and was buffered to a pH of 8.4. Each column was deemed spent after having been used for an enzymatic half life of the composite (about 30 days). Glucose isomerase activity was measured in IGIU.

Examples 1 through 7: In the experiments below, spent enzyme composites were regenerated by circulating varying amounts and concentrations of an aqueous NaOCl solution through the columns containing the spent immobilized enzyme. The amounts of composite in each column ranged from 15 g at start down to 4 g with the successive regenerations. In the regeneration steps, the goal is to provide a reusable carrier capable of as high an enzyme reloading as possible. To be economically feasible, glucose isomerase composites should demonstrate an in-use loading of 600 IGIU/g of composite. Hence, any regeneration of carrier which could assure such loading can be deemed successful.

It was found that to a limited extent, the amount of NaOCl solution used in the regeneration had some effect on subsequent enzyme reloading. Preferably, at least 5 ml of a 5% NaOCl solution is used per gram of carrier to be regenerated. As the amount of NaOCl solution was increased, it was found that there occurred an increase in enzyme reloading. For example, in one set of experiments 1800 IGIU of enzyme was offered per gram of carrier for adsorption.

After use in column (30 days) this material (15 g total) was treated in a fluidized bed reaction with 5 mg/g of NaOCl solution for 15 minutes at a flow rate of 60 ml/min. This carrier was then offered 2700 IGIU/g which resulted in an observed (in-column) activity of 800 IGIU/g as opposed to 700 IGIU/g for new carrier offered the same amount/g of enzyme. After operational use, the carrier was treated with 5 ml/g of 5% NaOCl solution and offered 2700 IGIU/g enzyme resulting in only 542 IGIU/g activity. However, after a subsequent treatment with 13.3 ml/g of 5% NaOCl, the activity following identical enzyme immobilization was 864 IGIU/g.

Table 1 shows the relatively high level of loading activity observed after spent composite had been regenerated with varying amounts of a 5% NaOCl solution. The original (unregenerated) carrier was initially offered 3500 IGIU/g of enzyme. After the regeneration step, the carriers were offered 2700 IGIU/g of enzyme. The regeneration step was accomplished by recirculating the indicated NaOCl solutions through approximately 15 g quantities of spent composite for about 120 minutes.

Table 1

Treatment (ml 5% NaOCl/g carrier)	Initial Activity Using Regenerated Carrier (IGIU/g)
5	651
10	709
13.3	809

After the above treatments, the carriers were rinsed in distilled water prior to adsorption of the enzyme. However, in a subsequent experiment, it was found that if enzymes were adsorbed to the carrier (no water rinse) after a 15-minute treatment with 5 ml/g of 5% NaOCl, the resulting activity of the composite having the regenerated carrier was 824 IGIU/g.

Table 2

Example No.	Time (min)	NaOCl (%)	Amount (ml solution/g)	Initial Activity (IGIU/g)
1	120	5	5	799
2	15	5	5	810
3	15	5	5	797
4	30	5	10	773
5	60	5	20	716
6	15	2.5	10	774
7	15	5.75	5	807

Examples 1 through 3 show that increasing the NaOCl treatment time did not increase effective enzyme loading. Examples 4 and 5 show that no advantage was found in additional increase of the amount NaOCl used. Examples 6 and 7 show that varying the concentration and source (using commercial bleach, as in Example 7) of NaOCl did not affect results.

USE OF OTHER INORGANIC SUPPORT MATERIALS

Porous Alumina, Titania or Zirconia as Supports

An immobilized enzyme composite is disclosed by *R.A. Messing; U.S. Patent 3,850,751; November 26, 1974; assigned to Corning Glass Works* having an enzyme adsorbed to the inner surface of a porous ceramic body having an average pore diameter at least as large as the largest dimension of the enzyme but less than 1000 A.

A critical feature of the carriers is the relationship between the average pore size and the dimension of the enzyme bonded to the carrier. The average pore size should be such that loading and half life of the bonded enzyme are maximized and that the substrate can readily diffuse into the carrier material and to the bonded enzyme. Preferred enzyme composites comprise an enzyme bonded to the inner surfaces of a porous, essentially nonsiliceous ceramic material consisting of agglomerated metal oxide particles selected from alumina, titania and zirconia. The porous ceramic carrier may comprise more than one of the above metal oxides.

The techniques for preparing the enzyme composites is as follows: First an essentially nonsiliceous porous ceramic body having an average pore size at least as large as the largest dimension of the enzyme but less than 1000 A is ground and sieved to the desired mesh. Preferably, the porous bodies have a substantially uniform pore distribution. Various nonsiliceous porous ceramic carrier particles having the desired average pore size can be used such as porous alumina, porous titania, porous zirconia, or combinations thereof and of other essentially water-soluble metal oxides. Once the porous bodies of a given average pore size and mesh size are acquired or prepared, they are preconditioned or hydrated with a suitable buffer for the subsequent bonding procedure to assure the formation and/or retention of maximum surface oxide or hydroxyl groups.

To bond enzymes by adsorption, the buffer system is first removed from the particles but the particles are kept wet prior to actual adsorption. The enzyme to be adsorbed is then added to the wet carrier particles at a concentration per unit weight of carrier which will permit optimum adsorption of the enzyme or maximum loading in view of the available surface area. The adsorption process is facilitated and mass diffusion of the enzyme into the pores is hastened by stirring, circulation, inversion, use of a fluidized bed reactor containing the carrier, or other known means for forming an agitated reaction environment.

Example: Glucose Isomerase Adsorbed to Porous Al_2O_3 – Glucose isomerase (largest molecular dimension about 100 A, molecular weight 18,000) was adsorbed on porous alumina bodies having the following physical characteristics.

Porous Al_2O_3 Carrier

Average pore diameter	175 A
Minimum pore diameter	140 A

(continued)

Maximum pore diameter	200 A
Pore volume	0.60 cm^3/g
Surface area	100 m^2/g
Particle mesh size	25-60

The enzyme used consisted of a crude glucose isomerase preparation containing 444 IGIU/g. The enzymatic activity of the glucose isomerase and the immobilized enzyme composites was determined via a modified cysteine-carbazole assay method. The immobilized glucose isomerase composite was prepared as follows: A solution of 28.7 ml of 0.1 M magnesium acetate was added to 5 g of the crude glucose isomerase preparation. The slurry was stirred for 25 minutes at room temperature and then filtered through filter paper. The residue on the paper was washed with 14.3 ml of 0.1 M magnesium acetate solution followed by 14.3 ml and then 4.3 ml 0.5 M $NaHCO_3$. The washes were collected directly into the original enzyme filter paper and the total volume of the enzyme wash solution was about 50 ml.

500 mg of the porous alumina were placed in a 50 ml round bottom flask. 10 ml of the glucose isomerase solution was then added. The flask was attached to a rotary evaporator. Vacuum was applied to the apparatus and the flask was rotated in a bath maintained at 30° to 45°C for 25 minutes. An additional 10 ml of glucose isomerase solution was then added and evaporation continued 35 minutes under the same conditions. The procedure was repeated two more times and then a final 7 ml aliquot of glucose isomerase was added and evaporation was continued for an additional 1 hour and 10 minutes at 45°C. The flask and contents were removed from the apparatus and placed in a cold room over a weekend. A total of 47 ml of glucose isomerase solution had been added to the porous alumina particles.

50 ml of buffer (0.01 M sodium maleate, pH 6.8 to 6.9, containing 0.001 cobalt chloride and 0.005 M magnesium sulfate) was added to the enzyme composite and the sample extracted over the next hour at room temperature. The 50 ml extract was saved for assay. The composite was then washed with 200 ml of water, followed by 10 ml of 0.5 M NaCl. The final wash was performed over a fritted glass funnel with 50 ml of water. The enzyme composite was then transferred to a 50 ml Erlenmeyer flask and stored in buffer at room temperature with periodic assays of the total sample over a 106-day period.

Over the assay period, the average assay value for the composites was 130 IGIU per 500 mg composite, the enzyme activity recovery on the composite being 6%. The average assay value was determined by making a total of 20 assays over the 106-day period with 16 of the assays showing an activity between 100 and 150 IGIU per 500 mg of composite (200 to 300 IGIU/g). The above results indicated a high degree of static stability for the composite at a relatively high enzyme loading value. The extract solution which had been saved was assayed and found to have 39.8 IGIU/ml, indicating the enzyme activity recovery in the extract was 90%.

Two or More Enzymes Immobilized on Porous Carrier

R.A. Messing; U.S. Patent 3,841,971; October 15, 1974; assigned to Corning Glass Works has found that immobilized enzyme composites having multiple enzymes can be prepared in such a way that the resulting composites promote two or more simultaneous reactions having a synergistic effect.

The synergistic effect of two or more enzymes adsorbed within the pores of the same carrier may be demonstrated with combinations of certain mixed enzyme systems. For example, it is known that the enzyme glucose oxidase catalyzes the oxidation of glucose to gluconic acid and hydrogen peroxide. However, as catalysis proceeds, the by-product hydrogen peroxide tends to oxidize the glucose oxidase, thus destroying enzyme activity. If catalase is adsorbed within the same porous carrier, the catalase acts on hydrogen peroxide removing it from solution to yield free oxygen needed by the glucose oxidase.

Examples of other synergistic enzyme systems effective in a similar manner are galactose oxidase and catalase and a mixed system consisting of D-amino acid oxidase and catalase. Examples of synergistic enzymes operating in a different manner are mixed systems consisting of the proteolytic enzymes chymotrypsin and trypsin or papain and *Streptococcus peptidase A*. Inasmuch as the above enzymes promote protein hydrolysis by breaking different bond sites in proteins, in combination the enzyme pairs can be used in a single step to yield polypeptide units which are smaller than those which could be obtained with each enzyme acting separately and alone.

A critical feature of the process is that the carrier must be porous and should have a high surface area to maximize the surface area available (mostly internal) for the loading of the enzymes. Generally, the surface area should be at least 5 m^2/g and preferably greater than 25 m^2/g. The materials which can be used as carriers include any essentially water-insoluble porous inorganic material having available surface oxide of hydroxyl groups necessary for the surface bonding of the enzymes through adsorptive forces. Typical carriers include porous glass particles or beads, dried silica gels, porous alumina, titania, or zirconia bodies, or their mixtures and like materials.

The average pore size of the carriers also must be at least equal to the largest dimension of the larger or largest of the enzymes to be immobilized to assure a composite containing all enzymes of a multiple enzyme system. This minimum average pore size is required to permit entry into the pores and mass diffusion of the enzyme through the pores of the carrier so that the largely internal surface area of the carrier can be fully used for adsorption of the larger or largest enzyme and the smaller enzymes. As a very practical matter many multiple enzymes can be advantageously adsorbed within carriers having an average pore diameter between 100 A and 1000 A.

Preparation of the Multiple Enzyme Composites: The Enzymes — The enzyme preparation used was a liquid preparation of the two enzymes stabilized in glycerol and standardized to contain 750 glucose oxidase units and 225 catalase units/ml. A 20 ml volume of this enzyme preparation was dialyzed against four charges of 3,500 ml distilled water in a 4-liter beaker over a 2-hour period with stirring at room temperature. The final volume of the dialyzed enzyme solution was 50 ml. This solution was centrifuged at 2,200 rpm for 15 minutes at room temperature. The insoluble particles were discarded. The estimated activity of the solution was about 300 glucose oxidase units/ml.

The Carriers — 300 mg quantities of the carrier were transferred to separate 10 ml cylinders. To each cylinder, 9.0 ml of 0.5 M sodium bicarbonate was added to precondition the carrier surfaces. The cylinders were then placed in a shaking water bath at 35°C for 3 hours. The cylinders were removed from the bath, the bicarbonate solutions decanted, and 9 ml distilled water was added to each cylinder. The cylinders were stoppered and inverted several times and the water washes were decanted. At this point each 300 mg sample of the porous bodies was ready for the adsorption and immobilization of the enzymes.

The Composites – The mixtures of glucose oxidase and catalase were adsorbed on the inner surfaces of 300 mg of the carrier as follows: An 8 ml volume of the dialyzed glucose oxidase-catalase solutions (containing 2,400 glucose oxidase units) was added to a cylinder containing a 300 mg carrier sample. The cylinder was then partially immersed in a shaking water bath at 35°C for 2 hours and 20 minutes. The simultaneous diffusion and adsorption of the enzymes was then allowed to continue overnight (about 15 hours) at room temperature without shaking. The enzyme solution was then decanted and 9 ml of distilled water was added. The cylinders were then stoppered and inverted several times and the water washes were decanted and discarded. The multiple enzyme composite was then washed with 9 ml of 0.5 M sodium chloride. The wash was decanted and the immobilized enzyme composite was exposed to 9 ml of 0.2 M acetate buffer, pH 6.1 as a further wash. The acetate buffer solution was then decanted and the composites exposed to a final wash with 9.0 ml of distilled water.

Examples 1 and 2: These examples illustrate the criticality of having the average carrier pore diameter of at least the size of the largest dimension of the larger enzyme (catalase, which has a largest dimension of about 183 A). The column of Example 1 contained the composites consisting of both enzymes adsorbed to 300 mg of an alumina carrier having an average pore diameter of 175 A. The column of Example 2 contained the composites consisting of both enzymes adsorbed to 300 mg of a titania carrier having an average pore diameter of 350 A. Into each of the above columns a 100 ml volume of the glucose-containing substrate solution was periodically introduced by circulation at intervals over a 38-day period. The stopcocks of the columns were adjusted to maintain a substrate solution head of ¾ inch above the top surfaces of the composites during the assay periods.

A flask containing 100 ml of substrate solution had an outlet tube feeding through a peristaltic pump into the top of the column and the inlet of the column fed back into the flask. Circulation of the substrate solution was at a flow rate of 145 ml/hr. The temperature for all assays in these two examples was 22°C except for the last (assay day 38) at which time the room and assay temperature was 19°C. The periodic assay results for both composites are shown in the following table for the first 38 days. The composite using the 350 A porous titania carrier (Example 2) was subjected to further periodic assays and after 165 days the composite quite surprisingly was found to have retained 82% of its original glucose oxidase activity.

Comparison of Glucose Oxidase Activities per Gram of Porous Al_2O_3 and Porous TiO_2

	Carrier	
	Example 1	Example 2
Assay Day	Porous (175 A) Al_2O_3	Porous (350 A) TiO_2
1	12.9	20.5
1	18.7	36.2
4	11.7	36.2

(continued)

Assay Day	. . . . Carrier . . . . Example 1 . . . . Porous (175 A) Al_2O_3	. . . Example 2 . . . . Porous (350 A) TiO_2
5	10.1	36.2
6	8.5	36.2
7	7.0	36.2
38	(*)	32.5
59	–	36.2
83	–	35.9
103	–	35.5
136	–	32.5
165	–	29.6

*Results too low to measure

The results in this table show that activity losses of the composites using the 350 A porous titania were essentially negligible over the first 38 days of storage. The composites using the smaller average pore size (175 A) alumina exhibited a continuous loss of activity over the same time span.

Use of a Porous Ceramic Matrix

M.H. Keyes; U.S. Patent 4,001,085; January 4, 1977; assigned to Owens-Illinois, Inc. has disclosed a process for immobilizing an enzyme by causing an aqeuous dispersion of the enzyme to flow through an inert, inorganic, porous, sorptive, dimensionally stable, fluid permeable supporting matrix to form a biologically active composite. It is important for the matrix to be sufficiently porous to be permeable to enzyme as well as the substrate that will be subsequently processed therein.

The driving force for the deposition is provided by the application of pressure to a reservoir of the dispersion of enzyme in contact with the porous matrix. It is important for the matrix to be sufficiently porous to be fluid permeable and pass the enzyme so that substantial amounts of enzymes are present in the permeate passing through the matrix. In this regard, the enzyme is not deposited on the matrix by a filtration process (e.g., ultrafiltration) because the matrix itself is enzyme permeable.

The mechanism or theory for this process is not understood, although the improved results have been clearly demonstrated. It is not known why the enzyme is immobilized more effectively by flowing an aqueous dispersion through the porous supporting matrix as compared with impregnating or soaking under static conditions. Any enzyme which can be prepared in aqueous dispersion can be used, including a wide variety of enzymes which may be classified under three general headings; hydrolytic enzymes, redox enzymes, and transferase enzymes.

It has been found that porous matrix having a volume porosity preferably of 15 to 50% are quite suitable for the present purposes. The pore size of the support is critical in that it should not be so small as to prevent or inhibit fluid permeability in the resulting biologically active membrane. Average pore diameters in the range of 0.01 to 10 μ are suitable for most applications with 0.01 to 2 being preferred for efficiency and economy.

The porous support can be formed by compacting and sintering refractory ceramic oxide powders such as alumina powder, zirconia powder, magnesia powder, silica powder, thoria powder, glass powder, powdered clay, powdered talc and the like.

Porous, inert, rigid dimensionally stable refractory supports can be prepared by compacting such refractory oxide powders to form a "green compact" of the desired configuration. The green compacts are then fired for a time and at a temperature sufficient for sintering to yield porous, inert, rigid, dimensionally stable, fluid permeable refractory support.

The porous matrix can be in any geometric shape such as rods, cylinders, discs, plates, tubes, bars, and blocks so long as it can be positioned in a flow-through pressure cell. In depositing and immobilizing the enzyme on the supporting matrix an aqueous dispersion of the enzyme is placed in a chamber equipped with the supporting matrix. The matrix is positioned in the chamber such that the aqueous dispersion of enzyme must pass therethrough when pressure is applied to the chamber. This type of device is called a "flow-through" deposition chamber.

Example 1: Porous supporting matrices in the form of discs are prepared from a fine alumina powder having an average particle diameter of about 1 μ (Alcoa A-16). A compaction mixture is formed by ballmilling 3% polyvinyl alcohol, 0.5% stearic acid, 25% H_2O, with the balance being alumina powder. The polyvinyl alcohol and the stearic acid serve as compaction aids. The milled powder is dried at 150°C under vacuum. The dried alumina powder is then compacted in a ram press into discs of ¾ inch in diameter and 55 mils thickness under 6,000 psi. The discs weigh 1 to 1.5 g each.

If desired at this point, the alumina powders can alternatively be extruded in the form of a porous alumina tube or other shape using a conventional extrusion or other processing equipment. The further processing is described in terms of the discs although tubes or other geometric shapes can be processed in the same fashion. The discs are sintered by heating gradually to 1500°C and then maintaining this temperature for 2 hours. The discs are then allowed to cool to room temperature over several hours.

The resulting discs are porous, fluid permeable, rigid, dimensionally stable and sorptive. Supporting matrices can also be formed in this manner from refractory materials, powdered titania, powdered zirconia, powdered thoria, powdered glass, and fine clay.

The faces of the resulting sintered discs are then ground so that they are substantially flat and parallel. The resulting discs have a porosity of 20 to 40% with an average pore diameter of 0.1 μ. Because of the chemical and thermal stability of the alumina support matrix, it can be reconditioned for reuse, if desired, after the immobilized enzyme has finally lost most of its activity. It is necessary only to place the disc in a furnace and heat at 800°C for 2 hours. This serves to remove enzyme and other entrapped residual organic materials as well as to dehydrate and reactivate the surface of the alumina. Thus, a single alumina support can be used repeatedly.

Example 2: A 0.01% by weight dispersion of chymotrypsin is prepared from enzyme material having an activity of 45 IU/mg, based on benzoyltyrosine ethyl ester (BTEE) conversion. The pH of the aqueous dispersion is adjusted to 6.5 with a 0.08 M tris(hydroxymethyl)aminomethane (TRIS) solution containing calcium chloride (0.1 M) as a stabilizer for the enzyme.

A porous disc-shaped alumina matrix prepared in Example 1, having an average pore size of about 0.04 μ and a porosity of 35%, is placed in a pressure deposition cell. Nitrogen pressure varying from 10 psi to 1,000 psi is applied to the cell while its contents are at room temperature and the enzyme dispersion permeates and flows through the matrix for several minutes. Enzymes are present in the permeate. After about 150 ml of permeate have been collected, the pressure is released and the enzyme has been immobilized in an active form on the porous, alumina matrix to form a fluid permeable, rigid, dimensionally stable biologically active composite membrane.

Electrodeposition of Enzymes on Porous Supports

Basically, the process developed by *M.H. Keyes; U.S. Patent 3,839,175; Oct. 1, 1974; assigned to Owens-Illinois, Inc.* involves electrodepositing a charged enzyme onto an inert, inorganic, porous, sorptive, dimensionally stable, fluid permeable, supporting matrix. Figure 1.1 is a schematic flow diagram of an electrodeposition cell for use in preparing membranes according to the process.

Figure 1.1: Electrodeposition of Enzymes

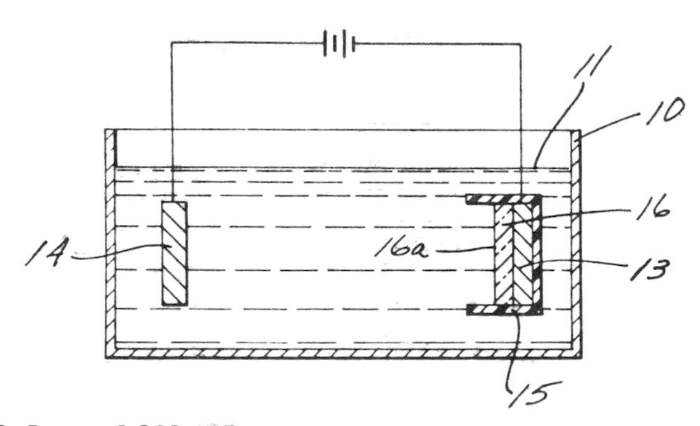

Source: U.S. Patent 3,839,175

Referring to Figure 1.1, cell **10** contains a bath **11** of an aqueous dispersion of ionized enzyme. Immersed in bath **11** are electrodes **13** and **14** which can be made of metal. Electrodes **13** and **14** are connected to a source of direct current potential which can be a battery, rectified alternating current or other electric power supply which provides a net DC potential. One such power supply can deliver voltages up to 600 volts and a current in the fractional ampere range at load resistances of up to 1 megaohm.

Immediately adjacent electrode **13** is mounted the inert, inorganic, porous, sorptive, dimensionally stable, fluid permeable, supporting matrix **16**, the dimensions of which are coextensive with the confronting dimensions of the electrode **13**.

The assembly of the electrode **13** and matrix **16** are mounted in housing **15**. Housing **15** envelops electrode **13** and matrix **16** so that only face **16a** of the matrix **16** is exposed to bath **11**. Thus, it is seen that matrix **16** is interposed in the electrical path defined between electrodes **13** and **14**. The support matrix used in this process is that prepared in Example 1 of the previous U.S. Patent 4,001,085.

Example: Electrodeposition of Enzyme — An electrodeposition cell like the one shown in Figure 1.1 is filled with an electrodeposition bath comprising 120 ml of a 0.1% by weight aqueous dispersion of ribonuclease. The pH of the solution is adjusted to 7 with sodium hydroxide. The activity of the enzyme is 1.1 International Units (IU) per mg, one unit being equivalent to the amount of RNase required to convert 1 μmol/min of cytidine-2',3'-cyclic monophosphoric acid to cytidine-3'-monophosphoric acid.

One of the porous alumina discs prepared in U.S. Patent 4,001,085, is mounted in the cell as matrix **16** and electrodes **13** and **14** are in the form of zinc discs of the same size as the alumina discs. Electrode **13** is made the cathode of the cell. A dc voltage is applied to initiate a current of 1.0 mA between the electrodes. The voltage is gradually lowered over a 30-minute deposition period so that the final current is 0.1 mA to prevent overheating of the bath. No external heating or cooling is employed. During the 30-minute deposition period, housing **15** is rotated at the rate of 30 rpm by means not shown in the drawing. At the end of this deposition period, ribonuclease has been electrodeposited and immobilized in an active form on the porous, alumina matrix to form the fluid permeable, rigid, dimensionally stable biologically active composite membrane.

For purposes of comparison a similar porous alumina disc is placed in the electrodeposition bath described above and allowed to adsorb enzyme under static conditions for several hours. Using identical assay techniques this disc is able to convert only 15% of the substrate to the product. The enzyme reaction using the biologically active composite membrane may be carried out in a conventional pressure reaction cell such as an Amicon Model 420 high-pressure ultrafiltration cell in which the ultrafiltration membrane has been replaced by the biologically active composite membrane prepared according to this process.

Dye Stabilizers for Alumina Supported Enzymes

E. Katchalski, Y. Levin and B. Solomon; U.S. Patent 3,873,426; March 25, 1975; assigned to Miles Laboratories have prepared water-insoluble enzymes where the enzyme is stably retained on an alumina carrier by adsorbing the enzyme and a dye on an activated alumina carrier at a pH below 7.

The activated alumina useful in this process is generally obtained by dehydration and calcination of aluminum hydroxide at 900°C in a carbon dioxide atmosphere which tends to coat the resulting alumina particles with a thin layer of aluminum oxycarbonate of approximate formula $[Al_2(OH)_5]_2CO_3 \cdot H_2O$. Alternatively, commercially available alumina can be activated by washing with concentrated hydrochloric acid. It is necessary that the alumina be in an activated condition in order to properly adsorb the enzyme and the dye. While the mesh size of the alumina is not critical for this process, it is preferred that the alumina have a mesh size of 14 to 48, as measured by the Tyler Standard Screen Scale when the insoluble enzymes are to be used in a column to treat liquid materials.

In carrying out the process the activated alumina is preferably contacted with an aqueous solution of the dye at a pH below 7 and the dye is conveniently and rapidly adsorbed on the surfaces of the alumina particles. If the pH is above 7, the amount of adsorption is minimal. Preferably the dye is contacted with the alumina at a pH of 4 to 5. The alumina is then washed with water to remove any unadsorbed dye. The dye-alumina combination is then adjusted to a pH below 7 by washing with appropriate buffer solution, the specific pH depending upon the enzyme being insolubilized. Amyloglucosidase and lactase prefer a pH of 4 to 4.5, fungal amylase a pH of 5, while bacterial amylase prefers a pH of 6. The dye-alumina carrier is then contacted with an aqueous solution of the enzyme at the preferred pH to allow the enzyme to be adsorbed and retained by the alumina.

Under these adsorption conditions below about pH 7, there is no covalent bonding resulting from chemical reaction between the dye and the enzyme that might otherwise take place at pH 7 to 9, for example. Even though there is no apparent chemical reaction between the dye and the enzyme, the presence of the dye stabilizes and improves the retention of enzyme activity on the alumina over the values obtained without the use of the dye. It is also understood that the enzyme can be adsorbed first by the alumina under the above pH conditions followed by adsorption of the dye at a pH below about 7.

Example 1: A 1 g quantity of alumina having a mesh of 60 to 100 was activated by washing it with 4 M HCl followed by washing with 0.02 M acetate buffer at pH 4.2. It was then suspended in 0.1 M acetate buffer at pH 5 and was contacted with Procion Brown MX5BR dye for about 20 minutes. The dyed alumina was then washed with 0.02 M acetate buffer to pH 4.2 and was mixed with 5 ml of an aqueous solution of amyloglucosidase containing 4.5 amyloglucosidase units, allowed to remain in contact with mixing overnight, and then separated from the liquid. The resulting insolubilized amyloglucosidase adsorbed stably on alumina contained 89% of the initial amyloglucosidase activity.

Example 2: The procedure of Example 1 was repeated with alumina having a mesh from 14 to 28. The resulting insolubilized amyloglucosidase contained 77% of the initial amyloglucosidase activity while the enzyme activity in the supernatant liquid increased as compared to Example 1. This is explained by the fact that the larger-sized alumina employed in this example contains a smaller overall surface area per unit weight than the alumina of Example 1. The total amount of adsorbed dye and adsorbed enzyme is thus reduced causing a reduction in retained amyloglucosidase activity.

Example 3: The procedure of Example 1 was repeated using an enzyme solution containing lactase activity. The resulting insolubilized lactase contained 7.6 lactase units which was a significant amount of the initial activity.

USE OF POLYMERIC SUPPORT MATERIALS

Porous Unsintered Fluorocarbon Polymers as Supports

In the process of *J.H. Fishman; U.S. Patent 3,843,443; October 22, 1974,* polypeptide materials such as enzymes are adsorbed directly to the surface of porous unsintered fluorocarbon polymers such as polytetrafluoroethylene.

The process is predicated on the discovery that porous unsintered fluorocarbon polymers will strongly adsorb protein and polypeptide polyelectrolytes. It had been expected that proteins and polypeptides would be adsorbed on, for example, unsintered polytetrafluoroethylene (PTFE) in the same manner that they are on sintered PTFE, namely, weakly bonded to the surface. Surprisingly, however, proteins are firmly adsorbed on unsintered PTFE to the extent that the adsorbate cannot be leached out of the fluorocarbon supporting matrix after repeated washings in water or organic solvents such as alcohol.

The nature and properties of the composite bound polypeptide material are a function of the fluorocarbon polymer matrix and the bound polypeptide. Although the precise mechanism by which the protein is bound to the unsintered fluorocarbon surface is not fully understood, it is, however, known that the supporting matrix must be sufficiently porous, at least at the surface to allow the protein to penetrate the material and become anchored to at least a portion of its surface; the term surface refers to the apparent external surface of the fluorocarbon polymer as well as the surface of the pores extending within the fluorocarbon support. Hence, the larger the polypeptide being used, the larger the required pore size of the supporting matrix. Consequently, high molecular weight polypeptides, such as the naturally occurring proteins, are more readily bound to a fluorocarbon having relatively large-sized surface pores of 0.1 to 1.0 micron.

This process for producing a bound polypeptide material comprises the steps of:

(1) Providing a porous unsintered fluorocarbon polymer having an atomic ratio of carbon to fluorine of 0.5 to 2.0,

(2) Flooding the unsintered porous polymer with a water-miscible organic solvent, and

(3) Contacting the flooded unsintered polymer produced in step (2) with an aqueous solution of polypeptide polyelectrolyte for a length of time sufficient to replace at least a portion of the organic solvent with the aqueous solution of polypeptide polyelectrolyte.

In an alternative embodiment of this process, the unsintered fluorocarbon polymer flooded with organic solvent in step (2) is subsequently flooded with water prior to contacting same with the solution of protein polyelectrolyte in order not to contaminate the aqueous polypeptide solution with an organic solvent.

The term polypeptide polyelectrolyte as used here refers to a linear polymer containing a plurality of polypeptide groups, specifically, natural and synthetic proteins including hemoglobin, albumin, gelatin, casein and a broad range of enzymes and hormones such as pepsin, ribonuclease, lysozyme, insulin and cytochrome C. The term unsintered as used here refers to the fact that the fluorocarbon polymer in question has not been heat-treated at or above its crystalline melting point.

Example: An experiment was performed to demonstrate the enzymatic activity of a bound polypeptide material of this process. A quantity of urease was dissolved in 100 ml of water and 5 ml of the resulting solution was added to a solution of urea comprising 5 g of urea in 95 ml of water. The mixture was agitated and maintained at 30°C. Ammonia was produced at the rate of 12 mg/min from the reaction mixture, thus providing a measure of the activity of the enzyme solution.

A 0.004-inch-thick unsintered PTFE film having an area of 25 cm^2 and a density of 1.69 g/cm^3, was flooded with ethyl alcohol, immersed for 1 minute in distilled water and then submerged in the urease solution for 24 hours. Two such samples were prepared (Samples A and B). Sample A was removed from the urease solution, rinsed with distilled water and air dried. Sample B was similarly treated except that after being rinsed with distilled water it was kept flooded by storing it in distilled water.

Sample B was thereafter submerged in the flooded condition, in urea solution containing 5 g of urea per 95 ml of water, at 30°C. Sample A was flooded with isopropyl alcohol, submerged thereafter in distilled water for 10 minutes and then transferred in the water-flooded condition to a urea solution identical to the one in which Sample B was submerged. The solutions were gently agitated.

The solution containing Sample A produced 14 mg of ammonia per minute; the solution containing Sample B produced 23 mg of ammonia per minute. This demonstrates that the bound enzyme was active.

Sample B was thereafter removed from the urea solution, placed in agitated distilled water for 6 hours and reinserted into the urea solution. Ammonia was produced at the rate of at least 21 mg/min for a period of at least 35 minutes.

Fatty Acid Esters of Polysaccharides as Supports for Phospholipases

A process for producing insolubilized enzyme having substrate specificity for glycerides or glycero-phosphatides has been disclosed by *Y. Horiuchi and S. Imamura; U.S. Patent 3,909,360; September 30, 1975; assigned to Toyo Jozo Company, Ltd., Japan.* It has been found that these enzymes can specifically be adsorbed by a carrier selected from fatty acid esters of water-insoluble polysaccharides or derivatives thereof containing hydroxyl groups and that the adsorbed enzymes are made insoluble in water without losing enzyme activity.

The glycerides or the glycero-phosphatides for which the insolubilized enzyme has substrate specificity are represented by the following formula:

$$\begin{array}{l} CH_2{-}OR_1 \\ | \\ CH{-}OR_2 \\ | \\ CH_2{-}OR_3 \end{array}$$

where R_1 and R_2 are hydrogen or an acyl group but R_1 and R_2 not being hydrogen at the same time, and R_3 is hydrogen, an acyl group, a residual group of phosphoric acid, phosphoryl choline, phosphoryl ethanolamine, phosphoryl serine, or phosphoryl inositol.

Examples of the enzyme having substrate specificity for glycero-phosphatides are phospholipases such as phospholipase A, B, C and D, phosphatidate phosphatase; and the like. The carriers for insolubilizing these enzymes are obtained by esterification of water-insoluble polysaccharides or their derivatives containing hydroxyl groups with reactive derivatives of fatty acids, particularly acid halides by known methods.

Examples of the water-insoluble polysaccharides or derivatives containing hydroxyl groups are natural vegetable fibers such as cotton, linen, jute or Manila hemp; cellulose fibers such as regenerated fibers (e.g., viscose rayon); cellulose derivatives such as carboxymethylcellulose, phosphocellulose, sulfomethylcellulose, sulfoethylcellulose, para-aminobenzylcellulose, aminoethylcellulose, diethylaminoethylcellulose, triethylaminoethylcellulose or guanidinoethylcellulose; crosslinked gel of dextran epichlorohydrin (dextran gel); dextran gel derivatives such as carboxymethyldextran gel, diethylaminoethyldextran gel or sulfoethyldextran gel; and agar.

As fatty acids to be used for esterification, saturated or unsaturated fatty acids having carbon atoms of 6 or more are advantageously used. These fatty acids may have either branched or straight chain structures. A mixture containing several kinds of fatty acids having different carbon atoms may also be used.

The enzyme having substrate specificity for glycerides or glycero-phosphatides is immobilized by merely contacting a solution containing the enzyme with a fatty acid ester of polysaccharide as described above. Treatment may be carried out in batchwise system or column at a temperature and pH which may be varied freely, so far as no detrimental effect on the enzyme activity results. It is especially preferred that the treatment be carried out at 20° to 30°C and pH of 7 to 8.

Example 1: Three grams of dry gauze and 10 ml of palmitic acid chloride were added into 80 ml of pyridine and the reaction was carried out at 30°C under stirring for 12 hours. To the reaction mixture was added 100 ml of chloroform-ethanol mixture (1:1) and the reaction product was collected by filtration. The product was washed again with 100 ml of chloroform-ethanol mixture (1:1), further washed with 100 ml of ethanol and dried to obtain 3.2 g of palmitoyl ester of gauze.

Two grams of the palmitoyl ester of gauze were immersed in 50% aqueous acetone solution, packed in a column of 1.2 x 9 cm and washed with water. Then, 40 ml (40 U/ml) of an aqueous solution of lipase was passed through the column, followed again by washing with water. The column contents were taken out and dried to obtain 2 g (730 U/g) of insolubilized lipase.

Example 2: By the use of the carriers as set forth below, insolubilized lipases were obtained according to the similar treatments as in Example 1. The carriers herein used were prepared similarly as in Example 1 by esterification of various water-insoluble polysaccharides or derivatives thereof containing hydroxyl groups with respective acid chlorides of fatty acids.

Carrier	Lipase Adsorption Ability (U/g)*
Palmitoyl ester of Sephadex G-25	720
Oleoyl ester of Sephadex G-25	840
Eraidoyl ester of Sephadex G-25	600
Palmitoyl ester of CM-Sephadex	640
Palmitoyl ester of SE-Sephadex	600
Palmitoyl ester of DEAE-Sephadex	560
Caproyl ester of defatted cotton	730
Palmitoyl ester of cellulose	940

(continued)

Carrier	Lipase Adsorption Ability (U/g)*
Caproyl ester of CM-cellulose	1,000
Lauroyl ester of CM-cellulose	1,010
Palmitoyl ester of P-cellulose	840
Lauroyl ester of DEAE-cellulose	930
Caproyl ester of DEAE-cellulose	880

*The amount of enzyme which liberates 1 μmol of fatty acids per minute at 37°C was defined as one unit.

Hydrophobic Derivatives of Cellulose or Glass

The development by *L.G. Butler, U.S. Patent 4,006,059; February 1, 1977; assigned to Purdue Research Foundation* provides a method for preparing simple hydrophobic derivatives of insoluble matrices, for use as strong adsorbents of enzymes.

This process is based on a property of proteins which has not been exploited as a means for binding. Electrostatic binding of proteins, by virtue of their charged groups, to ionic derivatives of cellulose, for instance, has long been important for protein purification by column chromatography. However, proteins are amphipathic (both hydrophilic and hydrophobic) macromolecules, and although the hydrophobic portion of the molecule may be largely buried in the core of the structure, it is recently being recognized that at least some proteins have a hydrophobic area at the surface, [*Biochemistry* 11, 3835, 3841 (1972)]. This site would presumably bind to hydrophobic materials supplied in an aqueous environment. It is likely that such hydrophobic areas are the sites of binding of the protein to a membrane under physiological conditions in the cell.

The preparations which have been found to be most useful have all been esters of hydrophobic acids with the corresponding amides apparently not binding proteins as strongly. The esters have all been prepared by reaction of the acid chloride of a suitably hydrophobic acid with a hydroxyl group in the support material.

In essence, strong noncovalent binding of proteins to hydrophobic derivatives is achieved with binding being sufficiently strong that the protein is effectively immobilized, yet the biological function of the protein is fully expressed.

Example 1: Phenoxyacetylcellulose — Whatman cellulose (CF11), 30 g, was suspended in 300 ml of 1:1 pyridine:dimethylformamide. A total of 21 ml of phenoxyacetyl chloride (Eastman) was slowly added with vigorous stirring resulting in the solution warming somewhat. After 1½ hours with occasional stirring, the resulting dark yellow suspension was heated to 70°C, then allowed to stand overnight without further stirring.

After decanting off as much solvent as possible, 95% ethanol was added and stirred. The product was recovered by filtration on a Buchner funnel and rewashed several times with ethanol. Since it is difficult to remove all of the solvent it was necessary to resuspend the pad of phenoxyacetylcellulose each time in fresh solvent rather than just washing it on the filter. The first few washes were quite yellow, but the final wash was water-clear. The pad of phenoxyacetylcellulose was partially dried on the funnel and completely dried by spreading it out on a sheet of plastic. From the 30 g of cellulose, 36 g of phenoxyacetylcellulose was recovered. No fines were discarded during the preparation and the material seemed to have a good flow rate when packed into a column. The material is stored dry and appears to be quite stable in this form. The phenoxyacetylcellulose has a maximum of ½ phenoxyacetyl group per glucose residue. It is slowly wet by water, expanding its volume considerably. On long storage in aqueous media it tends to hydrolyze to give phenoxyacetate and thus lose its capacity to bind protein.

The above phenoxyacetylcellulose was suspended in water and poured into a column 1.6 cm in diameter by 18 cm in height. The column was washed with several hundred ml of 0.05 M Tris, pH 7.4 and then enzyme was applied. The enzyme solution utilized was a preparation partially purified from bovine intestine, considerably enriched in alkaline phosphatase and phosphonate monoesterase activity; 42 ml of the enzyme preparation in the above buffer was applied.

The column was then washed with 50 ml portions successively of a large number of solutions which are described below. After each wash the eluate was tested for the presence of alkaline phosphatase activity and phosphonate monoesterase activity; in no case was any activity found.

The following solutions were utilized for washing the column in the order given (all were dissolved in the 0.05 M Tris, pH 7.4): no addition to buffer, then 1.2, 5, and 12 mM pyridine, 5, 10, 25, 50, 100, 200, and 500 mM sodium benzoate, buffer alone again, 3, 6, 7.2, 18, 36, 72, and 180 mM phenyl-1,2-ethanediol, 10% ethanol, and 20% ethanol. No enzymatic activity was eluted by any of these solvents. Passage of an assay mix over the column indicated considerable activity of both alkaline phosphatase and phosphonate monoesterase still bound to the column.

In control experiments, none of these solvents caused irreversible loss of catalytic activity of either enzyme. In other experiments these enzymes were not eluted from phenoxyacetylcellulose in high concentrations of salts (1.2 M ammonium sulfate) or 30% ethylene glycol. Other enzymes have been shown to be similarly strongly bound.

Alkaline phosphatase and phosphonate monoesterase can be completely eluted, with full activity, from phenoxyacetylcellulose by buffer solutions containing 0.1% Triton X-100. These are examples of enzymes which are stable in the presence of this detergent; other similarly stable enzymes may also be eluted in this way.

Example 2: Three different derivatives of glass were used. The chemical structure of these derivatives are shown below:

$$-\overset{|}{\underset{|}{Si}}-(CH_2)_3-NH_2$$

Alkylamine glass

$$-\overset{|}{\underset{|}{Si}}-(CH_2)_3-O-CH_2-\overset{OH}{\overset{|}{C}}H-CH_2OH$$

Glycerol glass

$$-\overset{|}{\underset{|}{Si}}-(CH_2)_3-O-\text{Dextran (MW = 1500)}$$

Dextran glass

Two 3.0 g samples of each glass derivative were placed in a small plastic bottle and 15 ml of pyridine were added to each. One sample of each type of glass was set aside as a control. To the remaining sample of each type of glass 1.0 ml of phenoxyacetyl chloride was added slowly with stirring. Heat was generated from all three samples; the alkylamine glass derivative turned yellow and the others turned dark reddish-orange. All samples were left overnight at room temperature and the next day were washed five times with approximately 25 ml each of 95% ethanol, along with the controls. After washing all samples twice with water, they were stored in the cold room. After treatment with enzyme solutions, the three treated glass samples retained approximately 90% of the activity of the applied enzymes.

All the derivatives described above have been obtained by using phenoxyacetyl chloride. Many other hydrophobic materials might also be suitable. Experiments to this effect have been carried out with lauryl chloride, benzoyl chloride, acetyl chloride and phenylbutyryl chloride. From the data, it would appear that the aromatic derivatives result in stronger binding than the aliphatic derivatives. In the aromatic series, phenoxyacetyl derivatives clearly bind proteins more strongly than do benzoyl derivatives and it seems likely that a certain distance of the aromatic group from the support is necessary for optimum binding.

Water-Insoluble Tannin Preparation

A water-insoluble tannin preparation is obtained by *I. Chibata, T. Tosa, T. Mori, T. Watanabe, R. Sano and Y. Matuo; U.S. Patent 4,090,919; May 23, 1978; assigned to Tanabe Seiyaku Co., Ltd., Japan* using covalent binding or physical adsorption of tannin onto a water-insoluble, hydrophilic carrier. The preparation has a specific affinity for proteins and can be used as an adsorbent for purification, isolation and/or separation of proteins (e.g., enzymes, albumin, globulin, hormonal proteins) from a mixture of compounds. Further, the water-insoluble tannin preparation having a catalytically active enzyme adsorbed thereon can be used as a heterogeneous catalyst to induce enzymatic reactions.

Tannin is the general term for astringent, aromatic, acidic glucosides of polyphenols found in various plants and trees and, depending on the structure, can be divided into two groups: (a) pyrogallol tannin (i.e., mono-, di- and/or trigalloyl monosaccharides, mono-, di- and/or trigalloyl disaccharides, and mono-, di- and/or trigalloyl trisaccharides); and (b) catechol tannin (i.e., polyhydroxyphenol condensates and polyhydroxy-flavan condensates). Examples of pyrogallol tannin include gallotannin, pentagalloyl glucose, Hamameli tannin and Acetannin. Examples of catechol tannin are polycatechol, polyepicatechol, polycatechin, poly(pistacia catechol) and poly(galloyl epicatechol). All of these pyrogallol and catechol tannins can be employed in the process.

The water-insoluble tannin preparation is prepared either by binding tannin and water-insoluble, hydrophilic carrier through at least one divalent functional group, or by simply contacting the tannin with the water-insoluble, hydrophilic carrier.

The term hydrophilic as used here means that the polymer is made wettable or swellable in water but not substantially soluble. Various polymers having these properties may be used in the process. Such polymers include, for example, polymers having hydroxy, amino, carboxyl, alkyl or phenoxyalkyl groups. Polysaccharides such as cellulose, agarose and crosslinked dextran (e.g., dextran crosslinked with epichlorohydrin or divinyl sulfone) are suitable as the hydroxy polymers.

Suitable examples of the amino-polymer include aminobenzyl-polysaccharide (e.g., p-aminobenzylcellulose), crosslinked polyacrylamide, crosslinked p-aminophenyl-polyacrylamide, crosslinked p-aminobenzamidoethyl-polyacrylamide (all crosslinked with N,N'-methylenebisacrylamide), p-aminobenzamido-porous glass, p-aminophenylalanine-leucine copolymer, p-amino-polystyrene, methacrylic acid-m-aminostyrene copolymer, aminoalkyl-scleroprotein (e.g., aminohexyl-wool, aminohexyl-silk) and crosslinked aminoalkyl-polymethacrylic acid (e.g., aminohexyl-polystyrene crosslinked with divinylbenzene).

Examples of polymers having carboxyl groups include crosslinked polymethacrylic acid such as polymethacrylic acid crosslinked with divinylbenzene; crosslinked carboxymethyl-polyacrylamide such as carboxymethyl-polyacrylamide crosslinked with N,N'-methylenebisacrylamide; and carboxyalkyl-polysaccharides such as carboxymethylcellulose, carboxyethylcellulose, carboxybutylcellulose, carboxyhexylcellulose, carboxymethyl-agarose, carboxyethyl-agarose, carboxybutyl-agarose or carboxyhexyl-agarose.

These polymers can be modified with the use of cyanogen halide, epihalohydrin, α,ω-bis(2,3-epoxypropyl)alkane or α,ω-bis(2,3-epoxypropoxy)alkane optionally in combination with at least one of alkylenediamine, aminoalkanol, aminoalkanoic acid, alkylenediol and haloalkanoyl halide. These reactants provide linkages between the polymer and the tannin consisting of $-CONH-$, $-CH_2CH(OH)CH_2-$, $-CH_2CH(OH)CH_2ACH_2CH(OH)CH_2-$, $-Y(CH_2)_nNHCO-$, $-Y'(CH_2)_nCONH-$, $-Y'-(CH_2)_nNHCH_2CH(OH)CH_2-$, $-Y'(CH_2)_nNH(CH_2)_mCO-$ and $-CONH(CH_2)_nOCH_2CH(OH)CH_2-$, wherein Y' is $-CONH-$ or $-CH_2CH(OH)CH_2NH-$, A is $-(CH_2)_q-$ or $-O(CH_2)_qO-$, each one of n and m is an integer of 1 to 16 and q is an integer of 1 to 6.

Example: 4 g of cellulose powder are immersed in 40 ml of an aqueous 25% sodium hydroxide solution. 10 ml of epichlorohydrin are added to the suspension, and the mixture is stirred vigorously at 60°C for 30 minutes. The epichlorohydrin-activated cellulose is collected by filtration, washed with water and then suspended in 80 ml of an aqueous solution (pH 10) containing 2.2 g hexamethylenediamine. The suspension is stirred slowly at 60°C for 2 hours. The precipitates are collected by filtration and washed first with an aqueous 0.1 M sodium bicarbonate solution and then with water. 10.5 g (wet form) of aminohexylcellulose are obtained.

10.5 g of chinese gallotannin are dissolved in 240 ml of water. The solution is adjusted to pH 10, and 0.3 ml of epichlorohydrin is added. The mixture is stirred at 30°C for 2 hours, whereby an epichlorohydrin-activated chinese gallotannin solution is obtained. The 10.5 g of aminohexylcellulose obtained above are added to the epichlorohydrin-activated chinese gallotannin solution, and the mixture is stirred at 45°C for 6 hours. After the reaction, the precipitates are collected by filtration and washed first with an aqueous 0.1 M sodium bicarbonate solution and then with water. 11.3 g (wet form) of a water-insoluble tannin preparation [i.e., chinese gallotannin covalently bound to cellulose through the linkage $-CH_2CH(OH)-CH_2NH-(CH_2)_6-NHCH_2CH(OH)CH_2-$] are obtained.

200 mg of aminoacylase (total activity: 4,000 μmol/hr) obtained from *Aspergillus orizae* are dissolved in 6 ml of an aqueous 0.2 M sodium chloride solution (pH 8.0) containing 0.2 v/v% of n-butanol. 500 mg (wet form) of the above water-insoluble tannin are added to the aminoacylase solution, and the mixture is stirred at 37°C for 30 minutes. The precipitates are then collected by filtration and washed with water. 530 mg (wet form) of an immobilized aminoacylase preparation (i.e., aminoacylase adsorbed by the water-insoluble tannin) are obtained. It shows an aminoacylase activity of 9,376 μmol/hr per 10 g (wet form) of the water-insoluble tannin preparation. (Aminoacylase activity is indicated in terms of μmol of L-methionine which is produced by reaction with N-acetyl-DL-methionine at 37°C at pH 7.0.)

Numerous permutations of the above examples with variation in the hydrophilic polymer and crosslinking agents used are covered by this process. It is also within the scope of the process to treat the tannin with the crosslinking agents and then react it with the polymer.

Reversible Immobilization on Polymers with Hydrophobic Residues

Y. Kosugi, H. Suzuki and A. Kamibayashi; U.S. Patent 4,013,512; March 22, 1977; assigned to Agency of Industrial Science & Technology, Japan have found that lipid hydrolyzing enzymes are adsorbed by the hydrophobic residue of an adsorbent, namely a hydrophobic organic compound, attached to an insoluble carrier.

The insoluble carriers usable for the process include agarose, dextran, polyacrylamide, cellulose, polystyrene, nylon and glass which are used as insoluble carriers for affinity chromatography. That is to say, any insoluble carrier may be used insofar as the carrier has a coarse structure such as to permit ready approach between the lipid hydrolyzing enzyme and the adsorbent. Particularly agarose is a desirable insoluble carrier because it is practically inert chemically and it has no possibility of causing unnecessary interaction with proteins.

It is the hydrophobic residue of the hydrophobic organic compound that provides specific adsorption of the lipid hydrolyzing enzyme. The term hydrophobic organic compound as used herein refers generally to an organic compound of the type having more hydrophobic residues than hydrophilic residues.

In the case of an aliphatic adsorbent, any of the aliphatic compounds having seven or more carbon atoms is used advantageously. Compounds including fats of even-numbered carbon atoms have been demonstrated to exhibit particularly desirable results. When an aliphatic compound having fewer carbon atoms is attached to an insoluble carrier, the distance between the carrier and the hydrophobic residue is so small that the expected interaction between the hydrophobic residue and the lipid hydrolyzing enzyme fails to proceed smoothly. These aliphatic compounds having fewer carbon atoms can be improved in affinity by incorporating a suitable spacer which is otherwise known as an arm.

Further, aromatic, alicyclic and heterocyclic hydrophobic compounds are also usable as adsorbents. For example, aromatic compounds such as benzene and chrysene, alicyclic compounds such as cyclohexane and deoxycholic acid and heterocyclic compounds such as pyridine are suitable for use.

Various methods are available for the attachment of the hydrophobic organic compound to the insoluble carrier: agarose or dextran is activated by use of cyanogen bromide and then allowed to react directly with a primary amine derivative as an adsorbent; an aminoalkyl acid is attached to an insoluble carrier activated with cyanogen bromide and the resultant insoluble carrier having the aminoalkyl acid attached is coupled with a primary amine derivative; a diamine is attached to an insoluble carrier activated with cyanogen bromide and the diamine compound on the carrier is coupled with an adsorbent containing a carboxylic group; an insoluble carrier of bromoacetyl is coupled with an adsorbent containing a primary amine and an imidazole compound; an adsorbent containing an amino group is coupled with a succinylamino propyl derivative; and an adsorbent containing a carboxylic group is coupled with a thiol derivative.

The desired attachment can be accomplished advantageously in about 50% of an organic solvent such as, for example, dioxane, ethylene glycol, methanol, ethanol or acetone. If an anhydrous alcohol or some other similar solvent is used, the dehydrating activity of this solvent may deprive the insoluble carrier of its coarse structure and consequently obstruct the required interaction between the adsorbent and the protein.

Example: In 5 ml of a 1:1 mixed solution consisting of ethanol and 0.1 M sodium hydrogen carbonate containing 0.5 M common salt and 10 mg of a different adsorbent shown in the table below, there was dispersed 3.4 ml of Sepharose 4B activated in advance with cyanogen bromide. The resultant dispersion was shaken at room temperature for 2 hours to effect a reaction. The reaction mixture was passed through a glass filter and the filter cake was dispersed in 5 ml of 1 M ethanolamine solution containing 0.5 M common salt (pH 8.0) and shaken at room temperature for 2 hours to cause a reaction, with the result that the preparation of a carrier was completed with the unimpaired active residue kept from the reaction.

Then, the prepared carrier was washed cyclically with a 1:1 mixed solution consisting of ethanol and 0.1 M sodium hydrogen carbonate solution containing 0.5 M common salt, 0.1 M acetate buffer solution containing 1 M common salt (pH 4.0) and a borate buffer solution containing 1 M common salt (pH 8.0) in the order mentioned to remove substances which were bound to the carrier by virtue of physical adsorbing activity. The washed insoluble carrier was equilibrated with tris(hydroxymethyl)aminomethane-HCl buffer (pH 8.0) having an ionic strength of 0.02 (hereinafter referred to as Tris buffer) and put to use in the test described below.

An inoculum of *Pseudomonas mephitica* var. *lipolytica* was aerobically incubated at 30°C for 24 hours in a medium incorporating 3% of defatted soy flour, 0.5% of K_2HPO_4, 0.03% of $MgSO_4 \cdot 7H_2O$ and 0.56% of oleic acid. At the end of the incubation, the incubated solution was centrifuged for 10 minutes at 6,000 *g*. The resultant supernatant represented a lipid hydrolyzing enzyme solution produced extracellularly by the microorganisms of genus *Pseudomonas.* This solution was adjusted to pH 4 to cause precipitation and the precipitate consequently formed was extracted by adjusting the solution to pH 8 to afford a crude lipid hydrolyzing enzyme solution.

The insoluble carrier prepared as described above was dispersed in 5 ml (208 units) of the crude lipid hydrolyzing enzyme solution, shaken for 2 hours at normal room temperature and then passed through a glass filter. The filtrate obtained was tested for enzymatic activity by the modified Nords' method. The table below shows the relation between various adsorbents used and corresponding amounts of lipid hydrolyzing enzyme adsorbed.

Adsorbent	Units
Laurylamine	1.62
Stearylamine	1.58
Oleamide	0.42
Palmitamide	1.40
Ethanolamine	-

Stabilization of Enzymes by Association with Synthetic Polymers

J.J. O'Malley; U.S. Patent 3,860,484; January 14, 1975; assigned to Xerox Corp. has found that sensitive enzymes may be stabilized by the presence of stabilizing synthetic polymers in solutions of the enzymes. The stabilizing effect is also produced in a dry form wherein the enzyme-polymer combination is freeze dried.

Enzymes of the oxidoreductase class which are used for biological oxidation and reduction, and with respiration and fermentation processes may be stabilized by this method. A particularly representative enzyme of this whole group is glucose oxidase which is an oxidase enzyme of fungal origin.

Stabilization occurs by the intimate association of sensitive enzymes, particularly enzymes of the above class with synthetic polymers and copolymers selected from groups consisting of polyvinyl pyrrolidone, polyethylene oxide, polyacrylamide, copolymers of polyvinyl pyrrolidone and polyvinyl acetate, and copolymers of polyvinyl alcohol and polyvinyl acetate. By intimate association, it is meant that the enzyme and stabilizer are sufficiently associated physically or chemically in nature so that the enzyme material is protected.

The process is effective for the stabilization of these enzymes when the weight ratio of polymer to enzyme in intimate association is from 5 to 1 to 50,000 to 1. A preferred and convenient weight ratio of polymer to enzyme is about 5×10^3 to 1.

Example 1: The enzyme glucose oxidase used was a purified, dry powder having a good degree of purity and a specific activity of approximately 20 U/mg. Stock solutions of the enzyme were prepared by mixing the enzyme with deionized water at room temperature so the final concentration of enzyme was 0.02 mg/cc.

Polyethylene oxide polymer having a molecular weight of about 20,000 was mixed with deionized water at room temperature so that the final concentration of polymer was 1 g/100 cc. One part of glucose oxidase stock solution was well mixed with 10 parts polyethylene oxide polymer solution and this mixture was placed in a freeze-drying flask and cooled to about -80°C in a 2-propanol-Dry Ice bath for approximately 1 minute. The sample was then vacuum evaporated at 10^{-3} mm Hg for 4 hours and then the dry mixture of polymer and enzyme was removed from the flask under a nitrogen blanket and placed in a clean stoppered tube. This procedure was simultaneously run in every detail except that the addition of polymer was omitted.

These two enzyme compositions were immediately tested for activity by mixing identical amounts of the samples in water, adding the standard testing reagent for glucose oxidase, and at the end of exactly 5 minutes, 5 N hydrochloric acid was added and mixed with the samples. Each sample was read in the Gilford 2,400 Spectrophotometer at 400 nm to determine the activity of the enzyme remaining prior to the addition of acid.

By using a control to determine the activity of the pure untreated enzyme in absolute units and taking this value to be as 100%, the freeze-dried enzyme had 42% of its activity remaining while the freeze-dried enzyme with polyethylene oxide had 75% of its activity remaining. This demonstrates the beneficial effect that polymers have for stabilizing enzymes during the freeze-dry process.

Example 2: The enzyme glucose oxidase used in Example 1 was prepared by the freeze-dry process of Example 1 with the following polymers: polyvinyl pyrrolidone having a molecular weight of 28,000 and polyvinyl pyrrolidone having a molecular weight of 180,000. Both samples were prepared, processed, and tested in the identical fashion as Example 1. In both cases, the enzyme had approximately 86% of its activity remaining after the freeze-dry process, thus demonstrating that molecular weight is not a factor in stabilizing the enzyme using the techniques of this process.

IMMOBILIZATION BY IONIC OR METAL BONDING

ION-EXCHANGE RESINS AS SUPPORTS

Quaternized Vinylpyridine Copolymer

Quaternized copolymers of vinylpyridine have been developed by *Y. Ishimatsu, S. Shigesada and S. Kimura; U.S. Patent 3,915,797; October 28, 1975; assigned to Denki Kaguku Kogyo KK, Japan* for use as carriers for preparation of immobilized enzymes. The overall process comprises four steps: (1) preparation of copolymers having vinylpyridine or derivatives substantially bonded within the molecules; (2) preparation of water-insoluble anion-exchangeable matrixes by quaternization of the pyridine rings of the copolymers; (3) preparation of immobilized enzyme compositions by combination of the matrixes with enzymes and microorganic cells; and (4) continuous conversion of substrates by catalysis with the immobilized enzyme compositions.

The term copolymers is used here to mean copolymers which are obtained by copolymerizing vinylpyridine or its derivatives with one or more other monomers or polymers by suitable methods of polymerization. The quaternization of the nitrogen atom of the pyridine rings in the copolymer is effected most advantageously by using the quaternization reaction of an amine compound by a halogenating agent, particularly by a halogenated alkyl reagent. Desirable halogenated alkyls are methyl chloride, methyl bromide, propyl bromide, methyl iodide, etc.

Besides these, sulfur compounds such as sulfur dioxide, sulfur trioxide, thionyl chloride and dimethyl sulfate, carbonyl compounds such as benzyl chloride, acid chlorides and acid anhydrides, metal halide compounds such as aluminum trichloride, copper chloride and cobalt chloride, and acidic high molecular compounds such as long chain alkyl sulfonates, polystyrene sulfonate and polyacrylic acid are also usable for quaternization.

The binding of the various enzymes and corresponding microorganic cells with the carriers can easily be effected by bringing the carriers into aqueous solutions of enzymes or aqueous suspensions of microorganic cells of any desired concentration in pH ranges in which the selected enzymes and microorganic cells can remain stable. This reaction of binding is completed in a short period of time. Desirably the temperature is in the range of from 5° to 20°C. These carriers can be obtained in any of a rich variety of forms by properly selecting the method of polymerization, conditions of polymerization and compositions of monomers. Accordingly, immobilized enzyme compositions may be obtained in any forms suited to the various reaction vessels employed.

Of the various possible reaction vessels, particularly effective is the packed bed column. In the case of this particular reaction vessel, the desired enzyme reaction can easily be carried out continuously and substantially automatically for a long time by merely packing the column with the immobilized enzyme composition produced in the form of beads or powder to a fixed capacity and feeding it with a given substrate at a fixed feed rate.

Example 1: Preparation of Quaternized 2-Vinylpyridinestyrene Block Polymer – A reactor having a volume of 1.5 liters and completely displaced with dry N_2 was charged with 800 cc of purified tetrahydrofuran and 55.2 cc (0.48 mol) of purified styrene (St) and cooled externally until the contents fell to a temperature of -20°C. Then, 2.0 mmol of n-butyl lithium were added to initiate polymerization. After 30 minutes, 50.4 cc (0.48 mol) of purified 2-vinylpyridine (2-VP) was added and the reaction was allowed to continue for 30 minutes. Next, a small amount of n-propanol was added to stop the polymerization and the reaction mixture was introduced in a large volume of water to separate the formed polymer. The polymer was then dried. The conversion was 100%.

Analysis of the copolymer by IR, NMR, etc. revealed that 2-VP and St were bonded substantially equimolarly in a pattern proper to block copolymerization. A glass column having an inside diameter of 3 cm was packed in a layered pattern with 20 g of the powdery (60 to 100 mesh) 2-VP-St block copolymer and methyl bromide in a vapor form was blown in at a feed rate of 100 cc/min for about 2 hours. The quaternized block copolymer thus obtained was found to reflect a weight increase of 7 g indicating that the corresponding amount of methyl bromide had been combined. This copolymer remained undissolved in tetrahydrofuran, methanol and water. It was found to have an anion-exchange capacity of about 3.5 meq/g.

Example 2: Preparation of Immobilized Aminoacylase Composition – 2 g of the quaternized block polymer (60 to 100 mesh) prepared in Example 1 were kept immersed overnight in 100 cc of 0.1 M phosphate buffer solution with pH 8.0, and the resultant bufferized polymer was filtered out to serve as a wet carrier. This wet carrier was suspended in a solution prepared by dissolving 1 g of a commercially available aminoacylase (activity units: 10,000 u/g) of a species of genus Aspergillus in 100 cc of 0.1 M phosphate buffer solution with pH 8.0. The suspension was agitated at 4°C for 2 hours. Thereafter, the suspension was centrifuged and the resultant sediment was washed several times with 0.1 M phosphate buffer solution with pH 8.0. The sediment thus obtained was finally lyophilized. The yield was 2.3 g.

The supernatant which occurred in the above reaction was analyzed for protein concentration by Lowry process [*J. Biol. Chem.*, 193, 26 (1951)], to show that absolutely no protein was present. This clearly indicates that the whole amount of aminoacylase used had been bonded to the carrier. The immobilized aminoacylase composition thus obtained was then tested for enzyme activity by the following method. A mixture consisting of 10 mg of test specimen, 1 cc of pure water, 2 cc of 0.1 M phosphate buffer with pH 8.0 and 1 cc of 0.1 M n-acetyl-DL-methionine (pH 8.0, containing 5.0×10^{-4} mols/l Co^{++}) was held at 37°C for 30 minutes by way of incubation. The liberated methionine was assayed by the ninhydrin colorimetric method.

Activity units were indicated by assuming the unity, 1 u/g, to represent production of 1 μmol of methionine from 1 g of a given specimen under a fixed set of conditions. The immobilized aminoacylase composition obtained in this example was found to have about 1,500 u/g of activity. The process may also be used to immobilize glucoamylase, glucose isomerase and glucose oxidase.

Aminoacylases on Polymers with Tertiary or Quaternized Amino Groups

Porous inorganic supports coated with a film of a crosslinked polymer containing tertiary amine or quaternary ammonium salt groups are used by *F. Meiller and B. Mirabel; U.S. Patent 4,132,596; January 2, 1979; assigned to Rhone Poulenc Industries, France* as supports for aminoacylase. These support-enzyme complexes possess very good mechanical properties, and may be used in a column, even under pressure; their volume does not undergo change; they are not biodegradable and may be sterilized. Moreover, they lose their activity only extremely slowly and may be regenerated very simply.

These complexes are enzymatically active, stable products consisting of aminoacylase adsorbed on a support, and characterized in that the support is a porous mineral substance having a grain size between 4 μm and 5 mm, a specific surface of the order of 5 to 150 m^2/g, a pore diameter of 500 to 2,500 A, a pore volume of 0.4 to 2 ml/g, and coated, in an amount of less than 10 mg/m^2, with a film of crosslinked polymer containing or bearing tertiary amine or quarternary ammonium salt groups.

As porous mineral support, there may be used metal oxides such as: titanium oxide, aluminas, and more especially silicas. These supports have average pore diameters of 500 to 2,500 A and preferably 600 to 1,500 A, a specific surface of 5 to 150 m^2/g and preferably 20 to 50 m^2/g, a grain size of 4 μm to 5 mm, and a pore volume of 0.4 to 2 ml/g. The functional groups, i.e., tertiary amines or quaternary ammonium salts, are represented by the formulas:

$$-CH_2-\underset{\substack{|\\ R}}{N}-CH_2-$$

or $-CH_2N^+(R)_3X^-$, in which R, which may be identical or different, represents an alkyl or hydroxyalkyl group having 1 to 4 carbon atoms and X represents an inorganic or organic anion, such as for example, chloride, sulfate, nitrate, phosphate and citrate. These functional groups form part of the chain of the crosslinked polymer, or are fixed to the crosslinked polymer which covers the entire surface of the support. The crosslinked polymers which cover the surface of the inorganic support are known products obtained by any conventional polymerization process. They are prepared from monomers capable of crosslinking between themselves, or with another monomer (copolymers), if necessary in the presence of a catalyst.

Among these monomers there may be mentioned: epoxide compounds which crosslink with polyamines as catalysts; polyamine-formaldehyde and phenol-formaldehyde mixtures, which crosslink without a catalyst; mixtures of vinyl monomers, e.g., vinylpyridine-ethyleneglycol diacrylate, vinylpyridine-vinyltriethoxysilane, styrene-divinylbenzene, and styrene-vinyltriethoxysilane, which crosslink with an initiator which liberates free radicals, such as organic peroxides and azonitriles.

In the case where the crosslinked polymer on the surface of the inorganic support does not have the functional groups, such as defined above in its chain, it is necessary to modify the polymer by any known process. In the coating operation of the inorganic support, the amount of monomer(s) used should be such that the amount of crosslinked polymer having functional groups and distributed on the surface of the inorganic support should be between 1 and 6 mg/m^2, so as to form a film which does not block the pores of the support.

The inorganic supports coated with crosslinked polymers having functional groups thus obtained, have an exchange capacity of between 0.3 and 1.2 meq/g. The aminoacylases adsorbed on the support are enzymes which may be of animal origin, e.g., extracted from pigs' kidneys, or even produced by microorganisms, such as *Aspergillus, Lactobacillus arabinosus, Micrococcus glutamicus,* and *Pseudomonas cruciviae.*

Example 1: Fixation of the Enzyme – The support used consists of a silica having a grain size of 40 to 100 μm, a specific surface of 37 m^2/g, a pore diameter of 1,100 A, and a pore volume of 0.97 ml/g, coated with a crosslinked polymer containing

$$-CH_2-\underset{\displaystyle C_2H_5}{\underset{|}{N}}-CH_2-$$

functional groups. The support has the following characteristics: carbon content, 8.8%; nitrogen content, 2.4%; amount of fixed polymer, 3.3 mg/m^2; and exchange capacity, 1 meq/g. Six ml of the support, i.e. 3 g, are added to a column 1 cm in diameter; 40 ml of a 0.5% by weight solution in distilled water of L-aminoacylase extracted from pigs' kidneys is passed through the column at a flow rate of 120 ml/h.

The support on which the enzyme is adsorbed is washed by passing 50 ml of distilled water through the column at a flow rate of 120 ml/h. The support-enzyme complex obtained has an activity of 40 units/g of complex, a unit being the number of micromols of L-amino acid obtained per minute at 55°C.

Separation of L-methionine – An aqueous solution of 0.1 M N-acetyl-DL-methionine, containing 5×10^{-4} M of Co^{++} ions, at a pH of 7 and a temperature of 55°C, is passed through the column at a rate of 60 ml/h. The concentration of L-methionine in the effluent is determined by the ninhydrin reaction. It is found that the conversion rate is 92 mol % after 3 hours' operation.

Example 2: Fixation of the Enzyme – A support consisting of silica having a grain size of 100 to 200 μm, a specific surface of 24 m^2/g, a pore diameter of 1,400 A, and a pore volume of 1.1 ml/g, and coated with a styrene-based crosslinked polymer having the functional groups:

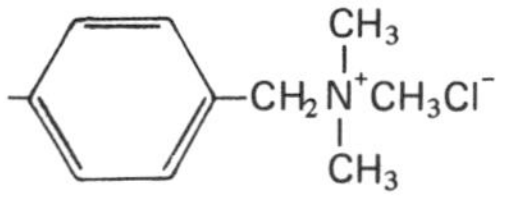

$$-C_6H_4-CH_2N^+(CH_3)_3Cl^-$$

is used; the support has the following characteristics: carbon content 4.8%; chlorine content, 2%; nitrogen content, 0.9%; amount of fixed polymer, 3.3 mg/m^2; and exchange capacity, 0.5 meq/g. 3 g of the support are contacted for 1 hour with 100 ml of a 0.5% by weight solution in distilled water of the same L-aminoacylase as in Example 1, the pH being adjusted to 7. The support-enzyme complex formed is then separated by filtration, and washed with distilled water. Its activity is 45 units/g.

Separation of L-methionine – The complex is added to a column identical to that of Example 1, through which a solution of N-acetyl-DL-methionine is passed under the same conditions as in Example 1. The molar conversion rate is 100%.

Porous Ion-Exchange Cellulosic Supports

Ion-exchange materials based on cellulose substrates have been prepared by combining a suitable compound containing an ionizable chemical group with natural or regenerated cellulose. The proposed materials are, however, costly because of their expensive preparation and have other disadvantages.

The process developed by *D.T. Jones, K.R. Rees and G.E. Jowett; U.S. Patent 3,905,954; September 16, 1975; assigned to Viscose Development Company Limited, England* enables ion-exchange activated regenerated celluloses to be obtained in a wide variety of physical forms, including many that have been difficult or impossible to attain. Further, the activated regenerated cellulose is generally obtained relatively economically and has superior absorption and/or ion-exchange properties for large ions. Examples of large ions for which the ion-exchange activated regenerated cellulose will generally be found suitable include ions derived from proteins such as enzymes and components of blood and tissue; carbohydrates, e.g., charged polysaccharides such as the mucopolysaccharides; nucleic acids, e.g., ribonucleic acids and deoxyribonucleic acids; dyestuffs, e.g., Congo Red; fatty acids, and quaternary ammonium compounds.

Suitable substances which may be used with cellulose to form ion-exchange materials include compounds containing amino, alkylamino, guanidino and quaternary amino groups for the preparation of anion-exchange materials, e.g., diethylaminoethyl chloride, di(hydroxyethyl)aminoethyl chloride, dimethylaminoethyl chloride, 1-(diethylamino)-2,3-epoxy propanol, p-morpholinoethyl chloride and salts thereof; and compounds containing sulfo, phosphoric and carboxyl groups for cation-exchange materials, e.g., chloromethane sulfonic acid, chloroethane sulfonic acid and 1,3-propane sultone.

Compounds are also known which can be attached to cellulose to form materials capable of fixing biologically active materials such as enzymes; and the use of such compounds is also included within the scope of the present process. Thus, the activating substance may be a compound containing pendant triazinyl groups. Preferably, in addition to such an activating substance, an activating substance is used in which the active component is an ionizable chemical group.

It will in most cases be preferable to crosslink the cellulose to some extent so as to obtain extra structural stability. Any of the crosslinking agents known in the art may be used for this purpose including, e.g., epichlorhydrin, dichlorhydrin, dibromoethane, dichloroethane, 1,2,-3,4-diepoxybutane, bisepoxy propyl ether, ethylene glycol bisepoxy propyl ether and 1,4-butanediol bisepoxy propyl ether. In this process a regenerated cellulose is prepared by the following steps:

(1) The alkali cellulose is prepared by any suitable method including the traditional one of soaking sheets of wood pulp in caustic soda, pressing out surplus caustic soda, and disintegrating the sheets to form crumbs of alkali cellulose. The alkali cellulose may also be prepared by grinding natural cellulose to a powder and impregnating the powder with a limited quantity of caustic soda solution.

(2) Treat with an ion-exchange activating substance, e.g., diethylchloromethylamine hydrochloride, and with a crosslinking agent such as epichlorhydrin.

(3) Treat with caustic soda.

(4) Treat with carbon disulfide.

(5) Dissolve the product in a solution of caustic soda.

(6) Precipitate in the required form by a known regeneration method, e.g., by the action of heat; by means of an acid; by treatment with a hot strong electrolyte; or by treatment with a strong electrolyte followed by an acid. One or more reinforcing agents and/or pore-forming materials may be added prior to regeneration. Examples of reinforcing agents include hemp, flax, cotton, viscose yarn, nylon and polyester. As pore-forming materials there may be mentioned sodium sulfate and sodium phosphate.

Example 1: Natural cellulose in the form of steeping-grade pulp normally supplied to the viscose industry was steeped in 18% (w/v) caustic soda solution, and the steeped pulp was drained and then pressed to a press ratio of 3.3. The resulting alkali cellulose, which comprised 27.5% cellulose, 15.5% caustic soda, and 57% water, was shredded in a Z-arm grinder to form crumbs. 363.5 g of the alkali cellulose crumbs were conditioned at 5°C in a Z-arm mixer, and 58.3 g N,N-diethyl-2-chloroethylamine hydrochloride were added to the cool crumbs.

The resulting mass was mixed for 30 minutes at 5°C, after which time the temperature was raised to 50°C and reaction allowed to continue for a further 60 minutes. The temperature was then reduced to 30°C and 65 g carbon disulfide were added to the reacting mass. The ensuing xanthation reaction was allowed to continue for 60 minutes at 30°C, giving diethylaminoethyl, (DEAE), cellulose xanthate having a γ number of 89. (The γ number provides a measure of the degree of substitution of the cellulose by xanthate groups.)

An activated regenerated cellulose sponge was produced from the DEAE-cellulose xanthate as follows. 153.5 g of 18.8% (w/v) caustic soda and 802 g water were added to 651.4 g of the xanthate to yield a viscose solution. 51.5 g cotton fiber as reinforcing agent and 5,500 g of $Na_2SO_4 \cdot 10H_2O$ (Glauber's Salt) as pore-forming agent were then added to the solution, yielding a paste. The paste was extruded into small cylindrical molds and was then regenerated in sodium sulfate solution at 95° to 100°C, the concentration of Na_2SO_4 being 25% (w/v). The resulting sponge had an ion-exchange capacity (for small ions, e.g., $C1^-$) of 0.95 meq/g.

Example 2: DEAE-cellulose xanthate was prepared as described in Example 1, and a viscose solution was prepared by treating 651.4 g of the xanthate with 385 g of 18.8% (w/v) caustic soda solution and 3,415 g water. The viscose solution was then diluted with water in the ratio of 1 part viscose to 5 parts water. After dilution, the viscose was sprayed by means of compressed air into a regenerating solution comprising 0.5% sulfuric acid and 10% sodium sulfate. The cellulose was regenerated in the form of a fine powder (200 to 400 mesh B.S.S.) which was washed to remove by-products and was then found to have an ion-exchange capacity of 1.13 meq/g.

Example 3: Powdered activated regenerated cellulose was prepared as described in Example 2. Crosslinking of the cellulose was effected by treating 100 g of the powder in a Z-arm mixer with a solution of 2.5 g epichlorohydrin, 30 g caustic soda, and 100 g water. The mass was raised to 50°C for 20 minutes, and was then washed free of by-products to yield a crosslinked regenerated DEAE-cellulose having an ion-exchange capacity of 1.11 meq/g.

Fibrous DEAE-Cellulose Embedded in a Hydrophobic Polymer

Fibrous ion-exchange cellulose has extremely high-loading capacities in regard to binding or immobilizing enzymes, especially glucose isomerase. In general, the fibrous cellulose materials determined most useful can be characterized as anion-exchange celluloses. Examples of such materials are the di- and triethylaminoethyl celluloses, such as DEAE-cellulose and TEAE-cellulose, and the cellulose derivatives of epichlorohydrin and triethanolamine, such as ECTEOLA-cellulose. The term fibrous refers to cellulose derived from natural sources which has been subdivided or fiberized by mechanical or chemical means and does not include cellulose or derivatives which have been subjected to chemical treatments which result in dissolution of the natural fibrous structure of the cellulose.

Although the high-loading capacity of DEAE-cellulose offers many advantages, especially for the isomerization of glucose to fructose with isomerase, such preparations suffer from the disadvantage of packing and, therefore, such are usually used in shallow beds to avoid problems due to excessive back pressure. Even when shallow beds are used, there is the possibility of channeling occurring whereby the substrate is not contacted to the desired degree with the bound or immobilized enzyme isomerase.

R.F. Sutthoff, R.V. MacAllister and K. Khaleeluddin; U.S. Patent 4,110,164; August 29, 1978; assigned to Standard Brands Incorporated have discovered that when fibrous ion-exchange cellulose is agglomerated with a hydrophobic polymer, such cellulose retains its capacity to immobilize or bind glucose isomerase. A number of polymers may be utilized in agglomerating the fibrous ion-exchange cellulose. Examples are melamine formaldehyde resins, epoxy resins, polystyrene and the like. The preferred polymer is polystyrene. The agglomerated fibrous ion-exchange cellulose composites also have the surprising property of being regenerated, i.e., after the activity of the immobilized glucose isomerase has decreased to a certain extent due to denaturation or other factors resulting from prolonged use, a solution of solubilized glucose isomerase can be brought into contact with the bed or column of agglomerated cellulose so that the glucose isomerase activity is increased again to the desired degree. Prior to regeneration, however, it is generally preferred to treat the composite with a solution of alkali to more readily make the ion-exchange sites of the fibrous cellulose more available to isomerase adsorption.

To form the agglomerated fibrous composite, the fibrous cellulose must be embedded in the hydrophobic polymer in such a manner that the cellulose is not completely encapsulated in the polymer. Otherwise, the capacity of the fibrous ion-exchange cellulose to adsorb enzymes would be substantially deleteriously affected. The greater the free surface of the cellulose, the greater the adsorptive capacity of the composite.

While a number of methods may be used to embed the fibrous cellulose in the cellulose in the hydrophobic polymer, the two which may be typically used involve dissolving the hydrophobic polymer and incorporating the other materials or heating the polymer to a plastic state and incorporating the other materials. The latter procedure is preferred since no solvent evaporation is necessary. The resulting material can then be reduced by grinding or the like and the granules classified on appropriate sized screens.

Example: The following materials were used to form the composite:

	Grams
Polystyrene (Polysar-510 high impact polystyrene)	265
Aluminum Stearate (utilized as a lubricant)	4.5
Fibrous DEAE-cellulose*	115
Alumina 10 μ	120

*Prepared by derivatizing C-100 chemical cellulose (International Filler Corp.) with diethylaminoethyl chloride hydrochloride. The ion-exchange capacity of the DEAE-cellulose was approximately 0.9 meq/g

The polystyrene was heated to a plastic state and the other ingredients mixed therein. The mixture was passed between the rolls of a two-roll heated mill. The rolls were 12 inches long and 6 inches in diameter and one roll was heated to about 180°C. The sheeted material was cooled, broken, ground in a Burr mill and screened to obtain a composite of 30 to 50 mesh granules using U.S. Standard Sieves. The fines were discarded. 12 g of the granules were added to a 196 ml solution of glucose isomerase containing approximately 20 IGIU ml^{-1} and buffered to a pH of 7.2. The suspension was stirred at 25°C for 7 hours.

The granules were separated by filtration, washed and the filtrate assayed for glucose isomerase activity. The granules adsorbed 330 IGIU g^{-1}db from solution by difference. 3.24 g of the immobilized granular enzyme was placed in a jacketed glass column 2.54 cm in diameter maintained at 65°C and a 50% dextrose solution was passed through the bed of granules in a continuous manner at a flow rate of 0.52 ml/min. The isomerized solution contained 33.1% fructose.

The pressure drop coefficient (K) of the granular immobilized enzyme was determined for column operation in the following manner. About 185 g db of the enzyme granules were added to a stainless steel column 5 cm in diameter to a height of 22.9 cm. The column was maintained at 50°C. A syrup containing 42% fructose and 50% dextrose at 70% ds, pH of 7.8, and 50°C was passed through the column at pressures of 394, 703, 1,758, and 4,218 g/cm^2 for 2 hours at each pressure. The flow rate and pressure drop were recorded at 30 to 40 minute intervals. The pressure drop coefficient of the packed column was determined to be 0.23 g min cm^{-1}ml^{-1}cp^{-1} at 394 g/cm^2 and at 4,218 g/cm^2 it was 0.35 g min cm^{-1}ml^{-1}cp^{-1} after 2 hours running time at each experimental pressure.

Crosslinked Cationic Starch Sponge

Another method of overcoming the packing of supported enzymes during continuous use in fixed bed columns has been developed by *R.E. Brouillard, U.S. Patents 4,033,820; July 5, 1977; and 4,029,546; June 14, 1977; both assigned to Penick & Ford, Limited.* A starch sponge modified to incorporate ion-exchange groups and optionally crosslinked is used as the enzyme carrier.

Starch in sponge form can be readily prepared by heating an aqueous starch solution until a colloidal dispersion results, freezing this dispersion, and then thawing the frozen dispersion to obtain the porous starch sponge material. The starch sponge is insoluble in water at temperatures up to those which begin to gelatinize the starch. The starch sponge may be prepared in large sheets, pads, or blocks. While it may be used in this form for packing the column, it is preferably used in subdivided form, which facilitates the filling and uniform packing of a fixed bed column. Exact sizing of the subdivided sponge material is not required. In general, the pieces of sponge material may range in size from 0.263 to 0.525 screen size (Tyler).

The starch used to form the sponge material is chemically modified to incorporate an effective amount of enzyme-immobilizing groups. Cationic groups are preferred but enzyme-immobilizing anionic groups are also known, and could be used for some purposes. The starch before conversion to sponge form may be modified to incorporate the di- and triethylaminoethyl groups, providing DEAE-starch or TEAE-starch respectively. Alternatively, the starch can be reacted according to known procedure with epichlorohydrin and triethanolamine to form ECTEOLA-starch, which is then formed into a sponge.

When certain enzymes are used, it is advantageous to crosslink the starch to prevent degradation of the sponge structure during use. This is particularly so when enzymes of the hydrolase group such as amyloglucosidase are used. Among the suitable crosslinking agents are formaldehyde, glyoxal, glutaraldehyde, and phosphorus oxychloride and various synthetic resins such as urea formaldehyde or melamine formaldehyde.

Example 1: Potato starch was slurried in water to a concentration of 22 Bé and a salt-caustic solution containing saturated sodium chloride and 30% sodium hydroxide was added under nongelatinizing conditions to yield 10% salt based on the weight of the water and 2.6% caustic based on the starch. An amount of 3-chloro-2-hydroxypropyltrimethylammonium chloride equal to 7% of the weight of the starch was added and the mixture reacted at 115°F for 24 hours. Thereafter, the slurry was adjusted to pH 1 with hydrochloric acid. Heating was continued until the starch derivative had been acid modified to a 5 g alkali fluidity of 17 ml [*Chemistry and Industry of Starch*, 2nd Ed., R.W. Kerr, Academic Press, N.Y., p. 133 (1950)].

Thereafter, the starch slurry was adjusted to pH 4 with soda ash. The product was filtered and washed by resuspending in water twice, filtering, and recovering the product each time. The final starch product was air-dried. An analysis indicated that the cationic starch product contained 0.35% added nitrogen. When stained with Fast Green SF dye solution, the granules became green, indicating strong cationic activity. A cationic starch sponge was prepared from the above product using the following procedure: 3 gal of a suspension containing 8% by weight of cationic starch in water was cooked at 195°F for 15 minutes.

After cooling to room temperature, the entire quantity in a 3 gal container was placed in a quick-freeze unit at 0°F and kept at this temperature for 72 hours. Afterward, the frozen cationic starch was removed and placed at room temperature for an additional 72 hours. At the end of this period, the starch sponge was placed in a wine press and excess water removed such that the wet sponge contained from 64 to 75% moisture.

Example 2: A quantity of 769 g of cationic starch sponge containing 200 g ds material was suspended in 400 ml of water and adjusted to pH 10 to 11 with calcium hydroxide. While maintaining the pH at 10 to 11, 60 g of phosphorus oxychloride was added dropwise with stirring over a period of 2 hours. Finally, the pH was adjusted to 4 with hydrochloric acid and the product washed with 4 volumes of water on the filter.

A portion of the final product was then placed in an oven at 50°C until dry, and another undried portion was further treated with enzyme and resin in the following manner: 148 g of damp starch sponge (50 g db) was mixed in a Hobart Blender with 50 ml of glucoamylase solution (Diazyme L-100), along with 25.0 g of Tybon 8003 in 25 ml of water.

The pH of the mixture was adjusted to 4.2. After stirring to insure uniform mixing, the mixture was placed in an air-dryer at about 50°C for 48 hours. Thereafter, the product was placed on a 20 mesh screen to remove any fines and the product thus prepared was ready for use in a column for the preparation of syrups.

Example 3: Stabilized (crosslinked) starch sponge containing bonded glucoamylase as prepared in example 2 was placed in a 1 inch diameter column, and 25% solids concentration syrup at 20% DE flowed through the column at a rate of 1.1 ml/min at pH 4.2 and 140°F. The DE of the effluent syrup was 84.9%. After 5 days of operation, the column was shut down while still producing 85.7% DE syrup.

Macroporous Anion-Exchange Resins

Insolubilized glucose isomerase wherein a large amount of isomerase is adsorbed and more than 40% of the activity is retained has been produced by *Y. Fujita, I. Kawakami, A. Matsumoto, T. Hishida, A. Kamata and Y. Maeda; U.S. Patent 4,078,970; March 14, 1978; assigned to Mitsubishi Chemical Industries Ltd., and Seikagaku Kogyo Co., Ltd., both of Japan.*

This process comprises contacting a macroporous anion-exchange resin having a porosity of 4.5 to 20% measured according to the aqueous dextran solution method and an exchange capacity of 0.035 to 0.1 meq/ml resin measured according to the polyanion decomposition method with glucose isomerase of 700 to 5,000 units per 1 ml of the resin at 40° to 75°C and a pH of 6.5 to 9 until at least 50% of the total isomerase and from 500 to 4,000 units of the isomerase per 1 ml of the resin have been adsorbed. Such macroporous anion-exchange resin is conveniently prepared by any known process, e.g., a monovinyl monomer and a crosslinkable monomer are copolymerized in the presence of any material which is removable by a solvent and does not take part in the reaction, such as polystyrene. After completion of the polymerization, the resin is treated with a solvent to extract the material, e.g., polystyrene, and then the anion-exchange group introduced.

The anion-exchange resin thus produced has, in general, macropores the radius of which ranges from 10^1 to 10^4 A. However, it should be noted that among the pores, smaller pores cannot adsorb glucose isomerase whereas much larger pores the internal surface of which is similar to the surface of the gel-type anion-exchange resin also do not take part in adsorption.

In general, although the pore size and the total pore volume of the macropores in an anion-exchange resin may vary depending upon the conditions under which the resin is prepared, such as the amount of a crosslinkable monomer and the amount and molecular weight of a polymer which is added initially in the polymerization mixture, it is difficult to determine the definite relationship of the pore size and the total pore volume to the preparation conditions. In this connection, it is desirable to select optimum conditions under which the most appropriate resin is produced by experimental procedures.

In order to introduce anion-exchange groups into a matrix resin, it is preferable that the matrix resin be subjected to treatment to introduce chloromethyl groups followed by treating with various amine compounds, e.g., an aliphatic amine such as trimethylamine, dimethylethanolamine, ethylenediamine, diethylenetriamine and triethylenetetramine, and a cyclic amine, such as pyrrolidine, morpholine and piperidine. However, it is appreciated that, if a weak basic amine is used, there is observed a tendency to a little release of the adsorbed isomerase from the resin, so it is desirable to use a tertiary amine, such as trimethylamine and dimethylethanolamine, to convert the resin into a quaternary ammonium type.

In order to assure that more than 50% of glucose isomerase supplied is adsorbed, it is required to contact the aqueous glucose isomerase with the anion-exchange resin for a long time. In the case of either immersing the resin in the aqueous isomerase solution or passing the solution upwardly through a column packed with the resin, the contact time may vary depending upon the type of resin employed, and in general, from 2 to 18 hours is satisfactory under the conditions abovementioned.

The anion-exchange resin may be a free type, although it is preferred to use a salt type, such as SO_4^{-2}, Cl^-, PO_4^{-3} or CH_3COO^- type prepared by treating an anion-exchange resin with the corresponding acids, since such salt-type resin can adsorb more glucose isomerase than a free type. The anion-exchange resin specified for this process is capable of adsorbing a large amount of glucose isomerase, but if too large an amount of glucose isomerase is adsorbed on the resin, the degree of the activity retention will be lower. Thus, the amount of the glucose isomerase to be adsorbed on the resin is from 500 to 4,000 units, preferably 1,000 to 3,000 units per 1 ml of the resin in wet state.

Adsorption of Glucose Isomerase on Various Resins

	Characteristics of Ion Exchange Resin						Insolubilized Glucose Isomerase			
							Degree of			
Ex.	Degree of Cross-linkage* (%)	Type of Anion Exchange Group	Total Ion Exchange Capacity (meq/ml)	Porosity (%)	Ion Exchange Capacity** (meq/ml)	Type of Matrix	Adsorption (%)	Activity Retention (%)	Activity Yield	Note
1	10	trimethyl ammonium	0.79	10.4	0.074	SO_4^{--}	99.9	91.0	90.9	–
2	10	trimethyl ammonium	0.95	10.0	0.088	SO_4^{--}	99.7	29.1	91.8	–
3	10	trimethyl ammonium	0.79	19.4	0.039	SO_4^{--}	93.7	88.2	86.6	–
4	15	trimethyl ammonium	0.57	19.4	0.061	SO_4^{--}	99.8	91.5	91.3	
5	25	trimethyl ammonium	0.67	7.5	0.070	SO_4^{--}	96.3	91.1	87.7	–
6	25	dimethyl ammonium	0.60	13.0	0.096	SO_4^{--}	99.9	88.5	88.4	–
7	10	trimethyl ammonium	0.79	10.4	0.074	PO_4^{---}	99.5	88.0	87.6	–
8	10	trimethyl ammonium	0.79	10.4	0.074	OH^-	95.0	80.2	76.2	–
9	10	trimethyl ammonium	0.79	10.4	0.074	Cl^-	100.0	90.3	90.3	–
10	10	trimethyl ammonium	0.79	10.4	0.074	CH_3COO^-	82.4	84.7	69.8	–
11	–	dimethyl ammonium	–	16.4	0.042	SO_4^{--}	72.3	61.7	44.6	Amberlite IRA-93
A	–	DEA***	1.01	8.7	0.028	SO_4^{--}	0	–	0	Amberlite IRA-910
B	–	DEA***	0.91	15.7	0.027	SO_4^{--}	0	–	0	Amberlite IRA-911
C	–	DEA***	0.53	0.7	0.034	SO_4^{--}	10	45	4.5	Amberlite IRA-938
D	8	trimethyl ammonium	>1.3	0	0.028	SO_4^{--}	0	–	0	Dianion SA-100
E	10	trimethyl ammonium	>1.3	3.0	0.030	SO_4^{--}	7	–	–	Dianion PA-320
F	–	trimethyl ammonium	1.03	16.4	0.038	SO_4^{--}	46.7	78.9	36.8	Amberlite IRA-900

*Percentage of the weight of the divinyl monomer to the total weight of the divinyl monomer and monovinyl monomer

**After adsorption.

***Dimethyl ethanol ammonium.

Examples 1 through 11: 100 g of glucose isomerase (NAGASE) produced from *St. phaeochromogenus,* was suspended in 700 ml of demineralized water and, after addition of 80 mg of crystals of egg white lysozyme, the suspension was agitated at 40°C for 45 hours and centrifuged to obtain an extracted solution of glucose isomerase. It was found that the activity of the extracted solution was 200 units/ml. Then, each 50 ml of the extracted solution (the total activity being 10,000 units) was mixed with each of anion-exchange resins listed in the table, the bed volume of which was 10 ml in wet state and the mixture was agitated at 50°C for 6 hours to effect adsorption of isomerase on the resin to obtain an insolubilized glucose isomerase.

The characteristics of the insolubilized isomerase are also given in the table. The table also deals with comparative examples referred to by letters. In Example 11 and Comparative Examples A through F, the anion-exchange resins used were commercially available. The anion-exchange resins used in examples other than Example 4 were produced by suspension polymerization using styrene, divinylbenzene, polystyrene and toluene in water and benzoyl peroxide as catalyst, followed by subjecting particles of styrene-divinylbenzene copolymer to chloromethylation and introducing each of the anion-exchange groups listed in the table.

The resin of Example 4 was produced by polymerizing a mixture of styrene, divinylbenzene and n-heptane using benzoyl peroxide as catalyst to obtain particles of styrene-divinylbenzene copolymer followed by chloromethylating and introducing anion-exchange groups. Resin types of SO_4^{-2}, PO_4^{-3}, OH^-, Cl^- and CH_3COOO^- listed in the table mean that the anion-exchange resins were treated with 2 N sulfuric acid, 1 N phosphoric acid, 2 N sodium hydroxide, 2 N hydrochloric acid and 1 N acetic acid, respectively.

CHELATING AND METAL BONDING OF ENZYMES

Metal Salt Modified Polymers as Supports

Active water-insoluble enzymes coupled with organometallic derivatives of a support such as polysaccharide, nylon or glass are disclosed by *A.N. Emery, S.A. Barker and J.M. Novais; U.S. Patent 3,841,969; October 15, 1974; assigned to Ranks Hovis McDougall Limited, England.* These insolubilized enzyme derivatives may be regenerated (after their activity has substantially decreased) by using a method similar to the original coupling for the regeneration. The organometallic derivatives of these supports may be obtained by reaction with a salt or other derivative of tin, titanium, zirconium or iron that gives a suitable organometallic derivative. The organometallic derivative is then reacted with the enzyme at pH 3 to 7 at 0° to 45°C for preferably 3 to 18 hours.

The weight of free enzyme applied per gram of titanium, tin, zirconium or iron derivative may be 20 to 1,000 mg, preferably 40 to 250 mg. The polysaccharide may be cellulose or a derivative which may be in the form of microcrystalline cellulose, diethylaminoethylcellulose, carboxymethylcellulose, sawdust, wood chips, filter paper or a membrane. The polysaccharide may also be crosslinked dextran or crosslinked starch. The enzyme may thus be chelated to a titanate, zirconate, stannate or iron complex of cellulose or a derivative thereof, still containing the requisite hydroxyl groups.

Thus cellulose may be reacted with solutions of $TiCl_4$, $TiCl_3$, $ZrCl_4$, $SnCl_2$, $SnCl_4$, $Ti_2(SO_4)_3$, $FeCl_2$, $FeCl_3$, or $FeSO_4$ or other salts of these metals and dried or stirred at 20°C to 60°C and washed free from unreacted salts. Microcrystalline cellulose or glass may be reacted with 12.5% w/v titanous chloride or 15% w/v titanic chloride and dried, e.g., for 18 hours at 45°C. Nylon may be reacted with 15% w/v titanic chloride and dried at 45°C.

Example 1: Titanous chloride solution (1 ml 12.5% w/v) was added to 100 mg of microcrystalline cellulose (Sigma cell type 19) in a 5 cm diameter watch glass and mixed well for 1 minute. The suspension was then placed in a 45°C oven overnight. The resulting solid was transferred to a test tube and then washed by stirring for 10 minutes with acetate buffer (0.02 M, pH 4.5) (10 ml) and centrifuging. Washing with acetate buffer was repeated a further 2 times after which the slurry was a cream color.

A solution of dialyzed commercial amyloglucosidase (e.g. *Aspergillus niger*) activity 67 units/mg protein) (1 ml) with 4 ml acetate buffer (0.02 M, pH 4.5) at 4°C was added to this slurry and stirred magnetically at 0° to 5°C for 18 hours. The water-insoluble amyloglucosidase derivative was subjected to 5 cycles of washing with acetate buffer (0.02 M, pH 4.5) (10 ml) and sodium chloride solution (1 M, 10 ml) in the same buffer. The amyloglucosidase derivatives were finally washed twice with acetate buffer (0.02 M, pH 4.5). Further washings failed to elute any enzyme from the water-insoluble amyloglucosidase derivative. The experiment was repeated using various concentrations of $TiCl_3$, namely (1) 12.5, (2) 6.25, (3) 2.5, and (4) 1.25% w/v. Results are given below.

	Enzyme Amount				
Run	Units/mg Free Protein	Units/mg Bound Protein	Units/g Dry Solid	Bound Protein, mg/g Dry Solid	Retention of Specific Activity, %
1	67	31.1	4,570	147	46.5
2	67	31.9	3,380	106	44
3	67	30.3	2,660	88.5	45.5
4	67	33.8	2,350	69.5	51.1

Note: 1 amyloglucosidase unit is taken as that which liberated 1 micromol of glucose at 45°C in 1 minute.

Example 2: The same procedure as in Example 1 was used except that in place of titanous chloride solution, samples (1 ml) of one of each of the following salts were taken at the various concentrations given: Titanic chloride (15, 7.5, 3, 1.5, 0.3 and 0.15%), titanous sulfate (15%), stannic chloride (1, 5, 10 and 20%), stannous chloride (10%), zirconium tetrachloride (1.5, 10 and 20%), ferrous chloride (0.1, 1, 5, 10 and 20%), ferric chloride (10%) and ferrous sulfate (10%). Enzyme activities and protein contents of the water-insoluble amyloglucosidase derivatives are given below.

			Enzyme Units		
Run	Coupling Salt	Salt Concentration (% w/v)	per g Solid	per mg Bound Protein	Retention of Specific Activity (%)
5	$TiCl_4$	15	3,300	34.5	52.0
6	$TiCl_4$	7.5	3,020	32.5	49.0
7	$TiCl_4$	3	2,800	30.5	46.0
8	$TiCl_4$	1.5	2,600	30.8	46.5
9	$TiCl_4$	0.3	775	29.2	43.5

(continued)

Run	Coupling Salt	Salt Concentration (% w/v)	. . . . Enzyme Units . . . . per g Solid	per mg Bound Protein	Retention of Specific Activity (%)
10	$TiCl_4$	0.15	665	48.0	72.5
11	$Ti_2(SO_4)_3$	15	389	32.8	51.3
12	$SnCl_4$	1	1,080	53.0	83.0
13	$SnCl_4$	5	1,900	34.0	53.0
14	$SnCl_4$	10	2,100	36.0	56.0
15	$SnCl_4$	20	1,570	30.2	47.5
16	$SnCl_2$	10	1,325	29.4	45.0
17	$ZrCl_4$	1	1,680	36.6	55.0
18	$ZrCl_4$	5	1,550	31.6	47.5
19	$ZrCl_4$	10	1,820	33.8	51.0
20	$ZrCl_4$	20	1,700	27.5	41.0
21	$FeCl_2$	20	1,720	32.2	48.5
22	$FeCl_2$	10	1,470	37.2	56.0
23	$FeCl_2$	5	1,025	32.8	49.0
24	$FeCl_2$	1	525	26.4	39.6
25	$FeCl_2$	0.1	400	49.5	74.0
26	$FeCl_3 \cdot 6H_2O$	10	1,430	41	61.5
27	$FeSO_4 \cdot 7H_2O$	10	526	24.6	31.5

Hydrous Metal Oxides or Hydroxides as Reversible Supports

A simple and inexpensive means has been found by *S.A. Barker, I.M. Kay and J.F. Kennedy; U.S. Patent 3,912,593; October 14, 1975; assigned to Gist-Brocades N.V., Netherlands* for the reversible immobilization of proteins (including enzymes), peptides, antibodies, antibiotics and whole cells and substances which contain protein, peptide and cyclic peptide moieties (sometimes termed, for convenience, nitrogen-containing organic substances) which involves the formulation of water-insoluble metal chelates.

The water-insoluble chelates are formed with a hydrous oxide or a hydrous hydroxide of a metal, e.g., tin, iron, vanadium, titanium or, preferably, zirconium. These metal chelates exhibit some aspect of the biological properties (e.g. catalytic, antibody or antibiotic properties) characteristic of the nitrogen-containing organic substances. Preferred chelates of the nitrogen-containing organic substances are those formed with the hydrous oxide of zirconium and to a lesser extent the hydrous oxides or hydroxides of tin [Sn(II)], iron [Fe(III)], vanadium [V(III)] and titanium [Ti(III,IV)]. Metal chelates of enzymes such as glucose oxidase, glucose isomerase, lactase, catalase, invertase, α- and β-amylase, pullulanase, penicillin acylase, bacterial protease, trypsin, chymotrypsin, glucoamylase, dextranase, glucosidase and Maxatase can be formed by this process.

In a preferred aspect of this process, an aqueous solution of a zirconium salt (e.g. chloride or sulfate), or other metal salt capable of forming a hydrous oxide or hydroxide able to chelate with a nitrogen-containing organic substance, or an aqueous solution of a mixture of such salts is added to a hydroxide, preferably ammonium or sodium hydroxide, until a pH of 3 to 8.5 (preferably 6 to 8) is attained and the hydrous oxide or hydroxide of the metal is formed. The suspension of the precipitated hydrous oxide(s) or hydroxide(s) is then mixed with the nitrogen-containing organic substance in an aqueous medium (preferably the organic substance is in aqueous solution) to yield a solid metal chelate of the nitrogen-containing organic substance. Alternatively, the chelate can be formed by coprecipitating in situ the hydrous metal oxide or hydroxide in an aqueous medium containing the nitrogen-containing organic substance and an appropriate metal salt by addition to the aqueous medium of a hydroxide (preferably ammonium hydroxide).

The process is usefully carried out at ambient temperature with the pH of the reaction medium at 4 to 8, and advantageously at pH 6 to 8. Mixing of the hydrous metal oxide or hydroxide with the nitrogen-containing organic compound in the aqueous medium can be continued for some hours but generally 0.5 to 3 hours is sufficient for maximum retention of biological activity of the organic compound.

Example 1: Optimum pH for Chelation of Glucose Oxidase with Hydrous Zirconium Oxide – To a 2.0 ml aliquot of zirconium tetrachloride solution (1.53 g/ml) was added slowly, with stirring, ammonium hydroxide (2.00 M) to precipitate hydrous zirconium oxide and give a pH of 4.5 to 8.5. To each of these samples was added 500 μl of glucose oxidase solution [0.0206 g of glucose oxidase (activity, 144 units/mg) in 10 ml phosphate buffer, pH 7.0]. The samples were stirred at 4°C for 18 hours, then centrifuged and washed with distilled water (6 x 5 ml). One milliliter of distilled water was added to each sample and stirred well to give a suspension. 0.5 ml of this suspension was taken and dried to constant weight at 120°C. A further 1.0 ml of the suspension was removed and diluted 100 times with distilled water.

The enzyme activity of this diluted sample was assayed as follows: 2.5 ml of 2,2'-azino-di(3-ethylbenzothiazoline-6-sulfonic acid) (ABTS) solution (0.5 mg/ml ABTS in 0.1 M phosphate buffer, pH 7.0), 10 μl of peroxidase (2 mg/ml in 0.1 M phosphate buffer, pH 7.0) and 500 μl of a 10% glucose solution were mixed and the optical density (OD) at 415 nm monitored for several minutes. 20 μl of the oxide suspension were added and the increase in OD at 26°C monitored continuously using a Beckman DB-G spectrophotometer linked to a chart recorder. The results are shown in the following table. (ABTS is a complex dye; ΔE is the change in optical density.)

pH of Chelation	ΔE/min at 415 nm	Wt of 0.5 ml Sample	Activity units/gram
4.45	0.00599	0.0201	60
5.02	0.0199	0.0226	176
5.54	0.0244	0.0226	215
5.92	0.0320	0.0236	271
6.47	0.0353	0.0258	273
7.49	0.0318	0.0254	250

Example 2: Chelation of Peroxidase with Various Metal Oxides – Peroxidase (1.0 mg) was chelated with the hydrous oxides of zirconium, tin(II), iron(III) and vanadium(III) at pH 6.0 as previously described for glucose oxidase. After washing, the activity of the enzyme chelate was assayed as follows. The suspension of solid phase peroxidase was diluted 100 times, and 25 μl added to 2.5 ml of ABTS (0.5 g/l) in 0.1 M phosphate buffer, pH 5.0 and 0.5 ml hydrogen peroxide solution (10 mM). The OD was monitored at 415 nm and 26°C. Standards were analyzed using 2.5 ml of ABTS, 25 μl of hydrogen peroxide of various concentrations (0 to 2.5 mM), and 0.5 ml of the original peroxidase solution (diluted 1,000 times).

Metal	Activity (units*/mg)
Zr (IV)	81
Sn (II)	69
Fe (III)	70
V (III)	36

*1 unit = 1 μM H_2O_2 oxidized per minute at 25°C, pH 5.0.

Precipitated Metal Oxides on Insoluble Supports

Insoluble supports are preactivated by precipitating a hydrous metal oxide in the presence of the insoluble support. The resulting composite materials are used by *R.W. Coughlin, M. Charles and B.R. Allen; U.S. Patent 4,115,198; September 19, 1978* as supports for immobilizing biologically active substances.

In the present process a hydrous metal oxide is precipitated from a solution of the metal upon an insoluble base carrier material suspended in the solution. The resulting composite material is then used as a support for immobilizing nitrogen-containing, biologically-active, organic substances. It appears that any suitable insoluble particulate material, organic or inorganic, metallic or nonmetallic, can be coated with hydrous metal oxide in this way. The method of coating by precipitation of the hydrous metal oxide from a solution of the metal in which the insoluble base carrier material is suspended is necessary to obtain the improved results of the present process.

Merely mixing base carrier material with already precipitated hydrous metal oxide does not result in composite materials upon which enzymes can be immobilized simply and conveniently to form immobilized enzymes of improved properties; neither does mere contact between the base carrier particles and the metal solution.

Apparently, by precipitation of the hydrous metal oxide in the presence of the insoluble base support material colloidal particles of the former are caused to adhere intimately, tenaciously and evenly upon particles of the latter, thereby forming a composite material having improved properties; e.g., the resulting material will have the approximate size and density of the insoluble base support material, but usually a greatly enhanced specific surface area, and in particular the resulting composite material will show a superior ability to bind biologically active, nitrogen-containing organic substances, presumably due to the coating of hydrous metal oxide.

It is not well understood why or how such hydrous metal oxides have superior ability to bind such biologically active materials but the mechanism has been attributed to chelation in U.S. Patent 3,912,593. The metal oxides employed in the present process are those insoluble in aqueous solutions of about the pH where the respective enzyme displays maximum activity. Ordinarily this will be within a pH of 3 to 11. Any metal salt can accordingly be used which will react to precipitate an oxide when the pH of its solution is adjusted to 3 to 11. Metals which are desirable in this regard are Ti, Zr, Sn, Fe, Al, V, Hf with Ti, Al, Sn and Zr preferred. Enzymes which can be immobilized by the present process include amyloglucosidase, lactase, glucose oxidase, invertase, trypsin, glucose isomerase, catalase, pronase, urease, lactate dehydrogenase, amino acid acylase, penicillin acylase, proteases and dextranase.

The insoluble support materials that may be used in this process are illustrated by the following table.

Porous Support Materials

Supplier	Material	Surface Area (m^2/g)	Particle Size (μ)
Carborundum	SAEH-33 (Al_2O_3)	4.0	-250 +149
Carborundum	SAEH-33 (Al_2O_3)	8.0	-177 +105
Houdry Chemical	Catalyst 9C79-12C (SiO_2)	unknown	-250 +149
Semi- or Nonporous Support Materials			
Glidden Metals	316-L-Si*	0.19	-149 +53
International Nickel	Sinter 90 (Ni-NiO)	unknown	-420 +180
Harshaw Chemical	Ni-1000 P (NiAl)	unknown	-100 +50

*Prealloyed stainless steel powder.

Example 1: To a reaction flask containing 10.0 g of alumina (SAEHS-33, 4m²/g) and 62.5 ml of chilled water, 6.3 ml of titanic chloride reagent was slowly added with vigorous stirring. The reaction flask was partially immersed in an ice bath to remove the heat generated by the addition of the chloride solution. The flask was then removed from the ice bath, heated to 80°C, and held at that temperature for 1.0 hour. The resulting titanium oxide-coated alumina support was then removed from the flask and washed thoroughly with tap water until a clear rinse was obtained. The coated alumina was then dried overnight at 110°C.

A solution of commercial amyloglucosidase was prepared by adding 1.0 g (dry wt.) of the enzyme powder to 50 ml of deionized water. Samples of both titanium oxide-coated and noncoated (untreated) alumina (5.0 g each) were contacted at 0° to 5°C for 18 hours with 25 ml of the enzyme solution. The alumina samples were then washed 4 to 6 times with deionized water before being assayed for enzyme activity. Amyloglucosidase activity of the immobilized derivatives was determined and expressed as units per gram of support (dry wt.). One unit of activity represents the production of 1.0 μmol of glucose per minute at 37°C.

Sample No.	Support Material	Immobilized Activity (u/g)	Binding Efficiency (%)
1A	Titanium oxide-coated Carborundum (4 m^2/g) Alumina (SAEHS-33)	88	24
1B	Noncoated Carborundum (4 m^2/g) Alumina (SAEHS-33)	26	7

Note: Binding efficiency equals the percentage of the soluble activity offered for binding which is recovered as immobilized activity.

Example 2: Samples of titanium-oxide-coated alumina (4 m^2/g) were prepared as described in Example 1 except that the liquid $TiCl_4$ reagent was added in differing proportions to the support materials slurried in water. The reaction time defined as the length of the reaction period at 80°C was also varied. The resulting coated samples were contacted at 0° to 5°C for 18 hours with solutions of equal concentrations of amyloglucosidase and then washed with deionized water before being assayed as in Example 1.

Sample No.	Water-to-Alumina Ratio (ml/g)	Volume Proportion of $TiCl_4$ Liquid Reagent to Water (ml/ml)	Reaction Time (min)	Immobilized Activity (u/g)
2A	6.1	0.1	60	88
2B	2.0	0.1	60	66
2C	2.0	0.1	15	41
2D	6.1	0.2	60	76
2E	6.1	0.2	180	51

Note: Support, Carborundum Co. (4 m^2/g) alumina.

Urease on Porous Titania Treated with Stannous Ions

An immobilized urease composite having improved stability and half life has been prepared by *R.A. Messing; U.S. Patent 3,910,823; October 7, 1975; assigned to Corning Glass Works* by reacting a porous titania carrier with a solution of stannous ions and then reacting the pretreated carrier with a urease solution.

The exact mechanism by which the pretreatment with a stannous salt solution enhances the bonding ability of the titania carrier for urease is not fully understood. However, it is possible that the salt, on dissociation in aqueous solution, forms a number of surface stannous bridges capable of reacting with available sulfhydryl groups on the urease molecule. Unlike many enzymes which are elongate, urease is substantially spherical; thus, it would be expected that the enzyme would offer few bonding sites which actually contact the titania surface. Hence, it is possible that available stannous ions somehow (e.g., by forming hair-like surface projections) provide a mechanism for bonding more sites on the urease than is possible by simple adsorptive forces.

In the example below, a stannous chloride solution is used as a source for stannous ions although it is thought other stannous salts which readily dissociate in water could also be used [e.g., $SnBr_2$, SnI_2, $SnSO_4$, $Sn(NO_3)_2$] to pretreat the carrier. The stannous salt treating solution should have a concentration of between 0.1 to 10% by weight, preferably 1% by weight. The surface should be treated for at least 5 minutes. The treated carrier is then reacted with a urease solution for at least 1 hour. To have a basis for comparing the stability and half lives of immobilized urease composites prepared by pretreating the carrier with a stannous salt solution, a urease composite is prepared according to the method of U.S. Patent 3,850,751.

Example: Urease Immobilized on Titania with a Stannous Pretreatment – A 500 mg sample of titania carrier was preconditioned with 20 ml of a 1% $SnCl_2 \cdot 2H_2O$ solution by placing both in a shaking water bath at 37°C for 45 minutes. The stannous chloride solution was then decanted and the carrier was washed with 3 aliquots containing 20 ml of distilled water. After the final wash was decanted, 20 ml of a 1% aqueous urease solution containing 80 Sumner Units was added to the carrier. The carrier sample and urease solution was shaken in a water bath at 37°C for 2 hours.

The mixture was then allowed to stand at room temperature for 22 hours, after which the enzyme solution was decanted and the enzyme composite was washed successively with 20 ml volumes of distilled water, 0.5 M sodium chloride solution and distilled water. The composite was then transferred to a small column where it was stored in water at room temperature and assayed periodically using a one molar urea solution (6%) to determine composite activity with time and the half life. The activities of the control composite and the composites of this example were measured to determine storage stability over a period of up to 62 days. The results are indicated in the following table.

Activity (Sumner Units per Gram)

Days of Storage	Control	Example
0	–	–
1	4.36	–
2	–	2.80
3	–	2.80
5	2.32	–
6	1.98	2.80
7	1.98	2.54
8	1.56	2.38
9	–	2.26
10	–	2.00
12	1.11	–
13	1.60	–
15	1.48	–
19	0.73	–
21	–	1.30
24	–	1.64
27	–	1.44
28	0.47	–
29	–	1.40
34	0.34	1.10
36	–	1.18
43	–	1.06
52	–	0.82
58	–	0.82
62	–	0.68

Reversible Bonding to Supports with Boronic Acid Groups

M.H.W. Phillipp, M.L. Bender and P.V. Valenzuela; U.S. Patent 3,912,595; October 14, 1975; assigned to the U.S. Secretary of Health, Education, and Welfare have found that serine-hydrolytic enzymes can be bound with high selectivity and with a minimum of manipulative steps by complexing the enzyme at pH near 7.5 with a boronic acid derivative covalently bonded to a water-insoluble granular support. It has further been found that a boronic acid derivative covalently bonded to a granular water-insoluble support material provides a suitable material for the separation of soluble serine-hydrolytic enzymes from other types of protein materials, including structurally similar proteins.

Soluble serine-hydrolytic enzyme means a soluble globular protein in which a serine group is located at the active catalytic site. Among the soluble serine-hydrolytic enzymes specifically separated by this method are proteolytic enzymes both of mammalian origin, such as chymotrypsin, and of bacterial origin such as subtilisin. Also separated by the process are serine-esterases, such as cholinesterase. The serine-hydrolytic enzymes may be separated from closely related protein materials, including the inactive enzyme precursors commonly known as zymogens and metalloenzymes, as well as from nonproteinaceous impurities.

Reversible complexes of the serine-hydrolytic enzyme and the boronic acid covalently bonded to a water-insoluble support are formed by passing a solution containing the enzyme at a pH from 7.0 to 8.5 over a column packed with the support to which is covalently bonded boronic acid. To assure removal of impurities from the serine-hydrolytic enzyme, the column containing the complexed enzyme is washed with a buffer solution at pH 7.0 to 8.5. The enzyme may be removed from the complex by using a buffer at a different pH. Preferred buffers for the aqueous solutions from which the serine-hydrolytic enzymes are complexed with the boronic acid covalently bonded to the support include phosphate buffers and volatile buffers such as triethylammonium bicarbonate (pH 7.5 to 8.5), triethylammonium acetate (pH 4 to 5) and triethylammonium formate (pH 3).

Although any pH from about 7.0 to 8.5 is operable for the complex formation step, the preferred pH range is from 7.2 to 7.8, especially 7.5. Liberation of the serine-hydrolytic enzyme from the complex with boronic acid can be done using a buffer in the same range as that in which the complex is formed. However, more rapid separation of the enzyme from the complex is achieved when a different pH is selected for this purpose.

Example 1: Binding of a Boronic Acid Group to Dextran – (A) Activation of Dextran with Cyanogen Bromide: Two hundred milligrams of dextran is combined with 8 ml of an aqueous solution containing 25 mg of cyanogen bromide per milliliter of water. The reaction takes place at 0°C while the mixture is stirred for about 30 minutes. During this time, the pH of the solution is kept between 10 and 11.5 by the addition of 2 N sodium hydroxide solution as needed. The product is washed with suction on a fritted glass filter using cold water and used directly for the ensuing steps.

(B) Attachment of the Boronic Acid Group to Activated Dextran – The product obtained above is contacted with 200 mg of triglycine in 2 ml of 0.1 M sodium bicarbonate solution and stirred for about 30 hours at room temperature. The product is filtered on a fritted glass filter and washed successively with 0.1 M sodium bicarbonate, 0.003 M hydrochloric acid, water and 0.5 M sodium chloride solution and then treated with a ten-fold molar excess of a 1:1 molar mixture of m-aminophenylboronic acid and N-cyclohexyl-N′-β-(4-methylmorpholinium)-ethylcarbodiimide at pH 4.5 to 6.0 for 6 hours at 0°C with stirring. The product is washed with water and used to pack a column.

Example 2: Subtilisin (Novo or BPN′ subtilisin) is suspended in triethylammonium bicarbonate buffer solution at pH 7.5 at an enzyme level of 0.01 M. The solution is passed through a column packed with the material prepared according to Example 1 and previously equilibrated with triethylammonium bicarbonate buffer at pH 7.5.

The column now containing complexed enzyme is washed with the same buffer at pH 7.5. The enzyme may be freed by washing the column containing complexed enzyme with acetate buffer at pH 5 to 6 and the product enzyme recovered from the buffer solution by freeze-drying.

FLOCCULATED ENZYMES

Flocculation of Whole Cell Enzyme Preparations

M.E. Long; U.S. Patent 3,935,069; January 27, 1976; assigned to R.J. Reynolds Tobacco Company has prepared flocculated whole cell enzymes in the presence of certain metallic compounds. The metallic compounds give improved hardness to the immobilized enzyme materials, particularly for use in continuous operations. The metallic compounds which are useful include the oxides, hydroxides, phosphates and sulfates of magnesium, calcium, iron and manganese. Preferred flocculating agents are synthetic anionic or cationic polyelectrolytes.

The effectiveness of the process is shown by results obtained in the treatment of microbial cells with magnesium oxide. Fermentation of *Arthrobacter* NRRL B-3728 was carried out using the procedure described in the example of U.S. Patent 3,821,086. A 19 liter portion of the resulting fermentation broth was treated with 7.6 liters of a 1% Primafloc C-7 solution (previously adjusted to pH 5) followed by 7.6 liters of a 1% Primafloc A-10 solution (previously adjusted to pH 7). Primafloc C-7 and A-10 are cationic and anionic flocculants, respectively.

Following addition of the flocculants to the broth, gentle agitation of the broth was continued for 5 minutes and the resulting flocculated cells were recovered by filtration. The wet flocculated cell aggregates were then extruded and the extrudate was dried at 55°C in a forced-draft oven for 18 hours. The dried material was ground in a Wiley mill and particles of 16 to 20 mesh were collected by sieving. These 16 to 20 mesh particles were assayed for glucose isomerase activity and found to have 62.7 milliunits per gram.

A second 19 liter portion of the *Arthrobacter* fermentation broth was stirred vigorously with 60 g of magnesium oxide (Magmaster No. 340) until the pH had increased from 5.7 to 8.5. Further processing of this magnesium oxide-treated broth was the same as above. The dried 16 to 20 mesh particles were found to contain 52.1 milliunits per gram of glucose isomerase activity.

The hardness of flocculated *Arthrobacter* NRRL B-3728 cells was evaluated by using an Instron Model 1132 system. The dried 16 to 20 mesh particles obtained in each case were immersed in water at room temperature and allowed to stand for 4 hours to induce any swelling of the particles.

The water covering the particles was then replaced with an equal volume of a 2 M glucose solution which had been buffered to pH 8.0 and the resulting immersed cells were maintained at 60°C overnight. Immersion of the particles in 2 M glucose was to simulate conditions to which the cell aggregate particles would be subjected when used for isomerizing glucose.

Measurement of hardness with the Instron system was done by randomly selecting 10 individual particles from the immersed sample, removing excess glucose solution from the particles by briefly contacting them with absorbent paper and immediately placing them on the flat measuring surface of the instrument so that the particles were grouped closely together in the central portion of the measuring surface but not touching each other. Using a drive speed of 5 cm/min, a chart speed of 5 cm/min and a force range of 50 kg full scale, the magnesium oxide-treated cell aggregates required a force of 9.5 kg while the control cell aggregrate particles required only 3.3 kg.

Evaluation of the flocculated *Arthrobacter* cells was also made under continuous column operating conditions to determine flow rates over extended periods of time. For this test 75 g of the dried 16 to 20 mesh particles obtained from each of the two flocculations were immersed in water for a period of time to permit any swelling of the particles.

The material was then washed with 8 liters of 0.004 M magnesium chloride that contained 0.1 M sodium bicarbonate as a buffer. The washed particles in each case were packed into 1-inch diameter jacketed columns that were heated to 60°C. A 50% by weight glucose solution was allowed to flow by gravity through each column and the flow rates were monitored hourly. The results shown in the table below, clearly indicate that the flocculated cells containing magnesium oxide exhibit a more gradual decrease in glucose solution flow rates than do the untreated cells.

Comparative Glucose Solution Flow Rates Through Beds of Flocculated *Arthrobacter* Cells

Bed Composition	Glucose Solution Flow - - Rate (liters/hour) - - Initial	Final	Total Volume of Glucose Solution (liters)	Total Operating Time (hours)
Cells flocculated with Primafloc C-7 and A-10, 75 grams of 16-20 mesh particles	21.9	6.0	349.8	44
Cells flocculated with Primafloc C-7 and A-10 and containing magnesium oxide, 75 grams of 16-20 mesh particles	21.0	13.8	681.75	44

In addition to magnesium oxide, other compounds have been found to be effective in producing flocculated *Arthrobacter* cell aggregates with increased hardness as determined by the Instron Model 1132 measuring system. Shown in the table below, are the results obtained in the flocculation of *Arthrobacter* NRRL B-3728 cells in the presence of various additives using Primafloc C-7 and A-10 flocculating agents used in accordance with the process.

Additive (Use Level: 2.5 g/l of broth)	Hardness of 16-20 Mesh Particles by Instron (kg)	Glucose Isomerase Activity (mU/g)
Control (no additive)	3.3	62.7
Magnesium oxide, CP	21.5	54.4
Magnesium hydroxide	10.5	43.1
Magnesium monohydrogen phosphate	7.4	45.4
Calcium oxide	7.5	37.2
Calcium hydroxide	8.0	35.0
Calcium triphosphate	4.6	54.6
Calcium sulfate	6.0	53.5
Manganese monoxide	6.7	40.4
Ferric oxide	5.4	46.5
Magnesium carbonate	2.9	59.4
Magnesium chloride	2.2	41.3
Calcium carbonate	2.8	49.6

Flocculation of Cell-Free Enzymes

C.W. Nystrom; U.S. Patent 3,935,068; January 27, 1976; assigned to R.J. Reynolds Tobacco Company has also produced flocculated cell-free enzyme preparations that are useful in effecting enzyme-catalyzed chemical transformations.

More specifically, cell-free enzyme preparations are treated with polyelectrolyte flocculating agents under conditions favoring formation of stable enzyme-containing aggregates. Inert support materials are preferably incorporated into the aggregates to produce a material that is particularly suitable for use in continuous enzymatic processes in which a substrate solution is passed through a bed of the aggregates.

It has been found that totally nonionic polyelectrolytes used alone are essentially ineffective as flocculants for cell-free enzyme preparations. However, nonionic or partially nonionic polyelectrolytes may be used in combination with anionic or cationic polyelectrolytes to give useful enzyme-containing aggregates.

Accordingly, the polyelectrolytes which may be satisfactorily used for this process are water-soluble anionic or cationic polyelectrolytes, preferably of synthetic origin, as well as combinations of anionic, cationic and nonionic polyelectrolytes. In a preferred embodiment of the process, an inert filter aid is added to the enzyme-containing medium prior to addition of the flocculant. This results in a uniform distribution of the filter aid throughout the enzyme-containing aggregates and improves substrate solution flow rates through a bed of the aggregates when the latter is used in a continuous enzymatic process.

By inert filter aid is meant a material which improves the filtration properties of the flocculated enzyme and which has no appreciable detrimental effect on either the activity of the enzyme or the stability of the substrate and product involved in the enzymatic process of interest. Typical of the filter aids which may be used are infusorial earth such as Celite and asbestos fibers. Also useful are water-insoluble polymeric adsorbents such as acrylic ester resins XAD-1 and XAD-4.

In another preferred embodiment of this process, a cationic and an anionic polyelectrolyte are used sequentially to effect immobilization of the cell-free enzyme preparation. In most cases a superior floc is obtained if the cationic polyelectrolyte is added to the enzyme-containing medium before addition of the anionic polyelectrolyte. A filter aid may be used in conjunction with this embodiment also but it is not necessary.

Example: Fermentation of *Arthrobacter* nov. sp. NRRL B-3728 is carried out as described in U.S. Patent 3,645,848 and the cells are harvested by centrifugation. A 10% by weight suspension of the harvested cells in 0.01 M Tris buffer (pH 8.2) and 0.001 M magnesium chloride is treated with crystalline lysozyme (0.05 mg/g of wet cells).

The cell suspension is incubated at 60°C for 2 hours with gentle agitation. The suspension is then cooled to 12°C and subjected to sonication for 5 minutes. Centrifugation yields a clear, cell-free supernatant containing isomerase. A 200 ml portion of this supernatant containing 2 g of Silflo filter aid is slowly agitated at about 25°C while adding to it 40 ml of a 1.5% solution of Primafloc C-7 cationic flocculant which has been previously adjusted to pH 8.0 by the addition of sodium hydroxide.

Agitation is continued for a few minutes before collecting the enzyme-containing flocculated material by vacuum filtration. A 1.3 g portion of this material is packed into a 1-inch diameter glass column and through it is passed a 50 weight percent solution of dextrose containing 0.004 M magnesium chloride and adjusted to pH 8.0. The column is maintained at 60°C and the flow rate of the effluent syrup is controlled to give 440 ml per day. After one day the effluent syrup contains 16.1% fructose and after nine days contains 14.2% fructose thus demonstrating the stability of the flocculated enzyme.

IMMOBILIZATION BY ENTRAPMENT

ENTRAPMENT OF ENZYMES IN MEMBRANES

Swollen Collagen or Zein Membranes

W.R. Vieth, S.S. Wang and S.G. Gilbert; U.S. Patent 3,843,446; October 22, 1974; and U.S. Patent 3,977,941; August 31, 1976; both assigned to Research Corporation have prepared enzymatically active protein-enzyme complex membranes by treating a swollen protein membrane with an aqueous solution of a compatible active enzyme. These membranes are used to effect enzymatic reactions.

A wide variety of synthetic polypeptides and natural proteins may be used in the process. Examples of suitable natural proteins include collagen, zein, casein, ovalbumin, wheat gluten, fibrinogen, myosin, mucoprotein, and the like. Examples of suitable synthetic polypeptides include polyglutamate, polyaspartate, polyphenylalanine, polytyrosine, and copolymers of leucine with p-aminophenylaniline. The selection of a particular synthetic polypeptide or natural protein, in modified or unmodified form, will be largely determined by the nature of the enzyme being complexed, the substrate to be treated, and the reaction environment to be encountered. Because of their inertness to a large number of enzymes, collagen and zein are preferred natural protein materials.

One suitable method for forming a collagen membrane is as follows. The collagen source is first treated with an enzyme solution to dissolve the elastin which encircles and binds the collagen fibers. Proteolytic enzymes, from either plant or animal sources, may be used for this purpose, although other types of enzymes are equally satisfactory. The collagen source is then washed with water and the soluble proteins and lipids are removed by treatment with a dilute aqueous solution of a chelating agent, such as ethylenediamine tetrasodium tetraacetate. The collagen fibers are then swollen in a suitable acid, such as cyanoacetic acid so as to form a collagen fiber dispersion. This dispersion can then be extruded or cast into a suitable membrane form. The dried collagen membrane is then annealed at 60°C, 95% RH for 48 hours.

The collagen fiber dispersion can also be electrodeposited to form suitable membranes. The collagen membrane is then prepared for complexing with the enzyme, generally by being swollen with a low molecular weight organic acid, or in some instances with suitable bases so that the pH ranges from 2 to 12. Suitable acids include lactic acid and cyanoacetic acid. If desired, plasticizers or other additives may be added during the swelling step. Swelling is accomplished by submerging the membrane in the acid bath for ½ to 1 hour, depending upon the particular conditions of the bath, generally at room temperature. In excess of this level will result in the conversion of the collagen to a soluble gelatin. The membrane is swollen by the acidity of the organic acid added and the use of the acid as a plasticizer. No other additive is needed. A change in water binding capacity results from the acid treatment.

As a second example of using natural protein to complex enzymes, zein film is prepared by casting a solution of zein. The same procedure was used to prepare protein-enzyme complexes as was done with collagen film, except that the swelling of the zein film was aided by adding plasticizers, such as 1,5-pentanediol. Zein is characterized by a relative deficiency of hydrophilic groups in comparison with most proteins. In fact, the high proportion of nonpolar (hydrocarbon) and acid amide side chains accounts for the solubility of zein in organic solvents and its classification as a prolamin.

Zein casting solutions are prepared by dissolving the protein in the organic solvent of choice by gentle stirring, at room temperature, for 1 to 2 hours until solution is complete. Examples of suitable solvents include 81% (w/w) isopropyl alcohol and 4% methyl Cellosolve (ethylene glycol monomethyl ether). The clear solutions which contained 20 to 30% by weight of dry zein, are of amber color. Curing agents, such as formaldehyde, and a plasticizer may be added shortly before film casting.

The zein membranes useful in this process generally have a thickness of 0.005 to 0.1 mm. The zein membrane is then prepared for complexing with the enzyme by swelling with a plasticizer, if plasticizer was not added before film casting. Suitable plasticizers include 1,5-pentanediol, glycerol, and sorbitol. This is accomplished by submerging the membrane in a bath of 2% (w/w) plasticizer in water for 10 hours at room temperature. The swollen membrane is then dried with tissue paper and soaked in an aqueous enzyme solution until complexing is completed. Usually, this requires a period of from 10 hours to 2 days. The temperature range during this period should be maintained within 4° to 20°C, depending upon the particular enzyme used.

Example 1: This example demonstrates the formation and use of a lysozyme collagen membrane complex. One cc of a 1 mil thick collagen film (postheated at 55°C, 90% RH for 48 hours) was swollen in a lactic acid solution (pH = 3) and then washed in running tap water for 5 minutes. The washed film was then soaked in an enzyme solution of 250 mg of lysozyme in 15 cc of water, and stored at 2°C for 14 hours. The soaked film was then layered on cellulose acetate, and dried at room temperature to yield a lysozyme-collagen membrane complex.

A solution of *Micrococcus lysodeictikus* (300 mg of dried cells per liter) was used to assay the enzymatic activity of the complex by measuring the decrease in optical density at 450 mμ of the bacterial solution. The dried membrane complex was washed with 10 liters of running water and its initial enzymatic activity measured. This was determined on the basis of the decrease in optical density divided by the optical density after a 30 minute reaction period. The complex was washed with 2 liters of water between individual experiments. Initial activity was 0.538, corresponding to 53.8% of the cells being lysed. The second repetition gave an activity of 0.420, corresponding to 42% lysis; the third repetition gave an activity of 0.489, or 48.9% lysis; the fourth experiment gave an activity of 0.533, or 53.3% lysis.

Example 2: This example describes the preparation and use of an invertase-zein complex. 2 cc of a 1.5 mil thick zein film (formaldehyde tanned) was swollen in 2% (w/w) 1,5-pentanediol. The film was then dried with tissue paper, soaked in an enzyme solution of 200 mg invertase in 15 cc of water, and stored in a refrigerator overnight. The soaked film was then layered on a cellulose acetate substrate, and dried at room temperature.

The dried film, which is a zein-invertase membrane complex, was used in the following experiment to hydrolyze sucrose. 400 cc of 6% sucrose solution was used as a substrate in assaying enzyme activity which was followed polarimetrically in a recirculation reactor system. After the complex was assayed for its activity, it was washed with 2 liters of water and reused. The results obtained are shown in the table below.

Number of Washings	Sucrose Inverted, %*
1	40
2	16
4	8
6	8
10	6
15	7

*In 2 hours of reaction time at 25°C and pH 6

The enzymatic activity of the invertase-zein complex decreased gradually after washing, finally reaching a stable limit.

Tanned Protein Membranes Containing Whole Cell Isomerase

Enzymatically active protein whole microbial cell complex membranes are prepared in the process disclosed by *W.R. Vieth, S.S. Wang and R. Saini; U.S. Patent 3,972,776; August 3, 1976; assigned to Research Corporation*. A wide variety of synthetic polypeptides and natural proteins may be used in forming the membrane portion of the complex. However, collagen and zein are preferred protein materials.

Suitable whole microbial cells which may be used in this process include: bacteria, particularly *Acetobacter; Lactobacillus; Clostridium;* and *Pseudomonas; Actinomycetes,* particularly *Streptomyces*; molds or fungi, particularly *Aspergillus* and *Penicillium*; and yeast, particularly *Saccharomyces* and *Torula*. These whole microbial cells may contain a wide variety of different enzymes which are membrane bound enzymes, intracellular or endo enzymes and essentially all the enzymes involved in cellular metabolism.

In preparing these complexes, it is necessary to "fix" the cells. That is, the whole cells should be maintained in a stationary phase such as by heat treatment at 60° to 80°C for one minute to two hours and/or by adding uncouplers such as 2,4-dinitrophenol, pentachlorophenol, other highly substituted lipid soluble phenols, gramicidin, and dicoumarol. When spores are used, as the whole cells, fixing can be dispensed with since spores are in a resting stage.

The whole microbial cells are dispersed in a suitable inert medium, such as water. Good results are obtained when the solids content of the dispersion is 0.5 to 10% by weight. The protein contained in the dispersion is from 0.1 to 5% by weight, depending upon the particular protein being used. For instance, with collagen, the suspension will become too viscous if more than 1% by weight is used. Preferably, the protein is used in amounts of 0.1 to 1.0% by weight.

The temperature of pretreatment of the cell suspension is important, since it is necessary to destroy the proteolytic activity of the cells. If the proteolytic activity is not destroyed, the cells will autolyze and destroy the structural integrity of the protein membrane. It has been found that effective destruction of proteolytic activity is obtained if the cell suspension is heated to 80° to 82°C for at least 10 minutes. Longer periods and higher temperatures are also effective, but if the temperature is raised above 90°C, there may be adverse effects on the quality of the cells.

The protein carrier is dispersed by any conventional technique. A suspension of whole cells is then comixed and complexed with it, after which a complex membrane is cast and dried. Good results are obtained when the thickness of the membranes used is 6 to 20 mils. The pH of the dispersion at the time of contact should correspond to the type of whole cell being used and can be 2 to 12. If the whole cells contain an alkaline enzyme, the pH of the dispersion should be alkaline, and if the enzyme is acidic, the pH of the dispersion should be acidic.

At the termination of the drying period, the whole microbial cells will have become bonded to the membrane. Bonding will be a multiple bonding mechanism which includes at least salt linkages, van der Waal forces and hydrogen bonding. One important aspect of this process is the fact that the microbial cells are bonded directly to the protein of the membrane. Unlike the previous techniques of bonding enzymes to protein materials, the whole cells are not prior diazotized, nor is the membrane diazotized, nor are the protein material or whole cells otherwise treated chemically to produce covalent linkages. In the process, the whole cells are contacted with the protein macromolecules and it has been found that bonding occurs directly.

After drying, the film can then be tanned by dipping in acidic chromium sulfate solution (11.67%, aged at least one month) for the case of an acid enzyme, or, in the case of an alkaline enzyme, alkaline formaldehyde solution (10% HCHO, 1% $NaHCO_3$) or glutaraldehyde solution. Tanning time is usually between ½ minute and ½ hour, depending on the stability of the enzyme. The tanned membrane is then washed thoroughly in running water for at least ½ hour. The membrane is then ready for use. The membrane can be stored either dry or wet. The thickness of the dried membrane is usually between 2 to 10 mils or 0.005 to 0.1 mm.

Example: To 20 g of frozen *Streptomyces phaeochromogenes* cells (batch number 6-7-72), distilled water was added to bring the volume of the suspension to 100 ml. The suspension was stirred for 10 minutes to disperse the cells. The cell suspension was heated at 80°C for 15 minutes with shaking (130 rpm) on a Gyrotory water bath shaker and the suspension was then cooled slowly to room temperature.

110 g of hide collagen was added to 100 ml of water and blended in a Waring blender. The pH of the blended mixture was adjusted to 11.2 by adding 1 N NaOH solution. The resulting cell suspension was then slowly blended into the collagen dispersion. The mixture was cooled intermittently to prevent a rise in temperature above 20°C. The mixed dispersion was cast onto a Mylar sheet and then dried to room temperature. Total dry weight of the complex was 12.5 g.

The dried complex was tanned in formaldehyde solution [10% (w/v), containing 1% sodium bicarbonate, pH 8.35] for one minute. It was then washed immediately for half an hour under cold running water. 1.23 g of the tanned complex membrane was cut into small chips which were assayed for glucose isomerase activity. The chips ¼" x ¼" in dimension were used to catalyze the isomerization of 5 ml of 18% glucose (containing 0.01 M Mg^{++} and 0.2 M pH 7 phosphate buffer) at 70°C.

Percent of glucose converted to fructose was noted at different reaction times. The table below summarizes the results. From the initial conversion data it was calculated that *Streptomyces phaeochromogenes* cells equivalent to 131 units of glucose isomerase activity were immobilized per gram of the complex.

Time, (hrs)	Percentage Conversion
0.5	9.0
1.0	16.6
1.5	23.5
2.0	30.1
4.0	42.1
5.5	45.0

Alkali Saponified Starch-Acrylonitrile Graft Polymer Membranes

Highly water-absorbent graft polymers have been developed by *M.O. Weaver, E.B. Bagley, G.F. Fanta and W.M. Doane; U.S. Patent 3,985,616; October 12, 1976; assigned to the U.S. Secretary of Agriculture* which may be modified and used to immobilize enzymes by entrapment or covalent linking. These polymers comprise water-insoluble alkali salts of saponified gelatinized starch-polyacrylonitrile graft polymers containing gelatinized starch (GS) and saponified polyacrylonitrile (HPAN) in molar ratios of 1:1.5 to 1:9 GS:HPAN. These materials are capable of absorbing in excess of 300 parts of water by weight per part of the water-insoluble solids.

These polymers are prepared by the steps of: (a) saponifying a gelatinized starch-polyacrylonitrile graft polymer (GS-PAN) in an aqueous slurry with an alkali in amounts such that the molar ratio of alkali to the acrylonitrile repeating unit of the GS-PAN is from 0.1:1 to 7:1 to form a water-soluble saponified GS-PAN (water-soluble GS-HPAN); and (b) optionally adjusting the pH of the water-soluble GS-HPAN to 3 followed by isolating the resulting water-insoluble acid form of GS-HPAN, and readjusting the pH of the water-insoluble acid form of GS-HPAN to 4 to 12 to reform the water-soluble GS-HPAN; and (c) drying the water-soluble GS-HPAN to a moisture level of 1 to 15% water by weight. The GS-HPAN finds use in materials requiring high fluid absorbing properties and can be used in many forms including films produced from water dispersions of the polymer.

The absorbency of a film prepared by this process can be reduced from 600 to 100 g of water per gram of polymer by 3 Mrad of gamma irradiation. Such control of film swellability can be used in the gel entrapment of enzymes, where a large initial swelling is desirable to allow the enzyme to penetrate into the film, but subsequent use of the film requires a lower degree of swelling to keep the enzyme trapped. Water-insoluble GS-HPAN film can also be treated with an aqueous solution of glucoamylase, then treated with a solution of calcium chloride. The calcium chloride shrinks the fluid swollen film sufficiently to entrap the enzyme, which remains active through five separate and consecutive reactions with starch. The amounts of enzyme, GS-HPAN, and calcium chloride would be different for each enzyme, depending upon the enzyme's molecular size and desired activity.

Any soluble mineral salt is suitable for shrinking the gel after it has absorbed the enzyme solution. Immobilized amylase enzymes were prepared by two methods. One, an α-amylase, was covalently bonded to the GS-HPAN composition, and the other, a glucoamylase, was absorbed according to the method described above. It was suprising that in neither of the two immobilizing enzyme compositions did the amylase degrade the starch moiety of the GS-HPAN, even though these enzymes are well-known degraders of starch.

Example 1: A stirred slurry of 27.0 g of cornstarch in 480 ml of water was heated for 30 min at 88°C and then cooled under nitrogen to 27°C. To the stirred mixture, 1.5 mmol of ceric ammonium nitrate solution in 1 N nitric acid was added. After 10 minutes, 31.8 g of acrylonitrile was added and the mixture stirred for 3 hours at 35°C. The starch-polyacrylonitrile graft copolymer was isolated by filtration and dried to yield 57.0 g of GS-PAN having a 1:33 ratio of starch to acrylonitrile. 50 g of the graft polymer was slurried in water to form a 6.25 wt % solution and NaOH was added to give a 1:3.1 molar ratio of alkali to acrylonitrile in the polymer.

The mixture was stirred at 90°C until the orange-red color, which formed on initial heating with alkali, had disappeared. The smooth, light yellow dispersion was diluted to 5% solids, cooled at room temperature, and sulfuric acid added to give a pH of 3.2. The water-insoluble acid form, which precipitated on acidification, was isolated by centrifugation and washed successively with water, 5:1 methanol:acetone, and acetone. The product was finally isolated by filtration and dried at 25°C, yielding 37 g.

To prepare the absorbent polymer, 5 g of the water-insoluble acid form of GS-HPAN were sifted into 500 ml of rapidly stirred water, and high speed stirring continued for 30 seconds. The pH of the dispersion was adjusted to 6 to 9 by addition of 1 N potassium. The resulting viscous solution was then spread on a tray and dried at 25° to 35°C to yield a continuous film of absorbent polymer. After tray drying, the film may be coarsely ground to a flake-like material which could absorb 648 g of water per gram of polymer.

Example 2: Enzymes may be rendered insoluble by entrapment in swollen water-insoluble films of GS-HPAN. A 0.25 g sample of tray dried film, prepared as described in Example 1, was swollen in 10 ml of water containing glucoamylase enzyme having approximately 160 glucoamylase units. The film absorbed all of the solution. A 10 ml portion of 1% calcium chloride solution was added to shrink the film, and the resulting film was then washed several times to remove untrapped enzyme.

Activity of the film was tested by placing it in a Petri dish and adding 5 ml of a 1% starch solution in 1% calcium chloride at pH 6.5. Five different starch solutions were used consecutively, and after incubation, solutions were tested for reducing sugar by the ferricyanide procedure as shown in the table below.

Trial	Reducing Sugar,* mg per 100 ml
1	288
2	214
4	340
5	520

*As glucose

Example 3: Films of GS-HPAN serve as insoluble supports for enzymes, and enzymes are readily covalently bonded to the support. A 0.25 g sample of tray dried film prepared as described in Example 1 was added to 10 ml of a 5% solution of 1-ethyl-3-(3-dimethylaminopropyl) carbodiimide and 5 ml of stock HT-1000 α-amylase solution, prepared by making a 10% solution and then filtering. The film was allowed to swell for 30 minutes and 10 ml of 1% calcium chloride solution was added. The film, which shrank considerably, was transferred to fresh calcium chloride solution and finally to distilled water. Gradual swelling of the film was observed on standing in water for 1 week. Water was finally decanted off, and the film reshrunk with 1% calcium chloride.

Enzyme activity was tested by reacting the film containing the bound enzyme, with a 1% starch solution in 1% calcium chloride. The film was placed in a Petri dish and 10 ml of the starch solution added. Incubation was for 1 hour at 40°C, and iodine solution was used to test for unreacted starch. The solution was then tested for reducing sugar by the ferricyanide procedure. The film was then washed with 1% calcium chloride solution, and the procedure was repeated with a fresh starch solution. The procedure was performed a total of three times, and the results are shown in the table below.

Trial	Starch-Iodine Color	Reducing Sugar,* mg per 100 ml
1	Colorless	114
2	Colorless	121
3	Slight blue	78

*As glucose

Charged Enzymes Entrapped in Porous Polymeric Membranes

The membrane of the process disclosed by *T. Miyauchi and S. Furusaki; U.S. Patent 4,051,011; September 27, 1977; assigned to W.L. Gore & Associates, Inc.*

has an enzyme which is electrically entrapped on the surface or in the fine pores or irregularities of the base material. An enzyme is dispersed in an appropriate solvent such as water, with a buffer solution to give an electrically charged colloidal solution. This solution becomes negatively charged at a pH higher than the isoelectric point (pH 4 to 8), and positively charged at lower pH. Electrophoresis of the charged enzyme solution takes place under the application of an electric field to the charged solution with the resulting accumulation of the colloidal particles at one electrode, either the anode or cathode depending on their electric charge.

When the electrophoresis of the enzyme solution is carried out in the presence of a suitable base material (the "carrier") which intersects the dc electric field applied to the colloidal solution, the colloidal enzyme particles collide with the carrier on their way to the electrode and adhere to it, and in due course, almost all the particles have adhered to the surface of the carrier to form a high density layer of enzyme.

The enzyme particles which adhere to the carrier cannot be stripped off easily while the electric field is applied. Therefore, the entrapped enzyme can be used in the same manner as a fixed enzyme. In this case, use of a carrier having the power of absorbing the enzyme after the removal of the electric field will eliminate the need for electric field application after the enzyme is once entrapped. Examples of such suitable carriers include porous materials, fibrous materials, or solid materials having a roughened surface with pore size or irregularity dimensions ranging from 100 A to 10 μ.

The shape of the carrier may be that of a plate, sheet (or film), and so on. One or more sheets of the carrier are placed in a colloidal enzyme solution in a spaced parallel arrangement and in a direction perpendicular to that of the electric field. Porous or irregular materials such as porous synthetic resin products, natural or artificial semipermeable membranes, felt, asbestos plate, sintered glass, sintered metal, or unglazed pottery are capable of entrapping an enzyme, and are effective as a carrier.

Among other materials, expanded porous polytetrafluoroethylene membrane has a fibrillated structure which allows the clinging of enzyme, excellent chemical stability, good mechanical strength, uniform and controllable pore size, easy availability in any size and shape, and thus, it is highly recommended for use as the carrier. Too weak an electric field applied to the colloidal enzyme solution prolongs the time of electrophoresis, and too strong a field produces an enzyme solution which aggregates and precipitates. An electrical field between electrodes should be preferably 2.5 to 3.5 V/cm.

A carrier having an electric charge opposite to that of a colloidal enzyme solution may be put in the electrically charged enzyme solution, the enzyme molecules strongly adhere to and are entrapped in the surface of the carrier. The carrier thus formed by the adherence of the enzyme may be used as a type of fixed enzyme. In this instance, the carrier consisting of an insulating material, can be electrically charged by the frictional method, corona discharge method, etc. However, the use of an electret is quite effective in the process. An electret is a permanently polarized piece of dielectric material such as: synthetic polymers like polytetrafluoroethylene (PTFE), polyvinylidene fluoride, polyvinyl fluoride, polyvinyl chloride, nylon, polyethylene; certain natural waxes; or inorganic insulation materials

The electret is produced by irradiating the material with a beam of electrons, or γ-rays, by applying a corona discharge, or by heating the material to its melting point and placing it in a strong electric field during cooling. The electric charge thus applied can last permanently, with a sufficiently high enough voltage (in some cases, the voltage is as high as 1,000 V). When an enzymatic solution having an electric charge opposite to that of the carrier is applied to the electretized carrier, the enzyme strongly adheres to the carrier to give a fixed enzyme effectively applicable to the enzymatic reaction.

Example 1: A colloidal enzymatic solution with a negative charge was prepared by dissolving 3 mg of catalase into 100 ml of phosphate buffer solution of pH 7. Two 1 cm wide by 6 cm long nickel plates having a porous PTFE membrane (pore size, 2 μ, maximum) adhered to their surface, one being for the negative and the other for positive electrode, were dipped into the colloidal enzyme solution with a parallel clearance of 1 cm, and an electrical voltage of 3 V was applied to the electrodes for about 6 hours.

Then, the positive electrode was taken out from the solution, lightly washed with water, and immersed into a 0.03 mol/l H_2O_2 aqueous solution (pH 7). At that time, oxygen gas (O_2) was generated vigorously from the porous PTFE membrane on the positive electrode, exhibiting effective enzymatic reaction. The negative electrode showed no signs of enzymatic reaction. Enzyme was not found in the solution after the reaction. That is, no enzyme elimination from the carrier occured.

Example 2: A carrier was made by placing a negatively polarized electret (a fluorinated ethylene-propylene resin, FEP, film) inside a porous PTFE tube (i.d., 2 mm; o.d., 4 mm; length, 30 mm; pore size, 2 μ maximum), and closing both ends by adhesion. The carrier thus prepared was put in the same type solution as used in Example 1, and allowed to stand overnight at room temperature. Then the carrier was removed from the solution, and immersed in a 0.03 mol/l H_2O_2 aqueous solution, and vigorous O_2 gas bubbling from the porous PTFE tube surface showed an effective enzymatic reaction. As in Example 1, the carrier exhibited strong enzyme absorption, but no enzyme mixed with the reaction solution due to elimination, and no drop in enzyme activity was seen.

Example 3: An electret carrier, made by the method of Example 2, was immersed in a pH 5.9 glucoamylase solution (10 mg/ml, negative charge), and left overnight. Then, the substrate was placed in a solution of starch buffered to pH 5.9 by a phosphate buffer, and the rate of starch decomposition was measured as the change with time of the starch concentration. The following table represents the results.

Glucose Concentration (mg/ml)	Time (min)
0.08	0
0.25	30
0.35	60
0.57	120
0.80	180
1.05	240

ENTRAPMENT OF ENZYMES IN FIBERS

Extruded Hollow Filaments of Gelled Polymers

The process developed by *M. Yoshino, Y. Hashino and M. Morishita; U.S. Patent 3,875,008; April 1, 1975; assigned to Asahi Kasei Kogyo KK, Japan* relates to hollow filaments encapsulating an enzyme, microorganism or both. In this process encapsulation means that an enzyme or microorganism is present and fixed in hollow spaces of hollow filaments. The terminal ends of these hollow filaments may either be open or closed. As the hollow filaments used in this process are very thin, the encapsulated enzyme or microorganism remains fixed even if the terminal ends are open.

The enzyme used in this process may be, for example, penicillin acylase, racemase, esterase, glucose isomerase and the like. The microorganism to be used may be any species, including, for example, cyan-utilizing microorganism, penicillin acylase producing microorganism, macrolide type antibiotic producing microorganism and the like.

Any polymer may be used so long as it is capable of forming semipermeable gel films excellent in substrate permeability when it is spun into water and producing hollow filaments which can endure the pressure employed. The polymers which are preferably used include polyacrylonitrile; polymethacrylonitrile; cellulose derivatives such as cellulose acetate, cellulose acetate phthalate, hydroxypropyl methylcellulose phthalate, hydroxypropyl methylcellulose trimellitate, nitrocellulose or ethylcellulose; vinyl chloride-vinyl acetate copolymer; polyurethane, polysulfone; and copolymers containing these polymers as principal components.

The organic solvents used for preparation of a spinning dope should have an excellent property for dissolving the polymer as well as an excellent miscibility with water. The solvent is suitably selected according to the polymer to be dissolved. Generally, however, N,N-dimethyl formamide, N,N-dimethyl acetamide, dimethyl sulfoxide, formamide, a concentrated aqueous solution of rhodan salts, and the like are used.

The water permeability of the hollow filament is preferably as large as possible. If it is less than 0.1 ml/cm^2/min/atm, the useful value of the hollow filament is very low. The internal and external diameters are preferably as small as possible, since surface area is increased as diameters decrease. But a hollow filament with an internal diameter of 0.06 mm or less is very difficult to manufacture, whereas a hollow filament with an external diameter of 1.5 mm or more is difficult to handle as the filaments become bulky and difficult to pack in a column. Therefore, a hollow filament having an external diameter of 0.1 to 1.5 mm and an internal diameter of 0.06 to 1.3 mm is preferably used.

Example 1: A polyacrylonitrile having an intrinsic viscosity of 1.2 (in N,N-dimethyl formamide) was dissolved in dimethyl sulfoxide to the concentration of 15 g per 100 ml, then the solution was filtered and defoamed. This polymer solution was extruded through the external portion of a double orifice, while a dispersion, wherein 1% (w/v) active charcoal is dispersed homogeneously in a tris-buffer liquid (0.05 M; pH 7.0) containing 10% (w/v) urease of jack bean, is extruded through the internal portion of the orifice. The whole extruded product was coagulated in a water bath. The double orifice used was a spinning nozzle having an external orifice diameter of 0.80 mm and an annular slit width of 0.1 mm. The hollow filament obtained, which contained encapsulated urease and active charcoal, had an external diameter of 0.20 mm and an internal diameter of 0.18 mm. The hollow filament was closed by knotting the ends and made into a unit with a length of about 20 cm. The units of hollow filaments were then packed in a column of 2 cm diameter and 10 cm height very carefully so that the hollow filaments might not be cut.

A phosphoric acid buffer liquid (0.05 M; pH 5.5) containing 1% urea was passed from the top of the column. Urea, passing through the walls of the hollow filaments, was decomposed by the encapsulated urease to produce ammonia which was adsorbed on the coencapsulated active charcoal. The percentage of the decomposition of urea was determined, by measuring the concentration of unaltered urea by microdiffusion analysis, to be 75%. The water permeability of this hollow filament was 2 ml/cm^2/min/atm.

Example 2: A cellulose diacetate having an intrinsic viscosity of 2.3 (in acetone) was dissolved in N,N-dimethyl formamide to the concentration of 10 g per 100 ml, then the solution was filtered and defoamed. This polymer solution was extruded through the external portion of the same double orifice as used in Example 1, while a solution, wherein lipase MY (Meito Sangyo Company) was dissolved in a phosphoric acid buffer liquid (0.02 N; pH 7.0) to the concentration of 15%, was extruded through the internal portion of the double orifice. The whole extruded product was coagulated in a water bath.

The hollow filament obtained, which contained encapsulated lipase, had an external diameter of 0.30 mm and an internal diameter of 0.25 mm. The percentage of ester hydrolysis of para-nitrophenyl butyrate by the use of the hollow filament obtained containing lipase was 60%. The hollow filament had a water permeability of 1 ml/cm^2/min/atm.

High Permeability Reduced Nitrocellulose Fibers

Fibers encasing an enzyme have previously been prepared by starting with polymer solutions adapted to give fibers and containing enzymatic compounds dispersed in the form of very small drops of the order of magnitude of emulsions. These emulsions may be wet or dry spun to produce a fiber having in its interior, very small cavities which contain the enzymes separated from the environment by a very thin membrane.

The known processes, however, have not produced enzyme-containing fibers having the desired permeability. *D. Dinelli, F. Bartoli and S. Gulinelli; U.S. Patent 3,947,325; March 30, 1976; assigned to Snam Progetti SpA, Italy* have produced highly permeable fibers using nitrocellulose as the fiber-forming material.

The starting material used consists of high molecular weight nitrocellulose with a nitrogen content (preferably higher than 5%) that renders the material soluble in organic solvents. It is dissolved in a solvent, immiscible with water, selected from n-butyl acetate, bis-butyl phthalate, methyl amyl ketone, ethyl amyl ketone and others which can be used as such or suitably diluted by aliphatic hydrocarbons such as pentane, hexane, heptane, octane, aromatic hydrocarbons such as toluene and xylene, or hydrocarbon mixtures such as ligroin. To the cellulose solution is added the aqueous solution containing the enzymes which may be selected from urease, invertase, lactase, acylase, transaminase, glucose oxidase, catalase, papain, penicillin acylase and others.

The emulsion is then spun according to known techniques by using known coagulants. Besides the process of emulsion dispersion of the enzyme into the polymer solution, other processes can be used to prepare fibers encasing the enzyme. For instance, the enzyme can be dispersed, as a powder, into the polymer which will then be spun according to known methods. The fiber is then reacted with a reducing agent which removes nitro groups and forms the cellulose fiber encasing the enzyme. A solution of ammonium sulfhydrate with or without excesses of ammonia or hydrogen sulfide may be used as reducing agent.

Example 1: 1,000 g of nitrocellulose (Snia Viscosa SpA) were dissolved into a solution constituted by 6,900 g of n-butyl acetate and 4,600 g of toluene. Then there was added 1,600 g of an aqueous solution containing the enzyme invertase. Under stirring, an emulsion was obtained which was spun through orifices having a 125 μ diameter by using a coagulating agent comprising a mixture of saturated hydrocarbons having boiling points ranging from 40° to 70°C. About 2,600 g of fiber encasing the enzyme were obtained.

1,000 g of this fiber, immersed in a 20% saccharose solution at a pH of 4.5 inverted 40 g of sugar per minute. 1,000 g of fiber were treated with a 2% ammonia solution, which had been saturated with hydrogen sulfide up to a pH of 8.5. After 6 reaction hours, this second fiber was washed and when immersed in a 20% saccharose solution at a pH of 4.5, it inverted 490 g of sugar per minute.

Example 2: 1,000 g of nitrocellulose (Snia Viscosa SpA) were dissolved into a mixture constituted by 7,000 g of n-butyl acetate and 3,750 g of toluene. Then 2,000 g of an aqueous solution of the enzyme penicillin acylase were added and the whole was stirred to form an emulsion. The emulsion was spun according to Example 1 and 3,000 g of fiber were obtained. 1,500 g of fiber were immersed in 37.5 liters of a 10^{-1} M potassium phosphate solution at pH of 8.0 containing 1,500 g of potassium penicillin G, at 37°C. The enzyme contained in the fiber catalyzed the hydrolysis of penicillin to 6-aminopenicillanic acid and phenylacetic acid.

During the reaction, the pH was kept constant by adding sodium hydroxide. In such a way, a 90% conversion of penicillin was obtained after 227 minutes. 2,000 g of an aqueous solution of the enzyme penicillin acylase were encased in 1,000 g of nitrocellulose according to the procedure described above and then treated with the ammonia solution saturated with hydrogen sulfide as described in Example 1.

A similar solution of penicillin acylase was dispersed in a solution formed by dissolving 1,000 g of cellulose triacetate in 13,300 g of methylene chloride. This emulsion was spun as described in Example 1, the coagulant being toluene. 1,000 g of the first fiber, at 37°C immersed in 25 liters of a 10^{-1} M potassium phosphate solution at pH of 8.0 containing potassium penicillin G at 10% hydrolyzed 90% of penicillin over 219 minutes. The second fiber, under the same conditions, hydrolyzed 90% of penicillin over 298 minutes.

Another 1,500 g of fiber were treated with the ammonia solution saturated with hydrogen sulfide as described in Example 1. Then, under the same conditions as the untreated fiber, they promoted the conversion of 90% of penicillin over 56 minutes.

Spun Fibers of Cellulose Triacetate Containing Cellulose Diacetate or Glucose Pentaacetate

Enzyme-containing fibers (or films) which can be treated to increase their permeability to the enzymes have been developed by *A.H. Emery, H.C. Lim and M.J. Kolarik; U.S. Patent 4,004,980; January 25, 1977; assigned to Purdue Research Foundation.* Basically, this process requires formation of a solid phase in fiber or film form and swelling of the fiber or film after formation.

In forming the solid phase, cellulose triacetate is blended with a second solid material and then dissolved in methylene dichloride and spun through an orifice into a toluene bath to form the basic fiber. Two kinds of solid materials were used, including glucose pentaacetate, GPA, (which was then dissolved out of the fiber by soaking in acetone and which produces a larger permeability than available with cellulose triacetate alone), and cellulose diacetate, CDA, (which has a higher affinity for acetone and water and thus is capable of being swelled to a higher extent). While the diacetate was not dissolved out of the fiber, it contributed to a higher degree of swelling and larger permeability than available with cellulose triacetate alone.

After formulation, the fibers were soaked in mixtures of acetone and water. The amount of swelling is a function of the solids used in the formulation and of the concentration of acetone in water. The soaking not only dissolves out constituents such as glucose pentaacetate, but also increases the permeability of the fiber by swelling it.

For the following experiments and examples, materials used included cellulose acetates, usually triacetate, in the flake form, as well as a secondary acetate powder, E-383-40 (Eastman Kodak). In addition, glucose pentaacetate was synthesized in two batches following a procedure designed to give the β-form, with the first batch having a melting point range of 130° to 131°C and used in work carried out with whole cells and the second batch having a melting point range of 129° to 130°C.

The solvents used were reagent quality acetone, toluene, and methylene chloride, while two sources of glucose isomerase enzyme were used, one of which was a whole cell preparation that had been treated to fix the enzyme to the interior of the cell and the other of which was a partially purified enzyme mixed with substantial amounts of filter aid and soluble. The buffer used in this work was 0.01 M succinate, pH 6.5.

Preparation of a Polymer-Enzyme Solution – In a beaker 10 g of cellulose triacetate were dissolved in methylene dichloride to bring the total volume to 100 ml. In a separate beaker 5 g of whole cells were slurried in 15 ml of distilled water. The whole cell solution was then poured into the acetate solution under vigorous agitation. The result was an emulsion of the aqueous phase in the organic phase. This mixture was stirred for an additional 5 to 30 minutes, then formed into fibers or films as described in detail below.

Preparation of Fibers – Fibers were formed from the polymer-enzyme solution by a wet spinning technique. A 10 ml hypodermic syringe was filled with the polymer-enzyme solution, and then was used to extrude the solution through a number 20 hypodermic needle into 100 ml of toluene held in a 100 ml graduated cylinder. The needle had been cut blunt for this use.

During the extrusion process, the coagulating fiber would form ringlets about 2 cm in diameter which slowly settled to the bottom of the graduated cylinder. The rate of extrusion was controlled to give a continuous strand of the above ringlets at a rate slow enough to keep them from sticking together at the bottom of the graduated cylinder. After the polymer-enzyme solution had been extruded into the toluene bath, the fibers were removed and air dried before use. The air dried fibers generally consisted of ribbons measuring 250 μ x 500 μ.

Example 1: 0.08 g of fibers were placed into each of a series of test tubes which contained 10 ml of solution consisting of various fractions of acetone and water. The fibers were allowed to swell overnight (8 hours) before the acetone-water solution was poured off and replaced with distilled water. After 4 to 6 hours, the fibers were assayed to determine the fraction of the added enzyme activity that could be observed. Figure 3.1a summarizes the results that were obtained with fibers that had been extruded through a number 20 hypodermic needle.

Example 2: Fibers were formed with 50% of the cellulose triacetate replaced by cellulose diacetate (CDA). Samples of the fiber were swelled and assayed as in the previous experiment. The results for the fibers containing 50% CDA are summarized in Figure 3.1b.

Example 3: Fibers were made with 50% of the CTA replaced by GPA. The fibers were swelled as in the swelling experiment, then assayed to determine the fraction of the added enzymatic activity that was observable. Figure 3.1c summarizes the results obtained with fibers containing 50% GPA.

Figure 3.1: Graphs Showing Activity of Swelled Fibers

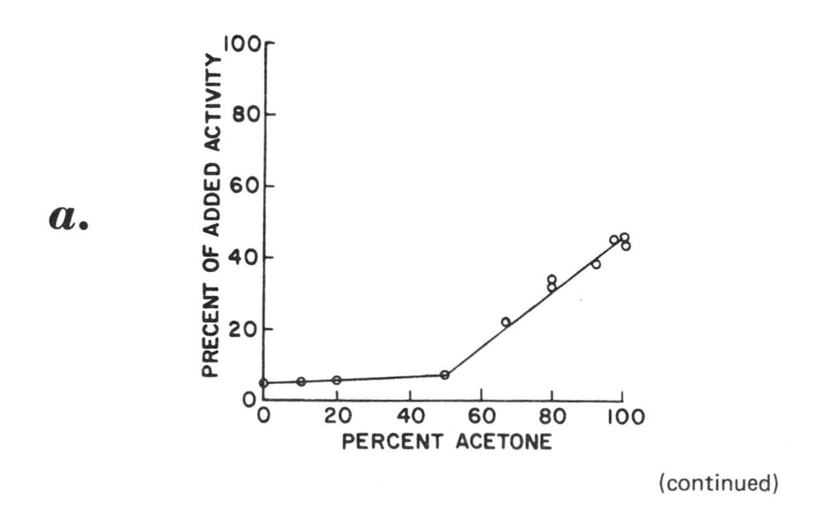

(continued)

Figure 3.1: (continued)

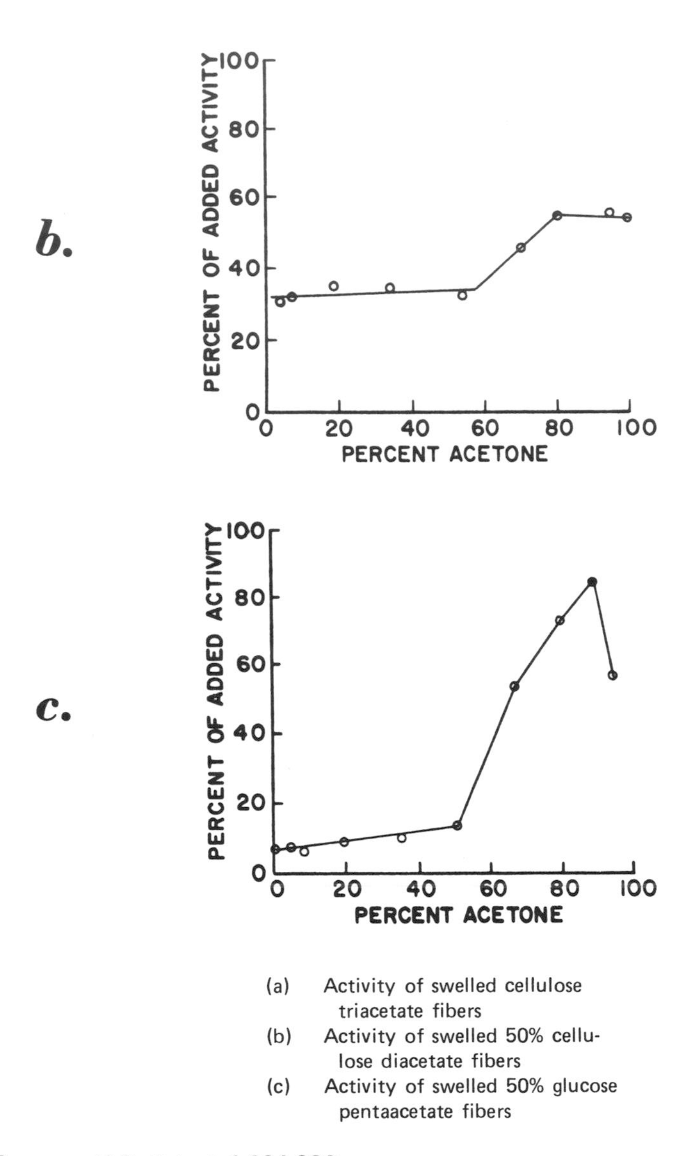

(a) Activity of swelled cellulose triacetate fibers

(b) Activity of swelled 50% cellulose diacetate fibers

(c) Activity of swelled 50% glucose pentaacetate fibers

Source: U.S. Patent 4,004,980

Films, or membranes, could also be formed by simply pouring some polymer-enzyme solution onto a flat glass plate and then spreading the solution into a thin layer which was allowed to air dry. A glass rod was used to spread the solution. A piece of paper at each end of the glass rod was used to adjust the thickness of the layer. The resulting films are generally 10 to 20 μ thick.

Enzymes and Coenzymes Embedded in Filaments

In the case of enzymes, or enzymic complexes which require a coenzyme of the NAD^+ type, it has been required that the coenzyme be added to the reaction mixture. Because of the low molecular weight of the coenzyme, it is difficult to immobilize it together with the enzyme on the same support so as to render the continuous addition of same coenzyme to the reaction mixture unnecessary. In addition, covalent bonding of the coenzyme to the support itself to which the enzyme is attached is unadvisable due to steric hindrances which prevent or drastically limit the interactions between the enzyme and the coenzyme, such interactions being required for activity.

As such coenzymes are comparatively expensive; adding them continuously to a reaction mixture has been a hindrance to the use of oxidoreductase systems in fiber form. It is likewise known that NAD^+ can be reacted at the 6-aminopurinic group and the functionalized NAD^+ can be attached to water-soluble, high molecular weight polymers.

G. Prosperi, W. Marconi, S. Giovenco and F. Morisi; U.S. Patent 4,100,029; July 11, 1978; assigned to Snam Progetti SpA, Italy have found that such polymers which have a coenzymic activity can be occluded in water-insoluble polymeric matrices, more particularly filamentary structure, together with an enzyme. The simultaneous occlusion of enzymes and coenzymes involves a stabilization of the oxidoreductase, which, when no coenzyme is present, undergoes equilibria of associations and dissociations of the subunits, the result being an activity loss.

This process permits the preparation of a polyenzymic biological reagent in which the polymeric derivative of NAD^+ is continuously cycled from the oxidized form to the reduced form, at the expenses of two oxidoreductases. The NAD^+ derivative can be cycled not only by the enzymic way, using two oxidoreductases, but also by the combined chemical and enzymic way, by using an oxidoreductase and a chemical compound. The recycle system is selected consistently with the product which one desires to obtain.

The way in which enzyme and coenzyme can be occluded comprises a fibrous or filamentous structural base of artificial or synthetic polymeric material with the required enzymes or enzymatic preparation encased and subdivided and partially enclosed in small alveoli or separated cavities. The applications of macromolecular coenzymes which can be embedded together with enzymes in polymeric matrices make up a wide variety. One application is in the syntheses of steroids or transformations of steroidal nuclei in preselected positions.

In addition, stereospecific syntheses can be carried out of amino acids starting from hydroxyacids or ketoacids. In the analytical field, it is possible to occlude enzyme-coenzyme systems in which the ultimate reaction of oxidation of the reduced coenzyme, or of reduction of the oxidized coenzyme, is carried out by a chemical substance, the absorption spectrum of which is a function of the reduced, or oxidized state in which it lies. The color of such a substance can easily be measured and correlated to the quantity of the substance sought for.

Example: In this example there is used an enzyme solution in buffer and glycerol (75:25) of polyethyleneimine-NAD^+ (PEI-NAD^+), lactic dehydrogenase (LDH, from rabbit) and alanine-dehydrogenase (Ala-DH, from *B. subtilis*), with the following specifications: PEI-NAD$^+$, 37.5 mg/ml (0.63 μmol NAD$^+$ per ml); LDH, 17.5 mg/ml (70 IU, NAD$^+$ as the substrate); and Ala-DH, 3.75 mg/ml. 10 g of cellulose triacetate are dissolved, with stirring in a reactor, in 133 g of methylene chloride. 20 g of the enzyme solution are added to the polymer solution and the emulsification of the two phases is encouraged by a vigorous stirring at 0°C for 30 minutes. The emulsion is poured in a small melting pot, kept at 0°C and spun under nitrogen pressure. The filament is coagulated in toluene at 0°C and collected on a bobbin frame.

Air-drying is carried out to remove the organic solvents. Two grams of the fiber, which correspond to about 1 g of dry polymer, are washed with a pH 8.0 bicarbonate buffer to remove the enzymes and PEI-NAD$^+$ which have been adsorbed on the surface, and then the fiber is placed in 10 ml of an aqueous solution having the following composition: ammonium L-lactate at 3% (w/v) at a pH 7.4.

Stirring at room temperature (approximately 22°C) is carried out. After 10 hours, by means of an amino acid autoanalyzer, the quantity of formed L-alanine is measured, which amounts to 0.163 g (96% of theory). Once the mixture has been discharged, the same fiber is contacted with a fresh solution of ammonium lactate for a fresh reaction cycle. After a 7 day run, the percentage of L-alanine is 94%, at 10 hours and after 30 days, still at 10 hours, it is 90%. L-alanine is separated by precipitation with ethanol at a pH of 6.2 $[\alpha]_D^{25} = +14.6°$ (c = 2.0 in 5 N HCl).

POLYMERIZATION OF MONOMER-ENZYME MIXTURES

Polymerization of Water-Soluble Acrylate Monomers

Stabilizing enzymes by entrapping in a polymeric gel has been disclosed by *K.F. O'Driscoll and M. Izu; U.S. Patent 3,859,169; January 7, 1975; assigned to Polymeric Enzymes, Inc.* Briefly the process comprises providing an aqueous reaction mixture containing: water containing 20 to 25% of an enzyme; at least one water-soluble polymerizable monomer dissolved in the water; at least one crosslinking agent for the monomers soluble in the water-monomer solution; at least one polymer soluble in the reaction mixture; and a free radical initiating system.

The mixture is polymerized under mild conditions in the substantial absence of oxygen to provide a gel entrapping the enzyme. This gel may be dried to a free-flowing powder, and subsequently may be swelled for use by the addition of water. The product has excellent enzymatic activity over a prolonged period of time, and exhibits substantially no mechanical degradation or breakdown when used in a column or stirred reactor. The gel may be provided in any convenient form, typically as a film which then is positioned and used in a suitable reactor.

The monomer used in the reaction mixture may be any desired water-soluble polymerizable monomer, and preferably is a polymerizable monomer containing glycol units which remain in the polymer. Examples of monomers which may be used include hydroxylated esters of acrylic and methacrylic acid, such as

2-hydroxyethyl methacrylate (HEMA). Other hydroxylated esters which may be used include monomethacrylates or acrylates of glycol, glycerol and other polyhydric alcohols, monomethacrylates or acrylates of dialkylene glycols and polyalkylene glycols. Many other derivatives of acrylic and methacrylic acid may be used in the reaction mixture, including dimethylaminoethyl methacrylate, piperidinoethyl methacrylate, morpholinoethyl methacrylate, methacryloglycolic acid and methacrylic acid itself, and the corresponding acrylic acid derivatives.

Diesters of methacrylic and acrylic acids also may be used such as triethanolamine dimethacrylate, triethanolamine trimethacrylate, tartaric acid dimethacrylate, triethylene glycol dimethacrylate, the dimethacrylate of bishydroxyethylacetamide, and the corresponding acrylate derivatives. Other water-soluble polymerizable monomers such as maleimide may be used. Typical crosslinking monomers which may be used include methyl acrylate, diethylene glycol dimethacrylate (DDMA), methyl methacrylate, acrylamide, methyl methacrylamide and ethylene glycol dimethacrylate (EDMA).

A further component of the reaction mixture is a polymer which is soluble in the reaction medium. Any convenient natural or synthetic polymeric material soluble in the system may be used. A mixture of polymers may be used, if desired. The polymers may be included in the reaction mixture to protect the enzyme during the polymerization of the reaction mixture. Additionally, the polymer may serve to increase the viscosity of the reaction mixture, and hence achieve a more ready polymerization of the mixture to the gel.

The polymeric material may be a polymer of the monomer or a polymer of a different material. Polymers which may be used include poly(2-hydroxyethyl methacrylate), polyvinylpyrrolidone, polyethyleneimine, agarose, dextran, albumin, polyacrylamide, polyacrylic acid and polyvinyl sulfate.

Free radical initiators are included in the reaction mixture to promote the polymerization of the monomer. The form of the initiator and the quantity used depend on the nature of the monomer used, and the reaction temperature required. For example, free radical initiators preferably operative at 15° to 25°C are used, although photochemical free radical initiators operative at temperatures down to -30°C may be used. Further details are given in the following examples.

Example 1: A water-soluble polymer of HEMA [s-poly(HEMA)] was prepared by polymerizing a 25% v/v solution of HEMA in dimethylformamide using benzoyl peroxide as an initiator at 100°C in the absence of air. The polymer was purified by precipitation in diethyl ether. In an ampoule, a solution was prepared containing 200 mg of s-poly(HEMA), 1 ml HEMA containing 2% EDMA, and 1 ml 0.05 M tris(hydroxymethyl)aminomethane (tris) buffer (pH = 8.0). To this solution were added 0.2 ml of a previously prepared solution of 10 mg of purified, crystalline trypsin dissolved in 1 ml of tris buffer and 0.2 ml of a previously prepared solution of 1 ml of di(sec-butyl) peroxydicarbonate (Lupersol 225) dissolved in 100 ml of methanol. Air was displaced from the ampoule by bubbling nitrogen for 5 minutes and the ampoule was then sealed and allowed to polymerize at 27°C for 14 hours.

The gel formed was somewhat opaque and was a hard mass which was reduced to small particles by shear action in a Waring blender. The gel was filtered on sintered glass and washed. After washing, the gel was dried at room temperature, pulverized to a fine powder with a mortar, washed with distilled water and stored at room temperature prior to use. The weight of the dry gel formed was 0.91 g.

Example 2: The procedure of Example 1 was followed, except that the 200 mg of a s-poly(HEMA) was replaced by 200 mg of poly(vinylpyrrolidone) (PVP) of 40,000 MW. The weight of the dry gel formed was 0.96 g.

Example 3: The procedure of Example 1 was followed except that 200 mg of s-poly(HEMA) was replaced by 200 mg of poly(ethyleneimine) (PEI) and 100 mg of acrylic acid monomer was also added. The weight of the dry gel formed was 1.01 g.

Example 4: Approximately 0.5 g of a hydrogel prepared as in Examples 1, 2 or 3 was suspended in a beaker containing 10 ml of tris buffer (pH = 7.9) at 25°C. The pH was adjusted to a stable value of 8.0 by the addition of 0.1 N NaOH solution and then 1 ml of a previously prepared ester solution was added. The ester was either p-toluenesulfonyl-L-arginine methyl ester (TAME) or benzoyl-L-arginine ethyl ester hydrochloride (BAEE) prepared as an 0.1 M solution in distilled water.

Additional amounts of 0.1 N NaOH were added from time to time to maintain the pH in the range 8.0±0.1. The rate of addition of the NaOH is a direct measure of esterolytic activity of the entrapped enzyme. (Gel without enzyme gave no activity.) The results of activity measurements for the hydrogels prepared as in Examples 1, 2 and 3 are summarized in the following table. The ratio of k_{TAME} to k_{BAEE} of the gels was 1.6 whereas the ratio of free trypsin was 6.8. (It is suggested that the apparent activity towards TAME was reduced because of its relatively slow diffusion through the gel matrix.)

Esterolytic Activity, k*

	BAEE	TAME
Example 1	0.022	0.033
Example 2	0.021	0.024
Example 3	0.012	0.019
Free trypsin	0.38	2.5

*The rate constant is defined as ml of 0.01 N NaOH added per minute per 0.5 g of gel or per mg of free trypsin.

Use of Crosslinking Methylenediacrylamide and N-Acryloylmorpholine Mixtures

R. Epton, C. Holloway, and J.V. McLaren; U.S. Patent 3,896,092; July 22, 1975; assigned to Koch-Light Laboratories, Ltd., England have prepared crosslinked polymers which are useful for gel permeation chromatography and the occlusion of enzymes. At least some of the repeating units of the backbone chains of the crosslinked polymers are bonded directly or indirectly to morpholine groups. These polymers have the advantage that they form gels in both water and a wide range of organic solvents. One preferred polymer can be prepared by copolymerization of N-acryloylmorpholine with N,N-methylenediacrylamide in aqueous solution, using a catalyst such as potassium persulfate, and purging with nitrogen. The molar ratio of N-acryloylmorpholine to N,N'-methylenediacrylamide in the reaction can be varied from 150/1 to 5/1.

The preparation of the polymers in bead form can be effected by forming a suspension of the reactants in an aqueous solution in paraffin mixed with suitable surfactants. This procedure is particularly satisfactory for preparing polymers from acryloylmorpholine. The morpholine compound is mixed with a crosslinking agent such as N,N'-methylenediacrylamide and dissolved in water. The mixture, preferably after degassing with an inert gas such as nitrogen, is mixed with a catalyst such as potassium persulfate. The mixture is then added to liquid paraffin mixed with a surfactant, and the whole stirred to produce a suspension in the paraffin. After polymerization, the resulting beads can be washed and shrunk by addition of alcohol.

The pore size of the polymers can be varied in a number of ways to suit different applications of the polymers. A first method is to change the ratio of the monomers that are to be reacted. The pore size can also be varied by macroreticular polymerization, that is, polymerization in a medium which is a good solvent for the initial reactants but is a poor solvent for the chains of the resulting polymer, so that as the polymer chains are formed, they agglomerate together resulting in larger pores.

For example, the polymerization can be carried out in water mixed with ethanol which is a nonsolvent for the constituent chains of the polymer. The polymers prepared by this method are xerogel-aerogel hybrids and a suitable molar ratio of polymerizable monomers to crosslinking agent is from 150:1 to 1.5:1. In the case of macroreticular polymerization using polystyrene and divinylbenzene, the optimum ratio is 3:1.

Another method of varying pore size is to vary the concentration of the reactants alone, providing that a ratio of morpholine-group-containing compound to crosslinker is chosen that ensures good mechanical stability. Such a ratio of morpholine-containing compounds to crosslinker is 10:1. It has also been found that the rate of stirring of the reaction mixture can affect the bead size and it is therefore important to use the correct rate to obtain the desired size of bead.

An additional feature of the process is that the mild conditions used in the polymerization reaction enable the polymers to be produced in which biologically active molecules, such as, for example, urease may be entrapped without the active molecules being destroyed. Such chromedia will have analytical and certain medical applications such as in enzyme chromatography and will be of use as catalysts in industrial biochemical reactors.

Another use of the chromedia is in desalting processes in which a mixture of the compounds to be desalted, for example, a starch-glucose mixture, is passed through the chromatographic columns. Polymers produced in this process can also be used to entrap catalytic water-soluble enzymes in order to insolubilize the enzymes.

Example: This example illustrates the immobilization of urease by molecular entrapment in copolymers of N-acryloylmorpholine and N,N'-methylenediacrylamide. Two samples of crystalline urease (12.5 mg) were weighed into stoppered test tubes. To one tube, a solution of acryloylmorpholine (0.617 g, 0.004 mol) and N,N'-methylenediacrylamide (0.071 g, 0.0005 mol) in phosphate buffer (0.025 m, pH 7.5, 5 ml) was added, when the enzyme dissolved. The second urease sample was similarly dissolved except that only 2.5 ml of buffer was used. Riboflavin (0.05 mg) was added to each tube after which nitrogen was blown over the surface of the liquid for 30 minutes. The tubes were then placed in a beaker of ice water and irradiated with a No. 2 photoflood lamp to bring about the polymerization.

The polymers were left at 0° to 5°C overnight to harden, after which they were broken by grinding in a mortar. After washing twice with phosphate buffer (0.025 mol, pH 7.5), twice with urea (150 mM) in the same buffer, and twice more with buffer, the polymers were finally resuspended in buffer (15 ml) and stored at 0° to 5°C. The activity of the immobilized urease derivatives were 333 units per gram and 78 units per gram for the 2.5 ml volume and 5 ml volume polymerization respectively. (One urease unit being that which liberated 1 mg NH_3 per hour from a solution of 0.15 M urea at pH 7.5 and 30°C.)

Radiation Polymerization of Frozen Monomer-Enzyme Mixtures

Water-insoluble enzymes are produced by *K. Kawashima and K. Umeda; U.S. Patent 3,962,038; June 8, 1976; assigned to Director of National Food Research Institute, Japan* using the following steps: freezing a solution of one or more monomers and enzyme; and polymerizing the resultant composition by irradiating with a dose of 40 to 100 krad of gamma rays. This method entraps the active enzyme within the resultant polymer lattice. Before freezing the solution "organic substance" can be incorporated which promotes the polymerization yield and protects the highly purified enzyme from radiation inactivation.

The following water-soluble monomers are used in this process: acrylic acid, acrylamide, bisacrylamide (N,N'-methylene bisacrylamide), sodium acrylate, potassium acrylate, acrylonitrile, acrylic acid esters and other acrylic acid derivatives, and propylene glycol. The following enzymes can be used in this process: amylase, protease, glucoamylase, lipase, acrylase, d-amino acid oxidase, catalase, various kinds of dehydrogenase, urease, ribonuclease, and the like.

The term "organic substance" means a substance which promotes the polymerization yield of monomer and which at the same time protects enzymes, especially highly purified enzymes in the solution from radiation inactivation. The existence of the substance is very important in the radiation polymerization of acrylamide and bisacrylamide. Examples of the organic substance are: substrate of the enzyme, starch, glycerine, sugar such as glucose, and protein such as defatted powdered milk.

Examples of suitable ionizing radiation are the gamma ray (^{60}Co, ^{137}Cs), x-ray, accelerated electron ray and the like. In this process, the freezing treatment is conducted below -5°C and preferable temperatures would be -20° to -100°C. The solution is placed in a suitable container and placed in a coolant, without bubbling nitrogen or argon, which completes the freezing technique.

When the quantity is large, a thick frozen layer is obtained and when the quantity is small, the layer obtained is thin. For the purpose of obtaining a thinner layer, the container with the solution is slowly rotated inside the coolant and after this, when the frozen layer is irradiated by ionizing radiation, a membranous, water-insoluble polymer is obtained. In regard to the coolant, a liquid coolant (dry ice-acetone) is preferred because of its easy heat conductivity. When the solution is frozen, the water forms small ice crystals and the monomer, enzyme and organic substance become encircled around these ice crystals. When these ice crystals are exposed to an ionizing radiation of 40 to 100 krad, the enzyme becomes distributed in the resultant polymer. During the irradiation process, the frozen mixture is placed in the coolant.

The resultant membranous water-insoluble polymer, by defrosting at room temperature, assumes a hollow appearance where the ice crystals were present and becomes a sponge-like mass, which maintains its structure even on immersing in water. Irradiation was carried out under aerobic conditions with gamma ray or x-ray. The critical dose for irradiation is 40 to 100 krad, if the dose is less than 40 krad, there is inadequate polymerization and if doses higher than 100 krad are used, the enzyme is bound to be inactivated.

If desired, drying under vacuum (lyophilization) is conducted immediately after irradiation and the activity (comparing with the total activity of the original enzyme) of the immobilized enzyme is increased. Lyophilization is carried out in the normal way. Water-insoluble enzyme prepared by this process is a whitish membrane. This membrane, if necessary, is finely cut or powdered. The preparation obtained through the lyophilization step can be powdered easily without any significant drop in enzyme activity.

Example: To 2 ml of a solution of 30 g of acrylamide and 1.6 g of bisacrylamide in 100 ml of 0.04 M acetate buffer at pH 4.6 were added 2.0 ml of 10% soluble starch and 1.0 ml of invertase solution (0.25 mg invertase/ml of the buffer). This mixture is placed in a 300 ml round bottom flask and immersed in the coolant (dry ice-acetone). The flask is slowly rotated while the flask is inside the coolant so as to obtain a thin layer on the wall of the flask.

After freezing, while maintaining the frozen state, the mixture was irradiated with 42.8 krad by ^{60}Co. This irradiated mixture is defrosted at room temperature to obtain a water-insoluble enzyme preparation. One-fiftieth of the above enzyme preparation, 5 ml of 0.3 M sucrose and 4 ml of 0.04 M acetate buffer at pH 4.6 were mixed and incubated at 40°C for 60 minutes. After termination of reaction, the enzyme activity was assayed by the modified Somogy method. The recovery of enzyme activity of the preparation was found to be 69.2% based on original activity.

K. Kawashima and K. Umeda; U.S. Patent 4,025,391; May 24, 1977; assigned to Director of National Food Research Institute, Japan have further modified this process to obtain bead-shaped particles of entrapped enzymes. These are obtained by adding a solution containing an enzyme and at least one water-soluble monomer or polymer to a water-insoluble or slightly soluble fluid to form a mixture containing enzyme-containing beads, freezing the mixture at -200° to -5°C and subjecting the frozen mixture to ionizing radiation under aerobic conditions to polymerize the monomer or polymer and produce polymer beads containing immobilized enzymes.

The polymer beads are strong enough to be packed in a column, and continuous enzyme reactions can be carried out with the entrapped enzymes. The enzyme can be combined in advance with a water-soluble or insoluble carrier such as water-soluble dextran, CM cellulose, an anhydrous maleic acid polymer, a polypeptide, a styrene derivative polymer, agalose, cellulose, dextran, starch and the like, to improve stability of the enzyme.

Water-soluble monomers which may be used include acrylamide, acrylic acid, methacrylic acid, sodium acrylate, potassium acrylate, calcium acrylate, magnesium acrylate, ferrous acrylate, cobalt acrylate, nickel acrylate, magnesium methacrylate, nickel methacrylate, acrylonitrile, propylene glycol, pyrrolidone, 1-vinyl-2-pyrrolidone and diacetone acrylamide. Further, water-soluble monomers such as diethylaminoethyl methacrylate and divinylsulfone can be used with the above monomers. Similarly, water-soluble polymers such as polyvinyl alcohol having a polymerization degree of 1,400, polyvinyl pyrrolidone and the like, can be used.

These monomers or polymers can be used singly or by combining two or more of them. Further, the concentration of monomer or polymer solutions is not critical, and thus, the concentration can be selected freely. Also, N,N'-methylenebisacrylamide and divinyl sulfone can be added as a crosslinking agent in suitable quantities.

Water-insoluble or slightly water-soluble fluids, oils and solvents can be used. Typical oils are vegetable oils such as soybean oil, sesame oil, olive oil; mineral oils such as volatile oils, kerosene and light oils. Typical solvents include aliphatic hydrocarbons such as n-hexane, petroleum ether; aromatic hydrocarbons such as toluene, xylenes; alcohols higher than propyl alcohols such as butyl alcohol, amyl alcohols; ethers such as ethyl ether; esters such as ethyl acetate; ketones such as methyl ethyl ketone. Further, a surface active agent can be added if desired.

A monomer or polymer solution containing an enzyme is added to the fluid while agitating the fluid or adding the solution dropwise by means of a nozzle; as a result, a bead-shaped suspension is formed. The size of the beads is influenced by the diameter of the nozzle, pressure of injection, speed of agitation and viscosity of the fluid. When a surface active agent is added to the fluid, beads with smaller size can be obtained easily. Generally, the size of the beads is preferably 1 to 10 mm.

When the monomer or polymer solution containing an enzyme is added to a fluid, the resulting bead-shaped solution is frozen, preferably at –20° to –80°C. For this purpose, dry ice-acetone, liquid nitrogen, dry ice-ethyl alcohol and $CaCl_2 \cdot 6H_2O$-ice mixtures and the like are used. Finally, polymerization is conducted by ionizing irradiation. A bead-shaped immobilized enzyme is obtained. As ionizing radiation sources ^{60}Co, ^{137}Cs, x-rays and accelerated electrons are applicable. The irradiation dose is 30 to 1,000 krad, preferably between 40 and 600 krad.

Example: Solutions (A), (B), (C), and (D) were prepared as follows: (A) acrylamide (30 g) and N,N'-methylenebisacrylamide (1.6 g) in 100 ml distilled water; (B) 30% aqueous solution of sodium acrylate; (C) 30% aqueous solution of calcium acrylate; and (D) 2 mg/ml solution of invertase enzyme. One ml of solutions (A), (B) and (D) were mixed with 2 ml of (C) and are referred to as enzyme monomer solution. 300 ml of soybean oil was taken in a large test tube (5 cm diameter) and immersed in a coolant of dry ice-acetone so that the oil in the lower part of the tube was frozen while the oil on the top of the tube was not frozen. The enzyme monomer solution was injected into the oil dropwise, and thus, was frozen in the oil in bead form. All of the oil was then frozen. In this state, the frozen enzyme monomer solution was copolymerized by ^{60}Co gamma rays, with a dose of 83.2 krad. Then at room temperature the mixture was melted, filtered, washed with detergent and water, resulting in bead-shaped immobilized invertase with diameters ranging from 1 to 10 mm.

For measuring the activity of the immobilized enzyme, $^1/_{50}$ of it was used. To this, 1 ml of 10% sucrose solution and 9 ml of citrate phosphate buffer (pH 3.8) were added, and reacted at 40°C for 20 minutes. The resulting glucose content was measured by the Willstatter-Schudel method.

For comparison purposes, 1 ml of native invertase (10 μg/ml) was reacted similarly, at pH 4.5, and resulting glucose was assayed. As a result, bead-shaped immobilized invertase of 1 to 2 mm in diameter showed a retained activity of 55.1% while those with a diameter of 3 to 7 mm had a retained activity of 35.2%.

Another example of polymerization of monomer-enzyme mixtures can be found in U.S. Patent 4,038,140 on page 101.

CROSSLINKING OF POLYMER-ENZYME MIXTURES

Curing of Resin Coated Supported Enzyme Particles

It has been disclosed by *R.D. Harvey and K. Ladenburg; U.S. Patent 3,849,253; November 19, 1974; assigned to Penick & Ford, Limited* that enzymes can be permanently bonded or held on or in an inert, insoluble material by polymerizing film-forming resinous material to the insoluble state while in contact with the enzyme and the insoluble material. Enzymes which are immobilized by this resin polymerization or curing still exhibit high activity even though sufficiently insoluble to survive prolonged continuous flow of substrate through a packed column. The process does not require a carrier having any particular chemical affinity for the enzymes. Inert, insoluble materials which are porous and capable of absorbing substantial amounts of water are especially suitable.

The use of carriers having high surface area per unit volume enable the inclusion of high proportions of enzymes. The solution of enzyme may be first absorbed on the carrier and then treated with a solution of polymerizable resinous material and polymerization of the resin accomplished under conditions which do not inactivate the enzyme. It is also feasible to mix the resin with the enzyme before absorbing on the carrier.

This process has several distinct advantages as compared with prior processes for immobilizing enzymes. It can make use of very low cost carriers and carriers which have a very high capability for absorbing enzyme solutions. Low cost resinous materials can also be used, for example, urea-formaldehyde resins. This allows a very advantageous cost with respect to enzyme activity per unit weight. The carrier should be an insoluble material which retains moisture when dispersed in water and drained. The carrier may advantageously be in the physical form of water-absorbing granules or flocs. It is desirable that the carrier wet readily with water and retain relatively high moisture content, but it need not be porous. Such hydrophilic carriers include, but are not limited to nylon, carbon, cellulose, granular absorbent resin, cellulose acetate, ion exchange resin, polyvinylpyrrolidone, and perlite filter aid.

The suitable resins are those which are soluble in water or a water-miscible solvent which polymerize or cure to form a water-insoluble film. Such resins include melamine-formaldehyde resins, methylated melamine-formaldehyde resins, urea-formaldehyde resins, cationic urea-formaldehyde resins, dimethylol urea resins, cationic polyacrylamide resins, alkyd resins, polyamide-amine-epichlorohydrin resins, and phenolic resins. A preferred subclass of resins for immobilizing soluble amylases is the formaldehyde-type resins of those listed, particularly the melamine and urea-formaldehyde resins. Mixtures of such resins can also be used where the blends are capable of polymerizing.

Preferably, the process can be practiced as follows. To a given amount of carrier, add enzyme solution. Volume and concentration can be adjusted to suit carrier and activity desired in the final product. Add resin, pH adjusted for proper curing. Resin may be added in a range from 10 to 200% on the dry solids weight of carrier. This will vary with type of resin, the carrier used, type and amount of enzyme used. However, the preferred range will normally be 30 to 100% on DS (dry solids) carrier weight.

The carrier-enzyme-resin mixture is then dried and allowed to cure (polymerize). Drying may take place as in forced air oven or vacuum oven with the temperature used being dependent upon the stability of the enzyme at the pH being used. The rate and degree of curing will vary as the drying conditions are varied and also vary considerably with the type and DP of the resin used. Curing for some resins is essentially complete on drying, while others may require up to 30 days aging to effect an adequate cure.

This process is especially useful for immobilizing starch hydrolyzing enzymes, such as the water-soluble amylases, and produces material that can be used for commercial manufacture of syrups of controlled sugar profiles, such as 92 to 98% glucose syrups substantially free of reversion products.

Example: This example illustrates the long half life of this immobilized enzyme product. Freshly precipitated nylon was filtered using a Buchner funnel. After filtration moisture content was approximately 80%. To a sample containing 36 g DS was added 800 ml of glucoamylase (Diazyme L-100) solution containing 75.0 IGU/ml. After thorough mixing the excess enzyme solution was drained off reclaiming 550 ml of a solution containing 60 IGU/ml. The lower enzyme activity of the reclaimed solution is due to dilution by the water on the wet carrier.

To the nylon-enzyme mixture was added 1 liter of resin solution (Pacific Resin N-1709-J-1, methylated melamine-formaldehyde), 20.6% DS (218 g) which had been adjusted to 4.2 pH. The enzyme-resin solution was drained off reclaiming 1,000 ml. Approximately 200% dry weight resin was retained on dry solid weight of carrier. The nylon-enzyme-resin mixture was dried overnight in a vacuum oven at 100°F. After drying, the product was washed several times in water and assayed. The product was found to have an activity of 20.0 IGU/g dry weight.

30 g dry weight was packed into a glass column 1" x 20". At a rate of 225 ml/hr, 30% DS, 20 DE syrup was passed through the column at 140°F, producing a product with average DE of 88.4 for 12 days with no noticeable change in conversion characteristics.

Irradiation of Polymer-Enzyme Solutions

The immobilization of enzymes by irradiating aqueous solutions of polyvinyl high molecular compounds containing enzymes with ionizing radiation has been disclosed by *H. Maeda, A. Yamauchi, H. Suzuki and A. Kamibayashi; U.S. Patent 3,933,587; January 20, 1976; assigned to Agency of Industrial Science & Technology, Japan.* This process is economical for it uses polyvinyl alcohol or polyvinylpyrrolidone which is an inexpensive organic high molecular material. In terms of operation, this method is simple because it only requires treatment of the enzyme-containing solution with ionizing radiation.

The preferred method of this process is as follows. First, a polyvinyl type high molecular compound such as polyvinyl alcohol or polyvinylpyrrolidone is dissolved in water heated in an inert atmosphere to produce a 2 to 40% aqueous solution. After the solution is cooled, an enzyme solution is added to the aqueous solution in an amount to give an enzyme content of 0.001 to 5% based on the weight of the aqueous solution of polyvinyl high molecular compound. This mixture is thoroughly stirred, placed in an ampoule and sealed hermetically under decreased pressure or in an inert atmosphere.

In this case, it is desired to have the mixture incorporate a highly diluted buffer solution prior to sealing so as to preclude possible variation of its hydrogen ion concentration (pH). Then, the sealed ampoule is irradiated with ionizing radiant rays, for example, gamma rays and accelerated electron rays. At the end of the irradiation, the gelled contents are removed from the ampoule and finely divided. The finely divided gel is used as desired on a given substrate as by being placed in a column, for example.

The results of immobilization obtained by this method are comparable with those previously obtained by immobilization with the use of acrylamide, and the gelled product is shown to be usable satisfactorily as an immobilized enzyme. While a slight degree of enzyme leakage is observed to occur in the immobilized enzyme prepared by the use of acrylamide, no enzyme leakage is observed at all in the product obtained by this process. This elimination of enzyme leakage may possibly be explained by postulating that the irradiation of the aqueous solution of polyvinyl high molecular compound with a large dosage of ionizing radiation gives a higher degree of crosslinking than by the conventional method where the enzyme is entrapped with a high molecular compound.

Example: In a nitrogen atmosphere, 10 g of polyvinyl alcohol was dissolved under heating in 90 ml of water saturated with nitrogen gas. The solution was cooled and then stirred thoroughly with 10 ml of added enzyme solution containing 3,061 units of glucoamylase (international standard units, pH 4.5, 40°C with maltose as substrate). Portions of the resultant solution, each 5 ml in volume and containing 139 units of the enzyme, were injected in ampoules, which were then hermetically sealed in a current of nitrogen gas. The sealed ampoules were exposed to several Mrad of gamma rays at a dosage rate of 6.54×10^4 rads per hour to form gel. After irradiation, the gelled products were removed from the ampoules, washed with water three times, drained, then weighed and tested for activity. The results are shown in the following table. In the case of a dosage of 5 Mrad, there was obtained about 6.5 g of gel which was found to retain a total of 26.2 units of activity. This means that the gelatin lowered the activity from the initial value of 139 units to 26.2 units, the recovery ratio of activity being 19%. According to this method of enzyme entrapping, substantially all the glucoamylase added was entrapped and immobilized and absolutely no enzyme leakage was observed.

Dosage (Mrad)	Weight of Formed Gel (g)	Activity of Added Enzyme (A) (units)	Activity of Gelled Enzyme (B) (units)	Ratio of (B)/(A) (%)
3	8.4	139	26.5	19
4	10.1	139	22.5	16
5	6.5	139	26.2	19
6	6.1	139	22.3	16
7	6.2	139	21.7	16

The glucoamylase activity was determined by the following procedure. A total of 10 ml of the solution of substrate adjusted to have 1% maltose content and a pH value of 4.5 was shaken at 40°C for 30 minutes to induce the enzymatic reaction. At the end of the period, the reaction system was thermally inactivated and assayed for liberated glucose content. The degree of enzyme leakage from the gelled product was determined as follows. A 1 to 2 g sample of the gel was added to a solution consisting of 20 ml of 2% maltose solution and 20 ml of 0.01 M acetic acid buffer solution (pH 4.5). The resultant mixture was shaken at 40°C for 60 minutes to induce a reaction. At the end of the reaction, the reaction system was filtered. The filtrate was subjected overnight to dialysis and the dialyzate was tested for residual enzymatic activity.

The procedure described above was repeated, except the ampoules were sealed under a pressure decreased to less than 1 mm Hg instead of in the current of nitrogen gas. Consequently, there were obtained substantially the same results as before. Other enzymes which may be immobilized by this process include glucose isomerase, invertase, β-galactosidase, glucose oxidase and catalase.

Curing and Foaming Polyurethane Prepolymers

A hydrophilic polymeric foam in which an enzyme is immobilized by entrapment in the poly(urea-urethane) foam matrix has been produced by *J. Klug; U.S. Patent 3,905,923; September 16, 1975; assigned to Minnesota Mining and Manufacturing Company*. This process involves the reaction of an isocyanate terminated urethane prepolymer, the prepolymer being an isocyanate terminated polyoxyalkylene glycol containing at least 50 mol percent oxyethylene in the glycol, with an aqueous solution containing enzymatic material. The resulting enzyme-hydrophilic foams form a catalytic agent which is especially compatible with aqueous substrate mixtures.

In this enzyme-foam composition, the polymeric foam immobilizes the enzyme and retards leaching of the enzyme from the foam material, while simultaneously containing and supporting the enzyme in an active configuration, thereby maintaining a high degree of catalytic activity. The enzyme-hydrophilic foam material is easily recovered from reaction media thereby allowing recycling of the enzyme. In addition, the hydrophilic foam is self-supporting and flexible, providing broad latitude in reactor design. A wide variety of enzymes are capable of being immobilized in the poly(urea-urethane) foam carrier including papain, amyloglucosidase, carboxypeptidases, thrombin, ficin, pepsin, trypsin, chymotrypsin, keratinase, bromelain, kallikrein, pronase, urease, amylase, maltase, aldolase, penicillin-amidase, amino acylases, lysine decarboxylase, lipase, cholinesterase, glucose oxidase, galactose oxidase, catalase, alcohol dehydrogenase, lactate dehydrogenase, tyrosinase, phosphoglucomutase, ribonuclease, α-amylase, and peroxidase, etc.

These enzymes need not be purified, but may be a more or less crude preparation containing an enzyme or a number of enzymes. Fillers can also be incorporated in the foam to reduce the cost, increase the mechanical strength of the foam and otherwise improve the properties of the foam. Suitable fillers include cellulose fibers, wood chips and fibers, synthetic fibers, natural and/or synthetic polymers, nonwoven web, woven mats and scrims, cotton, wool, glass particles, polymeric films, metal screen or rods, metal particles, paper, cork, rubber as well as other materials which will not interfere with the catalytic activity of the entrapped enzyme.

To form this enzyme-hydrophilic foam structure, enzyme-containing material is first dispersed in a large amount of water, about 0.15 to 15 parts by weight of the enzyme-containing material per 100 parts water usually being sufficient. The amount of water present is at least several times that necessary to convert all the isocyanate present into urea links. The resulting aqueous suspension or solution is then reacted with the isocyanate terminated prepolymers, most of the prepolymer reacting with the water present to form the enzyme-hydrophilic foam. Much of the excess water present will be trapped in the foam.

Generally, the isocyanate terminated prepolymers used in this process are the reaction product of a polyoxyalkylene polyol and a polyisocyanate. The OH equivalent weight of the polyoxyalkylene polyols should range from 300 to 4,000 and have at least 50 mol % oxyethylene to provide a prepolymer with good hydrophilicity. Particularly good results are obtained with polyol precursors having an OH equivalent weight of 450 up to 1,200. A suitable commercially available series of polyoxyethylene glycols is available under the trade name Carbowax. Other materials useful in making prepolymers include pol polyoxyethylene diamines with equivalent weights of 1,000 as well as polyoxyethylene dimercaptans and the like.

The isocyanate components of the prepolymers are derived from an aliphatic, aromatic or aralkyl polyisocyanate, preferably a diisocyanate, such as tolylenediisocyanate (TDI), xylene diisocyanate (XDI), naphthalene diisocyanate, 4,4'-diphenylmethane diisocyanate (MDI), phenylene diisocyanate (PAPI), etc.

The isocyanate-water reaction is carried out at 0° to 70°C, a temperature range which will not deleteriously affect the catalytic properties of the enzyme being entrapped. A preferred range of reaction temperatures is 0° to 3°C, reaction mixtures at these temperatures having a longer reaction time and allowing slower and more thorough mixing. It is important however, to keep the temperature of the reaction mixture above 0°C since freezing enzymes sometimes tend to reduce or destroy their catalytic activity.

The reaction mixture will cure to an immobilized enzyme poly(urea-urethane) foam in 30 seconds to 10 minutes, the longer reaction times being associated with the lower reaction temperatures, and vice versa.

Example: An aqueous enzyme solution was formed by adding 1.5 g papain to 40 ml of distilled ice water and the aqueous solution stirred with magnetic stirring bar until the enzyme was dissolved. An isocyanate terminated prepolymer solution was formed by reacting stoichiometric amounts of 80/20 2,4-/2,6-tolylenediisocyanate and a 1,000 MW polyoxyethylene polyol (Carbowax 1000) at 135°C for 5 hours using a 2-ethylhexanoic acid and tin octoate catalyst and diluting the resulting isocyanate terminated urethane prepolymer to 90% solids with acetone.

To 24.5 g of the acetone solution of isocyanate terminated prepolymer was added 0.22 g of Tween-80 (a polyoxyethylene derivative of sorbitan fatty acid ester having a molecular weight of about 1,309) and the viscous mixture stirred by hand for approximately 30 seconds. 20 ml of the aqueous enzyme solution was added to the prepolymer solution and the resulting pasty mixture stirred to homogeneity. The remaining cold aqueous enzyme solution was then added and the reaction mixture again stirred to homogeneity. The reacting mass was then poured into an aluminum pan 16 cm x 24 cm and allowed to stand at ambient temperature for 5 minutes at which time the reaction was essentially complete.

The resulting white spongy, wet, semirigid, enzyme-polymer foam was 0.4 to 1.0 cm thick. The foam was removed from the pan and stored in a refrigerator at 0° to 5°C, the foam remaining damp during storage. A strip of the papain foam structure containing the equivalent of 78.1 mg of papain was tested for biological activity against α-N-benzoyl-L-arginine ethyl ester hydrochloride and found to contain approximately 53% of the original activity.

OTHER ENTRAPPING PROCESSES

Enzymes Immobilized in Emulsified Particles

The process developed by *N.N. Li, D.R. Brusca and R.R. Mohan; U.S. Patent 3,897,308; July 29, 1975; assigned to Exxon Research and Engineering Co.* relates to immobilized enzymes which comprise an enzyme-containing solution emulsified in immiscible liquid. Preferably, the enzyme-containing solution is aqueous and will comprise from 0.1 to 10 wt % enzyme. A surfactant is provided in the immiscible liquid to stabilize the emulsion, the surfactant comprising from 0.01 to 90 wt %, preferably 0.5 to 20 wt % of the immiscible liquid.

The resulting immobilized enzyme, i.e., enzyme emulsion, may be conveniently used in chemical processing by dispersing the emulsion in a reactant containing feed stream for a time sufficient to obtain the desired chemical reaction. The emulsion is designed so as to be immiscible with the reactant containing feed stream. That is, when the feed stream is aqueous, a hydrocarbon liquid may be used for preparing the enzyme emulsion. Further, the enzyme emulsion is designed so as to selectively allow the reactants of the feed stream to permeate through the immiscible liquid into the interior phase of the emulsion wherein the reactants are contacted with the enzyme. The reactants are converted to reaction products in the interior phase of the emulsion, and the reaction products may permeate back through the immiscible liquid into the feed stream. Thus, in effect, the immiscible liquid, i.e., the exterior phase of the emulsion, forms a liquid membrane around the enzyme-containing solution.

The enzymes subject to this process may be selected from hydrolases, oxidoreductases, transferases, lyases, isomerases and ligases. The preferred enzymes are the hydrolases, including proteolytic enzymes which hydrolyze proteins, e.g., papain, ficin, pepsin, trypsin, chymotrypsin, bromelin, keratinase, carbohydrases which hydrolyze carbohydrates, e.g., cellulase, amylase, maltase, pectinase, chitinase, esterases which hydrolyze esters, e.g., lipase, cholinesterase, lecithinase, alkaline and acid phosphatases; nucleases which hydrolyze nucleic acid; e.g., ribonuclease, desoxyribonuclease; and amidases which hydrolyze amides, e.g., arginase, asparaginase, glutaminase, and urease; and the oxidoreductases including glucose oxidase, catalase, peroxidase, lipoxidase, and cytochrome reductase, as well as those mentioned specifically above.

The surfactant utilized for preparing the immobilized enzymes of the process may be selected from the group consisting of anionic, cationic, or nonionic surfactants. Nonionic surfactants are the preferred surfactant type for this process. A useful group of nonionic surfactants include the polyethenoxyether derivatives of alkyl phenols, alkylmercaptans, and alcohols, e.g., sorbitol, pentaerythritol, etc. Particular nonionic surfactants for use in the process include compounds having the formula

$$R_{10}-C_6H_4-O-[CH_2CH_2O]_m-CH_2CH_2OH$$

where R_{10} may be C_8H_{17}, C_9H_{19}, or $C_{10}H_{21}$ and m is an integer varying from 1.5 to 8. The most preferred nonionic surfactant is Span 80, a fatty acid ester of anhydro sorbitol condensed with ethylene oxide. In general, surfactants will comprise 0.01 to 10 wt % of the immiscible liquid, more preferably, from 0.1 to 5 wt %.

In general, the enzyme-containing solution being aqueous, the immiscible liquid will be an organic immiscible liquid, e.g., hydrocarbons, including halogen, nitrogen and oxygen derivatives thereof. For example, immiscible liquids are fluorinated, chlorinated, amino, alkoxy, etc, substituted hydrocarbons containing up to 50, preferably 30, carbon atoms. The preferred immiscible liquids are the fluorocarbons, including the chlorofluorocarbons and the perfluorocarbons, and the highly branched isoparaffins since compounds of these classes generally do not react with and do not substantially denature enzymes.

The immiscible liquid, which functions as a liquid membrane, is further designed to allow the permeation of selected reactants present in the feed stream. For example, when the reactant is phenol, an immiscible liquid comprising a nonionic surfactant and a hydrocarbon oil may conveniently be used in preparing the immobilized enzymes.

Example: The catalyst, polyphenol oxidase, is contained in an aqueous surfactant membrane. Phenol in cyclohexane solution permeates from the outside phase through the membrane, in which it is oxidized. The oxidized product is recovered in the inside cyclohexane phase. The aqueous surfactant membrane was prepared by combining 0.4 g saponin, 10 mg mushroom polyphenol oxidase (Worthington Biochemical Corp.), 100 g glycerol, and 99.6 g 0.05 M tris buffer, pH 7.0. This membrane was stirred at 500 rpm for 20 minutes with 200 g spectrophotometric grade cyclohexane to produce a stable emulsion.

100 ml of this emulsion was combined with 100 ml of a phenol solution in cyclohexane (0.1 g/200 g) in a 250 ml separatory funnel. The removal of phenol from the outside phase was monitored by sampling this phase after 3, 6, 12, 18 and 24 hand shakes of the separatory funnel. The absorbence difference between 265 mμ and 240 mμ was used as a measure of phenol concentration.

Number of Shakes	Phenol in Outside Phase (A_{265} - A_{240})
0	1.85
3	1.39
6	0.95
12	0.63
18	0.44
24	0.35

The results indicate that phenol is rapidly removed from the outside phase. Likewise, the appearance of oxidized product in the inside phase was confirmed by comparison of the ultraviolet spectrum of that phase, following centrifugation at 10,000 rpm for 20 minutes with the spectrum of known oxidized product. The oxidized product in cyclohexane is characterized by the relatively high A_{240}/A_{265} ratio of 0.56 as compared with 0.08 for the initial phenol reactant in cyclohexane.

Isomerase Immobilized Within the Cells by Use of Citrates

N. Tsumura and T. Kasumi; U.S. Patent 4,001,082; January 4, 1977; assigned to Director of National Food Research Institute, Japan provide a simple and effective method of treating microbial cells containing glucose isomerase and using these cells in either a batch or continuous type process for isomerization. As the microorganisms containing glucose isomerase, those belonging to genus of *Aerobacter, Lactobacillus, Brevibacterium, Bacillus, Arthrobacter, Streptomyces, Nocardia, Micromonospora, Microbispora, Microellobospora, Streptosporangium, Leuconostoc, Pasteurella* and *Actinoplanes* are known.

Any of these microorganisms can be applied to the process aggregated by various protein denaturants such as an aldehyde, acid, salt of heavy metal, organic solvent, etc. Some of them denature and inactivate an enzyme and are harmful from a hygienic viewpoint. In developing this process, microbial cells were treated with various organic and inorganic acids under proper pH conditions. For instance, hydrochloric and sulfuric acids as inorganic acids, and oxalic, lactic, succinic, maleic, tartaric, aspartic, tannic, citric acids and gum arabic, etc, as organic acids have been tested as treating reagents. As a result, it has been found that citric acid is particularly advantageous. That is, by this process, enzyme retentivity of microbial cells at reaction time is significantly improved by dipping the cells in a solution containing citric acid and/or a citrate of a specified metal (defined as "citrate solution") and then, if required, drying these microbial cells. As metals of a citric acid salt, monovalent and divalent cations and a mixture thereof (for instance, lithium, sodium and potassium as monovalent metals, and magnesium, cobalt and calcium as divalent metals) are used.

The method of treating microbial cells containing glucose isomerase comprises dipping the cells in a citrate solution at a pH range wherein no significant damage to the enzyme occurs, and then, if required, drying the cells at a temperature at which the enzyme is not significantly inactivated. This method is applied to both intact and heat treated cells of bacteria such as *Actinomycetes.* In the case of intact cells, the method is applied to the cells separated from a cultivation broth. As the effect of treatment is more significant with heat treated cells than with intact cells, it is preferred to use heat treated cells. The heating is done at a temperature killing intact cells, and usually a temperature of from 60° to 85°C is preferred.

With regard to concentration of citric acid and/or a citrate in a citrate solution, the effect is observed at 0.1% or more calculated as citric acid, but a concentration of 1 to 8% as citric acid gives favorable results. The volume of this solution is preferred to be one in which microbial cells are dispersed readily with agitation, and specifically, 40 to 150 times by weight of dispersed dried cells.

The effect of citrate treatment on improvement of enzyme retentivity of microbial cells varies somewhat depending on the kind or the state of the cells. For example, it has been observed that a heat treatment of the cells, the retention period after harvesting of the cell and also a pretreatment of the cell with a flocculating agent, etc, influence somewhat the effect of citrate treatment. The frequency of dipping microbial cells in a citrate solution is arbitrary. It is effective to repeat dipping and separation more than twice when a citrate solution of low concentration is used.

A citrate solution is used at a broad pH range without significant loss of glucose isomerase activity, though at pH below 5 the enzyme is apt to be inactivated. However, the tendency that the enzyme-retaining effect of treated microbial cells is lowered as pH rises from acidic to alkaline is recognized, and so it is preferred to adjust pH to 5 to 6.

The effect of citrate treatment on the cells at a broad range of temperature is recognized. The cells are dipped in a citrate solution at a temperature not inactivating glucose isomerase significantly, i.e., from 0° to 40°C, and retained for a sufficient time for full immersion of the cells in the solution, and then recovered by centrifuge or filtration. In addition to this, drying of the cells after dipping treatment reinforces the effect of the citrate treatment.

As drying methods, air-drying at room temperature, drying under reduced pressure, heat-drying, dehydration and drying with organic solvents, lyophilization, etc, can be effectively applied. In the case of heat-drying, it is possible to dry at any temperature which does not cause significant inactivation of the enzyme, but preferably a temperature below 50°C is used.

This process can be applied not only to intact cells or heat treated cells, but also effectively to cells aggregated in advance by a pretreatment with a flocculant.

Example: 2 g of commercial frozen cells of *Actinomycetes* containing glucose isomerase (belonging to genus *Streptomyces* samples) heat treated with about 60% moisture, were dipped in 50 ml of solutions containing 0 to 16% citrate (adjusted to pH 6.0 with sodium hydroxide), stored at 4°C overnight and then centrifuged. The sludge of the microbial cells was air-dried on a Petri dish for 5 days.

40 mg of dried cells (as solid) thus obtained were dispersed in 6 ml of reaction solution, sealed in 10 ml volume test tube and shaken for 20 hours at 60°C for isomerization. The reaction solution is composed of glucose, 40% (w/v), magnesium ion 0.02 M and phosphate buffer solution 0.025 M (pH 7.5).

After reaction, microbial cells were recovered by centrifugation and repeatedly used for second and third reactions with the addition of 6 ml of fresh reaction solution. The amount of fructose produced in each reaction mixture was determined, and the remaining enzyme activity of microbial cells (enzyme retentivity) was expressed as the percentage of the amount of fructose produced in second and third reactions related to that of the first reaction. The amount of fructose produced by each sample was compared with one another in the initial reaction, and it served as relative activity. The results of these determinations are shown in the table below.

	Sample No.						
	1	2	3	4	5	6	7
	(%)						
Concentration of Citrate	0	0.1	1	4	8	12	16
Cell moisture	13.8	12.0	11.0	11.4	13.9	14.9	15.5
Relative activity	100	97	95	98	96	95	95
Remaining activity A*	41	52	63	77	81	73	74
Remaining activity B**	11	37	53	68	71	64	62

*In second reaction.
**In third reaction.

Enzyme-Containing Marumes of Reduced Tack

The disclosure by *D.M. van Kampen; U.S. Patent 4,016,041; April 5, 1977; assigned to Lever Brothers Company* relates to spheronized enzymes with reduced stickiness and improved sievability, and a method for preparing them. A special granular form of enzyme preparations (marumes) has been introduced. These are solid granules containing homogeneously distributed enzymes and are prepared by spheronizing an enzyme-containing extrudate in an apparatus which comprises a cylinder with a smooth wall and having positioned therein a roughened, horizontal rotatable table.

The extrudate, which is obtained by extrusion of a suitable mixture of enzymes and organic and/or inorganic extrudable material, in mixture with small amounts of binding or lubricating agents, for example, noodles, is fed into this apparatus, and by the centrifugal forces exerted upon the noodles by the rotation of the table, the noodles are transformed in spheronized granules, called marumes.

The apparatus known as a Marumerizer is described in U.S. Patent 3,277,520. Marumes are prepared wherein a water-containing extrudate is spheronized to form solid small sphere-like bodies, marumes, in an apparatus as described above. These marumes can then be coated with a suitable coating material, e.g., nonionic surface-active agents such as polyethylene glycols, condensation products of ethylene oxide with alkyl phenols, fatty alcohols and the like compounds.

During the spheronizing a dusting powder, such as anhydrous sodium sulfate or titanium dioxide can be used to prevent caking of the marumes. The marumes are subsequently sieved to remove small particles. While, for example, anhydrous sodium sulfate may indeed provide free-flowing, nonsticky marumes, the sievability of the treated marumes is unsatisfactory. It has been found that if the marumes are treated during spheronization with magnesium oxide, either alone or in mixture with magnesium carbonate as dusting powder, nonsticky, free-flowing marumes are obtained which are very easily sievable.

The MgO or $MgO/MgCO_3$ mixture used in the process has a particle size predominantly ranging from 40 to 250 μ. Technical MgO often contains a minor proportion of $MgCO_3$ (in a ratio of 10:1) and is equally applicable in the process. In general, the amount of dusting powder necessary ranges from 0.1 to 5%, usually from 1 to 3%, based on the weight of the marumes to be treated.

Although the process is applicable to marumes, prepared from water-containing extrudates, it is also applicable to water-free extrudates. In the latter case a final drying step, usually necessary in the former case, is normally not necessary. The marumes obtained are particularly suitable for incorporation in detergent compositions to formulate particulate enzymatic detergent compositions. The marumes contain proteolytic and/or amyloytic enzymes, preferably bacterial proteases and/or amylases, prepared by submersed fermentation of strains of *B. subtilis*.

Although the process is in particular applicable to enzyme-marumes, it is in principle also applicable to marumes containing other detergent adjuncts, such as low temperature bleach precursors, germicides or bactericides, fluorescers, antisoil redeposition agents and coloring agents.

Example: Enzyme marumes were prepared by mixing 63% NaCl, 18% secondary C_{11-15} linear alcohol condensed with 9 mols of ethylene oxide, 4% citric acid and 15% of an enzyme concentrate (a proteolytic bacterial enzyme Alcalase) in a mixer, extruding this mixture through a convenient extruder to form noodles, and spheronizing these noodles in a Marumerizer. During the spheronizing operation, which took place for 2 minutes, 1% by weight of one of the dusting powders listed in the table below was added in each experiment.

The flow characteristics of these treated marumes were visually assessed after the dusting treatment (by visually assessing the caking and flowing characteristics of a sample in a glass tube); the sievability was assessed by sieving the treated marumes.

The results of these tests are reported in the table on the following page. The $MgO/MgCO_3$ had the following particle size distribution: 89.9% $< 250\ \mu$; 8.9% $> 250\ \mu$; and 1.3% $> 1,000\ \mu$.

Dusting Powder	Flow Characteristic After Dusting	Sievability
Highly voluminous silica (Aerosil)	very sticky	big lumps on 12 mesh
Highly voluminous silica (Gasil)	very sticky	big lumps on 12 mesh
Highly voluminous alumina (Alusil)	sticky	big lumps on 12 mesh
Carboxymethylcellulose	free flowing	lumps on 44 mesh; blinded
Sodium sulfate	free flowing	lumps on 44 mesh; blinded
Sodium chloride	free flowing	lumps on 44 mesh; blinded
Sodium tripolyphosphate	free flowing	lumps on 44 mesh; blinded
Calcium phosphate	free flowing	lumps on 44 mesh; blinded
Calcium carbonate	free flowing	lumps on 44 mesh; blinded
Calcium sulfate	free flowing	lumps on 44 mesh; blinded
Titanium dioxide	very sticky	big lumps on 12 mesh
Titanium dioxide/magnesium silicate (4:1)	very sticky	big lumps on 12 mesh
Magnesium silicate	very sticky	big lumps on 12 mesh
Talcum powder	slightly sticky	lumps on 44 mesh; blinded
Magnesium oxide	free flowing	good; no lumps, no blinding
Magnesium oxide/magnesium carbonate (10:1)	free flowing	good; no lumps, no blinding
Control (no dusting powder)	slightly sticky	lumps on 44 mesh; blinded

IMMOBILIZATION OF MODIFIED ENZYMES

CROSSLINKING SUPPORTED ENZYMES

This method of preparing water-insoluble or immobilized enzymes comprises adsorbing the enzyme on a substrate and then crosslinking the enzyme, e.g., by the action of a water-soluble dialdehyde. The enzyme is mainly crosslinked to itself rather than bonded to the substrate by covalent linkages.

Aldehyde Crosslinking of Adsorbed Penicillin Acylase

Previous formulations of immobilized and crosslinked acylases have not had the mechanical stability and enzyme activity needed for economical processing. Improved water-insoluble enzyme complexes are provided by *T.A. Savidge and L.W. Powell; U.S. Patent 4,001,264; January 4, 1977; assigned to Beecham Group Limited, England* which comprise a penicillin acylase adsorbed on a water-insoluble polymer or copolymer of methacrylic acid and crosslinked with glutaraldehyde, glyoxal or formaldehyde. The complexes are useful in the production of 6-APA from penicillins G and V.

The supports for use are polymers or copolymers of methacrylic acid. They therefore contain free carboxylic groups which give the polymer an acidic function to a degree about optimum for good adsorption of penicillin acylase enzymes without denaturation. Macroporous polymers and copolymers of methacrylic acid are to be preferred to gel polymers and copolymers of methacrylic acid. The methacrylic polymers are required to be water-insoluble and are usually in the form of crosslinked copolymers, for example copolymers of methacrylic acid, with divinylbenzene or a diester of a glycol with methacrylic acid, for example by the use of ethylene glycol bis-methacrylate.

One advantage of the process is that it permits the use of polymer supports that are already commercial products for other purposes, particularly as cationic exchange resins of a weakly acidic nature. Particularly suitable is the methacrylic acid/divinylbenzene copolymer, Amberlite IRC-50, and the methacrylic acid copolymer, Zeokarb 227.

The latter is a divinylbenzene/methacrylic acid copolymer which also contains some benzene sulphonyl groups.

The polymer supports for use are preferably in the form of finely-divided particles or beads of particle size such that they will pass a 100 A.S.T.M. sieve, i.e., of particle size diameter below 0.1 mm. However, the resulting enzyme complex must not be so finely divided that it cannot be separated from the reaction mixture by a mesh filtration process or used in a column reactor. Thus the polymer should have a particle size in excess of 0.01 mm, i.e., it should be substantially retained on a 800 A.S.T.M. sieve.

The acylase enzyme to be contacted with the polymer should be an aqueous solution and have been dialyzed until its ionic conductivity has been lowered to within the range of from 0.1 to 5 mmho. The pH of the enzyme solution should desirably be between 4.5 and 7.0 and the polymer should be contacted with the enzyme solution for a sufficient period to ensure maximum enzyme adsorption: this residence time is usually between 2 and 16 hours.

After its adsorption on the support, the enzyme is crosslinked in situ by treatment with a crosslinking agent selected from glutaraldehyde, glyoxal and formaldehyde. The use of glutaraldehyde or glyoxal is preferred, glutaraldehyde especially resulting in enzyme complexes of most advantageous properties when used in the production of 6-APA. The crosslinking agent is normally used in aqueous solutions at a concentration preferably of 0.5 to 5.0% by weight. After completion of the crosslinking reaction, it is desirable to ensure that any of the crosslinking agent that has not reacted is removed or rendered innocuous. For instance, excess agent may be removed by washing with water or with a solution of an amine compound and urea has been found to be very effective for this purpose.

Example: Aliquots (20 to 25 or 50 g) of commercial cationic and anionic ion-exchange resins as stated below were suspended in distilled water (ca 100 to 500 ml) and adjusted to pH values between 4.4 and 6.3 in the case of the cationic resins and to pH values between 6.5 and 9.0 in the case of the anionic resins by the addition of sodium hydroxide or hydrochloric acid with vigorous agitation. The resins were recovered by filtration, washed well with distilled water and were resuspended in a solution of partially purified penicillin acylase (60 to 100, 250 or 500 ml; specific activity, 3.85 to 6.75 μmol/min/mg protein; conductivity <1 mmho).

The amount of enzyme adsorbed on the resin was varied between 71 and 308 μmol/min/g of resin. Enzyme was allowed to adsorb with gentle agitation for ca 16 hours when the resins were recovered by filtration, resuspended in a solution of glutaraldehyde in water (100 ml, 0.825 to 3.3% w/v) and allowed to react for ca 16 hours. The resulting enzyme-resin complexes were recovered, washed 3 times with distilled water, resuspended in water or 0.2 M phosphate buffer pH 7.8 and adjusted to pH 7.8, treated for 1 hour with an aqueous solution of urea (100 ml, 0.1 M, pH 7.8) and finally washed 3 times with distilled water.

Each enzyme complex so prepared was used to prepare 6-APA from benzylpenicillin under standard conditions at pH 7.8 and 37°C. The activities of the enzyme complexes are given in the following table. The activities refer to complexes prepared at the optimum pH for enzyme adsorption and retained enzyme activity.

Activities of Enzyme Complexes

Experiment	Nature of Matrix and Functional Groups in the Resin	Resin	Physical Form of the Resin	Specific Activity of Enzyme Resin Complex at Optimum pH .... μmol/min/g .... Damp Weight	Dry Weight
A	styrene-quaternary ammonium	Amberlite IRA-938	macroporous	10.4	32.4
B	styrene-quaternary ammonium	Amberlite IRA-401	gel	0	0
C	acrylate-quaternary ammonium	Amberlite IRA-458	gel	4.71	10.1
D	polystyrene-polyamine	Lewatit MP-62	macroporous	15.1	30.7
E	styrene-polyamine	Amberlite IR-45	gel	0	0
F	acrylate-polyamine	Amberlite IRA-68	gel	0	0
G	styrene–SO_3H	Lewatit SP-120	macroporous	19.3	39.2
H	styrene–SO_3H	Lewatit S-100	gel	0	0
I	styrene–SO_3H	Zerolit 325	gel	0	0
J	styrene–SO_3H	Amberlite IR-120	gel	0	0
K	phenol-formaldehyde–SO_3H	Bio-Rex 40	gel	4.01	8.3
L	styrene–SO_3H	Bio-Rad AG/MP/50	macroporous	9.9	19.8
M	styrene–PO_3H_2	Bio-Rex 63	gel	0	–
N	styrene–CH_2–N(CH_2–COOH)$_2$	Chelex 100	gel	7.16	30.1*
O	methacrylate–COOH	Amberlite IRC-50	macroporous	37.0	88.8
P	acrylate–COOH	Amberlite IRC-72	macroporous	~8.0	~24.0**
Q	acrylate–COOH	Amberlite IRC-84	gel	7.0	13.9
R	acrylate–COOH	Zerolit 236	gel	~8.0	~46.5**
S	***–COOH	Lewatit CHP-80	macroporous	15.2	36.5
T	***–COOH	Lewatit CHP	macroporous	0	0
U	phenol-formaldehyde {–OH, –COOH}	Zerolit 216	gel	0	0
V	polymethacrylate {–COOH, –SO_3H}	Zeokarb 227†	gel	25.9	57.0

Note:
Resins A–C are strongly anionic; D–F are weakly anionic; G–L are strongly cationic and O–U are weakly cationic.

*Mesh size 100–200 cf 14.50 of other resins.
**Results very variable, quoted values are mean of several determinations.
***Polymer structure not known.
†Resin not available commercially.

It will be noted that the advantages of the process are limited to the cases where the substrate is a polymer or copolymer of methacrylic acid.

Crosslinking by Disulfide Rearrangement

The process developed by *M.H. Keyes; U.S. Patent 4,008,126; Feb. 15, 1977; assigned to Owens-Illinois, Inc.* provides for the polymerization of enzymes by in situ reaction on an inert support to form active composite having prolonged service life.

This process comprises the steps of selecting a protein where the sum total of one-half of the cystine amino acid residues plus the cysteine amino acid residues equal at least 14 per mol of protein, depositing the protein on an inert support to form a protein/support composite and maintaining the composite at a pH which facilitates polymerization of protein and maintaining the composite at a temperature and for a time sufficient to polymerize the protein and immobilize the protein in situ on the support.

It could be theorized that the immobilization reaction involves a disulfide rearrangement within the protein molecule. Either the cystine or cysteine amino acid residues can be zero so long as the sum total of the two fulfills the ratio set forth above. However, when no cysteine is present, it may be necessary or desirable to add a trace of a compound containing sulfhydryl groups (e.g., an amount such that the reaction mixture is 10^{-8} to 10^{-5} molar in sulfhydryl groups) such as β-mercaptoethanol or dithiothreitol, to initiate the disulfide interchange reaction. The term amino acid residue is used in its conventional sense and refers to the residues of the amino acids which chemically combine in forming the protein molecule.

The cysteine amino acid residues contain the –SH group and the cystine contains the –S–S group, and under the pH condition specified in this process these groups are theorized to undergo a disulfide rearrangement which results in the polymerization of the protein. This disulfide rearrangement is acid and base catalyzed and does not proceed at an acceptable rate at a pH of 6.6 to 9.4. The following table sets forth the cysteine and cystine proportions for several enzymes useful in the process.

Enzyme	Molecular Weight	Cysteine Residues plus One-Half Cystine Residues per Mol
Urease (Jack Bean)	500,000	85
Histidase	210,000	17
L-amino acid oxidase	135,000	14
Alcohol dehydrogenase (horse liver)	40,000	28

Other enzymes which are suited for this process are glucose-6-phosphate dehydrogenase ED 1.1.1.49, glucose-6-phosphate dehydrogenase, glutamate synthase, glutamine phosphoribosylpyrophosphate amidotransferase P-300 monomer, glutaminyl-tRNA synthetase, glutathione synthetase, glyceraldehyde-3-phosphate dehydrogenase, glycerol-3-phosphate dehydrogenase, EC 1.1.1.8, glycine N-methyltransferase, glycinin, glycogen synthetase and the acid pH which facilitates polymerization is preferably in the range of 4.5 to 6 for efficiency and economy. A basic pH which facilitates polymerization is in the range of 9.5 to 11.

In carrying out the process the enzyme or other protein and support are brought into contact in an aqueous solution of enzyme and the enzyme thoroughly wets and permeates the support. The pH of the enzyme solution is maintained as indicated above for time and temperature sufficient to polymerize the enzyme for immobilization in situ on the support. This usually requires time periods ranging from a few minutes to several (e.g., 100) hours depending on the temperature, concentration, enzyme and other factors. Time periods of 1 to 50 hours are suitable for most applications. The temperature is usually maintained below 50°C to prevent denaturing the enzyme and temperatures in the range of 0° to 20°C are satisfactory for most applications.

The composition of the support is not particularly critical as long as it is inert, dimensionally stable, and provides sufficient surface area for retention of enzyme. The support can be porous, fluid-permeable membranes or porous particulates. It has been found that the porous materials having a volume porosity preferably in the range of 15 to 50% are quite suitable for the present purposes. The pore size of the support is critical in that it should not be so small as to prevent immobilization of the enzyme thereon. Average pore size diameters of either fluid permeable membrane or porous particulates in the range of 0.01 μ to 10 μ are suitable for most applications with 0.01 to 2 being preferred for efficiency and economy.

Example: A—20 g of particulate porous alumina are washed with 4 liters of distilled water by swirling in a flask and decanting the cloudy supernatant liquid. The particulate alumina has a particle size of from -80 to +100 mesh (U.S. Sieve) and an average pore size diameter of 0.1 to 0.2 μ with a volume porosity of 10 to 20%. The alumina is then washed on a vacuum filter with 1 liter of distilled water. Finally the alumina is deaerated by adding 300 ml of a 2 x 10^{-3} molar, pH 8.5, tris(hydroxymethyl)aminomethane-maleic acid buffer (tris-maleate) to the alumina and evacuating the flask continuously with a water aspirator vacuum for about 1 hour. This buffer is discarded and 300 ml of fresh tris-maleate buffer are added.

B—An aqueous urease solution is prepared by dissolving 0.56 g of urease (Miles-Servac, Batch 11) in 120 ml of 2 x 10^{-3} M tris-maleate buffer. This solution is centrifuged for 20 minutes at 10,000 gravity. The resulting supernatant liquid is decanted from the precipitate and the supernatant is added to the alumina in tris-maleate buffer of Part A. This mixture of urease and alumina in buffer is mixed for 35 minutes at 0° to 5°C. The urease supernatant solution is decanted from this alumina and is transferred to the second, fresh sample of alumina (like that of Part A except that the particle size is -60 +80 mesh) which had been pretreated as described above in Part A. The resulting mixture is mixed for about 30 minutes at 0° to 5°C.

C—The pH of the solution above the alumina is 8.2. The pH is then adjusted to pH 5.0 by the slow addition of a 0.02 M sodium acetate-acetic acid buffer solution having a pH of 4.5. The buffer is prepared by titrating 0.02 M sodium acetate against 0.02 M acetic acid to a pH of 4.5. A total volume of 98.2 ml of this buffer is required for this pH adjustment of the enzyme-alumina mixture.

After the addition of the acetate buffer is completed, the resulting mixture is shaken for one hour at 0° to 5°C. During this period the urease polymerizes in situ and the resulting polyurease polymer becomes immobilized and entrapped in and on the alumina support to form a biologically active composite. The resulting "poly" urease/alumina composite is filtered and washed with about 1.5 liter portions of 3.0 M sodium chloride, 5×10^{-3} M ethylenediaminetetraacetic acid (EDTA) and 2.0×10^{-3} M tris-maleate buffer (pH 7.0). The poly urease/alumina composite is analyzed to have an activity of 212 U/cc of composite and is effective in prolonged use.

Activated Esters as Crosslinking Agents

The disclosure by *M. Rubinstein, S. Simon and R. Bloch; U.S. Patent 4,101,380; July 18, 1978; assigned to Research Products Rehovot Ltd., Israel* relates to several classes of multifunctional compounds which can serve as crosslinking agents for proteins and especially enzymes. These crosslinking agents fall into two groups comprising: (a) water-soluble poly(ethylene oxide) derivatives of the formula

$$(Z)_m\text{–Y–}(CH_2CH_2O)_nCH_2CH_2\text{–Y–}(Z)_m$$

where both n and m are at least 1; Y is a covalent bond or is an –R– or –RO– radical in which the oxygen is bound to the poly(ethylene oxide) and R is selected from methylene, ethylene, propylene, o-, m- and p-phenylene, o-, m- and p-phenylene carbamate optionally substituted by one or more alkyl, aryl, halo, nitro, oxo, carboxy, hydroxy, thio, sulfonate and phosphate groups; and Z is a reactive group selected from haloisocyanato-, isothiocyanato-, tosylate, acyl halides, acyl azides, aryl diazonium salts, acyl imidoester salts, activated esters of acyl residues and 2,4-dichlorotriazines.

The second group comprises (b) activated esters of di- and polycarboxylic acids where the acids and the alcohol moieties are water-soluble, and where the alcohols forming the activated esters are selected from N-hydroxysuccinimide, 1-hydroxybenzotriazole, 8-hydroxyquinoline or 2- and 4-thiopyridine and the acids are selected from tartaric, citric, malic, dimethoxysuccinic and trimethoxyglutaric acid.

Preferred examples of such crosslinking agents which can be prepared by known methods include: di-N-hydroxysuccinimide ester of poly(ethylene oxide)disuccinate, di-N-hydroxysuccinimide ester of poly(ethylene oxide)diglycolate, diacid chloride of poly(ethylene oxide)diglycolate, diazide of poly(ethylene oxide)diglycolate, dihydrochloride salt of diimido methyl ester of poly(ethylene oxide) diglycolate, poly(ethylene oxide)-di(2-, 4-, or 6-methyl-3-isocyanatophenyl carbamate) and di-N-hydroxysuccinimide ester of d-tartaric acid (TDN).

These crosslinking agents can be used in several ways for the crosslinking of proteins, e.g.:

(a) The pure enzyme or protein is reacted with the crosslinking agent to form three-dimensional species which are completely insoluble in water, or complex oligomeric soluble derivatives;

(b) The enzyme is crosslinked in the presence of a second

protein, to form a cocrosslinked derivative;

(c) The enzyme is first adsorbed on an insoluble, surface-active support and then crosslinked intermolecularly with a multifunctional reagent; or

(d) The enzyme can be crosslinked to a protein matrix.

Example 1: The following is the preparation of chymotrypsin membranes with various crosslinking agents. A chymotrypsin solution containing β-phenyl-propionate as inhibitor (3.3 mg chymotrypsin and 7 mg of inhibitor/cc) in 0.05 M phosphate buffer pH 7.0 was passed through a millipore membrane (0.1 μ). The membrane was washed briefly and immersed in TDN dissolved in acetone (10 mg/0.1 cc). After two hours it was washed with HCl 10^{-3} M (to break up the enzyme-inhibitor complex) and buffer until no UV and no enzymatic activity were found in the wash water. The membrane was tested for its enzymatic activity at R.T. using benzoyl-tyrosine-ethyl-ester in 30% ethanol tris buffer pH 7.8 (0.05 M) as substrate and measuring the amount of product formed in the effluent.

A chymotrypsin membrane was prepared as described above using N-hydroxysuccinimide ester of poly(ethylene oxide) disuccinate (Mx-54) instead of TDN in the crosslinking solution. The membrane was washed and analyzed above. A chymotrypsin membrane was prepared as described above using N-hydroxysuccinimide ester of poly(ethylene oxide) diglycolate (Mx-170) instead of TDN in the crosslinking solution. The membrane was washed and analyzed as above. The enzymatic activity of the membranes prepared above are given in the following table.

In the table the chymotrypsin membrane is Millipore MF-VC, pore size 0.1 μ. (The membrane parameters are: section, 10.7 cm^2; thickness, 150 μ; porosity, 74%.) The substrate is N-benzoyltyrosine ethyl ester, 2.5 x 10^{-3} molar concentration. The enzymatic activities when the various crosslinking agents are employed are given in μmol/min cm^2 at the indicated pressures and flow rates. The conversion percent is also recorded.

				Conversion Percent
Pressure, atm	0.5	1.2	2.7	
Flow rate, cc/min/member	3.0	6.6	13.0	
Activity, μmol/min cm^2				
TDN	0.61	1.17	1.7	50
Mx-54	0.81	1.2	1.78	46
Mx-170	0.34	0.37	0.53	60

Example 2: The following is the preparation of asparaginase particles using Mx-170 as a crosslinking agent. 1 cc of sepharose (4B, particle size 40 to 190 μ) particles (55 mg dry weight) were impregnated with asparaginase (6 mg/ml) in borate buffer (0.05 M, pH 8.5). The particles were then suspended with stirring in a solution of Mx-170 (10 mg/ml) in phosphate buffer (0.05 M, pH 7) for 2 hours. The particles were rinsed with water and borate buffer until no UV absorption and no enzymic activity were found in the rinse water. The particles were packed into a column (diameter 1.7 cm, height 0.3 cm) and asparagine (0.01 M) in borate buffer (0.05 M, pH 8.5) flowed through it. NH_3 was measured in the effluent. The activity of the particles increases with flow rate.

The activity of a 100 mg of particles (dry weight) is 7.0 μmol/min.

Flow Rate (ml/min)	Product Concentration (μmol/ml)	Reaction Rate (μmol/min)
0.25	4.6	2.3
1.13	3.2	3.7
2.1	1.8	3.87
3.6	1.1	3.86

IMMOBILIZATION BY POLYMERIZATION OF MONOMER MODIFIED ENZYMES

The majority of processes for the immobilization of enzymes on synthetic polymers involve the reaction of the polymer or modified polymer containing an enzyme-reactive group with the enzyme. In the following four processes, the enzyme is chemically coupled with a polymerizable monomer and the immobilized enzyme formed by polymerization of the monomer-enzyme intermediate.

Enzymes Coupled with Monomers Containing Reactive Groups

The method developed by *J.J. Hamsher; U.S. Patent 3,925,157; December 9, 1975; assigned to Pfizer Inc.* comprises reacting an enzyme with a polymerizable, ethylenically unsaturated monomer containing a reaction group, and then polymerizing or copolymerizing the covalently bonded enzyme-monomer in the presence of a crosslinking agent and an initiator. Suitable polymerizable ethylenically unsaturated monomers are those having the formula

$$CH_2{=}\underset{\displaystyle R_1}{\underset{|}{C}}{-}W$$

where R_1 is selected from hydrogen, methyl and chloro; W, the reactive group, is selected from $-COR_2$, $-SO_2Cl$ and

$$-C_6H_4-SO_2Cl$$

R_2 is selected from (A) hydroxy, halide, azido, 2,3-epoxypropoxy, 2,3-epithiopropoxy, N-(2,3-epoxypropyl)amino, N-[(p-diazonium chloride)phenyl] amino, acryloyloxy, lower alkoxy carbonyloxy, and benzenesulfonyloxy; and (B) $-X-(Y-X')_n-Z$ where X and X' are selected from $-O-$ and $-NR_3$ where R_3 is selected from hydrogen and alkyl containing 1 to 6 carbons, Y is selected from alkylene containing 2 to 3 carbons, n is an integer from 1 to 2; and Z is selected from hydrogen and Z' where Z' is selected from haloacetyl, 2-(4,6-dichloro)-s-triazinyl, p-toluenesulfonyl, p-(halomethyl)benzoyl, and cyano; with the proviso that when Z' is p-toluenesulfonyl, X' is $-O-$.

Each of the monomers of the above formula where R_2 is a member of group (A) contains, except for those where R_2 is hydroxy, a reactive group capable of combination with an enzyme. However, when R_2 is hydroxy the monomer acids are potentially reactive in that the carboxylic group is readily activated by reaction with an appropriate carboxylic group-activating reagent, such as carbodiimide, for example, dicyclohexyl carbodiimide or ethylmorpholinocarbodiimide; N-ethyl-5-phenylisoxazolium-3'-sulfonate (Woodward's Reagent K); ketenimines such as pentamethyleneketene cyclohexylimine; acetylenic ethers, for example, ethoxyacetylene; hexahalocyclotriphosphatriazines; N-hydroxyphthalimide or N-hydroxysuccinimide, and other reagents used to form a peptide bond. Monomers where R_2 is from group (B) are also activated with the exception of compounds in which Z is hydrogen. Such monomers, however, are transformed into reactive monomers by reaction with halo-Z' according to standard procedures.

The monomer-enzyme product can be polymerized with an addition polymerizable monomer in the presence of a crosslinking agent and an initiator. Suitable addition-polymerizable monomers (comonomers) are, for example, acrylic, α-chloroacrylic, methacrylic acids and the glycidyl, lower alkyl esters, N,N-(disubstituted)-aminoalkyl esters, amides, lower alkyl substituted amides, methylol substituted amides, N-monosubstituted aminoalkylamides and N,N-disubstituted aminoalkylamides thereof, styrene, butadiene and isoprene.

Polymerization or copolymerization of the monomer-enzyme product is conducted in the presence of a crosslinking agent to impart a three-dimensional network character and insolubility to the final polymer. A wide variety of crosslinking agents can be used, for example, acrylic monomers or olefin compounds.

Representative of such agents are 1,3-butylene diacrylate, ethyleneglycol dimethacrylate, 1,3-butylene dimethacrylate, 1,6-hexamethylene diacrylate, ethylene diacrylate, diethylene glycol dimethacrylate, N,N'-methylene-bisacrylamide, neopentyl glycol dimethacrylate, 1,1,1,-trimethylol ethane trimethacrylate, divinylbenzene, and the like. The preferred crosslinking agent is N,N'-methylene-bisacrylamide since it imparts desirable hydrophilicity to the final product.

The polymerization and copolymerization reactions are initiated by free radicals. The choice of initiator is determined largely by the mild conditions necessitated by the thermal sensitivity of the enzyme moiety. Suitable free radical initiator systems are those which generate free radicals at a suitable rate for polymerization or copolymerization at a temperature below 40°C. Redox initiator systems are preferred for this reason.

Example: Penicillin Acylase Coupling to Cyanogen Bromide Activated Hydroxyethyl Methacrylate—To a solution of 20 g of 2-hydroxyethyl methacrylate in 80 ml of water is added a solution of 10 g of cyanogen bromide in 80 ml of water. The pH is adjusted to 11.0 with 30% w/v NaOH and maintained at this pH until stabilized, keeping the temperature between 20° to 30°C. The pH is brought to 6.5 with HCl, and 20 g (8 units/mg activity) of penicillin acylase from *Proteus rettgeri* ATCC 9250 in 200 ml of water is added with stirring. The mixture is stirred at room temperature for 1.5 hours, then 10.0 g of N,N'-methylene-bisacrylamide is added and the mixture stirred for an additional 30 minutes. The reaction mixture is placed in an ice bath and cooled to 4°C under nitrogen. A catalyst system consisting of 3 ml of dimethylaminopropionitrile and 375 mg of ammonium persulfate is added with stirring.

The ice bath is removed and the reaction mixture is allowed to come slowly to room temperature under nitrogen. Polymerization occurs gradually over a 30 to 60 minute period until a thick gel results.

The resulting solid is broken up, 500 ml of water is added, and the mixture stirred to break up the thick gel into a granular solid. The solid is filtered, washed with water, and may be air-dried or vacuum dried to give 37 g of immobilized enzyme material which, based on activity assay, contains about 50% of the total initial activity of the free enzyme.

The immobilized penicillin acylase is capable of performing over 30 reaction runs hydrolyzing benzylpenicillin to 6-aminopenicillanic acid over a 12 week period without significant loss of activity. The conversion of benzylpenicillin to 6-aminopenicillanic acid under optimum conditions using this immobilized penicillin acylase system is greater than 95%. Numerous other examples of the complete patent cover the preparation and use of penicillin acylase and glucose oxidase coupled to various monomers and monomer mixtures activated with trichlorotriazine, bromoacetyl bromide, or a diimide, or directly to monomers containing reactive groups.

Copolymerization of Enzymes Acylated or Alkylated with Monomers

A mild process for binding enzymes to insoluble carriers, which gives stable products having high activity and activity yield is used by *D. Jaworek, M. Nelböck-Hochstetter, K. Beaucamp, H.U. Bergmeyer and K.-H. Botsch; U.S. Patent 3,969,287; July 13, 1976; assigned to Boehringer Mannheim GmbH, Germany.*

In this process, compounds capable of acylating or alkylating enzymes and which contain a polymerizable group are coupled with the enzyme. The coupled enzyme is then copolymerized with other monomers to form a polymer bound enzyme.

It is especially preferred to use a coupling compound which contains at least one double bond capable of copolymerization, especially a carbon-carbon double bond, as the further functional group capable of producing a bond with a carrier. As carrier substances, there can be used water-insoluble solid materials which, via the further functional group of the coupling compound, can be coupled in aqueous solution under mild conditions. Preferably, there are used carrier substances which are hydrophilic, easily swellable, substantially charge-free and also stable towards microorganisms. The carrier substance can be introduced as such into the aqueous solution for the production of the bond with the intermediate product but preferably the carrier substance is itself produced in the aqueous solution by the polymerization of water-soluble monomers.

In this preferred embodiment the reaction of the enzyme with the coupling compound can take place either in the presence of the polymerizable monomer or monomers, whereafter the polymerization is carried out with the polymerizing in the presence of the coupling compound-protein intermediate product, or the polymerizable monomer or monomer mixture is first added to the solution after the reaction has taken place between the protein and the coupling compound whereafter the polymerization is initiated.

Typical examples of preferred coupling compounds include maleic anhydride and its homologues in which the hydrogen atoms of the carbon-carbon double bond are replaced by alkyl radicals containing up to 6 carbon atoms, allyl halides, especially allyl bromide and its homologues, acryloyl chloride and its homologues in which one or more hydrogen atoms are replaced by one or more lower alkyl radicals, maleic acid and fumaric acid chlorides and their homologues corresponding to the above definition in the case of maleic anhydride, maleic acid azide and ethyleneimine compounds, such as 1-allyloxy-3-(n-ethyleneimine)-propan-2-ol and the like.

As monomers used for copolymerization with the coupled enzyme-monomer it is preferred to use the water-soluble derivatives of vinyl alcohol, acrylic acid or methacrylic acid, for example, the amides, nitriles or esters of these compounds, especially good results being obtained with the use of acrylamide. The compounds can also be substituted by alkyl radicals so long as the water-solubility of the compounds is not reduced too much. However, such compounds with reduced water-solubility are of advantage if, subsequently, the carrier-bound enzyme is to be used in a system which is not entirely aqueous, for example, in an aqueous-organic system. The corresponding derivatives of maleic and fumaric acids can also be satisfactorily used.

Alternatively, there can also be water-insoluble monomers. In this case, the polymerization is carried out in suspension and not in solution. Suspension polymerization is of advantage if a finely-divided pearl-like matrix which does not swell in aqueous systems is desired.

Depending upon the desired consistency of the end product, crosslinking agents, i.e., compounds containing more than one polymerizable group, can be added to the monomer. Examples of crosslinking agents of this type include N,N'-methylene-bis-acrylamide and ethylene diacrylate. These are preferred in the case of working in aqueous solution. If the polymerization is carried out in a heterogenous phase, i.e., as a suspension polymerization, there can also be water-insoluble crosslinking agents, for example, divinylbenzene or ethylene dimethacrylate.

Example 1: 100 mg glucose oxidase (GOD; 220 U/mg) were dissolved in 10 ml 1 M triethanolamine buffer (pH 8.0) at 10°C under an atmosphere of nitrogen. 0.03 acryloyl chloride in 3 ml ether were then added and the reaction mixture was stirred for 30 minutes. Subsequently, it was dialyzed overnight against 2 liters 0.01 M triethanolamine buffer (pH 8.0) and then the precipitate was centrifuged off and discarded. The enzymatic activity amounted to 16,000 U.

The solution was then mixed with 0.4 ml 5% dimethylaminopropionitrile and 0.4 ml 5% ammonium peroxydisulfate at 5° to 10°C (enzymatic activity 14,500). Thereafter, 3 g acrylamide and 0.015 g N,N'-methylene-bis-acrylamide in 9 ml of water were added under an atmosphere of nitrogen. The polymerization which commenced immediately led to a gel-like solidification of the mass. The product obtained was granulated by forcing through a 0.4 mm metal sieve and then washed with 2 liters 0.2 M phosphate buffer (pH 7.5). The enzymatic activity removed with the wash water was 600 U.

The polymer was lyophilized to give 3 g of dry product with an enzymatic activity of 1,500 U. When the process was repeated but without the use of acryloyl chloride, the inclusion polymerization gave 3 g of a product with a total activity of 330 U.

Example 2: 300 mg trypsin (1,500) were dissolved, under an atmosphere of nitrogen, in 10 ml 0.5 M phosphate buffer (pH 8.0) at 10°C. The solution obtained was mixed with 0.1 ml acryloyl chloride in 10 ml ether and the reaction mixture was stirred for 30 minutes.

There were then added 0.4 ml 5% dimethylaminopropionitrile and 0.4 ml 5% ammonium peroxydisulfate and the reaction mixture stirred for 30 minutes. Subsequently, 3 g acrylamide and 0.015 g N,N'-methylene-bis-acrylamide in 9 ml water were added under an atmosphere of nitrogen at a temperature of 5° to 10°C and these conditions were maintained until a gel-like solid mass had formed. This mass was granulated by forcing through a 0.4 mm metal sieve and washed with 3 liters of 0.2 M phosphate buffer (pH 7.5). 23 U of enzyme activity were found in the wash water. The washed produce was lyophilized. There were obtained 3 g lyophilizate with a specific activity of 12.9 U/g. A repetition of the process under the same conditions but without the addition of the acryloyl chloride gave a specific activity of 0.5 U/g of lyophilizate.

Copolymerization of Enzyme-Monomer Products Within a Molecular Sieve

According to the process of *D. Jaworek, K.-H. Botsch, G. Weimann, M. Nelböck-Hochstetter and H. Determann; U.S. Patent 4,081,329; March 28, 1978; assigned to Boehringer Mannheim GmbH, Germany* carrier-bound macromolecular compounds are prepared in which the macromolecular compounds are covalently bonded onto the surface of a polymer body which penetrates a molecular sieve material body.

Macromolecular compounds include proteins, especially biologically-active proteins, such as enzymes, hormones and antibodies, nucleic acids, peptides, porphyrins, hemins, phosphatides, cerebrosides, ganglyosides, glycosides and the like. It is important that the macromolecule is so large that it cannot penetrate into the interstices of the molecular sieve material used. Therefore, the minimum size of the macromolecules which can be bound depends upon the size of the interstices of the molecular sieve material used.

Compound B which, in this process is coupled with the macromolecular compound A possesses at least one function capable of coupling with the latter and at least one further function capable of polymerization. If, for example, the macromolecular compound A contains amino acid residues, then the function capable of coupling can be, for example, an oxirane group, an ethyleneimine group, a halide group, an acid halide group, an acid azide group or an acid anhydride group. Apart from the coupling function, the compound B has at least one further functional group which can participate in a polymerization reaction. This group may be one suitable for an addition polymerization but also one suitable for a condensation polymerization.

Specific examples of compounds B which can be used include acrylic acid 2,3-epoxypropyl ester, but-2,3-ene oxide, 1-allyloxy-3-(N-ethyleneimino-propan-2-ol), maleic anhydride, allyl bromide, acryloyl chloride, maleic acid azide, methacrylic acid, 2,3-epoxypropyl ester, maleic acid 2,3-epoxypropyl monoester, fumaric acid epoxy-(2,3-epoxypropyl diester), 1-(allyloxy)-2,3-epoxybutane and the like.

Molecular sieve materials which can be used are those with a mesh size which excludes the macromolecular compound A but which allows the polyfunctional coupling compound B to pass through. There are preferably used gel-like molecular sieves based on crosslinked polymers, such as polyacrylamide gels and crosslinked carbohydrates, such as crosslinked dextran, agar and the like.

The molecular sieve material must be wholly or at least partially in unswollen state. This is necessary since the polymerization or copolymerization of the polymerizable function of the coupling compound B is to take place within the molecular sieve. Therefore, the swelling of the molecular sieve is carried out in such a manner that the polymerization system is drawn in through the swollen molecular sieve material and is present substantially in the interior of the molecular sieve material. In the case of the swelling of the molecular sieve, the solution is thereby drawn in which the coupling of the macromolecular compound A with the compound B has taken place so that, in principle, only the macromolecular part of the coupling product AB remains behind on the surface of the molecular sieve body.

The volume of the reaction solution is preferably so chosen that it is insufficient for the complete swelling of the molecular sieve, which ensures that the subsequent polymerization takes place exclusively within the molecular sieve body.

Example: 100 mg glucose oxidase (220 U/mg) were dissolved in 5 ml 0.5 M triethanolamine buffer (pH 8.0) and cooled to 10°C in at atmosphere of nitrogen. To this reaction batch were added 0.03 ml acryloyl chloride in 2 ml methylene chloride and the reaction mixture was then stirred for 30 minutes. In this enzyme solution preincubated with acryloyl chloride, were dissolved 0.8 g acrylamide and 0.05 g N,N'-methylene-bis-acrylamide. The polymerization was then initiated with 0.05 ml 5% N,N-dimethylaminopropionitrile solution, as well as 0.05 ml 5% ammonium peroxydisulfate solution. Immediately after this addition, 5 g of molecular sieve crosslinked dextran (Sephadex G-25, medium) were added to the reaction batch and this was then stirred in an atmosphere of nitrogen.

The volume of the reaction solution was so calculated that it was not sufficient for the complete swelling of the molecular sieve so that a uniform polymerization block was not formed but rather the polymerization took place exclusively within the molecular sieve particles. The material obtained was suspended in 500 ml distilled water, vigorously slurried with a stirrer, filtered off with suction and lyophilized. There were obtained about 5 g (referred to the lyophilizate) of material with a specific activity of 160 U/g lyophilizate.

Grafting of Enzyme-Monomer Couples on Activated Polysaccharides

A disadvantage of the binding of active proteins onto activated polysaccharides, such as cellulose, crosslinked dextran, agarose and the like, is that, after completion of the protein binding, some of the groups are hydrolyzed and thus charged. Indeed, some of these carriers are so strongly charged that the greater part of the enzymes to be immobilized are absorptively held and the bound protein, depending upon the polyvalent character of the carrier, undergoes a shift of the pH optimum. A further disadvantage is the strong absorption of the substrate and the reaction product on such charged carriers.

The disclosure by *D. Jaworek, J. Maier and M. Nelböck-Hochstetter; U.S. Patent 4,038,140; July 26, 1977; assigned to Boehringer Mannheim GmbH, Germany* provides for binding biologically-active proteins with an activated polysaccharide graft-polymerized with a hydrophilic graft copolymer.

In one embodiment of the process, proteins, activated polysaccharide and hydrophilic monomer or monomers to be grafted on and possibly also crosslinking agents are brought together in aqueous phase, whereafter the graft copolymerization is carried out. There are obtained inclusion-fixed biologically active proteins with good stability and improved accessibility for substrates.

In another embodiment, the protein, together with a bridge-building compound which contains at least one copolymerizable group and at least one group acylating or alkylating protein in aqueous solution, is reacted and the reaction product is brought together with the activated polysaccharide and the hydrophilic monomer or monomers to be grafted on and possible crosslinking agents in aqueous phase, whereafter the graft copolymerization is carried out. This process variant gives covalently bound proteins with especially high activity yields.

Examples of biologically active proteins which can be used include enzymes, enzyme associations, protein and peptide hormones, antigens, antibodies and the like. Polysaccharides which can be used include starch, cellulose and cellulose derivatives, such as cellulose acetate, cellulose butyrate, allyl cellulose, carboxymethylcellulose, deoxythiocellulose, methylcellulose, hydroxyalkylcellulose, regenerated cellulose, allyl starch, carboxymethyl starch, dialdehyde starch, the corresponding dextran derivatives, polyglucosides and the like.

The activation of the polysaccharide necessary for the graft polymerization can take place by the introduction of a double bond, for example, by reaction with a bifunctional compound which contains a copolymerizable olefinic double bond and a group which is reactive with hydroxyl groups, for example, an epoxide group, episulfide group, cycloimine group or lactam group, or by ionizing irradiation with the formation of long-life radicals or also by radical formation on the hydroxyl groups of the starch with a one electron acceptor, such as Ce(IV).

Especially preferred difunctional compounds for the activation of the polysaccharide are activated allyl derivatives, such as allyl halides, for example, allyl bromide, and activated acrylic or methacrylic acid derivatives, such as the acid chlorides, anhydrides, azides and the like, as well as activated dicarboxylic acid derivatives, such as chloromaleic acid. Other known alkylation and acylation monomers can also be used.

Examples of comonomers which can be used for the grafting include acrylamide, acrylonitrile, vinyl acylates, such as vinyl acetate, propionate and phosphate, acrylates, methacrylates, allyl citrate, polyalkyleneglycol acrylates and methacrylates, N'-vinyl-lactams, such as N-vinyl-pyrrolidone, and N-substituted acrylamides and methacrylamides. The monomers preferably possess at least one carboxyl, aminocarbonyl, sulfo or sulfamoyl group, acrylamide being especially preferred.

Difunctional crosslinking agents can also be used which must possess at least two olefinic double bonds capable of copolymerization and which are preferably also hydrophilic. Examples of crosslinking agents include diacrylates, dimethacrylates, diacrylamides, such as N,N'-methylene-bis-acrylamide, the corresponding methacrylic compounds and the like.

According to a further preferred embodiment of the process, the graft copolymerization does not take place in the presence of the unmodified biologically active protein, but in the presence of a protein reaction product with a bridge-building compound. Bridge-builders of this type, which are also called coupling compounds, have at least one copolymerizable olefinic double bond or another function capable of forming a covalent bond with the grafted on side chain in aqueous solution and a group which acylates or alkylates proteins in aqueous solution.

The following table shows, for three different enzymes as biologically active proteins, the activity yields and specific activities obtained in the case of binding onto a modified carrier of this process and onto a carrier comparable therewith but crosslinked in another way according to the two-stage process, with the use of a bridge-builder.

Enzyme	Comonomers in Carrier (g/g)	Ratio of Polymer: Enzyme* (g/g)	Ratio of Enzyme: Bridge Builder** (mg/μl)	Activity Yield (%)	Specific Activity (U/g)
D-hydroxynitrile lyase from almond powder	acrylamide:starch allyl ether 1:0.33	1:0.23	1:0.05	20	41
D-hydroxynitrile lyase from almond powder	acrylamide:N,N'-bis*** 1:0.057	1:0.25	1:0.05	10	27
Trypsin	acrylamide:starch allyl ether 1:0.2	1:0.125	1:0.25	33	30
Trypsin	acrylamide:N,N'-bis*** 1:0.1	1:0.16	–	14	12
Yeast hexokinase	acrylamide:starch allyl ether 1:2	75:1	1:1	7	125
Yeast hexokinase	acrylamide:N,N'-bis*** 1:0.2	75:1	1:1	2.5	46

*Protein used for the immobilizing.
**Acrylic acid chloride was used as bridge builder.
***N,N'-methylene-bis-acrylamide.

Example 1: Mechanical Inclusion—The acrylamide (3.0 g), starch allyl ether (2.0 g) and acylase (600 mg, specific activity 20 U/mg) were dissolved in 40 ml phosphate buffer at pH 7.5. The solution was mixed with 2 ml 5% ammonium peroxide disulfate and 2 ml 5% 3-dimethylaminopropionitrile and flushed with nitrogen. After 5 minutes, gelling commenced. The gel was forced through a sieve of 0.4 mm mesh size and eluted in a column with 5 liters 0.3 M aqueous sodium chloride solution. The product had a specific activity of 26 U/g. The yield was 5 g.

Example 2: Copolymerization with Bridge-Builder—The acrylamide (6.0 g) and starch allyl ether (2.0 g) were dissolved in 15 ml phosphate buffer (pH 7.5, 0.2 M) to form Solution 1. The hydroxynitrile lysase was dissolved in 35 ml of the same phosphate buffer, cooled to 4°C and mixed with a similarly cooled solution of 0.1 ml acrylyl chloride in 5 ml diethyl ether. The mixture was intensively stirred for 30 minutes and subsequently added to Solution 1. It was flushed with nitrogen until the greater part of the ether had been removed. The starter solutions, 2 ml 5% ammonium peroxydisulfate and 2 ml 5% 3-dimethylamino-propionitrile, were then added.

After about 5 minutes, the reaction mixture begins to solidify. After the polymer has stood for at least 3 hours, it was forced through a sieve with a mesh size of 0.4 mm and eluted in a column with 5 liters of 0.5 M aqueous buffered sodium chloride solution. The yield was 8.4 g. The specific activity was 41 U/g (lyophilizate) = about 345 U = 18% activity yield.

CHEMICAL ACTIVATION OF ENZYMES FOR COUPLING

In the following process, the enzyme is chemically modified other than by coupling with a polymerizable monomer to provide means of covalent coupling to a support. Related processes include the conversion of enzyme disulfide groups to thiol groups prior to coupling to enzymes as in U.S. Patent 4,048,416 on page 204.

Oxidized Glycoenzymes Immobilized on Aminostyrene Polymers

The development of *O.R. Zaborsky and A.I. Laskin; U.S. Patent 3,970,521; July 20, 1976; assigned to Exxon Research and Engineering Company* provides for immobilizing glycoenzymes. In this process, a glycoenzyme is first treated with an oxidizing agent at conditions whereby carbonyl groups or their precursors, e.g., acetals, are formed in the carbohydrate portion of the glycoenzyme. The oxidized product is then contacted with an amino-containing material under conditions whereby a conjugate of the amino material and the glycoenzyme is formed.

In a preferred method, a glycoenzyme, e.g., glycose oxidase, is contacted with an aqueous periodic acid or sodium periodate solution at a pH of 2.5 to 7.5 and 5° to 40°C for a time to convert a portion of the carbohydrate to an oxidized product, i.e., one containing carbonyl or acetal groups. The oxidized glycoenzyme is then contacted with a water-insoluble polymer containing an amino group to form a water-insoluble conjugate of the glycoenzyme and the polymer. Preferably the amino-containing material is a water-insoluble amino-containing polymer, e.g., poly-p-aminostyrene, aminoethylcellulose, carboxymethylcellulose hydrazide, Biogel P-2 hydrazide, which is a derivative of polyacrylamide, amino-Sepharose which may be prepared according to the procedure described by Cuatrecasas, *J. Biol. Chem.* 245, 3059-3065 (1970), or an aminoalkylated glass.

It has been found that the glycoenzyme poly-p-aminostyrene conjugate of this process shows activity substantially equivalent to the glycoenzyme in its native state, i.e., in aqueous solution. More surprisingly, it has been found that many of the glycoenzyme amino polymer conjugates show enhanced thermal stability when compared to the native enzymes.

Example: Oxidation of Glucose Oxidase with Periodic Acid—To a stirred solution of glucose oxidase (40.2 mg, 2.68×10^{-7} mol, in 20 ml 50 mM acetate buffer, pH 5.60) in a thermostated vessel at 25°C protected from light was added 0.4 ml of a periodic acid solution (9.12 mg, 4.00×10^{-5} mol). The yellow solution was stirred for 4 hours upon which 0.025 ml of ethylene glycol, 4.48 $\times 10^{-4}$ mol, was added to react with excess periodate and stirred for an additional 0.5 hour. The solution was transferred to an Amicon Model 202 ultrafiltration cell equipped with an XM-50 filter at 50 psi N_2 pressure and dialyzed with 50 mM acetate buffer, pH 5.59, until no further dialyzable material came forth.

The conversion of periodic acid after the 4 hours of reaction with glucose oxidase, based on the absorbance change at 223 nm, Dixon, J.J. and Lipkin, D. (1954) *Anal. Chem.* 26, 1092-1093, was 61.6%. The oxidized enzyme was stored at 5°C.

Coupling of Oxidized Glucose Oxidase to p-Aminostyrene—To a 5 ml solution of oxidized glucose oxidase (10 mg) was added 250 mg finely powdered p-aminostyrene. The suspension was adjusted to pH 9 with 0.1 N NaOH, stirred at 25°C for 1 hour and then filtered with a Millipore 0.45 μ filter. The solid was washed with 1 liter of H_2O and 1 liter of 1 M NaCl in 50 mM phosphate buffer, pH 6.4. After several ml of the H_2O wash, no further activity was detected in the wash. The original filtrate exhibited high activity; the NaCl wash exhibited no activity.

Coupling of oxidized glucose oxidase with p-aminostyrene resulted in an active enzyme-polymer conjugate. It is presumed that the enzyme is bound to the polymers through an imine linkage. However, when the amino-containing material is a hydrazide, the linkage is a hydrazone. The protein loading (mg of enzyme/g of enzyme-polymer conjugate) with the p-aminostyrene is 5 to 8 mg. The activity of the immobilized enzyme is equivalent to the native and oxidized enzymes. See Table 1 below. It is likely that this is due to immobilizing the enzyme via catalytically nonessential carbohydrate residues for no amino acid residues were found to be oxidized in the enzyme.

Table 1: Specific Activities of Glucose Oxidases

Glucose Oxidase	Specific Activity (u/mg protein)
Native	91.9*
Oxidized	92.5*
p-Aminostyrene-bound	93.1**

*Based on protein as determined by anaerobic spectral titration with glucose.
**Based on protein as determined by amino acid analysis. Protein loading was 5.87 mg enzyme per gram of enzyme-polymer conjugate.

The conjugate was stored for six weeks in 100 mM phosphate buffer, pH 6.31. The conjugate retained its activity and the buffer remained inactive, thus showing no desorption. Table 2 shows the thermal stability of the glucose oxidase. It is noted that the native and oxidized enzymes have similar stabilities (the oxidized enzyme exhibiting slightly more stability), whereas the water-insoluble conjugate has definite enhanced stability.

Table 2: Thermal Stability

	-------- Relative Activity of Glucose Oxidases (%) -------		
Time (hr)	Native	Oxidized	p-Aminostyrene-Conjugate
0	100	100	100
0.25	63.4	76.0	76.8

(continued)

Table 2: (continued)

Time (hr)	Relative Activity of Glucose Oxidases (%) Native	Oxidized	p-Aminostyrene-Conjugate
0.50	47.1	59.1	60.0
0.75	36.8	49.3	59.4
1.0	27.4	38.1	56.6
1.5	15.7	30.0	48.1
2.0	8.7	21.0	41.4
3.0	3.5	11.0	34.5
4.0	1.8	6.4	28.6
5.0	1.1	4.0	24.6
6.0	0	2.7	24.3
48.0	0	0	12.7

COVALENT BONDING OF ENZYMES GENERAL

PHOTO-ACTIVATED COUPLING PROCESSES

Aryldiazosulfonates as Coupling Agents

There is an acute problem with coupling high molecular weight naturally occurring molecules such as proteins and sugars to like or different molecules. Molecules, such as proteins, sugars, and the like will have a plurality of hydroxyl, mercapto or amine functionalities, which are reactive to a wide variety of reagents. Therefore, in any type of reaction, one obtains a relatively random reaction of the functionalities and a complex mixture of products.

Desirably, one would wish to react the first polyfunctional compound to a desired degree to provide a new compound which could then be reacted with the second compound. By controlling the ratio of reactants, one could minimize or preclude self-reaction.

A substituted aryldiazosulfonate is used for this purpose by *T. Burkoth; U.S. Patent 3,843,447; October 22, 1974; assigned to Syva Company* to couple or link an enzyme to another compound having a reactive functional group. In this process a first polyfunctional compound or composition is combined with a substituted aryl anti-diazosulfonate so as to substitute the first compound or composition with at least one anti-diazosulfonate group, to form an anti-diazosulfonate-substituted polyfunctional product.

The anti-diazosulfonate-substituted polyfunctional compound is combined with a second compound, usually polyfunctional, having at least one functionality susceptible to reacting with a syn-diazosulfonate, by causing the light induced isomerization of the anti-diazosulfonate (inactive) to the syn-diazosulfonate (active).

The polyfunctional compounds will for the most part have a molecular weight of at least 125, and may be as high as 10 million molecular weight or more. In most instances, at least one of the polyfunctional compounds will be a lower order or higher order polymer, having at least three recurring units, usually at least six recurring units, and may have recurring units of 500,000 or greater. (For proteins, individual amino acids will be considered recurring units.) Of particular interest are polyamino acids having a molecular weight of 1,000 to 500,000, more usually 2,000 to 200,000. For the most part, these compounds will be the polyamino acids, either as polypeptides or proteins.

There will be two types of polypeptides of particular interest. The first type will generally be of 5,000 to 600,000 molecular weight, more usually 10,000 to 200,000 molecular weight, and will be enzymes. Therefore, it will generally be of particular interest to join an enzyme or enzymes to a different polyfunctional compound or composition.

For the most part, the linking compounds will have the following formula: $YArN_2SO_3^-M^+$, where Ar is an aromatic group of 5 to 12 carbon atoms, more usually of 6 to 10 carbons, and having 0 or 1 heteroatom, i.e., nitrogen; and Y is an active substituent which is capable of forming a stable link to a functionality present in the first polyfunctional compound, wherein the functionality may be bonded directly to an annular member or be separated from an annular member by an aliphatic chain.

Usually, the aromatic ring will be unsubstituted except for the specified functionalities. The sulfonate will usually have ammonium or an alkali metal, particularly sodium or potassium, counterion as the cation M^+. Preferred compounds have the following formula:

$$R'OC(=O)C(=O)-C_6H_4-N_2SO_3^-M^+$$

where R' is hydrocarbyl, usually alkyl of from 1 to 8 carbon atoms, more usually of from 1 to 6 carbon atoms; and M^+ has been defined previously.

Example 1: To a suspension of 4.1 g (30 mmol) of p-aminobenzoic acid in 300 cc of water was added 4.5 cc of concentrated hydrochloric acid. The suspension was warmed slightly to hasten solution, at which time an additional 9 cc of concentrated hydrochloric acid was added and the solution cooled to 0° to 5°C.

To this stirring solution was added at once a precooled solution of 2.1 g of sodium nitrite in 6 cc of water. After 15 minutes at 0°C a test with starch-iodide paper indicated excess nitrous acid. The pH was raised to the pH range of Congo red by the addition of saturated aqueous sodium acetate.

To this solution was added at once 6.3 g of sodium sulfite in 15 cc of water. A certain amount of color was produced at this step and some precipitation was observed as well. The solution could be assayed for active diazonium salt or the reactive syn isomer by touching a drop to filter paper which had been treated with β-naphthol in aqueous alcoholic carbonate solution.

An instantaneous red color signaled the presence of the reactive species. After one hour at room temperature, this spot test showed no active diazonium species remaining. The aqueous solution was treated with decolorizing charcoal and filtered. The addition of solid sodium chloride to saturation caused the precipitation of the desired anti-diazosulfonate as a yellow crystalline solid which is shown by spectroscopic techniques to be the monohydrate.

Example 2: To an ice-cold suspension of 500 mg of the acid-diazosulfonate monohydrate (Example 1) in 10 cc of dry dimethylformamide was added by syringe 1 cc of isobutyl chloroformate followed by 1 cc of triethylamine. The mixture was stirred for 3 hours at 0°C under nitrogen, then sealed and placed in a refrigerator at 5°C for 12 hours.

During this time the amount of undissolved material decreased significantly, but a crystalline solid remained. The reaction mixture was stirred at 40°C and 1 mm Hg to remove about 5 ml of clear liquid, which contains some dimethylformamide and the excess isobutyl chloroformate. The resulting solution suspension was adjusted in volume to 5 ml with dry dimethylformamide and used without further treatment.

Example 3: A solution of 100 mg (0.0069 mmol) of lysozyme in 10 cc of water was adjusted to pH 9.5 with 0.05 N NaOH. After cooling to 0°C a volume of stock dimethylformamide solution of Example 2 calculated to contain 0.05 mmol was added dropwise with stirring. The pH was maintained at 9.3 to 9.5 throughout the addition by adding 0.05 N NaOH. After addition was complete, the pH remained stable for 1 hour at 0°C. The pH was then adjusted to 7.5 and the cloudy solution was centrifuged. The supernatant was dialyzed against 4 portions of 3 liters of water for 6 hours each change.

The precipitate was dissolved with urea and likewise dialyzed. Each dialysis resulted in a yellow clear solution. The redissolved material was centrifuged again to remove a small amount of insoluble material. Either solution could be irradiated at pH 9 in the presence of excess β-naphthol to give a red solution. The red color could not be removed by extensive equilibrium dialysis. Both solutions were found to contain active lysozyme when assayed using the normal procedure of measuring the rate of clearing of a suspension of lyophilized bacteria.

Example 4: A stock, buffered solution of morphine was prepared by dissolving 15 mg of morphine in 7.5 cc of 0.2 N sodium carbonate, then diluting to 10 cc volume with 0.2 N sodium bicarbonate. The final pH was 10.3. To 4 cc of this solution was added 400 μl of a $\sim 2 \times 10^{-4}$ M lysozyme conjugate solution of Example 3. The final concentration of morphine was 5×10^{-3} M and that of lysozyme conjugate was $\sim 2 \times 10^{-5}$ M.

This solution was cooled and irradiated with visible light of wavelength longer than 390 nm for one-half hour. The solution was only slightly darker in color after this irradiation. The solution was dialyzed against 4 liters of distilled water in a small hollow fiber device. The solution was found to have active enzyme by lysis of a bacterial suspension. The rate was carefully measured and a clearing rate of 190 optical density units in 48 seconds was recorded. Addition of antibody to morphine reduced the rate to 140 units in 48 seconds for an inhibition of 25%. This result is indicative of photobonding of morphine to lysozyme.

Bifunctional Arylazides as Coupling Agents

A bifunctional agent is used in the process disclosed by *P.E. Guire; U.S. Patent 3,959,078; May 25, 1976; assigned to Midwest Research Institute* for producing immobilized enzymes. Preferably a solid support material is treated with an activating agent containing a thermochemical functional group and a photochemical functional group. The activating agents are reacted thermochemically with a solid support to form an "activated solid matrix" which is then subjected to photochemical reaction with an enzyme to covalently bind the enzyme.

In a preferred embodiment of this process, a solid support material is treated with a thermochemical-photochemical bifunctional agent such as 1-fluoro-2-nitro-4-azidobenzene (4-fluoro-3-nitrophenyl azide), 1-fluoro-2,4-dinitro-5-azidobenzene (5-fluoro-2,4-dinitrophenyl azide) and other derivatives of arylazides containing substituents which allow activation of the azide by visible light and substituents which react thermochemically with support material. Thus, the fluorine may be replaced with alkylamino, alkylcarboxyl, alkylthio, alkylmethylimidate, alkylisocyanate, alkylisothiocyanate, alkylaldehyde, alkylhalide and other such groups which react with functional groups of support materials. While the azide on a nitrophenyl ring provides advantages in degree of dark stability and susceptibility to activation by visible electromagnetic radiation, other functional groups which generate nitrenes and/or carbenes can be used as the photochemical group for the thermochemical-photochemical bifunctional reagent.

Some examples of such useful photochemical groups are alkyl azides, acyl azides, α-keto diazo compounds (α-diazo ketones, esters, etc.), diazirines and diazoalkanes. Such photochemical reactive groups can be placed on the same molecule with chosen thermochemical reactive groups. The resulting bifunctional activating agents react thermochemically in the dark with solid support material containing a functionally available reactive group, the reaction occurring between the thermochemical group of the activating agent and the reactive group of the solid support material.

Conditions such as temperature, pressure, pH and the like are not critical insofar as the reaction is concerned. Generally, however, it is preferred to conduct the reaction at a temperature of 25° to 40°C, atmospheric pressure, and at a pH of 7 to 10. The time of reaction will depend upon the temperature, pH and reactivity of the functional groups and in general the reaction can be completed in periods of 2 to 24 hours. By effecting the above reaction with minimum exposure to visible light, the photochemically functional azide moiety is substantially unaffected and is available for subsequent covalent binding with an enzyme.

Examples of suitable solid support materials are the ion exchange glass beads, celluloses, dextrans, agaroses, polyacrylamides, polymethacrylates, silicone rubber elastomers, partially hydrolyzed polyamides and polyesters, collagen and the like. The nitroazidophenyl derivative of the solid support material, the "activated matrix," is stable in the absence of light and can be reacted with an enzyme to form a solid stable and immobilized enzyme. To prepare such a solid stabilized enzyme system, the "activated matrix" is contacted with an enzyme and irradiated with visible light.

Example: About 100 mg of white, fine granular, dry alkylamine glass beads [(AAG), (ENCOR, 550A, 40 to 80 mesh, porous glass beads, Corning Biomaterials)] were added to 1 ml of 0.9 M sodium borate buffer (pH 9.1) in an amber glass bottle. 1-Fluoro-2-nitro-4-azidobenzene (FNAB), about 16 mg, was dissolved in 2 ml ethanol in dim light and added to the reaction bottle. This mixture was stirred at 37°C for 20 hours in the dark. The solid product was separated and washed with ethanol, with 3 M sodium chloride, and with 0.02 M potassium phosphate buffer, pH 8. The washed solid nitroazidophenyl-aminoalkyl glass product (fine pink granules) was dried under vacuum at room temperature and stored at this temperature. The preceding operations were performed in the dark or dim light.

Still under dim light, about 75 mg of this nitroazidophenyl-aminoalkyl glass (NAP-AAG) derivative and about 7 mg of L-asparaginase (AsNase) from *Escherichia coli* B were placed in a colorless glass round-bottom flask of about 5 ml capacity. To this was added about 0.5 ml water. This mixture was exposed with stirring under refrigeration (0° to 15°C) to light from a focused 40 watt tungsten lamp passing through about 0.5 cm of 1 M aqueous sodium nitrite solution for about 16 hours. The solid glass-bound enzyme product (AsNase-ANP-AAG) was washed sequentially with assay buffer (0.2 M phosphate, pH 8), 3 M sodium chloride, and assay buffer, then dried under vacuum at room temperature. This dry pink glass-asparaginase was found by a standard enzyme assay (in 0.02 M phosphate buffer, pH 8, at 37°C, with 15 mmol L-asparagine) to exhibit 1,450 IU enzyme activity per gram of product (1 IU is defined as the production of 1 μmol product per minute under the assay conditions).

The pink granular powder from the above procedures comprises a stabilized immobilized enzyme system having high catalytic efficiency and specificity, capable of acclerating the amido-hydrolysis of L-asparagine without extremes of pH or temperature. This enzyme system can be used to convert L-asparagine to L-asparaginic acid and ammonia under very mild reaction conditions with very low energy requirements and minimal side reactions or undesirable by-products. This form of the catalyst can be readily recovered for subsequent use in either batch or flow processes for conversion of dissolved L-asparagine to aspartic acid and ammonia. Also, bound to an ammonium-ion-sensitive electrode membrane, the glass-bound enzyme can be used to quantitatively measure L-asparagine concentration in physiological and other fluids.

The production of asparaginase and invertase each bonded to treated aminoethylcellulose, and invertase, peroxidase and glucose oxidase each bonded to treated alkylamine glass are disclosed in other examples of the complete patent.

Use of Photosensitizers to Immobilize Enzymes

Prior attempts to bond polypeptides to carriers using photochemical processes have resulted in low yields and reduced enzyme activity. It was, thus, surprising that polypeptides could, by a photochemical technique developed by *D. Kraemer, K. Lehmann and H. Plainer; U.S. Patent 4,039,413; August 2, 1977; assigned to Rohm GmbH, Germany* be bound quantitatively, with extensive retention of their biological activity and under protective conditions to various carrier substances.

This process comprises bonding a polypeptide to a macromolecular carrier by irradiating an aqueous solution of the polypeptide, in the absence of oxygen and in the presence of a carrier compound, with ultraviolet light in the optional presence of a photosensitizer or with visible light in the presence of a photosensitizer and an organic peroxide.

Because of the low specificity of the photochemical reaction, a great variety of nonpeptide-like compounds come into consideration as macromolecular carrier substances. In the majority of cases, natural or synthetic organic polymers are involved. Nevertheless, inorganic carriers can also be used to the extent that they are sufficiently transparent to the activating radiation, have a sufficient internal surface area, and have, on this surface photochemically activatable organic groups. As examples of inorganic carriers of this type can be mentioned porous glass, swellable silicates, particularly clay minerals, or highly dispersed silicic acid, which are modified with siloxanes containing alkenyl-, toluyl-, or other groups capable of photochemical addition.

The organic carrier substances used in the majority of cases have a molecular weight greater than 10,000. They can be water-soluble, noncrosslinked, materials but for most uses water-insoluble substances are employed. Among these, are the preferred predominantly hydrophilic monomer units which are water-insoluble only because of their crosslinked structure.

Among the water-soluble or water-swellable carrier substances which are basically suitable for use in this process those characterized by a high photochemical reactability are formed completely or partially from highly hydrophilic monomer units tending toward free radical formation. This is true for macromolecular compounds having carboxylic acid amide, carboxylic acid ester, lactone, semiacetal, and cyclic ether groups. The homopolymers and copolymers of acrylamide or of vinyl pyrrolidone, as well as polysaccharides, have proved to be very reactive.

The photochemical reaction between the polypeptide and the carrier substance takes place in an aqueous solution of the polypeptide. This also encompasses solutions of polypeptides in mixtures of water with lower alcohols such as methanol, ethanol, propanol, or butanol, or with acetone, inter alia. The concentration of the polypeptide in the solution is suitably from 100 μg/ml to 10 mg/ml. The carrier substance is generally introduced in such an amount that the polypeptide, on complete binding to the carrier, amounts to 1 to 50% by weight, preferably about 10% by weight thereof. The photosensitizers used should have a triplet energy between 50 and 85 kcal/mol. A number of such materials are tabulated below with their triplet energies.

Compound	Triplet Energy, kcal/mol in hydrocarbon solvent
Propiophenone	74.6
Xanthone	74.2
Acetophenone	73.6
1,3,5-Triacetylbenzene	73.3
Isobutyrophenone	73.1
1,3-Diphenyl-2-propanone	72.2
Benzaldehyde	71.9
Triphenylmethyl phenyl ketone	70.8
Carbazole	70.1
Diphenylene oxide	70.1

(continued)

Compound	Triplet Energy, kcal/mol in hydrocarbon solvent
Triphenylamine	70.1
Dibenzothiophene	69.7
o-Dibenzoylbenzene	68.7
Benzophenone	68.5
4,4'-Dichlorobenzophenone	68.0
p-Diacetylbenzene	67.7
Fluorene	67.6
9-Benzoylfluorene	66.8
Triphenylene	66.6
p-Cyanobenzophenone	66.4
Thioxanthone	65.5
Phenylglyoxal	62.5
Anthraquinone	62.4
Phenanthrene	62.2
α-Naphthoflavone	62.2
Flavone	62.0
Ethylphenylglyoxalate	61.9
4,4'-Bis(dimethylamino)benzophenone	61.0
Naphthalene	60.9
β-Naphthyl phenyl ketone	59.6
β-Naphthaldehyde	59.5
β-Acetonaphthone	59.3
α-Naphthyl phenyl ketone	57.5
α-Acetonaphthone	56.4
α-Naphthaldehyde	56.3
5,12-Naphthacenequinone	55.8
Biacetyl	54.9

The excited photosensitizer, by an exchange effect with the polypeptide of the carrier molecule, produces a carbon radical. Typical reactions of this type occur for the glycine group (1) of a polypeptide, for a glucose unit (2) of a polysaccharide, for an acrylamide unit (3) of a vinyl polymer, and for a carrier polymer having a toluyl group (4), as depicted below, where C* designates a carbon radical.

(1)

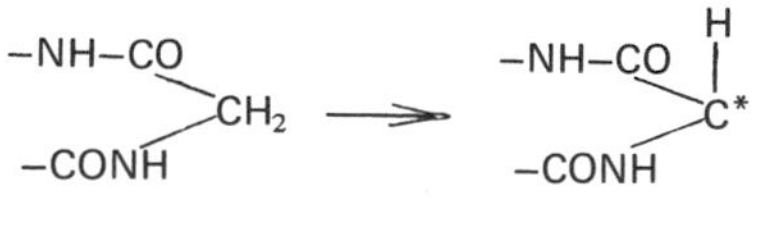

```
 -NH-CO                     -NH-CO  H
        \                          \ |
         CH2     ---->              C*
        /                          /
 -CONH                      -CONH
```

(2)

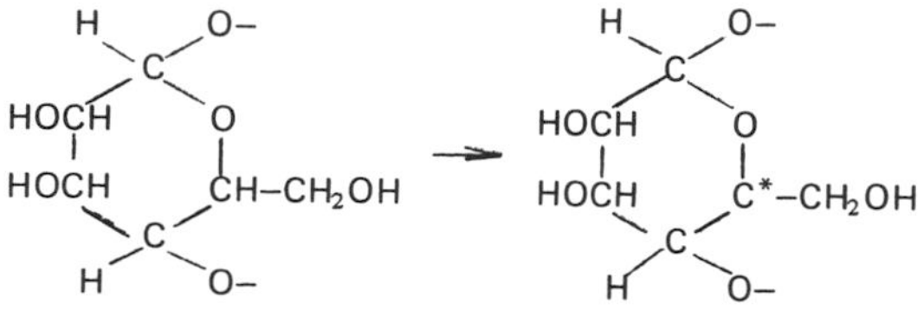

```
   H     O-                     H     O-
    \   /                        \   /
      C                            C
     / \                          / \
 HOCH    O                    HOCH    O
   |     |         ---->        |     |
 HOCH    CH-CH2OH             HOCH    C*-CH2OH
     \  /                         \  /
      C                            C
     / \                          / \
   H     O-                     H     O-
```

(3)

```
        H                           *
 -CH2-C-          ---->      -CH2-C-
       |                           |
      CONH2                       CONH2
```

(4) $CH_3 \rightarrow \overset{*}{CH_2} \leftrightarrow CH_2^{*}$

In principle, the initiation of photochemical reactions is not dependent on the presence of a photosensitizer. It is quite possible to excite a polypeptide or a carrier molecule by a sufficiently energetic irradiation, i.e., that is to convert it into a free radical. However, the direct excitation of polypeptides often leads to inactivation, i.e., the molecular structure is irreversibly disturbed. For best yields, this process is, preferably, carried out with radiation which is not sufficient for a direct activation of the polypeptide, but which nevertheless does excite a photosensitizer present. Radiation well suited for this process is ultraviolet light of a wavelength of more than 209 nanometers.

The period of irradiation depends on the strength of the radiation source and the efficacy of the photosensitizer and can be between 10 minutes and 50 hours. When conventional high pressure mercury immersion lamps are employed, i.e., sources having an output of 125 to 500 W, the radiation times are in general between 1 and 30 hours. After the light source is extinguished, the reaction is immediately ended. Unreacted carrier material, protein, and the protein-carrier compounds are stable and, in contrast to other bonding processes, there is no danger of the hydrolytic cleavage of the photochemically-produced compounds. The photochemical reaction can also be initiated by irradiation with light in the visible region if, in addition to a photosensitizer excitable by visible light (such as biacetyl), a peroxide such as ditertiary butyl peroxide is employed.

An important prerequisite for the bonding of the polypeptide on a carrier is the absence of oxygen. Oxygen, as is known, is very readily photochemically activated in the presence of sensitizers and initiates a variety of undesired side reactions. Thus, it is necessary to carry out the reaction in an oxygen-free protective gas atmosphere, for example, in nitrogen or argon.

Example: Acetone-Photosensitized Bonding of Trypsin on Agarose — 100 to 200 mesh agarose pearls (Bio-Gel A-150) were suspended several times in water, washed, and centrifuged. 10 g of the moist gel so obtained, corresponding to 100 mg of agarose, were suspended in a 90:10 water/acetone mixture containing 50 mg of dissolved trypsin (crystalline 4220 NF/mg). The material was put into a Schenck irradiation apparatus comprising a Pyrex immersion opening, a Pyrex cooling opening, and the reaction vessel. The material was then gassed for 1 hour with oxygen-free nitrogen which was led through 10% aqueous acetone.

Subsequently, the material was irradiated for 18 hours with good stirring and with water cooling with a high pressure mercury lamp (Phillips HPK 125 W, Model 57203B/00). During the irradiation, the apparatus was wrapped with aluminum foil. After irradiation, the material was recovered by centrifugation, washed several times with water and 1 N NaCl, and centrifuged.

The moist gel was washed five more times with water and lyophilized. The nitrogen content was determined by the Kjeldahl method and protein content calculated. Protein content: 21%; yield of bound protein: 63%. After use of the material several times in the hydrolysis of casein, no decline in activity was determined which is a criterion for the presence of covalent bonding.

CHEMICAL COVALENT COUPLING PROCESSES

Formation of Bridges via Isonitrile Groups

In the process developed by *R. Axen, P. Vretblad and J.O. Porath; U.S. Patent 3,847,745; November 12, 1974; assigned to Exploaterings A/B TBF, Sweden* biologically active materials such as enzymes are coupled by an isonitrile-group-containing compound to insoluble organic, hydrophilic polymeric carriers.

Broadly, this process relates to a method of preparing polymer products having a molecular weight over 1,000, especially adsorption material and polymer-bridged enzymes by covalent coupling of two or more substances, whereby at least one is a polymer acting as a carrier, the carrier polymer, and is characterized in that each of the above substances contains at least one of the following functional groups: isonitrile, aldehyde or ketone anion, primary or secondary amines, and that the reaction occurs in a reaction mixture containing these substances, possibly supplemented with additional substances so that all the functional groups in question are simultaneously present at the beginning of the reaction.

The process is based on the tendency for isonitriles to undergo the α-addition of suitable pairs of functional groups. The functional groups can also be chosen so that the addition product is transposed to a stable chemical structure. Substances containing such functional groups can thus be coupled together via isonitriles to stable conjugates.

Thus, the process works with four functional groups, $-NH_2$, $-CHO$ or $>C{=}O$, $-COO^-$ and $-NC$. The protein (enzyme) normally contains amino- and carboxyl groups. The polymer is selected and can thus be provided with $-NH_2$, $-CHO$ or $>C{=}O$, $-COO^-$ or $-NC$ functions, or two or more of these functions. Various examples of this process are shown in the table on the following page for the bonding of chymotrypsin to various carriers.

The polymeric carriers used are obtained in the following way: CM-Sephadex is epichlorohydrin-crosslinked dextran substituted with carboxymethyl groups. CM-agarose has been synthesized by treating 4% agarose gel with chloroacetic acid. Agarose-lysine-ethyl ester has been synthesized by treating 4% agarose with cyanogen bromide at pH 11 and then fixing lysine-ethyl ester in 0.5 M sodium bicarbonate solution.

Amino ethyl polyacrylamide is obtained by treating 1 g of crosslinked polyacrylamide (Bio-Gel P-300) with 15 ml ethylene diamine for 3 minutes at 90°C. The product is washed with a 0.1 M common salt solution. Enzacryl AA is a crosslinked polyacrylamide polymer substituted with aromatic amino groups. Aldehyde-agarose is obtained by periodate oxidation at pH 3 with 0.05 M or at pH 5 with 0.1 M periodate.

Chymotrypsin Fixed to a Series of Polymeric Carriers with Isonitrile—Catalytic Properties of the Conjugate

Carrier	Functional Carrier Group	Mg Chy in Reaction Mixture to 50 mg Carrier	Amount of Isonitrile plus Additional Reagent, µl plus µl	Reaction Time, hr	Amount of Fixed Chy in mg/g Conjugate	Activity, mol ATEE***	Activity, mol Casein
CM-Sephadex C-50	$-OCH_2-COOH$	5	25 plus acetaldehyde 25	1	55	13	0
CM-Sephadex C-50	$-OCH_2-COOH$	20	25 plus acetaldehyde 25	1	175	10	0
CM-Sephadex C-50	$-OCH_2-COOH$	35	25 plus acetaldehyde 25	6	345	7	0
CM-Sephadex C-50	$-OCH_2-COOH$	50	125 plus acetaldehyde 125	1	395	5	0
CM-Agarose	$-OCH_2-COOH$	35	25 plus acetaldehyde 25	1	60	10	5
Polymethyl methacrylate (partially hydrolyzed)	$-COOH$	10	25 plus acetaldehyde 25	6	-	(*)	0
Crosslinked amino ethyl-polyacrylamide	Aliph $-NH_2$	8	25 plus acetaldehyde 25	6	45	10	5
Enzacryl AA	Arom $-NH_2$	35	25 plus acetaldehyde 25	6	135	25	2-3
Agarose-lysine ethyl ester	Aliph $-NH_2$	20	25 plus acetaldehyde 25	6	15	50	20
Periodate-oxidized Sepharose 4B (oxide at pH 3)	$>C=O$	25	25	6	6	60	40
Periodate-oxidized Sepharose 4B (oxide at pH 5)	$>C=O$	35	25	6	17	80-90	40-60
Epichlorohydrin crosslinked agarose oxidized with dimethyl sulfoxide plus acetic acid anhydride	$>C=O$	15	10	6	40	46	45
Sheep wool	$-NH_2$ and $-COOH$	35	25 plus acetaldehyde 25	6	-	(**)	-

*0.8 µmol/min/mg.
**10 µmol/min/mg.
***N-acetyl-L-tyrosine ethyl ester.

Source: U.S. Patent 3,847,745

Polymethyl methacrylate (Plexiglass) in the form of filings is partially hydrolyzed with 2 N caustic soda for 24 hours at 50°C and is then washed with 10^{-3} M hydrochloric acid. Sheep wool is washed with acetone and 1 N caustic soda. The coupling reactions were carried out in the following manner.

Example: (A) The coupling to aldehyde polymer is as follows. Varying amounts of chymotrypsin are chemically coupled by isonitrile to a series of different polymers. Different amounts of isonitrile and acetaldehyde are added to the reacting suspension of 50 mg polymer and 2 ml water. The acetaldehyde was removed when the coupling to the aldehyde polymer occurred. Fixing was done at pH 6.5 with the help of a pH meter.

(B) The coupling to keto-polymer is as follows. Epichlorohydrin-crosslinked agarose is oxidized with a mixture consisting of 12 ml dimethyl sulfoxide and 5 ml acetic acid anhydride at 40°C for 90 minutes. After it has been washed on a glass filter with acetone, the polymer is treated with 15 ml hydrochloric acid (0.4 M) at 35°C for 2 hours. The final washing is done on a glass filter with water. 50 mg of the keto-group-containing polymer was reacted with 15 mg chymotrypsin and 10 μl 3-dimethylaminopropyl isonitrile at pH 6.5 for 6 hours at room temperature.

Diamines plus Glutaraldehyde for Bonding Penicillin Acylase to Supports

In the process by *T.A. Savidge, L.W. Powell and K.B. Warren; U.S. Patent 3,883,394; May 13, 1975; assigned to Beecham Group Limited, England,* water-insoluble enzyme preparations are formed from a penicillin acylase enzyme; a water-insoluble absorbent material; a water-soluble dialdehyde, e.g., glutaraldehyde; and an aliphatic diamine, preferably an α,ω-diaminoalkane having 2 to 10 carbon atoms. The resulting enzyme preparations have improved activity in the production of 6-aminopenicillanic acid and can be recovered for reuse.

Suitable absorbent materials include cellulose powder and various cellulose derivatives, ion exchange resins, silica gel and other polymeric materials. Particularly suitable are crosslinked polyacrylate or polymethacrylate resins, preferably when these are in macroreticular form, with a large ratio of surface area to weight.

The water-insoluble absorbent material is preferably a finely-divided polymeric material so that there is sufficient surface area to give water-insoluble enzyme preparations of high specific activity. Thus the surface area available for absorption of the enzyme increases as the particle size of the absorbent material decreases.

The order with which the four components of the enzyme preparations are mixed is often of importance in producing a preparation with optimum properties. In one method, the water-insoluble absorbent material is first contacted with the acylase enzyme optionally in the presence of the diamine and then consecutively or concurrently with the dialdehyde, and aliphatic diamine if not previously added. In such methods the enzyme is preferably absorbed in the presence of 0.1 to 5.0, preferably 0.2 to 2.0 mmol of the diamine per gram of absorbent material and 0.25 to 10.0, preferably 2.5 to 5.0 mmol of the dialdehyde per gram of absorbent material.

To ensure that any excess aldehyde groups are reacted after completion of the coupling reaction, the enzyme resin is washed with either solutions of sodium metabisulfite, aliphatic diamine, urea or glycine. Using the above conditions, the water-insoluble enzyme preparations can be sufficiently stable to enable at least 25 successive splittings of penicillin in a batch reactor to be carried out without an appreciable decline in the efficiency of the reaction.

An alternative coupling procedure which results in water-insoluble enzyme preparations of even greater stability is the pretreatment of the absorbent material with a mixture of the diamine and dialdehyde followed by absorption of the acylase enzyme. Important parameters in this preferred procedure are the quantities and molar ratios of aliphatic diamine and dialdehyde.

The diamine and dialdehyde should be present in concentrations of 0.01 to 20.0, preferably 0.1 to 10.0 mmol/g of absorbent material, and the molar ratio of diamine to dialdehyde should be 1:0.1 to 1:40, preferably 1:1 to 1:10. The pH at which pretreatment of the resin is carried out is between 4 and 9. Under optimal conditions, the stability of these enzyme preparations can undergo 50 or more successive splittings of penicillin.

In the examples the Amberlite XAD-7 resin is a commercially available cross-linked methacrylate polymer in macroreticular form; and the activity given for each enzyme preparation is in micromols (μmol) of 6-aminopenicillanic acid produced at pH 7.8 and 37°C per minute per gram of preparation.

Examples 1 through 7: An amount of a diamine as stated below was dissolved in 1,500 ml of 0.5 M citrate phosphate buffer at pH 5.0. The pH was adjusted to 5.0 and the stated amount of penicillin acylase was added. 100 g of Amberlite XAD-7 resin was then added and the whole was stirred slowly for 18 hours at room temperature.

100 ml of 50% aqueous glutaraldehyde was then added and the pH was adjusted to 5.0. After 18 hours further agitation, the solid was recovered by filtration and washed with water, 5% sodium metabisulfite, 1 M sodium chloride and 0.5 M phosphate buffer at pH 7.8. The activity of the preparation was as stated below.

An amount (as stated below) of some of the preparations made as described above was used to carry out a number of successive deacylations of potassium benzylpenicillin (6%) in a 1-liter stirred reactor at pH 7.8 and 37°C. In each of the experiments the conversion yield is stated after 5 hours.

Example	Diamine of Formula $H_2N(CH_2)_nNH_2$ n	Amount	Acylase Amount	Activity of Preparation (μmol/min/g)	Use for Deacylation Amount	No. of Successive Deacylations	Conversion (%)
1	6	15 g	5 g	16	200 g	15	90
2	3	11 ml	5 g	24	100 g	25	90
3	2	10.2 ml	100 ml	17.2	100 g	10	85
4	4	14 g	100 ml	40	-	-	-
5	5	13 g	150 ml	10	-	-	-
6	8	18.7 g	100 ml	12.8	-	-	-
7	10	22.2 g	100 ml	15.2	-	-	-

Polymers Containing Hydrazide or Azide Groups

According to *H.D. Orth, W. Brummer, M. Klockow and N. Hennrich; U.S. Patent 3,959,080; May 25, 1976; assigned to Merck Patent GmbH, Germany* it is possible to introduce not only a single reactive group, but selectively different reactive groups into a biochemical immobilizing matrix. Thus, the covalent linkages between the matrix and one or more chemically active substances can be effected via different (basic, acidic, aromatic) functional groups; for example, the optimal immobilization reaction conditions which are different for each individual substance can thus be more readily fulfilled in order to maintain the activity of the active substance.

According to the process, a water-soluble polymer of sufficiently high molecular weight so as to be capable of crosslinking to a water-insoluble polymer is used. A polymer containing, e.g., hydrazide or acid azide groups is partially crosslinked by reaction of these groups with a bifunctional crosslinking reagent, and any residual acid hydrazide or acid azide present uncrosslinked in the product simultaneously or subsequently converted at least partially into further different reactive groups which can covalently bind with the biochemically active substances. The polymers used as the matrix are water-soluble hydrazides of CMC (carboxymethylcellulose) or of polyacrylic acid and their copolymers.

The starting material is preferably a water-soluble, highly substituted CMC hydrazide containing 3 to 3.5 meq of acid hydrazide groups per gram of dry CMC. The average degree of substitution, i.e., the average number of acid hydrazide groups present on one glucose unit, ranges between 0.65 to 1.0. The average degree of polymerization is preferably about 1500 glucose units per CMC molecule. It is also possible to use uncrosslinked, water-soluble hydrazides of polymeric acrylic acid, preferably as a random, block or graft copolymer with acrylamide. In this connection, the mol ratio of acrylic acid hydrazide:acrylic acid amide can be preferably of 1:1 to 1:5. The content of acid hydrazide groups in these copolymers then ranges from 6.3 to 0.34 meq/g. Suitable polyacrylic acid homopolymers and copolymers preferably have a molecular weight of 10,000 to 30,000.

Suitable crosslinking agents include dialdehydes, dicarboxylic acid halides, diepoxides, diamines, diisocyanates and dirhodanides; halocarboxylic acid halides, e.g., chloroacetyl chloride, bromoacetyl chloride, or iodoacetyl chloride; halogenated epoxides, e.g., epichlorohydrin or epibromohydrin; and halides of aldehydic acids, e.g., glutaraldehydic acid chloride. Especially preferred crosslinking agents are glutaric dialdehyde and hexamethylenediamine.

By the crosslinking reaction, respectively two strands of the polymer are linked by groups of the formula –CO–NH–Z–Y–Z'–NH–CO– where Y is a single covalent bond or an alkylene hydrocarbon chain of up to 18 carbon atoms, optionally interrupted by up to 4 oxygen atoms and Z and Z' are each independently a single covalent bond, –N=CH–, $-NHCH_2-$, $-NHCH_2CHOH-$, –NHCO–, –NH–CO–NH–, or –NH–CS–NH–, preferably where the two groups Z and Z' are alike.

The partial crosslinking of the polymeric acid hydrazide is normally effected in an aqueous solution and at appropriate pH values dependent on the crosslinking reagent. For example, the reaction with dialdehydes is advantageously conducted in a slightly acidic solution (pH about 4.5). The reaction temperatures are preferably 15° to 30°C. At least 10 mol % crosslinking of the hydrazide groups is generally required; for complete conversion into a water-insoluble, highly swelled gel, crosslinking 15 to 20 mol % of the hydrazide groups present is generally sufficient, and preferably at least 70 to 80% of the hydrazide groups are uncrosslinked.

Any uncrosslinked functional groups, especially acid hydrazide groups, still present in the crosslinked product can be converted, at least partially, into other reactive groups capable of reacting with the, e.g., enzymes by activation. A preferred form of activation is the conversion of the hydrazide groups into azide groups with nitrous acid, suitably with sodium nitrite in 0.01 N hydrochloric acid at temperatures of -10° to +10°C.

The residual acid hydrazide groups can also be activated by reaction with a crosslinking reagent, as listed above. Only one of the reactive groups reacts with the acid hydrazide groups, whereas the other reactive group is preserved and available to react with the amino group of an enzyme. Thus, for example, with dialdehydes the monoacylhydrazones are obtained, the free aldehyde groups of which can be reacted with the amino groups of enzymes. If desired, the C=N bonds of the thus produced Schiff bases can be reduced to CH–NH– bonds, e.g., with $NaBH_4$.

The activation can also take place with bifunctional reagents which are poorly suitable for crosslinking. For example, it is possible to react acid hydrazides with nitrobenzaldehydes, e.g., p-nitrobenzaldehyde, thus producing the corresponding acylhydrazones of the nitrobenzaldehydes. The nitro groups can then be reduced to amino groups; the latter can be diazotized and coupled with aromatic groups present in the biochemically active substances, e.g., with the tyrosine portion of albuminous or other proteinaceous substances.

If it is desired to interpose a side chain spacer, the partially crosslinked polymeric acid hydrazides and/or azides can also be reacted successively with several reagents. In this process, the length of the interposed side chains can be varied arbitrarily dependent on the type of spacer reagent employed and on the number of successive chain-extending reactions.

Example 1: 2 g of CMC hydrazide (with 3.2 meq/g acid hydrazide groups) is dissolved in 100 ml of water. The solution is adjusted to pH 4.5 with HCl and combined with 0.6 ml of a 25% aqueous solution of glutaric dialdehyde. The mixture immediately assumes a high viscosity, and after 2 to 3 minutes, a gel-like polymer is precipitated. A homogeneous suspension is produced by agitation, cooled to 0° to 2°C, and the pH is lowered by adding 5 N HCl to a value of 1.2. 10.4 ml of a 5% aqueous $NaNO_2$ solution is then added to the reaction mixture.

After 15 minutes, the polymer which contains acid azide groups, is filtered, washed with water, and immediately made into a homogeneous suspension with 1 g of crystalline trypsin dissolved in 100 ml of 0.2 M triethanolamine/HCl buffer (2 mM $CaCl_2$; pH 8.5), which solution was precooled to 2°C. After 2 hours, the protein which has not been bound in a covalent linkage is removed by repeated washing with 0.3 M phosphate buffer (pH 8.0), as well as 1 M NaCl solution. Matrix-bound trypsin is obtained having the schematic formula $R-O-CH_2CONH$–trypsin (R is a partially crosslinked cellulose strand). Of the provided enzyme, 52% is bound to the matrix by a covalent linkage. The bound trypsin is still enzymatically active to an extent of 80 to 90% against N-benzoyl-arginine ethyl ester hydrochloride (BAEE).

Example 2: 2 g of a water-soluble copolymer of acrylic acid hydrazide and acrylamide (2.7 meq/g acid hydrazide groups) is suspended in 100 ml of 0.2 M triethanolamine/HCl buffer (pH 8.0) and reacted for 3 hours with 100 ml of 5% aqueous glutaric dialdehyde solution at 25°C. The excess dialdehyde is washed out with water, and the activated copolymer is immediately homogeneously suspended in a solution, precooled to 5°C, of 1 g of crystallized trypsin in 160 ml of 0.2 M triethanolamine/HCl buffer (0.01 M $CaCl_2$; pH 8.5).

After 18 hours, the protein which has not formed a covalent linkage is washed out as described in Example 1, thus obtaining matrix-bound trypsin of the partial formula $R-CONHN{=}CH(CH_2)_3CH{=}N$–trypsin (R is a partially crosslinked $CH-CH_2$ strand). Of the enzyme employed, 10% forms a covalent linkage with the matrix. The bound trypsin has an enzymatic activity of 80% against BAEE.

Mixtures of Alkane Dihalides with Alkane Diamines as Coupling Agents

A method for chemically immobilizing enzymes on a support to form an active composite by using a preformed reaction solution of an alkane dihalide and an alkane diamine has been disclosed by *M.H. Keyes; U.S. Patent 3,933,589; January 20, 1976; assigned to Owens-Illinois, Inc.*

Suitable alkane diamines include branched and straight chain alkane diamines such as diaminomethane, diaminoethane, diaminopropane, diaminobutane, diaminopentane, diaminoisooctane, diaminohexane, diaminoheptane, diaminooctane, diaminoisobutane, and diaminoisohexane. The position of the amino group on the alkane structure has not been found to be critical. Usually the alkane groups have 1 to 10 carbons, preferably 1 to 6 carbons for ease in forming the reaction solution.

Suitable alkane dihalides include dibromo, diiodo, and dichloro branched and straight chain alkane dihalides such as dibromomethane, dibromoethane, dibromopropane, diiodopropane, dibromopentane, dichloroethane, dibromobutane, diiodopentane, diiodomethane, dichloromethane, and other dihalo alkanes having 1 to 10 carbon atoms. The position of the halide group on the alkane structure also has not been found to be critical.

The composition of the support is not particularly critical as long as it is inert, dimensionally stable, and provides sufficient surface area for retention of enzyme. The support can be porous, fluid-permeable membranes as in U.S. Patent 3,839,175, porous particulates as in U.S. Patent 3,850,751 or natural or synthetic fibers. When porous supports are used, they should be sufficiently porous and sorptive enough to retain enough enzyme to form a biologically active composite.

In the usual practice, the preformed reaction solution is formed by mixing the alkane dihalide and alkane diamine in an aqueous solution until the components are thoroughly dissolved to the limits of their solubility. In some instances an organic phase can be present in addition to the preformed reaction solution. This condition is usually not desired when the residual organic phase has a tendency to denature the enzyme or detract from its performance.

As soon as the components are dissolved, the reaction solution is ready for use in immobilizing enzymes. Such dissolution readily takes place at temperatures from 20°C (i.e., room temperature) to the boiling point of the mixture in a time period of from a few seconds to an hour or longer. Temperatures of 20° to 50°C for a time from one minute to one-half hour are convenient and practical.

The molar proportion of alkane diamine to alkane dihalide has not been observed to be critical in this process. Empirical observations have confirmed the molar ratio of alkane diamine to alkane dihalide of 0.005:1 to 1000:1 provides satisfactory results with such ratio of 0.1 to 20 being practical for many applications. The pH of the resulting preformed reaction solution is not particularly critical although a pH of 2.5 to 11 is practical depending on enzymes used and pH obtained in the final solution. Although it is known that the alkane dihalide and alkane diamine react to form complex polyalkylene compounds, it is not known whether this reaction takes place in the process.

Example: Immobilized Glucose Oxidase on Alumina Powder — 1 g of particulate alumina is washed thoroughly with distilled water. The particulate alumina has a particle size in the range of -40 to +70 mesh (U.S. sieve screen) and an average pore size diameter of 0.1 to 0.2 μ. 50 mg of glucose oxidase (having a reported activity of 140 IU/mg) is added to the wet particulate alumina in 40 ml of an aqueous solution which has been buffered to pH of 5.5 with standard buffer comprising a mixture of potassium dihydrogen phosphate and disodium hydrogen phosphate. The resulting mixture is stirred gently for one-half hour at 6° to 8°C to sorb the enzyme.

To this mixture is added a preformed crosslinking reaction solution formed by mixing 20 ml methanol, 10 ml distilled water; 0.15 ml concentrated hydrochloric acid, 0.08 ml diaminopropane, and 0.02 ml dibromoethane at room temperature for 15 minutes to form a solution. This corresponds to a 4 to 1 mol ratio of diaminopropane to dibromoethane. The combined alumina, glucose oxidase and crosslinking reaction mixture is stirred gently with a magnetic stirrer at 6° to 8°C overnight to immobilize the glucose oxidase on the alumina. The resulting immobilized glucose oxidase/alumina composite is washed with about 2 liters of distilled water and stored in distilled water. The catalytic activity is calculated from the measured rate of oxidation of β-D-glucose to gluconic acid by parabenzoquinone in the presence of this composite.

Using this assay procedure the glucose oxidase/alumina composite is found to have an activity of 1.0 x 10^3 International Units of glucose oxidase per cm^3 of glucose oxidase/alumina composite. Ten days later, after being stored in distilled water at 4°C, the activity is 8.3 x 10^2 International Units of glucose oxidase per cm^3 of glucose oxidase/alumina composite.

Coupling Agents Containing Phenyl Azide and s-Triazine Moieties

The asymmetric bifunctional linking compound which may be used in the process disclosed by *N.L. Smith, III; U.S. Patent 4,007,089; February 8, 1977; as-*

signed to Nelson Research & Development Company must have at least one phenyl azide moiety and at least one s-triazine moiety. Linking compounds which satisfy the foregoing criteria include those compounds having the structural formula

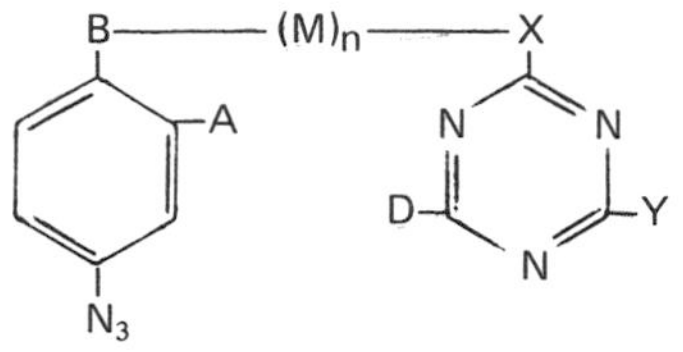

B—$(M)_n$—X
A
N N
D Y
N
N_3

where B is NR, O, S,

—N N—

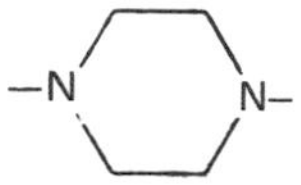

or CH_2; $(M)_n$ is a saturated or unsaturated hydrocarbon chain; X is NR, O or S; A is SO_2R, CN, NO_2 or H; Y is NR_2, N_3, halogen or SH; R is H and/or a lower alkyl group; D is N_3 or halogen and n is 1 to 12.

A preferred linking compound has the following structural formula

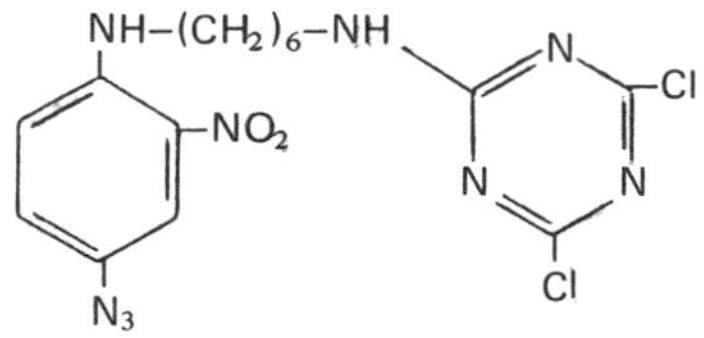

$NH-(CH_2)_6-NH$
N
Cl
NO_2
N N
Cl
N_3

Carriers which may be used in this process include polymeric and nonpolymeric carriers. Polymers which may be used are water-insoluble organic polymers of synthetic or natural origin. These polymers all react with the phenyl azide moiety of the linking compound. Examples of suitable carriers include polyethylene (conventional and linear), polypropylene, polymethylpentene, ethylene propylene copolymer, polystyrenes, polycarbonate, polyvinyl chloride, etc.; organic polymers from biological origin, such as, for example, cellulose, starch, pectin, etc., and proteins such as enzymes, lipoproteins, mucopolysaccharides.

Examples of carbonyl polymeric carriers which may be used include those produced according to any known procedure from such aldehyde monomers as acrolein; α-alkyl acroleins, e.g., methacrolein, α-propylacrolein; crotonaldehyde; 2-methyl-2-butenal; 2,3-dimethyl-2-butenal; 2-ethyl-2-hexenal; 2-decenal; 2-dodecenal; 2-methyl-2-pentenal; 2-tetradecenal and the like, alone or in admixture with up to 95% by weight based on the total weight of the copolymer of each other and/or such other copolymerizable monomers known to react therewith.

Exemplary of the enzymes which may be used include proteolytic enzymes, hydrolases, amylases, dehydrogenases, kinases, oxidases, deaminases, amidases, enzyme anticoagulants, etc.

Generally, the reaction is carried out in two separate steps. In step one, the phenyl azide moiety of the bifunctional linking compound is linked to a polymeric carrier by photochemical reaction. To carry out the reaction, a bifunctional linking compound, as described above, is reacted with a polymeric carrier in a suitable solvent and exposed to light at a wavelength of preferably 420 to 500 nm and for 0.5 to 2 minutes at temperatures between 0° to 25°C. Upon completion of the reaction, unbound linking agent is removed from the reaction mixture, e.g., by washing with a suitable solvent.

In step two, the polymeric carrier with covalently bound linking reagent is incubated with a biologically active compound, e.g., an enzyme, for 0.5 to 2 hours at 0° to 30°C and at a pH of 4.5 to 6.5. During this reaction, the s-triazine moiety reacts with the biologically active compound. The phenyl azide-polymer bond is not affected by this reaction. Unbound biologically active compound is then removed from the reaction mixture, e.g., by washing with a suitable solvent. The resultant complex consists of a biologically active compound such as an enzyme bound through the bifunctional linking compound to a polymeric carrier.

Example 1: Method of Immobilizing Glucose-6-Phosphate Dehydrogenase to Polycarbonate – 2 mg of the linking compound 1-N-(2-nitro-4-azidophenyl)-6-N-(4,6-dichloro-sym-triazinyl)diaminohexane having the structural formula

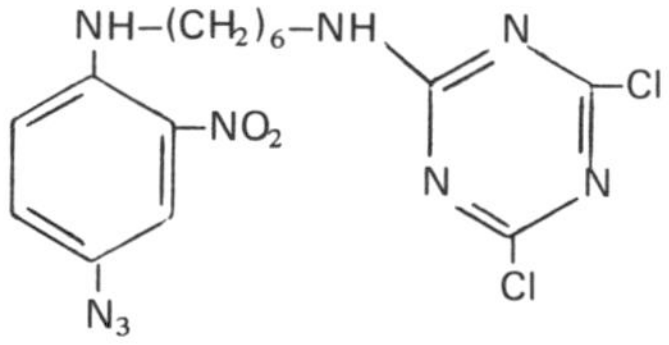

$NH{-}(CH_2)_6{-}NH$ ... N ... Cl

NO_2 ... N ... N

N_3 ... Cl

was dissolved in 100 ml ethyl alcohol and 2 ml of this solution was placed in a polycarbonate test tube and exposed to sunlight at ambient temperature for about 20 minutes. The test tube was washed with ethanol to remove unbound linking compound and an enzyme solution, made by adding 10 μg glucose-6-phosphate dehydrogenase to 2 ml of 0.05 M acetate buffer having a pH of 5.5, was added to the test tube and incubated for 2 hours at ambient temperature. The enzyme solution was then removed and the test tube was twice rinsed with 0.1 M Tris buffer to remove any unbound enzyme. Subsequent enzyme assay of the test tube confirmed the presence of bound enzymes. The enzyme-polycarbonate preparation showed no significant decrease in enzyme activity over a period of 60 days after forming the complex.

Example 2: Method of Immobilizing Hexokinase to Polycarbonate – Example 1 is repeated, except hexokinase is used in the place of glucose-6-phosphate dehydrogenase. Comparable results are obtained.

Example 3: Method of Immobilizing Glucose-6-Phosphate Dehydrogenase to Polystyrene – The method of Example 1 is used, except polystyrene discs are used in place of polycarbonate and an ultraviolet lamp is used. Comparable results are obtained.

Use of Diazotized m-Diaminobenzene

A water-insoluble solid containing an adsorbed diazotized m-diaminobenzene is used by *S.A. Barker and C.J. Gray; U.S. Patent 4,043,869; August 23, 1977; assigned to Koch-Light Laboratories, Ltd., England* to immobilize enzymes.

The material which can be used as the water-insoluble solid is generally any material which can be dyed or redyed with diazotized aromatic diamines, such as solids containing free hydroxyl or polyamide groups. Particular examples include silica, i.e., sand, polysaccharide, crosslinked dextran, polyacrylamides, polyamides, polycarbonates, polyesters, glass (particularly porous glass), nylon, diatomaceous earth, natural or regenerated cellulose such as paper, viscose rayon, Sigmacell 38, Whatman CC31, Whatman CF11, Celite, Neosyl, carboxymethylcellulose, diethylaminoethylcellulose, Bioglas-1000, Biogel P-6, Enzacryl gel K_2, cellulose acetate polyurethane, Sephadex G-200, as well as inorganic hydrous oxides such as titanium oxide, zirconium oxide, aluminum oxide, and iron oxide. Generally, all polar surfaces are dyeable as are the surfaces of natural products such as the cells of wood, e.g., balsa wood.

The preferred aromatic diamine for use in the process is diaminobenzene, and the preferred isomer is m-diaminobenzene. The biologically active molecules which can be used in the process include proteins such as enzymes, coenzymes, lectins, and antibodies. Examples include, B-glucosidase, dextranase, amylase, glucamylase, catalase, glucose oxidase, thermolysin, N-acetyl amino acid amido hydrolase, peroxidase, chymotrypsin, uricase, pepsin, urease, pronase, lactate dehydrogenase, cholinesterase, glucose isomerase, isoamylase and pullulanase, which may be used alone or in any combination.

With certain enzymes such as catalase and dextranase both the optimum pH for enzyme activity and the pH stability of the enzyme are modified considerably when the enzymes are insolubilized in the biologically active material. This effect is believed to be due to the residual amino groups of the diazotized aromatic diamine which have not been diazotized. This allows two enzymes in solution having different pH values to be used together since one enzyme can have its pH value reduced or increased to that of the other by bonding the enzyme to a diazotized aromatic diamine adsorbed on the water-insoluble solid.

The effectiveness of the enzymes depends to a large extent upon its method of preparation. The absolute and relative amounts of aromatic diamine, sodium nitrite and hydrochloric acid, the area and type of water-insoluble solid surface and the pH values used are important and the optimum values of these can be found by experimentation for each water-insoluble solid.

In order to achieve maximum enzyme insolubilization and efficiency, each carrier should be "optimized;" in other words, the best reaction conditions and concentrations of reactants for preparing the carrier and adsorbed diazotized aromatic diamine should be found.

The attachment of the biologically active molecules to the diazotized aromatic diamine molecules depends upon the concentration and pH value of the solution containing the biologically active molecules to be attached to the diazotized aromatic diamine molecules and thus the optimum concentration of active molecules and pH should also be found for each enzyme.

It is believed that the diazotized aromatic diamine molecules when adsorbed on the surface of the carrier exist as oligomers or high molecular weight species and that these molecules attach themselves to and possibly surround the carrier surface.

Example: (A) Coupling of β-Glucosidase to Cellulose via Diazotized Diaminobenzene by the Standard Procedure — m-Diaminobenzene (50 mg), cellulose (100 mg) and hydrochloric acid (1.0 N; 2.0 cm^3) were combined in a stoppered test tube equipped with a magnetic stirrer, and cooled to 0°C. Precooled (0°C) aqueous sodium nitrite (2% w/v; 2.0 cm^3) was then added and the suspension stirred at 0°C for one-half hour. The solid was then washed three times with ice cold acetate buffer (0.2 M, pH 5.0; 3 x 50 cm^3) and β-glucosidase (0.2 mg/cm^3; 5.0 cm^3) added. The suspension was allowed to stir for sixteen hours at 4°C. β-Naphthol (saturated in 2 cc saturated sodium acetate) was then added to the stirring suspension, and the suspension allowed to stir for a further one-half hour at 4°C.

A cellulose control was prepared following the same procedure but omitting the diaminobenzene. The suspensions were then centrifuged, the supernatants discarded and the solids subjected to an exhaustive washing procedure. A cycle of ten washings was employed, the wash liquids being acetate buffer (0.2 M, pH 5.0; 5.0 cm^3) and sucrose-salt in acetate buffer (0.2 M, pH 5.0, 1 M in sucrose and 1 M in sodium chloride; 5.0 cm^3). Three further washes with acetate buffer were carried out and the solids were then suspended in acetate buffer (0.2 M, pH 5.0; 5.0 cm^3) prior to storage at 4°C.

(B) Coupling of β-Glucosidase to Cellulose by the Standard Method Using a More Concentrated Solution of Enzyme — The procedure previously described was followed. β-Glucosidase (10 mg/cm^3 in acetate buffer, 0.2 M, pH 5.0; 0.5 cm^3) was added to the diazotized diaminobenzene/cellulose preparation. Use of more concentrated enzyme solution was found to facilitate washing; the cycle of sucrose-salt/buffer washes described above was found to be sufficient to remove all visible color from the supernatants.

Enzymic activities on the diaminobenzene preparation and on a control solid (prepared without the addition of diaminobenzene) were determined in the usual manner. Results are given in the table below.

	Volume Assayed (cm^3)	Change in OD per Minute	Enzymic Activity Detected (%)
Example A			
Test (0.2 mg/cm^3)	1.0	0.070	20.6
Control (0.2 mg/cm^3)	1.0	0.003	0.9
Example B			
Test (1.0 mg/cm^3)	0.35	0.296	24.2
Control (1.0 mg/cm^3)	0.35	0.008	0.6

This process also provides methods for reactivating immobilized inactivated enzymes prepared with a diazotized aromatic amine. One method comprises (a) adsorbing additional molecules of a diazotized aromatic diamine on the surface, and (b) attaching fresh enzymes to the additional molecules of diazotized aromatic diamines. Another method comprises (a) treating the material with sodium dithionite to regenerate amine groups, (b) diazotizing the amine groups, and (c) attaching biologically active enzymes.

MISCELLANEOUS COUPLING METHODS

Bonding Enzymes to Support in Presence of Substrate

It is the primary object of the process developed by *G.P. Royer; U.S. Patent 3,930,950; January 6, 1976; assigned to The Ohio State University Research Foundation* to immobilize an enzyme by chemically bonding to a support in such a manner that the bonded enzyme exhibits higher activity than it would have exhibited had it been bonded by conventional techniques.

This process comprises providing an active support capable of chemically bonding with an enzyme and then contacting the active support with an enzyme-substrate complex which has been formed by mixing an enzyme and a specific substrate while minimizing the transformation of the substrate to product, whereby the enzyme component of the enzyme-substrate complex becomes chemically bonded to the support. If desired, the substrate may be removed from the enzyme-substrate complex by conventional means, such as water washing, with the resultant product being an active enzyme which is chemically bonded to the support.

The support used in this process can be any material capable of reacting with an enzyme so as to cause the enzyme to become chemically bonded thereto. Among those organic support materials suitable are carbohydrates, vinyl polymers, amino acid polymers, derivatives of amino acid polymers, nylon, polystyrene structures, and phenol-formaldehyde resins. In addition, among those inorganic supports suitable are porous glass (arylamine derivatives of porous glass), nickel screen, and alumina-silicate structures.

Example: An arylamine derivative of porous glass (550 A) was provided as the insoluble support member. The support member was activated by treatment with a HCl-$NaNO_2$ solution of 0°C to convert the amine to a diazo structure; this process is described in detail in an article entitled "Immobilized Pronase" published in *Biochemical and Biophysical Research Communication,* Vol. 44, No. 2 July 1971. The insoluble support member consisted of a plurality of glass beads. The enzyme used was trypsin (Lot TRL2DA). The substrate used was N-α-benzoyl-L-arginine ethyl ester. The enzyme-containing solution was buffered by the use of 0.05 M Tris, pH 8.0 containing 25 mM $CaCl_2$. The enzyme solution (200 ml, 1 mg/ml) and substrate solution (200 ml, 10^{-2} M) were placed in separate gradient bottles connected by a T-tube. The remaining opening of the T was connected to a column containing the activated glass beads.

The solutions were mixed and pulled through the column by a peristaltic pump. The flow rate was 5 ml/min and the column contained 1 g of activated glass. A temperature of 0° to 4°C was chosen to minimize depletion of substrate and take advantage of the negative enthalpy change for substrate binding. The resultant product, i.e., the chemically bonded (to the support) enzyme-substrate complex, was then subjected to a water washing treatment whereby the substrate material was removed therefrom. The activity of the immobilized enzyme was about 15%. While washing with water is listed as the preferred material, other relatively inert washing fluids may be used. Subsequently, for the purpose of comparison, unmodified trypsin was immobilized by chemically binding it to a similar type of insoluble support material by conventional means. The activity of the enzyme bonded to the support member without substrate was 10%.

Pressure Application of Enzymes to Membranes of Mixed Polymers

Pressure-driven enzyme-coupled membranes are prepared in the process disclosed by *H.P. Gregor; U.S. Patent 4,033,817; July 5, 1977*. These matrix membranes can be prepared from either homopolymers, from copolymers or from interpolymer mixtures. When prepared from homopolymers they must have an appropriate chemical and mechanical stability so that they can be used for periods of months in the presence of a solvent (usually water) without suffering significant changes in their porosity (volume fraction which is pores), pore diameter or dimensions. There are a limited number of homopolymers which are useful for these purposes. Cellulose is one homopolymer, and polyvinylbenzyl chloride is another.

A preferred method of preparing matrix membranes is to use the procedures prescribed by H.P. Gregor in U.S. Patent 3,808,305 where membranes are cast from interpolymer mixtures. In a typical example, two parts of polyacrylic acid are dissolved in a suitable solvent such as dimethylformamide with one part of a film-forming matrix polymer such as polyvinylidine fluoride (Kynar, Pennwalt Co.) together with a suitable epoxide crosslinking agent and this is then cast in the usual manner.

In order to make these membranes of suitable porosity, while the cast membrane is partially dry, it is then coagulated by the introduction of a vapor which causes a partial coagulation but one which does not cause a loss of the structural integrity of the membrane. A suitable solvent in this case is water or another substance which causes the ionogenic polymer to become ionic, such as ammonia or triethylamine, following which final drying and curing by crosslinking of the film renders a membrane which has pores of appropriate dimensions while still containing a high concentration of carboxyl groups to which coupling reactions can take place. Similarly, amines such as polyvinylimidazole or poly-N-methyleneimine can be used in this formulation.

Polymer mixtures which include polystyrene sulfonic acid provide for a highly polar negative charge within the membrane, and can be used together with a group capable of coupling such as polyacrylic acid. Accordingly, it is evident that by the judicious selection of matrix polymers and polymers to which coupling can take place, one can prepare a wide range of suitable matrix membranes. Similarly, a wide range of known coupling chemistries are available. In some cases, activation is required as in the treatment of cellulose by cyanogen bromide or the treatment of a polyvinyl-alcohol-containing membrane with cyanuric chloride.

In order to make appropriate use of the pressure-driven enzyme-coupled membrane concept, it is important that the matrix membrane have certain characteristics. First, its pores must be of molecular dimensions at least twice the size of the enzymes to be accommodated and preferably 3 to 10 times the diameters of such enzyme molecules, but less than 20 times these diameters so that the membranes can have a high internal surface for coupling. Membranes with pores of 15 to 200 A in diameter are preferred. The matrix membranes must be capable of withstanding the pressure gradient across them without collapse. It is not necessary that the membrane withstand the pressure unsupported because the available membrane technologies allow one to support thin and fragile membranes on a variety of support materials. It is an important criterion, however, that the membrane pores remain substantially constant in size under conditions of use. An auxiliary requirement is that a substantial part of the interior pore surface be available for the coupling of enzyme molecules. Having a large number of very fine pores which are permeable to water but not to the enzyme molecule is not desirable.

Example: A membrane was prepared from a 1:1 mixture of Kynar and cellulose in MSO-paraformaldehyde solution, allowed to dry for 2 minutes in air, covered and allowed to equilibrate in the solvent vapor for 1 hour and then coagulated with water vapor and washed to have a thickness of 20 μ, a water content of 90% and a hydraulic permeability of 4.5 liters per hour at 50 psig pressure for an 11.3 cm^2 area. Activation by cyanogen bromide (40 mg CNBr in pH 11 water) was performed at 50 psig for 25 minutes. The membrane was washed for one minute in a 0.1 M phosphate buffer at pH 7.5 and then a solution consisting of 50 mg of the enzyme horse liver alcohol dehydrogenase in 100 ml of the same buffer was passed through at 50 psig. The membrane was then stored in this effluent for 24 hours at 4°C, then washed with water and its protein content was found to be 18% of the dry weight of the membrane.

The activity of this enzyme was determined using the standard procedures described in the *Worthington Biochemical Corporation Manual* (Freehold, N.J., 1972). The free solution activity of this horse liver alcohol dehydrogenase was determined at pH 7.5 in the phosphate buffer employing NAD as a coenzyme and with ethanol as substrate. The free solution activity of this enzyme was 0.31 units/mg and at 100 psig substrate pressure the activity of the coupled enzyme was 0.19 units/mg.

Fixation of Enzymes Using Nonaqueous Media

The process developed by *G. Durand and P. Monsan; U.S. Patent 4,119,494; October 10, 1978; assigned to Rhone-Poulenc Industries, France* permits the rapid fixation of enzymes on supports. The fixing is effected in anhydrous organic medium at a temperature of more than 60° to 120°C.

The classes of enzymes which can be fixed include: oxydoreductases, transferases, hydrolases, lyases, isomerases, and ligases. The mineral, organic or organo-mineral support used is insoluble in the organic medium and must be active, that is to say, possess one or more functional group reactive with groups such as amine, carboxyl, sulfhydryl, hydroxyl, by which the enzymes can be fixed. When the support does not in itself possess the functional groups, it must be modified.

The support can be made of brick, silica, alumina, the clays, sand, agarose, starch, polydextrane, cellulose, polymers such as polybutadiene, polystyrene whether or not crosslinked, copolymers of methacrylic acid and copolymers of maleic anhydride and ethylene. The modification or activation consists in treating the support so as to supply it with the functional group or groups reacting with the enzyme. One modification might be the reaction of the support with a sulfuryl, cyanuryl, cyanogen or thionyl halide; another possibility is grafting of the support, especially the polymers, by hydrolyzable polyfunctional silanes of carbonyl groups. These supports are generally used in the form of grains, the granulometry of which, in most cases greater than 100 μ can be greatly varied. They must be free from all products which inhibit or denature the enzyme and free from all traces of water.

The organic medium, in which the enzyme is insoluble, is selected from the aliphatic, cycloaliphatic or aromatic hydrocarbons, possibly containing chlorine atoms or hetero atoms, such as atoms of oxygen and/or sulfur. They are represented by hexane, benzene, xylene, chloroform, carbon tetrachloride; dichloroethane, trichloroethylene, perchloroethylene, dioxane, tetrahydrofuran, and dimethylsulfoxide. These hydrocarbons are used alone or in mixtures.

According to this process, the active support and one or more enzymes are dispersed simultaneously or successively in the hydrocarbon, then the reaction medium is brought to the reaction temperature, preferably the boiling temperature of the medium, for the time necessary for fixing the enzyme. The desired fixation at a rapid rate without excessive loss of any of enzymatic activity can be achieved at reaction temperature of above 60°C up to 120°C. It is believed that the ability to use such high temperature for rapid fixation of the enzymes depends upon the reaction being carried out in the nonaqueous organic liquid medium since the reaction at such temperatures in aqueous medium would denature the enzyme and destroy its enzymatic activity.

The amount of enzyme used is in excess in relation to the functional groups of the support. As regards the hydrocarbon, the quantity used must be sufficient to obtain a good dispersion which permits contact between the active support and the enzyme. It is surprising that such high temperatures permit the fixing of the enzymes without the enzymatic activity being considerably reduced, since it is heretofore accepted that temperatures higher than 40° to 50°C denatured the enzymes. Contrary to the known processes, the pH value of the fixation medium has no influence upon the properties of the enzyme fixed. The reaction time is generally between several minutes and 2 hours, which represents a considerable gain in time in comparison with the known processes.

After fixing, the hydrocarbon is easily separated and support-enzyme complex obtained is washed with the aid of a buffer solution which is variable with the enzyme, and such that its pH value and its composition have no denaturing action upon the complex. This washing permits separation of the enzyme which has not reacted and that which is simply adsorbed upon the support. This separated enzyme is recovered and can be reused.

The obtained support-enzyme complexes are constituted by one or more enzymes fixed on the support. In the case where the complex contains several enzymes, these are fixed either simultaneously or successively, or the complex is obtained by mixing of two complexes prepared in accordance with the process. The quantities of enzyme fixed by this covalent bonding process, are a function of the nature of the enzyme and of the nature and structure of the support and are clearly greater than the quantities of enzyme fixed according to the process in aqueous medium.

These support-enzyme complexes can be used at higher temperatures than the complexes prepared in aqueous medium. Moreover they are advantageously usable continuously for relatively long times without loss of activity or with a slight loss of activity of the enzyme.

Example: Fixing of Pectinase on Brick — Brick crushed and screened so as to have a diameter between 0.5 and 0.8 mm, is washed with distilled water and then with an aqueous solution of N hydrochloric acid and finally with distilled water. After 15 hours at 700°C in an oxygen atmosphere, the impurity-free brick is activated: 2 g of brick are introduced into 40 ml of a 10% solution by volume of sulfuryl chloride in anhydrous benzene (BP 80°C). The obtained suspension is agitated and brought to the boiling temperature of the mixture, which temperature is maintained for 20 hours. The active brick formed is then separated. 0.5 g of commercial pectinase, then 2 g of active brick are dispersed by sonic vibration in 50 ml of anhydrous benzene. The obtained dispersion is heated to the boiling temperature and kept at this temperature for 1 hour.

After cooling, the solid is filtered under vacuum, washed by placing in contact for 15 hours at 4°C with 20 ml of a solution of M sodium chloride, centrifuged and dried. This washing operation is carried out seven times. It permits desorbing the enzyme adsorbed on the support, which can be reutilized. After each washing of the solid constituted by the support-enzyme complex, in which the enzyme is fixed on the support by covalent bonding, the enzymatic activity is determined as follows.

1 g of the complex is dispersed in 10 ml of a 0.4% by weight solution of polygalacturonic acid in 0.01 M acetate buffer at pH 4. The dispersion is heated to 35°C for 30 minutes. After cooling and decanting, 5 ml of the solution are extracted, then clarified by addition of 0.3 ml of a 9% by weight solution of zinc sulfate in water and 0.3 ml of an N aqueous solution of soda. After centrifuging, the reducing compounds present in the surface-floating portion are determined by the dinitrosalicylate method.

By way of comparison, the same test is repeated but the fixing is effected in aqueous medium with a solution of 0.5 mg of pectinase in 50 ml of water, 2 g of active brick, at 4°C for 7 hours. The results obtained are summarized in the table below where they are expressed in percentages of activity of the complex in comparison with the activity of the enzyme before fixing.

	Number of Washings with M NaCl						
	1	2	3	4	5	6	7
	Activity (%)						
Fixing in benzene medium	69	50	47	47	47	47	47
Fixing in aqueous medium	44.5	39	36	32	28	27.5	24

By comparison it is observed that the fixing of the enzyme in benzene medium is more rapid than in aqueous medium.

COVALENT BONDING USING INORGANIC SUPPORTS

SILANIZED SUPPORTS

Silanized Calcined Attapulgite Clays

R.A. Burns; U.S. Patent 3,953,292; April 27, 1976; assigned to Engelhard Minerals & Chemicals Corporation has disclosed an insolubilized mineral-supported enzyme composite having outstanding catalytic activity and mechanical stability. This improved composite comprises an enzyme covalently attached to silanized porous, attrition-resistant granules of heat-activated attapulgite clay.

An essential feature of the process resides in the selection of attapulgite clay as the clay support and the use in the heat-activated (calcined) form. Uncalcined attapulgite clay is not suitable for the reason, among others, that its granules disintegrate in water and the requisite granular form is not preserved. Bentonites and kaolins, which constitute the most common clays, are also unsuitable and are therefore outside the scope of the process. The clay particles used in this process are obtained by calcining naturally-occurring attapulgite clay (Georgia-Florida fuller's earth) at 600° to 1300°F for a time sufficient to reduce the volatile matter to a value below 18% by weight.

To prepare the heat-activated clay, the raw clay is crushed and nonclay impurities may be removed. If desired, the raw clay may be extruded in known manner after being mixed with a small amount of water. The raw clay, extruded or not extruded is calcined in suitable equipment, usually a rotary calciner. The calcined crushed clay must be placed in the form of fairly uniform sized granules. For example, the discharge from the calciner may be processed in a corrugated roll mill and the milled clay classified to segregate particles of desired size. Alternatively the clay can be sized before heat activation.

Generally, the finer the particles of the calcined attapulgite clay support, the higher the activity of any given enzyme attached thereto. However as the particles become finer the excessive pressure drop becomes a problem. Consequently, the ultrafine particles, e.g., particles finer than 44 microns, or 325 mesh are undesirable.

To attach the enzyme covalently to the granules of heat-activated attapulgite clay through an intermediate coupling agent, the clay is silated in known manner to introduce functional groups which may then be linked to the enzyme by crosslinking agents, also known in the art.

Of the substituted organosilanes, hydrolyzable aminosilanes, especially omega-aminoalkyl and aminoaryltrialkoxysilanes exemplified by gamma-aminopropyltrimethoxysilane, aminopropyltriethoxysilane and aminophenyltriethoxysilane are preferred. Examples of preferred crosslinking agents which form a covalent bond with a reactive group on the enzyme that is not essential to the enzyme activity and which also form a covalent bond with a functional group of the aminosilanes are aldehydes including, e.g., formaldehyde, glyoxal, glutaraldehyde and acrolein. Also suitable are bispropiolates and disulfoxyhalides.

Active enzymes useful in this process include: the redox enzymes that catalyze oxidation or reduction reactions, e.g., glucose oxidase and catalase; hydrolytic enzymes which hydrolyze proteins and transferase enzymes. Any inert medium may be used to bond the enzyme, the crosslinking agent and the organosilane which is reacted with sites on the activated clay. Usually an aqueous medium is used. The pH and temperature used are selected to avoid or minimize deactivation of the enzyme. With most enzymes, the temperature is room temperature or below. Generally the enzyme is dissolved in a buffer solution at appropriate pH and temperature and the solution is usually assayed. The silanized clay is added and the glutaraldehyde or other immobilizing agent is added to the slurry.

The resulting slurry is maintained at a desired temperature for a suitable time, generally 1 to 72 hours, while maintaining pH at desired level. The bonded enzyme may then be assayed. The bonded enzyme may be stored in water or in a buffered solution at room temperature or below. In some cases the bonded enzyme may be dried mildly prior to use. Insolubilized enzyme composites of this process are of special value in industrial enzymatic processes carried out in continuously operated columns or stirred tank reactors.

Example: The clays used were commercial granular grades of heat-activated attapulgite clay. One was an RVM grade and the other an LVM grade, these designations referring respectively to regular and low volatile matter contents. Samples of these clays were ground in a hammer mill and sieved to different particle size ranges. Typical properties of these clays are as follows:

	RVM-AA	LVM-A
Free moisture, wt % loss at 220°F	3–7	0
Volatile matter, wt % loss at 1832°F	12–16	6–10
Surface area, BET, m^2/g	125	125

Glucose oxidase was bonded to samples of the sieved activated attapulgite clay via a gamma-aminopropyltrioxysilane coupling agent with glutaraldehyde employed as the immobilizing agent. The procedure used is described by Herring, Laurence and Kittrell, "Immobilization of Glucose Oxidase on Nickel-Silica Alumina," *Biotechnology and Bioengineering,* Vol. XIV, pages 975-984 (1972). This procedure was modified by using glutaraldehyde as the immobilizing agent in a conventional manner. The activities for catalysts prepared with the two different samples of activated attapulgite were essentially the same at all mesh sizes. Consequently, the catalytic activity values of glucose oxidase coupled to particles of heat-activated attapulgite clay, at -26°C and pH 5.5, are summarized below in table form and represent average values in mols of oxygen removed per second per milligram of support x 10^{-12} for the two different grades of activated attapulgite clay.

Particle Size of Support, microns	Activity
500/1,000 (granular)	111
250/500 (granular)	270
125/250 (granular)	546
0/125 (powdered)	893

Porous Matrix Containing Dispersed Silanized Fillers

Enzymes are immobilized in the process developed by *B.S. Goldberg; U.S. Patent 4,102,746; July 25, 1978; assigned to the Amerace Corporation* on a microporous binder or matrix with finely divided filler particles dispersed throughout the binder. The proteins are coupled to the filler particles, and the microporous member has a relatively large surface area and a large number of available protein coupling sites. Enzymes coupled to the microporous member have a relatively high reaction efficiency when used to act on a substrate.

Suitable microporous material comprises a normally hydrophobic polymeric matrix (e.g., polyvinyl chloride), finely divided normally hydrophilic filler particles (e.g., silica) dispersed throughout the resinous matrix, and a network of interconnected micropores formed throughout the material.

The network of interconnected micropores, in turn, consists of micropores formed between adjacent or neighboring particles of the dispersed inorganic filler, between particles of dispersed filler and the resinous matrix, and in the resinous matrix itself, with the size distribution of the micropores typically ranging over a relatively broad range from 0.01 μ to 100 μ, and the mean pore diameter of the micropores typically being in the range of 0.1 μ to 0.2 μ as determined porosimetric lly by the Mercury Intrusion Method. Furthermore, the total porosity of such material is typically within the range of 50 to 70%.

Other microporous materials may also be used in the present process. Thus, for example, in lieu of the thermoplastic binder constituent of the above microporous material, synthetic or natural thermosetting rubber polymers or copolymers may be used. In broad aspect, the present process uses the finely divided filler particles dispersed throughout the binder or matrix of the microporous material as active sites to which enzymes may be coupled.

Due to its porous construction and the dispersion of the filler particles throughout the matrix or binder, such microporous material has a relatively large surface area, typically on the order of about 80 m^2/g, and the number of available enzyme coupling sites is relatively large; hence, the loading factor or amount of enzymes which may be coupled per unit volume of such microporous material has been found to be correspondingly large. It is preferred that the microporous material be treated in such a manner as to effect a chemical bond between the catalytically active enzyme and the insoluble microporous support material.

In one preferred embodiment aliphatic primary amine functionality may be imparted to the microporous starting material by covalently bonding directly to the dispersed filler particles in the microporous material a bridging agent in the form of an organosilane such as gamma-aminopropyltriethoxysilane, whereas in another alternatively preferred embodiment aliphatic primary amine functionality may be imparted to the microporous material by irreversibly chemiadsorbing directly to the dispersed filler particles in the microporous material a bridging agent in the form of a macromolecular polyelectrolyte such as polyethyleneimine (PEI). Enzymes may then be covalently bonded or crosslinked to the chemically modified microporous material and more specifically to the aliphatic primary amine residues imparted to the surface of the material by the aforementioned bridging agents.

Example 1: Preparation of Untreated Enzyme Support Member – A sheet of microporous material was prepared by first dry blending 20.0 lb of Conoco 5385 polyvinyl chloride resin having a particle size of about 80 mesh, and 40.0 lb of Hi-Sil 233, a precipitated hydrated silica, in a Patterson Kelley low shear liquids-solids blender for approximately 3 minutes. Thereafter, and during continued agitation, 54.6 lb of solvent (cylohexanone) were added over a 20 minute period by means of a pump. Water in an amount of 59.0 lb was then added to the mix in the agitating blender over a subsequent 20 minute period to form a damp, stable, free-flowing powder.

The powder was then introduced into a screw extruder having a barrel temperature of approximately 120°F, and the extrudate passed between the rolls of a calender to obtain a substantially flat sheet having a thickness of 0.02 inch (0.5 mm). The sheet was then passed through an extraction bath of water at 170°F, and subsequently dried in a hot-air oven at 225°F for 6 minutes. The finished microporous sheet had a relatively wide pore size distribution extending from about 0.01 μ to 100 μ, and a mean pore diameter in the range of about 0.15 μ to 0.25 μ as determined by the Mercury Intrusion Method.

In addition, the total porosity of this material is approximately 65% by volume and the dispersed filler content (e.g., silica) comprises approximately 56% by weight. Liquid water soaked rapidly into the material without any applied pressure indicating that the micropores are substantially interconnected from surface to surface. From the resulting substantially flattened, semi-rigid, microporous sheet a plurality of untreated support members 5 x 5 cm in size were cut and heat-sterilized by immersion in a steam bath for 1 hour and allowed to cool and dry in open air.

Example 2: Chemical Modification by Chemiadsorption – Another untreated support member prepared in accordance with Example 1 was incubated in a 5% w/v aqueous solution of 50,000 MW branched chain polyethyleneimine (PEI) at room temperature for 1 hour. The treated support was flushed with water and 1 M NaCl to remove any unadsorbed PEI. Assay by the trinitrobenzene sulfonic acid test gave an intense orange trinitrophenylamine derivative on the surface of the treated support member demonstrating substantial aliphatic amino functionality.

The nitrogen loading on the treated support member was quantitated by elemental analysis and was 1.25% nitrogen by dry weight vs 0.02% nitrogen by dry weight of an untreated support member. The chemiadsorption of PEI on the treated support member appeared virtually irreversible since it could not be removed by incubation with high ionic strength solutions (e.g., 1 M NaCl or

1 M K_2HPO_4/KH_2PO_4) at pH values between 3 and 9. Only in the case of strong acidic conditions (incubation in 1 M HCl for 2 hours) was there evidence of partial desorption amounting to 50% of the nitrogen content as indicated by elemental analysis. The surface of the treated support by standard BET procedure was 55.4 m^2/g vs 81.1 m^2/g for the control. The support member treated with PEI displayed identical flow properties compared to an untreated support member irrespective of the buffer or ionic strength used.

Example 3: Enzyme Coupling Reaction (Glucose Oxidase) – The support member treated in accordance with Example 2 was incubated for 1 hour in 10% v/v aqueous solution of glutaraldehyde at pH 7. The support member was then rinsed with water and incubated for 1 hour in a solution of glucose oxidase which had been purified to homogeneity from *Aspergillis niger.* The conditions of the enzyme coupling reaction were as follows: glucose oxidase concentration 20 mg/ml in 0.1 M K_2HPO_4/KH_2PO_4 buffer pH 6.0 at ambient room temperature.

Directly pumping the enzyme solution through the support member under a positive hydraulic pressure did not improve enzyme loading relative to that obtained by simple incubation. The temperature of the coupling reaction was not found to be critical with the only requirement being that it did not exceed the thermal inactivation region of 50°C for glucose oxidase. The support member was then extensively washed with water and 1 M NaCl to remove unreacted enzyme.

Quenching of electrophilic residues on the support member surface was achieved by incubation of the immobilized enzyme composite with 0.1 M ethanolamine at pH 7.0 containing 50 mmol $NaCNBH_3$. The immobilized enzyme appears to have an indefinite shelf life when stored at 4°C in 0.1 M K_2HPO_4 buffer pH 6.0.

Porous Metallic Compounds Grafted with Haloalkyl Silane Groups

Grafted mineral carriers for use as enzyme supports have been developed by *H. Mazarguil, F. Meiller and P. Monsan; U.S. Patent 4,034,139; July 5, 1977; assigned to Rhone-Poulenc Industries, France.* These grafted carriers comprise oxides, hydroxides or other porous, insoluble mineral compounds, and are characterized in that they carry substituted or unsubstituted haloalkylsilane groups whose alkyl residue comprises 3 to 11 carbon atoms. These grafted carriers are produced by reacting a mineral carrier having hydroxyl groups with a haloalkylsilane having 1 to 3 groups which are reactive with the hydroxyl groups of the carrier.

The mineral carrier used must have hydroxyl groups, a grain size range of 40 μ to 5 mm, a specific surface area of 2 to 600 m^2/g and preferably 20 to 70 m^2/g, when it is intended for fixing enzymes, a pore diameter of 50 to 10,000 A and a pore volume of 0.5 to 1.8 ml/g. The following are illustrative of carriers which have these characteristics: aluminas, brick, glass, mineral silicates, metal oxides and more particularly silicas. The compound to be grafted is a haloalkylsilane having the formula:

$$X{-}(CH_2)_n{-}\overset{\displaystyle A}{\underset{\displaystyle C}{Si}}{-}B$$

in which: X represents a halogen atom – chlorine, bromine or iodine; n is an integer of from 3 to 11; A, B and C, which are similar to or different from each other, represent a Cl atom, or a methoxy, ethoxy, methyl or ethyl group, with the condition that at least one of the substituents A, B or C is capable of reacting with an OH group of the mineral carrier.

The following are compounds suited for grafting: triethoxybromopropylsilane, trimethoxyiodopropylsilane, triethoxyiodopropylsilane, triethoxychlorobutylsilane, trichlorochloropropylsilane, dimethylchlorochloropropylsilane, methyldichloroiodoundecylsilane. The grafting reaction can be carried out by any known processes in a solvent or aqueous medium, at atmospheric pressure or under pressure, and generally at elevated temperature. These grafted carriers can be used as industrial catalyst carriers, in chromatography, affinity chromatography, and more particularly for fixing enzymes.

In the latter case, a very large number of enzymes can be fixed, represented in particular by: oxidoreductases, such as glucose oxidase; hydrolases, such as lipase, pectinase, trypsin, urease, glucoamylase and α-amino-acylase. Fixing of the enzyme is effected by any known processes, either cold in an aqueous solution which is buffered in accordance with the pH value compatible with the enzyme, or hot in hydrocarbons or chlorinated solvents. The particular process will depend on the nature of the enzyme. Likewise, the choice of the carrier and the halogenated graft depends on the specificity and the actual reaction characteristics of the enzyme.

The enzymes, which are fixed in this way, are particularly stable and resistant to denaturation factors, pH and temperature. The following examples illustrate the process.

Example 1: 100 g of a silica, in the form of microballs, characterized by a grain size of from 100 to 200 μ, a specific surface area of 44 m^2/g, a pore diameter of 660 A and a pore volume of 0.9 ml/g, is dried at 150°C under vacuum for a period of 4 hours. The dried silica is introduced into 200 ml of a solution of 20 g of triethoxybromopropylsilane in xylene, and the mixture is heated for 8 hours at 140°C. After cooling, the grafted silica is drained, washed with acetone and dried under vacuum. The product contains 0.4% by weight of bromine.

Example 2: Operation is as in Example 1, but with 20 g of trimethoxyiodopropylsilane dissolved in 200 ml of xylene at boiling temperature. The grafted product contains 0.45% by weight of iodine.

Example 3: Glucoamylase is fixed in aqueous medium on the grafted silica of Example 1 (Test A) and on the grafted silica of Example 2 (Test B). One gram of carrier is added to 20 ml of a 0.1 M acetate buffer, pH 4.5, solution, containing 20 mg of glucoamylase, and the resulting dispersion is stirred for 18 hours at 4°C. After decantation, the carriers are washed 3 times with distilled water, and their activity is measured. The carrier-enzyme complexes produced are introduced into 10 ml of a 3% by weight starch solution in a 0.1 M acetate buffer, pH 4.5, and left in contact with stirring for 10 minutes at 40°C.

After separation, a measurement of the reducing sugars liberated is carried out on the liquid phase, by colorimetry with 3,5-dinitrosalicylic acid. Several washing operations and activity measuring operations are thus carried out in succession. Between each measurement operation, the complexes are preserved in

2 M NaCl solution. The results related to the glucose and expressed in μmol of glucose liberated/min/g of carrier are as follows:

Activity	Test A	Test B
Initial	8.5	1.4
After 5 days	5	1
After 10 days	0	0.7

It is found that the silica carrying a bromine-containing graft fixes more enzyme than the silica with an iodine-containing graft.

OTHER COUPLING COMPOUNDS WITH INORGANIC SUPPORTS

Carriers with Diazide Couplers

In the process developed by *R.A. Messing; U.S. Patent 3,930,951; January 6, 1976; assigned to Corning Glass Works,* enzymes can be coupled to water-insoluble inorganic carriers by reacting the carriers with 4,4'-bi(methoxybenzenediazonium chloride) (BMBD) to form a surface of diazo groups which are subsequently reacted with an enzyme. In preferred embodiments, the inorganic carriers are highly porous, have an average pore size of less than 1000 A, and consist of 4 to 200 mesh porous particles of materials selected from glass, silica, alumina, and mixtures of silica and alumina.

In the example, unless otherwise indicated, all composites were assayed using standard reagents and assay techniques. A control composite, prepared by using the same enzyme and carrier and adsorbing the enzyme to the internal surface of the carriers provided a comparison for the overall utility of the BMBD method.

Example: Alkaline Bacillus subtilis Protease Coupled to Porous Silica – In this example, the porous inorganic carrier consisted of 60 to 100 mesh porous SiO_2 particles having a surface area of 50 m^2/g, a pore volume of 0.8 cc/g and an average pore diameter of about 510 A with a maximum pore size of 680 A and a minimum pore size of 350 A. The solution of 4,4'-bi(2-methoxybenzenediazonium chloride was 0.01% aqueous solution prepared by transferring 1 mg of the BMBD salt to a 10 ml cylinder, diluting to the 10 ml mark with distilled water and then sonicating the slurry at room temperature in an ultrasonic bath until the salt was dissolved (about 5 minutes).

The alkaline *Bacillus subtilis* protease used was Alcalase. Two examples each of composites prepared by adsorption and coupling with the BMBD were prepared and compared as to enzyme activity over the same period of time. All four samples were prepared essentially simultaneously and assayed simultaneously over the same time span. The average activities of the two samples in each set at each indicated time period are shown below.

The immobilized enzyme composites were assayed with a 1% casein solution in 0.1 M phosphate buffer, pH 7.8, at 37°C in a shaking water bath. The results of these assays were compared to a standard curve produced by relating the optical density at 280 mμ of a trichloroacetic acid extract of the reaction mixture to the number of milligrams of free enzyme utilized in the assay. From this, one could obtain the number of milligrams of active enzyme remaining on the carrier over various periods of time.

Immobilization by Adsorption – To each of two 500 mg samples of 60 to 100 mesh porous silica in separate 25 ml Erlenmeyer flasks, 9 ml of 0.5 M sodium bicarbonate was added. The flasks were placed in a shaking water bath at 37°C and shaken for 1½ hours. The sodium bicarbonate was decanted and the porous bodies were washed with water. After decanting the water wash, 10 ml of enzyme solution containing 2 g of the Alcalase in 0.10 M phosphate buffer pH 7.8 was added to each sample of porous bodies. The flasks containing the carrier and the enzyme suspensions were placed in the 37°C shaking bath and shaken for 3 hours.

The samples were then removed from the bath and allowed to stand at room temperature overnight. The enzyme solutions were then decanted from the porous bodies and the immobilized enzyme was washed first with water and then with 0.5 M sodium chloride and finally again rewashed with water. The samples were stored in water at room temperature between assays for the half life determination period.

Immobilization by Direct Coupling via BMBD Residue – To each of two 500 mg samples of 60 to 100 mesh porous silica in separate 25 ml Erlenmeyer flasks were added 2 ml of the 0.01% diazonium coupling solution and the coupling solution was allowed to react at room temperature with occasional shaking. The porous silica bodies turned red during this period of time. At the end of 20 minutes the coupling solution was decanted and the porous bodies were washed with distilled water. The distilled water was then decanted and 10 ml of enzyme solution containing 2 g of the Alcalase in 0.1 M phosphate buffer pH 7.8 was added to each of the flasks containing the 500 mg sample of pretreated porous silica.

The flasks containing the enzyme and carrier were placed in a shaking water bath at 37°C and reacted with shaking at this temperature for 3 hours. The flasks were then removed from the bath and allowed to stand at room temperature overnight. The samples were then washed with water followed by 0.5 M sodium chloride and finally with water. The samples were stored in water at room temperature between assays over the half life determination period. All four composites were assayed over a period of 103 days to determine the amounts of active enzyme calculated to mg of active enzyme per gram of carrier for each composite. The average of activity determinations for each set of composites is shown in the table below.

	Mg of Active Protease/Gram Carrier	
Days	Adsorbed	Coupled
3	7.20	7.52
6	7.28	6.96
14	6.64	6.72
26	5.52	6.44
45	3.92	5.60
70	2.48	3.92
82	2.20	3.84
95	2.28	3.80
103	1.84	3.08

Papain was also coupled to porous glass particles in other examples of the complete patent.

Use of o-Dianisidine as Coupling Agent

Proteins such as enzymes or antibodies are immobilized in the process disclosed by *R.A. Messing and G. Odstrchel; U.S. Patent 3,983,000; September 28, 1976; assigned to Corning Glass Works* on various inorganic supports via an intermediate residue of o-dianisidine. The carriers which can be reacted with the o-dianisidine solution include any essentially water-insoluble inorganics having available surface oxide or hydroxyl groups capable of reacting with the o-dianisidine. The exact mechanism whereby the surface inorganic reacts with the organic compound is not fully understood although the bond formed was found to be quite strong. For example, the bond could not be washed off with 0.5 M sodium chloride, boiling water, or urea.

For reasons of economy, convenience, and control of surface area, and, where desired, porosity, it is preferred to use siliceous support materials, consisting mainly of silica. Silica per se, glass, and porous glass or fritted glass are all useful supports and can be readily surface activated and coated with o-dianisidine from either an aqueous or organic solvent.

After the inorganic support has been reacted with the o-dianisidine solution to form a surface coating of o-dianisidine residues, the coated support can be reacted directly with a protein solution or, where desired to introduce certain functional groups, indirectly with the solution after some further modification. For example, in some of the examples below, glutaraldehyde was found to be an excellent modifier between the o-dianisidine residue and the proteins of the protein solution, thus permitting a relatively mild protein bonding step.

Where it is desirable to space the protein away from the support a coupling arm of o-dianisidine and modifiers can be extended by simply alternating the reaction of o-dianisidine and modifiers such as glutaraldehyde. The general methods of treating the carriers are outlined below.

In the aqueous solution preparation, the carrier material consisted of porous silica particles (45 to 80 mesh) having a surface area of 40 m^2/g and an average pore diameter of 425 A. The minimum pore diameter was 270 A and the maximum diameter was 475 A. The o-dianisidine solution consisted of 5 g of o-dianisidine plus 4 ml HCl diluted to 500 ml with water.

A 100 g sample of the silica was transferred to a coarse, 350 ml fritted glass funnel. To this sample, 200 ml of the o-dianisidine solution was delivered at a rate of 1,000 ml per hour. The o-dianisidine was trapped at the top surface of the silica and the solution came through clear. After delivery of the 200 ml was completed, the silica cake was drained. An additional 200 ml of the o-dianisidine was delivered to the silica. Again, after passage through the silica, the solution was colorless.

After the second delivery, isopropanol was delivered to the derivative to wash away excess o-dianisidine. The first 100 ml of alcohol delivered was intensely colored brownish-red; however, after a second 100 ml volume of the alcohol was delivered, the wash solution began to lose color. The support was then washed with 300 ml of water followed by another 100 ml of isopropanol and, finally, washed with 100 ml of acetone and then air-dried with aspiration for 1 hour. The carrier derivative was transferred to a glass bottle and used subsequently for preparing the immobilized proteins.

In the organic solution preparation, the inorganic support consisted of silica particles having a surface area of 370 m^2/g and an average particle size of about 4 μ (Syloid 72). The o-dianisidine solution consisted of 1 g of o-dianisidine diluted to 100 ml with methanol. The carrier was prepared by mixing 20 g of the silica with 100 ml of the o-dianisidine solution in a 150 ml beaker for 10 minutes at room temperature. The slurry was transferred to a fine frit 600 ml funnel and filtered with aspiration on a filter flask. The solution came through the funnel with about 50% of the color of the original solution. The silica was then washed on the funnel with aspiration with 100 ml of methanol. The filtrate was clear.

The carrier derivative was then washed with 50 ml of acetone and air-dried. Samples of this carrier were used for the immobilization of antibodies, described below. This carrier can be further derivatized for aldehyde function as follows: To 3.64 g of the above-described treated carrier, 18.2 ml of a 2.5% aqueous glutaraldehyde solution was added in a 30 ml beaker and stirred for about 10 minutes. This derivative was then immediately filtered with aspiration on a Buchner funnel containing S and S576 filter paper. The derivative was then washed with 100 ml of distilled water and finally washed with 50 ml of acetone and air-dried.

Example: Lactase derived from *Escherichia coli* was immobilized on the derivatized carrier prepared in aqueous solution by placing 1.0 g of the carrier in a column and circulating through the column at 400 ml/hr 12.5 ml of an aqueous slurry of the enzyme containing 417 lactase activity units. The reaction was carried out at 20°C with circulation for 22 hours. It was found that the amount of enzyme immobilized yielded between 92 and 98 lactase units per gram. It is significant to note that in this preparation no further functionalization of the surface o-dianisidine residue was required.

In a subsequent experiment where the same carrier was further functionalized with glutaraldehyde, an enzyme loading of 261 lactase units per gram was achieved. When the carrier having the o-dianisidine residue was further functionalized to form a reactive diazonium group, the lactase loadings were 197 lactase activity units per gram.

Use of Polyisocyanates for Bonding to Porous Metallic Oxides

Enzymes can be immobilized in the process developed by *R.A. Messing and S. Yaverbaum; U.S. Patent 4,071,409; January 31, 1978; assigned to Corning Glass Works* on polyisocyanate modified inorganic supports. The process comprises reacting an organic solution of polymerized isocyanate compounds having a molecular weight of at least 250 with a high surface area inorganic material having surface hydroxyl or oxide groups to form a polymeric surface having reactive isocyanate groups, and then reacting the treated support with a dispersion of the proteins under conditions sufficient to assure bonding of the protein in a biologically active state to the surface of the support via the isocyanate groups. The polymeric isocyanates useful for bonding the proteins may be represented as follows.

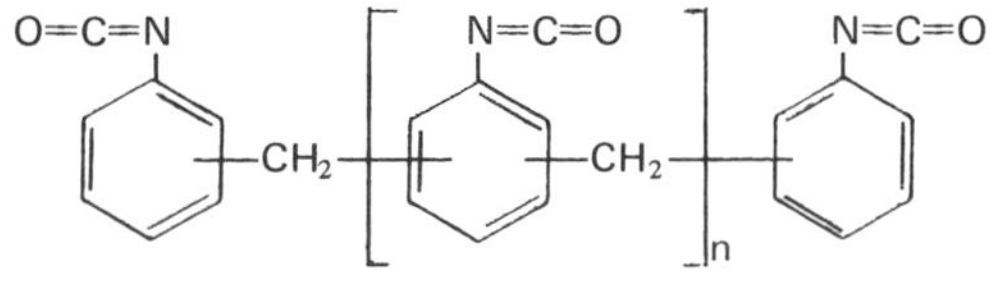

O=C=N N=C=O N=C=O

CH_2 CH_2

n

In the above formula, n is an integer having a value of at least 2. Such polymeric isocyanates are available commercially (Papi). A very effective polymeric isocyanate for use with porous titania, silica, alumina is Papi-901. All of these polymeric isocyanates are insoluble in water and they must be applied from an organic solvent that does not react with the polyisocyanates. The most effective solvent is acetone. The reaction of the polymeric isocyanate coupling agent with the carrier (e.g., titania) results in the formulation of a metal carbamate. The isocyanate groups that have not reacted with the carrier remain available for coupling to the enzyme in the following manner:

Reaction 1 (under alkaline conditions) yields substituted ureas:

$$-\phi-N{=}C{=}O + RNH_2 \longrightarrow -\phi-NH-\overset{\overset{O}{\|}}{C}-\overset{\overset{H}{|}}{N}-R$$

Reaction 2 (under mildly acidic conditions) yields urethanes:

$$-\phi-N{=}C{=}O + ROH \longrightarrow -\phi-NH-\overset{\overset{O}{\|}}{C}-O-R$$

For efficient and practical protein loading, the inorganic supports should have a high surface area (e.g. >0.2 m^2/g) and may be porous. A preferred average pore size range is 100 to 2500 A. Although any essentially water-insoluble inorganic having surface hydroxyl or oxide groups capable of reacting with the polymer may be used, a preferred support consists of porous titania particles, 30 to 80 mesh, having an average pore size of 320 to 1500 A.

Although TiO_2 is preferred as a support, inorganic supports consisting of silica, alumina, and porous glass may also be used successfully to bond biologically active proteins via the polymeric isocyanate. Among the biologically active proteins successfully used in this process are the enzymes papain, urease, glucoamylase, and lactase.

Example: Papain Coupled to Porous TiO_2 and SiO_2 – The polymeric isocyanate consisted of a 1% (v/v) solution of the Papi-901 polymer. To prepare this solution, 0.25 ml of the Papi-901 was diluted to 25 ml with acetone. The papain was 0.2 Hoover Ball Milk Clot. The titania support consisted of porous TiO_2 particles (30/80 mesh) having an average pore diameter of about 420 A and a surface area of 35 m^2/g. The porous silica particles (25/60 mesh) had an average pore diameter of 500 A and a surface area of 50 m^2/g.

Preparation of Carrier Surfaces – Two 500 mg quantities each of titania and silica support were placed in four 25 ml Erlenmeyer flasks. To each flask was added 1 ml of the coupling solution which was allowed to react with the carrier at room temperature for 5 minutes. The titania immediately reacted with the coupling agent to form an orange-coated porous body. The silica was somewhat slower to react but a yellow coating of coupling agent formed on the silica bodies. The color due to the coupling agent was completely removed from solution in the case of the titania porous bodies. The residual acetone solvent was decanted from the flask and the porous bodies were then ready for reaction with the enzyme.

Coupling of the Enzymes – Series 1: Immobilization was from alkaline, pH 8.0, enzyme solution. One gram of the papain was diluted to 10 ml with 0.2 M Na_2HPO_4. To one flask containing isocyanated titania and one flask containing isocyanated silica was added 10 ml of the above enzyme solution and the flasks were placed in a 23°C water bath with shaking for 7 hours. The flasks were then removed from the shaking water bath, the unreacted enzyme solution was decanted, and the immobilized enzymes were washed first with water, followed by 0.5 M sodium chloride and finally followed by water again.

Series 2: Immobilization was from mildly acidic solutions, pH 5.2. One gram of the papain was diluted to 10 ml with distilled water. To each of the remaining two flasks, one containing the isocyanated titania and one containing the isocyanated silica, were added 10 ml of papain solution and the flasks were placed in a shaking water bath at 23°C. The flasks were shaken for 3 hours, removed from the bath and the unreacted enzyme solutions were decanted. The immobilized enzymes were first washed with water and then 0.5 M sodium chloride solution, followed by a final wash with water.

The Series 1 and 2 immobilized enzymes were stored at room temperature in water between assays. The immobilized enzymes were assayed with 0.2% egg albumin solution in acetate buffer containing cysteine and EDTA at pH 5.0 at 37°C. The optical density of the trichloroacetic acid extract of the reacted substrate was determined at 280 nm and compared to a curve for the free enzyme to give the milligrams of active papain per gram of carrier. The results are recorded below.

Days of Storage	Series 1 SiO_2	Series 1 TiO_2	Series 2 SiO_2	Series 2 TiO_2
	(mg active papain/g)			
1	73.4	58.6	115.7	63.8
4	–	39.2	28.9	53.4
14	21.6	32.0	43.2	39.6
21	19.0	39.0	30.0	34.0

In previous immobilization studies that were preformed with porous silica and porous glass, it was noted that most of the activity (99% or more) of papain was lost during the first 21 days of storage. In addition, on a per gram basis of carrier, the loadings were usually under 30 mg of papain. It is significant to note from the results above that there was little difference between the loading and stability on the titania of the alkaline and acid series. It would appear as though higher loadings were achieved under the acidic condition in the silica series.

Use of p-Phenylenediamine as Coupling Agent

Arylamine surface derivatives useful for the immobilization of enzymes can be prepared by the method of *M. Lynn; U.S. Patent 4,072,566; February 7, 1978; assigned to Corning Glass Works* by adsorbing p-phenylenediamine onto the surfaces of water-insoluble inorganic carriers having a relatively low surface isoelectric point.

Briefly, this process comprises the steps of:

(a) reacting a high surface area, insoluble, inorganic support material having a relatively low surface isoelectric point with a solution of p-phenylenediamine (1,4-diaminobenzene) under conditions sufficient to adsorb at least a portion of the p-phenylenediamine onto the surfaces of the support material;

(b) separating the supporting material from the solution of the p-phenylenediamine; and

(c) reacting the separated support material with a solution of proteins under conditions sufficient to assure the covalent bonding of at least a portion of the proteins to the support material without significant loss of biological activity.

In various preferred embodiments, the surface derivatized support of step (a), after removal from the p-phenylenediamine solution, is further functionalized to reaction with the proteins. For example, the free amino groups of the derivatized surface can be readily modified via a diazotization step or Schiff base reaction (reaction with an aldehyde).

The main requirements for the inorganic carriers are that they have a relatively high surface area ($>10\ m^2/g$) to assure relatively high loading of proteins; be relatively water-insoluble; and, very importantly, have a relatively low surface isoelectric point. The isoelectric point of the surface is the pH at which the electrostatic surface potential (zeta potential) is zero. Isoelectric points of surfaces are described by G.A. Parks, *Advances in Chemistry*, Series 67, 121-160 (1967), and G.A. Parks, *Chemical Reviews,* 65, 177-198 (1964).

The purpose of having a relatively low surface isoelectric point (less than pH 7.0) is to enhance the formation of strong electrostatic bonds between the acidic inorganic surface and the basic organic amine. Typical examples of inorganic carriers having low isoelectric points are siliceous materials such as glass and silica particles. Other suitable inorganics are stannic oxide, titania, manganese dioxide and zirconia.

Example 1: A series of arylamine glass derivatives was prepared in the following manner. Solutions of the diamine were prepared by dissolving 1 g, 5 g and 10 g of p-phenylenediamine in 100 ml of ethanol. To each of the diamine solutions was added 10 g of porous glass particles (40 to 80 mesh) having an average pore diameter of 550 A. The reaction mixtures were shaken for 30 minutes at 60°C. The reaction solutions were decanted and the products were washed with ethanol until no more diamine could be removed (as indicated by color in the washes). The products were then washed with 0.5 M NaCl solution until the washings were colorless, followed by additional washing with water. All three products gave positive tests for arylamine with the β-naphthol test. The products were analyzed for reactive amine loading by titration with perchloric acid.

Reactive Amine Loadings

Sample	Reaction Solution (g/100 ml)	N (%)	meq Amine/g
A	1	0.29	0.21
B	5	0.33	0.23
C	10	0.46	0.33

Example 2: Papain was immobilized on the arylamine supports in the following manner. To 1 g samples of the arylamine supports described above was added 10 ml of 2 N HCl and the reaction mixture was cooled in an ice bath. To the mixtures was added 2.5 ml of 4 M sodium nitrate solution, then the reaction mixtures were evacuated for 30 minutes to remove gas bubbles from the pores. The reaction solutions were decanted, and the products were washed several times with water.

Papain was coupled to the supports by adding 1 ml of a solution containing 100 mg of papain in pH 8.5 phosphate buffer to 1 g of diazotized support. The reactions were allowed to continue for 2 hours in an ice bath, then the reaction solutions were decanted. The immobilized enzymes were washed for 15 minutes with 6 M urea solution, then with water, then for 30 minutes with 0.5 M sodium chloride solution, then several more times with water. The immobilized enzymes were stored under water in a refrigerator.

The immobilized enzymes were assayed with 1% casein solution in phosphate buffer at pH 7 containing cysteine and EDTA at 37°C. The enzyme loadings, reported as mg of active papain per gram of support are listed below. The immobilized enzymes were assayed again after storage for 83 days, and these results are also listed below.

Immobilized Papain Activities

Carrier	Enzyme Loading, mg Papain/g Support	
Sample	Initial	83 Days Storage
A	31.0	–
B	20.5	32.9
C	19.2	28.8

Immobilization of Sulfhydryl Oxidase on Glass Beads Using a Diimide

H.E. Swaisgood; U.S. Patent 4,087,328; May 2, 1978; assigned to Research Triangle Institute has found a method of isolating and purifying sulfhydryl oxidase enzyme from milk which consistently yields preparations of greater than 3,000-fold purification over skim milk. Sulfhydryl oxidase in a substantially purified form has been found to catalyze the oxidation of sulfhydryl groups in both small compounds and proteins using oxygen as an oxidant. The enzyme in a substantially purified as well as immobilized form has been found useful in treating milk to remove cooked flavor.

As an initial step, whole raw milk is treated to remove the fat from solids and obtain skim milk. Whey is obtained from the skim milk by coagulation of the casein fraction with rennin. The whey is removed, cooled and added to a half-saturated solution of ammonium sulfate to precipitate the crude enzyme. The crude enzyme fraction obtained as above is dissolved in a dilute neutral buffer solution (e.g., a phosphate buffer) in order to keep the enzyme stable. The enzyme in solution is then allowed to equilibrate (period to allow for dissociation of the enzyme) preferably under refrigerated conditions.

The solution is then subjected to a separation treatment (e.g., centrifugation) to separate the enzyme fraction from molecularly larger materials. Upon centrifugation of the solution, the enzyme fraction is removed with the supernatant liquid.

The supernatant liquid is then concentrated to a 2½ to 3½% protein concentration by an ultrafilter. However, other means such as addition of dry molecular sieves, or vacuum evaporation appear to be equally suitable. A concentration of at least 2½% appears to be necessary in order to increase the enzyme molecular size by reassociation. The enzyme fraction is recovered in pellet form (e.g. by centrifugation) thereby achieving separation from molecularly smaller materials. The active enzyme fraction isolated in the resulting pellet possesses a specific activity at least about 1,400 times greater than that of skim milk.

Increased specific activity of the enzyme is obtained by dissolving the pellets in a neutral buffer solution at about twice the dilution (liquid volume) and subjecting the resulting solution to separation to obtain an even purer form of the enzyme having as much as a 3,000-fold increase in specific activity over that of skim milk. The isolated substantially purified sulfhydryl oxidase may be stored while refrigerated at 4°C without significant loss in activity. It is preferred to immobilize the isolated substantially purified sulfhydryl oxidase. Otherwise, the enzyme becomes an integral part of the reaction mixture and cannot be recovered following completion of the reaction.

A preferred means of immobilizing the isolated sulfhydryl oxidase according to the present process is to attach the enzyme to glass beads. A suitable method is described in *Biochim. Biophys. Acta,* 243–265 (1974). Since immobilized sulfhydryl oxidase is activity sensitive to bacteria, heat, pH and dryness, it is essential that in storage the enzyme be prevented from drying out. Preferably, the immobilized sulfhydryl oxidase is maintained in a neutral buffer solution (phosphate buffer) under refrigeration until ready for use.

Example: Immobilization of the purified sulfhydryl oxidase was achieved by affixing the enzyme to glass beads. Succinilate glass beads were prepared from γ-amino-propyl glass beads (40 to 60 mesh, 2000 A pore diameter, Corning Glass) which were then washed with distilled water and equilibrated for 24 hours in a 0.2 M phosphate buffer at pH 4.75. The beads were degassed and to 0.5 g of the beads crystalline 1-ethyl-3-dimethylaminopropyl carbodiimide (EDC) was added and the reaction was carried out for 20 minutes at 25°C and a constant pH of 4.75.

The beads were then washed with distilled water and cold neutral phosphate buffer solution to remove excess EDC. The isolated enzyme (0.1% w/v) in a phosphate buffer solution was then contacted with the succinilate beads for between 16 and 24 hours after which the beads were washed with 0.1 M phosphate buffer of pH 7. This resulted in covalently coupling the isolated enzyme to the glass beads yielding an immobilized sulfhydryl oxidase. As previously noted, the immobilized enzyme when not in use should be prevented from drying out and subjected to extreme temperatures by storage under refrigeration in a phosphate buffer solution of pH 7.

ENZYMES BONDED TO DENSE PARTICLES FOR FLUID FLOW PROCESSES

An improved fluid flow process described by *R.W. Coughlin; U.S. Patent 3,928,143; December 23, 1975* and *R.W. Coughlin and M. Charles; U.S. Patent 4,048,018; September 13, 1977* uses enzymes bound to a solid supporting carrier. A liquid stream containing the reactants is passed through a catalyst bed, made up of enzymes bonded to small, dense particles of carrier or support materials, in a manner which causes the bed to expand or fluidize. This chemical reaction is simultaneously free from limitations due to plugging and excessive pressure drop and which also has the advantage of the high mass transfer rates than can be realized between the liquid and small particles.

It is possible to practice the present process using various types of dense particulate material to which the enzyme of choice is then bound. By proper selection of these materials it is possible to vary the density of the insolubilized enzyme catalyst particles over a wide range; for example it is possible to use glass (density 2.5 g/cm^3), nickel (density 8.9 g/cm^3), iron or preferably stainless steel (density 8 g/cm^3), silver (density 10.5 g/cm^3), molybdenum (density 10.2 g/cm^3), tantalum (density 16.6 g/cm^3), gold or tungsten (density 19.3 g/cm^3), rhodium (density 21.4 g/cm^3), platinum (density 21.5 g/cm^3) or various alloys of these metals. When it is possible to form a tough, tenacious, insoluble and otherwise suitable oxide coating on metallic particles of the proper density, the enzyme of choice can be bound to the oxide by the use of known coupling agents.

The use of dense materials for the particles which carry the enzyme permits good fluidization while using particles of small size; using such small particles increases mass transfer rates of reactants to the particles and therefore also increases the rate of chemical reaction. Furthermore, the use of dense, enzyme-carrier particles has the added advantage of permitting high liquid flow rates through the expanded- or fluidized-bed without entrainment and loss of the particles from the bed.

The superiority of suspended or fluidized-bed operation of liquid plug flow reactors using as catalysts enzymes bound to small, dense particles is demonstrated by design calculations for a reaction, the global or net rate of which is about equal to the rate of mass transfer of the reactant from the bulk solution to the surface of the solid, enzyme-bearing particles, i.e., for a situation in which the rate of chemical reaction is controlled by mass transfer to the catalytic particles.

These computations have been carried out for reaction of a dilute aqueous solution of ethanol containing excess nicotinamide adenine dinucleotide (NAD) using the Ergun equation for pressure drop in flow through beds of solid particles [see *Unit Operations of Chemical Engineering,* by McCabe and Smith (2nd ed) p 161, equation 7-26, McGraw-Hill Book Company (1967)], correlations for mass transfer coefficients in fixed and in packed beds of particles [see *Chemical Engineering Kinetics,* J.M. Smith, pp 380–383, equations 10-36 through 10-39 and Figure 10-2 on p 364, McGraw-Hill (1970)], a smoothed correlation of particulate fluidization [see *Fluidization and Fluid Particle Systems,* by Zener and Othmer, p 236, Figure 7.7, Reinhold Publishing Co. (1960)], and the common, well-known design equation for plug flow reactors.

Assuming that the reactant is ethanol in aqueous solution in which the diffusivity of ethanol is about 10^{-5} cm^2/sec (i.e., a dimensionless Schmidt Number equal to 1,000 as would be the case for a dilute aqueous solution of ethanol reactant), that the reactor effects a conversion of the ethanol reactant equal to 99.9%, that the liquid reactant solution is passed in plug flow through beds of approximately spherical particles of approximately uniform diameter in each type of reactor, and assuming further that the flow rate of the liquid reactant solution is 70,000 $lb/hr/ft^2$ of reactor cross section and that the liquid reactant solution has viscosity equal to 1 cp and density equal to 1 g/cc, the following results are computed for different particle sizes, different particle densities, and two different types of reactor:

(1) a fixed-bed reactor in which the particles of enzymic catalyst remain packed in an immobile bed, and

(2) fluidized-bed reactors in which the particles of enzymic catalyst are suspended and agitated by the motion of the liquid through the reactor.

Example: Conversion of ethanol by reactor equals 99.9%, Schmidt Number is 1,000, viscosity of reactant feed equals 1 cp, density of reactant feed is 1 g/cc, mass flow rate of reactant feed equals 70,000 lb/hr/ft^2 of reactor cross section. It is to be noted from the results of the example, that the reactor length necessary to carry out the specified reaction is an order of magnitude larger when a fixed bed reactor packed with catalyst particles of about 5,300 μm or 0.53 cm diameter is employed in contrast to the reactor lengths required for fluidized-bed reactors containing catalyst particles of about 500 μm or 300 μm diameter. With regard to these same results it is also noteworthy that the pressure drop required for the fluidized-bed reactor for this reaction is between 6 and 12 times lower than for the fixed-bed reactor.

Reactor Type	Particle Diameter (μm)	Particle Density (g/cm^3)	Bed Porosity	Reactor Pressure Drop (psia)	Reactor Length (ft)
fixed bed	5,325	immaterial	~0.36	95	49.5
fluid bed	500	8.6	~0.8	7.9	4.8
fluid bed	500	16.6	~0.7	12.8	2.8
fluid bed	500	19.3	~0.65	14.8	2.6
fluid bed	300	16.6	~0.8	7.7	2.9
fluid bed	300	19.3	~0.75	8.7	2.4

COVALENT BONDING TO PROTEINS AND CELLULAR MATERIALS

BONDING ENZYMES TO NONENZYME PROTEINS

Crosslinked Enzyme-Protein Particulate Compositions

Water-insoluble enzyme compositions in particulate form of desired sizes, which can be used in column fillings, bed reactors, stirred reactors, etc. have been developed by *A.G. van Velzen; U.S. Patent 3,838,007; September 24, 1974; assigned to Gist-Brocades N/V, Netherlands.* The water-insoluble enzyme preparation is obtained by a process involving the steps of suspending a nonproteolytic enzyme in an aqueous solution of gelling protein, combining the mixture with an organic liquid poorly miscible or immiscible in water to produce a suspension having the protein in particulate form, treating this suspension to effect gelation of the protein and contacting the gelled protein particle with a bi- or polyfunctional protein reagent to crosslink the proteins present in the particles.

Enzymes which are used in the process are those having no proteolytic activity because proteolytically active enzymes would decompose the gelling protein even after the crosslinking step and give physically very unstable products. Examples of enzymes which may be used are invertases, amyloglucosidases, lactases, maltases, amylases, ureases, lipases, esterases, glucose isomerases, glucose oxidases, dehydrogenases and penicillinases. Mixtures of enzymes may also be used so that the resulting particles can be used for carrying out two or more enzymatic reactions. Also, insoluble enzymes may be used when they are in an unfavorable form, e.g., a powder and even microorganisms and spores containing enzymes may be used in the process. Complexes of enzymes and hydrated oxides such as hydrated aluminum oxide may be employed too.

The enzyme-gelling protein mixture which is used as a starting material may be prepared by dissolving or suspending the enzyme in an aqueous gelling protein solution. The temperature of the aqueous gelling protein solution should be such that an active mixture is obtained in a liquid form. Therefore, a temperature of 20°-35°C is preferred with the maximum temperature depending on the enzyme being generally 60°-65°C, but higher temperatures may be used in exceptional cases.

The concentration in the solution is dependent on the specific gelatin used and may vary within limits of 0.1 to 25% by weight based on the water used, preferably within limits of 5 to 10% by weight. The enzyme concentration depends on the purpose for which the insolubilized enzyme is to be used as well as on the activity thereof. The pH of the solution is preferably that at which the enzyme possesses its greatest stability under the circumstances involved, but as gelatin is also present, the pH should be preferably 3 to 10 to allow the gelatin to gel. For other gelling proteins, the pH range may differ.

The organic liquid used to bring about the particulate form of the aqueous enzyme-gelling protein solution is an organic liquid poorly miscible or immiscible in water and compatible with the nonproteolytic enzyme and gelling protein. It must be at a temperature below its boiling point or range at the pressure employed, and generally at a temperature below 65°C. Examples of suitable organic liquids are aliphatic alcohols with four or more carbons such as butanol, esters of alcohols and lower fatty acids such as ethyl acetate, butyl acetate and ethyl propionate, branched or straight chain aliphatic hydrocarbons such as paraffin oil and petroleum ether, aromatic hydrocarbons such as benzene and its homologs, chlorinated hydrocarbons such as methylene chloride and trichloroethylene, and mixtures of two or more of the abovementioned liquids. One or more surface-active agents may optionally be added to the organic liquid as they generally enable a better control of the dispersion state of the enzyme-gelling protein suspension to be obtained in the organic liquid.

The crosslinking step is carried out with a bi- or polyfunctional protein reagent forming covalent bonds with the enzyme and/or the gelling protein. Examples of suitable bifunctional protein reagents are aldehydes such as glutaric dialdehyde, acrolein or crotonaldehyde, esters such as chloroformic acid esters, acid halides such as acid chlorides, epoxides such as epichlorohydrin, derivatives of dimethyladipic acid, carbodiimides, phenol-2,4-disulfonyl chloride, bromocyanide, activated agents such as bromocyanide-activated compounds of acid halides, or mixtures of two or more of those compounds. Preferably, an aqueous solution of glutaric dialdehyde is used.

Example: 2 grams of invertase were suspended in 100 ml of an aqueous solution containing 7.5 grams of gelatin at a temperature of about 45°C and the pH was adjusted to 7. This aqueous enzyme-gelatin suspension was added with stirring to 400 ml of n-butanol whose temperature was 50°C and then the suspension obtained was rapidly cooled to 10°C whereby enzyme-gelatin-containing particles were formed. The butanol was decanted off, and after the enzyme-gelatin-containing particles were dehydrated by the addition of 400 ml of a 96% (v/v) aqueous ethanol solution, the liquid was subsequently removed. The particles obtained were crosslinked by reaction with a solution of 3.75 g of glutaric dialdehyde in 200 ml of water and after removal of the liquid, the particles were washed with water until the odor of butanol could no longer be detected. The particles obtained had good physical stability and were insoluble in water.

The use of the enzyme-gelatin-containing particles was demonstrated by filling a plug flow column reactor (height 22 cm; diameter 4 cm) with the enzyme-gelatin-containing particles and passing a 10% (w/v) aqueous solution of saccharose at pH 4.5 through the reactor at varying flow rates. The amount of invert sugar formed was calculated from the glucose content of the effluent, and was expressed as the percentage of saccharose converted. The percentage conversion decreased with increasing flow rate. The reactor was kept in operation continuously for 12 weeks at room temperature. During the 9th week of operation, however, the conversion rate of the column reactor dropped which was caused by microorganisms growing on the enzyme particles. No precautions were taken to maintain the sterility until this 9th week.

However, the original conversion rate was restored by thoroughly washing and sterilizing the enzyme-gelatin-containing particles in the column by percolating a 2% (w/v) solution of glutaric dialdehyde in water through the column. The following results are reported in the table below and show the high stability of the insolubilized invertase used.

	- - Percent Conversion of Saccharose- -	
Flow Rate, ml/hr	Initially	After 12 Weeks
200	100	100
300	–	80
400	80	–
500	70	68
700	–	50
800	50	–
1,500	30	30

Natural Sponge as Support

In *Biopolymers* 4, 441-448 (1966), the use of collagen sponge is proposed for binding papain. However, collagen sponge has the disadvantage of dissolving at a pH of 3 to 4 and gelatinizing in water at elevated temperatures. It is also reported that mechanical disintegration of the collagen sponge-papain conjugate occurs after addition of cysteine and EDTA, and the conjugate becomes partly soluble upon activation of the bound papain.

J.J. Dahlmans and T.A.J. Meijerink; U.S. Patent 3,919,048; November 11, 1975; assigned to Stamicarbon B.V., Netherlands have unexpectedly found that insoluble enzyme complexes having outstanding properties can be prepared by binding one or more enzymes to an insoluble carrier comprising natural sponge.

Natural sponge is particularly suitable as an enzyme carrier for the following reasons. (1) Because of the great elasticity and porosity of sponge, the enzyme-sponge complex is extremely suitable as a column filling. (2) Sponge is strongly hydrophilic. (3) Sponge contains reactive groups of the same type as those present in various enzymes, as a result of which linkage reactions are possible under mild conditions. (4) As a skeleton substance, sponge is especially resistant to microbiological and mechanical disintegration.

The term sponge generally includes any member of the invertebrate phylum Porifera characterized by a porous skeleton of interlocking spicules, glasslike rods, or horny fibers. Commercial sponges, which are the type used, comprise the skeletons surrounding the numerous cells of the sponge animal, and are usually obtained from the genera *Spongia* or *Euspongia,* and *Hippospongia.* In commercial sponges, the horny skeletal fibers are composed of spongin, a plastic-like material, which is an organic substance and consists mainly of proteins.

Among the known methods for bonding the enzyme to the sponge, particularly preferred methods include the azide method, the carbodiimide method, bonding with anhydrides, the cyanogen bromide method, bonding by means of cyanuric chloride or its derivatives, bonding of tyrosine-containing enzymes by diazotization of various aminophenyl derivatives and bonding by means of isothiocyanates or suitable fluoronitroaryls, such as 2,4-dinitro-1-fluorobenzene.

A wide variety of industrial enzymes can be used to complex with the sponge. The more important enzymes that can be used include trypsin, papain, chymotrypsin, lactase, urease, acylase, invertase, catalase, α-amylase, penicillin-amidase, β-amylase, lipase, cellulase, glucose oxidase, amyloglucosidase, protease, gelatinase, hemicellulase, pectinase, lysozyme and pepsin, as well as glucoamylase.

Example: Sponge-Papain Complex – To 100 mg of ground sponge in a double-walled reaction vessel connected to a cryostat and provided with stirrer are added 25 ml of water. The temperature is maintained at 5° to 7°C. The pH value is then adjusted to 3.9 with the aid of 5 N H_2SO_4. Subsequently, 85 mg (0.2 mmol) of N-cyclohexyl-N'-[β-(N-methyl-morpholino)ethyl]-carbodiimide-p-toluenesulfonate (CMECDI) are added and the mixture stirred for 15 minutes, while the pH value is maintained at 3.9. The carbodiimide reacts with the carboxyl groups of the sponge.

After the pH has been adjusted to 7, 250 mg of technical papain are added. While keeping the pH value constant at 7 (using, for example, an autotitrator with 0.1 N NaOH), the reaction mixture is stirred for 2 hours, during which time the amino groups of the enzyme react with the carbodiimide-substituted carboxyl groups of the sponge, whereby the urea derivative formed from the carbodiimide is eliminated. The pH value is then reduced to 3.9 and, once again, 85 mg of CMECDI are added, followed by 15 more minutes of stirring at this pH. After the pH reaches a value of 7, stirring is continued for 14 hours at a constant pH of 7 and a temperature of 5° to 7°C. The product is washed 20 times, alternately with 0.1 M NaCl and water.

The product thus obtained is freeze-dried to yield 63 mg of insoluble papain sponge complex having a content of 4% (gram of enzyme per gram of sponge) and an enzymatic activity of 125% referred to the free enzyme. An enzymatic activity greater than 100% can be due to any one or more of the following reasons: (1) The substrate concentration at the carrier surface is larger than the substrate concentration in solution. This phenomenon occurs when the carrier and the substrate have opposite charges; (2) The starting enzyme is contaminated by other proteins or compounds which are not bound to the carrier. The starting enzyme, therefore, is purified by becoming transferred to the carrier. (3) Because of the enzyme being coupled to the carrier, a change in the conformation of the enzyme occurs, as a result of which the activity of the enzyme increases.

Other examples cover the bonding of trypsin, papain and γ-chymotrypsin to succinic-acid-anhydride-modified sponge, plus sponge modified with iminobispropylamine, glutaraldehyde, and p-aminophenyl-β-D-thiogalactopyranoside.

Inactive Proteins as Supports

A process of crosslinking active proteins together with at least one inactive protein, with or without preexisting support or carrier has been developed by *S. Avrameas, G. Broun, E. Sélégny, and D. Thomas; U.S. Patent 4,004,979; Jan. 25,*

1977; assigned to Agence Nationale de Valorisation de la Recherche (ANVAR), France. The active protein, e.g., an enzyme, preferably comprises 1 to 10% by weight of the total protein mixture while the crosslinking agent comprises 1 to 25% by weight of the entire mixture being treated.

In one embodiment of the process, the carrier itself may be an inactive protein, which may be coreticulated together with the biologically active protein substance. It has surprisingly been found that crosslinking of biologically active proteins, such as enzymes together with an inactive protein, such as human or animal plasma, albumin or plasma proteins, ovalbumin, fibrinogen or hemoglobin, by means of a crosslinking agent, i.e., a bifunctional or polyfunctional agent, eventually in the presence of a suitable carrier, and preferably with the active protein to whole proteinic substance ratios above specified, needs smaller amounts of crosslinking agent and provides higher activity ratios than the previous known techniques where such high amount of inactive proteins was not used. This process is a relatively simple one which is carried out in one step and can be performed in a short time.

It is believed that the process involves a competition of active and inactive proteins for the crosslinking agent, thus restricting the number of amino groups of each molecule to be involved in the crosslinking step. The active proteinic molecules are only slightly modified and few active sites are affected by steric hindrance or by denaturation. Their activity is well preserved, for example in a macromolecular structure where the framework is crosslinked albumin.

The active protein substances which can be used in the process may be natural or synthetic products, and they can be used in the crude state or after prior purification. These active protein substances are selected from the group consisting of enzymes, antibodies, antigens, allergens, hormones, microbes and viruses.

The inactive protein substances which can be used are human or animal plasma albumin, or plasma proteins, ovalbumin, fibrinogen or hemoglobin, or any mixtures thereof.

The active protein and inactive protein containing solutions are usually dissolved in aqueous buffer mediums. The buffers are most frequently inorganic buffers, containing alkaline or alkaline-earth phosphates. In the process, any suitable carrier compatible with the active protein substances can be used, that is to say, one that is not liable to denature the substance.

The crosslinked active and inactive protein articles can be soluble in an aqueous medium, and therefore be in the form of an aqueous solution, or be suspended in an aqueous medium. The articles can be gels, solid masses such as granules, pills, tablets, or a plate, cake or other molded mass. The carriers used can also provide the article with its final form such as a film, membrane or an inert, porous material, for instance. The following macromolecular carriers may be used: cellulose, regenerated cellulose (Cellophane), amylose, alginates, dextran, collagen, polyvinyl alcohol, polysilanes, polyacrylamide and their substitution products. The carrier may, further, be an aqueous medium or aqueous solution.

As a general rule, the crosslinking agents used are bifunctional compounds, but they may also be polyfunctional compounds. So bodies having multiple, identical or different functions can be used as crosslinking agents, such as: bisdiazobenzidines, bisdiazo-o-anisidine, biepoxides, chloro-5-triazines, diisocyanates, dialdehydes including glutaraldehyde, bismaleimides, ethylchlorocarbonates, carbodiimides.

Example 1: Several 0.05 mm thick sheets of Cellophane (Rhone Poulenc's 550 PTOO) were 50/50 impregnated with a solution in water of 2 mg/ml of glucose oxidase and 25 mg/ml of a mixture of albumin and fibrinogen for about 5 minutes. They were then dried by ventilation in a cold chamber. The operation was repeated 1 to 5 times to obtain sheets having different activities. The sheets were then impregnated with a 5 mg/ml solution of bisdiazo-o-dianisidine buffered to pH 6.8 by means of a 0.02 M solution containing 3 parts of monobasic sodium phosphate to one part of dibasic sodium phosphate. The glucose oxidase molecules which had not been immobilized were removed after the sheets had been passed through several agitated rinsing baths. The sheets were then ready to be used as membranes adapted to many uses.

The enzymic yield, that is the percentage of the activity introduced within the medium which was retained by the final structure, was found to be 65%.

The technique of the preceding examples was repeated using carbonic anhydrase, chymotrypsin and trypsin instead of glucose oxidase; comparable results were obtained. Such enzymes were also attached to other carriers such as cellulose, dextron and polyvinyl alcohol and similar results were obtained.

Example 2: 25 mg of glucose oxidase were solubilized in 0.7 ml of phosphate buffer at pH 6.8 and 50 mg of albumin were also solubilized in 0.7 ml of the same buffer. The two solutions were mixed and agitated until a homogeneous mixture was obtained. A solution of 2.5% by weight of glutaraldehyde was added, drop by drop with agitation. The solution so obtained was placed in a flat bottomed glass mold with 40 cm^2 surface. After one hour a 0.1 to 0.2 mm thick film was obtained and was kept in water to prevent it drying out. The enzymatic yield was found to be 80%.

When a carrier impenetrable to protein solutions and solutions of bridging agents was used, a detachable film was obtained that could be used alone or applied to another carrier.

Polymeric Membrane Coated with Inert Proteins

A microporous polymeric membrane coated with a thin layer of an inert protein is used by *C.J. Lai and S.M. Goldin; U.S. Patent 4,066,512; January 3, 1978; assigned to Millipore Corporation* for the immobilization of enzymes.

The membrane filters used in the process are thin porous structures composed of pure and biologically inert cellulose esters or similar polymeric materials. They are produced commercially in many distinct pore sizes, from 14 micrometers to 25 nanometers (0.025 micrometer), and in discs ranging from 13 to 293 mm in diameter. They can of course, be shaped in any desired configuration.

The pores in these membrane filters are extraordinarily uniform in size. For example, the total range of pore size distribution in one commercially available membrane filter with a mean pore size of 0.45 micrometers is plus or minus

0.02 micrometer. Each square centimeter of membrane filter surface contains millions of capillary pores which occupy approximately 80% of the total filter volume.

The biologically active agent, e.g., enzyme is bound to the porous polymeric filter by a material such as a water-insoluble protein. Two useful inactive proteins for binding enzymes and antibodies to the polymeric membrane are zein and collagen. Zein is the prolamin (alcohol-soluble protein) of corn. It is the only commercially available prolamin and one of the few readily available plant proteins. Collagen is a hydroxyproline, glycine-type protein, which is the chief organic constituent of connective animal tissue and bones. It can be obtained in good yields from a wide variety of mammal and fish parts.

A wide variety of enzymes can be bound to polymeric filter membranes with natural proteins such as collagen and zein. Compatible combinations of enzymes and multienzyme systems can also be complexed with the collagen in this manner. Examples include hexokinase with glucose-6-phosphate dehydrogenase and hexokinase with glucose oxidase.

Example 1: A zein solution was prepared by mixing 18 cc of ethanol, 34 cc of n-butanol, 8 cc of water, 3 cc of Cellosolve solvent, and 7.9 g of zein. The zein dissolves readily in these solvents and the solution can be prepared at room temperature (20°C). The solution is mixed for a sufficient amount of time to enable all the zein to be dissolved.

Thereafter, a microporous polymeric filter is soaked in the zein solution. A filter known as type SS filter which has an average pore size of 3.0 micrometers was used. The filter remained in the solution for approximately 24 hours at room temperature. The coated filter is then air dried.

Example 2: A collagen-coated filter is prepared from a collagen solution consisting of 0.7 g of acid-soluble collagen in 80 ml of 0.1 N acetic acid. A microporous polymeric membrane filter is soaked in the collagen solution for 24 hours. Prior to being complexed with an enzyme or an antibody, the collagen-coated filter is then air dried.

The procedure for complexing a biologically active agent such as an enzyme on to the protein-coated polymeric filter is quite simple. After the protein-coated filler has been prepared for complexing in accordance with the above procedure, the protein-coated filter is contacted with a solution of the enzyme for 48 hours. During this period, it is believed that secondary bonds that were previously formed between protein molecules (collagen-collagen bonds or zein-zein bonds) form between the enzyme and the inactive protein coating (collagen-enzyme bonds or zein-enzyme bonds). However, the exact mechanism by which the enzyme is bound to the filter is not precisely known.

The bond between the enzyme and the inactive protein coating is, however, strong. No chemical reactions that are detrimental to enzyme activity occur during complexing. Thus, an enzyme is firmly bonded to the filter with a minimum loss of activity.

Example 3: Six pieces of zein-DNase-complexed filter were prepared from a DNase solution of 5 mg/ml and stacked together. The amount of DNase leached out upon distilled water rinse leveled off rapidly as shown below:

10 cc Volume Aliquot of Distilled Water Passed Through (cc)	Amount of DNase Leached Out (mg)
10	4.20
20	0.90
30	0.40
40	0.12
50	0.01
60	–

A DNA solution (0.82 mg/cc) was then passed through the filter stack. The result is summarized below (no DNA was retained by the filter).

Flow Rate of Solution Through Filter (ml/min)	Percent of Phosphodiester Bonds Cleaved
0.250	54
0.167	78*
0.250	52
0.250	54

*Due to longer residence time

Reduced Keratin as Support

Difficulty has been encountered in preparing immobilized sulfhydryl-containing enzymes such as urease. Chemical crosslinking agents have a deleterious effect on the enzyme. For example, attempts at insolubilizing urease with such conventional agents as glutaraldehyde, cyanogen bromide, and substituted agarose gels were unsuccessful in that they substantially reduced the urease activity.

Insolubilized but active sulfhydryl-containing enzymes are prepared by *W.L. Stanley, G.G. Watters and B.G. Chan; U.S. Patent 4,069,106; January 17, 1978; assigned to the U.S. Secretary of Agriculture* by mixing an aqueous solution of the enzyme with reduced-keratin-containing material.

The keratin not only contributes to insolubilization of the enzyme but also provides useful physical properties to the product. In particular, the keratin acts as a support or carrier so that the insolubilized enzyme product forms a column through which water and other liquids can percolate readily. This is in sharp contrast to known insolubilized enzymes many of which are generally amorphous materials that cannot be used directly in a column because liquids will not flow therethrough. These known products require the addition of a carrier such as diatomaceous earth, crushed firebrick, or the like to provide a liquid-permeable mass.

A most important advantage of the process is that no crosslinking agent is used. The enzyme is reacted directly with the reduced keratin to produce the insolubilized enzyme product. Consequently, the activity of the enzyme is not destroyed or appreciably decreased.

Reduced keratin is prepared by treating keratin-containing material with a reducing agent. The keratin is dissolved in an aqueous alcoholic (ethanol, propanol, and the like) solution containing 0.25 to 3.0 parts of alcohol per part of water (volume to volume). Additionally, the aqueous alcoholic solution contains 1 to 10%, based on the amount of keratin, of a sulfur-containing, reductive disulfide-splitting agent, such as mercaptoethanol. The temperature of the reduction reaction is 50° to 110°C and the duration of contact is 20 to 80 minutes. Water is added gradually to the alcoholic solution to precipitate the reduced keratin in a granular form. Usually, about 0.5 to 3 parts of water are used per part of alcoholic solution.

During this precipitation it is desirable to vigorously agitate the material; thus, the solution should be stirred, shaken, rocked, etc., during the addition of water to achieve this agitation. It may be necessary to break up larger pieces of the granular product to pass through a 20 to 80 mesh screen. The product is washed thoroughly with water and collected by filtration, decantation, centrifugation, and the like.

Preferably, precipitation of the reduced keratin in granular form may be accomplished by cooling the hot alcoholic solution slowly with gentle agitation. In general, the hot alcoholic solution should be cooled to ambient temperature over a period of 1 to 2 hours. The so-precipitated material can then be treated as described above.

The keratin-containing material used in the process can be selected from a wide variety of sources, namely, poultry feathers, cattle hoof, hog hoof, animal hair, animal horn, animal hide, snake skin, etc.

The aqueous dispersion of the starting enzyme is added to the reduced keratin. Generally, 0.01 to 1 part of crude enzyme per 10 parts of reduced keratin are used. The mixture is gently agitated by conventional means such as shaking, stirring, or the like while being held for 1 to 48 hours at 1° to 25°C to cause the enzyme to react with the reduced keratin.

It is also possible to place the reduced keratin in a column and to percolate or pump the aqueous enzyme solution therethrough. The enzyme becomes immobilized on the reduced keratin and the column is then ready for use in enzymic reactions.

This process can be applied to all kinds of enzymes which contain sulfhydryl groups. Illustrative examples are urease, papain, yeast alcohol dehydrogenase, aspartate kinase, phosphofructokinase, carbamyl phosphate synthetase, fructose diphosphatase, ribulose diphosphate carboxylase, citrate cleavage enzyme, methionine-S-RN_4 synthetase, and the like. The starting enzyme need not be a purified substance but may be a preparation containing an enzyme.

Example: Preparation of Urease Immobilized on Poultry Feather Keratin – Powdered poultry feathers (2 g) were suspended in 100 ml of 50% aqueous 2-propanol solution and mercaptoethanol was added until the concentration was 2% by volume. The mixture was heated to 80°C with gentle stirring and held at that temperature for 30 minutes. Undissolved material was removed by filtering the solution through glass wool on a heated funnel. The opaque filtrate was allowed to cool to ambient temperature over a period of 1 hour with constant, gentle mechanical stirring. The reduced keratin precipitated in granular form, and the precipitate was washed with water and collected by filtration. Larger pieces were broken up to pass through a 20-mesh screen.

A 15 gram portion of the reduced keratin was packed with light tamping into a column 1.2 cm in diameter and 18 cm in length. A solution of 150 mg of urease in 50 ml of distilled water was prepared. The solution was made 10^{-3} M in sodium ethylene diamine tetraacetate. The mixture was centrifuged at 12,000 g for 20 minutes and the precipitate was discarded.

The supernatant solution (UV absorbance of 1.6 at 260 nm) was pumped through the above column until 30 ml (over one bed volume) had been used. The UV absorbance of the eluate was 1.3 at 260 nm. The column was washed with 50 ml of 0.1 M potassium chloride solution containing 10^{-3} M sodium ethylene diamine tetraacetate.

An aqueous urea solution (0.2 M in urea and 10^{-3} M in sodium ethylene diamine tetraacetate) was passed through the column at room temperature at a rate of 60 ml per hour. Five ml eluate fractions were collected. The extent of urea hydrolysis was determined by analyzing 1- or 2-ml aliquots of each fraction for ammonia with an automatic titrimeter. The results are summarized below:

Eluate fraction	1	2	3	4	5	6	7	8	9	10	11	12	13	14	15
Hydrolysis, %	2.7	25.7	79	85	73	67	63	62.5	57	57.5	59	55	53	53	59

Partially Hydrolyzed Collagen as Support

P.-M. Lin, J.R. Giacin, S.G. Gilbert, and J.G. Leeder; U.S. Patent 4,092,219; May 30, 1978 have found that the amount of enzyme complexed to a protein membrane is increased by treating the protein before or after forming the membrane with a proteolytic enzyme. Preferred proteolytic enzymes are pepsin, trypsin and pronase. In particular, the protein collagen has been found to be such a protein which is particularly susceptible to the process.

Specifically, a proteolytic enzyme which does not attack the well-ordered crystalline or helical region of the protein, but only the amorphous or telopeptide region thereof, is used.

It has been discovered that the enzyme binding capacity of the so-treated protein can be significantly increased, both with respect to freshly cast protein membranes and most significantly with respect to protein which includes substantial crosslinking, either by chemical or natural means, such as aging.

In a preferred embodiment, the protein collagen, which is a protein including a well-ordered crystalline region and a nonordered or amorphous region, is used, and various proteolytic enzymes other than the enzyme collagenase are used for treating the collagen. That is, collagenase will attack the ordered or crystalline region of collagen. In this embodiment, various other noncollagenase proteases may be employed, including pepsin, trypsin, pronase, etc.

The protein (collagen) membranes useful in the process will preferably be from 0.03 to 0.04 mm thick. The collagen membrane is thus prepared for complexing with enzyme, generally by being swollen with a low molecular weight organic acid, or in some instances with suitable bases so that the pH ranges from 2 to 12. Suitable acids include lactic acid and cyanoacetic acid. Swelling is accomplished by submerging the membrane in the acid bath for between ½ hour and 1 hour, depending upon the particular conditions of the bath, generally at room temperature.

Following the swelling treatment, the swollen collagen membrane is washed thoroughly with water until the pH level of the membrane is within the acceptable range for the particular enzyme being complexed.

Either prior to casting of the film, or preferably at this point in the membrane preparation, the protein is treated with the proteolytic enzyme, such as pepsin. Preferably, this step is carried out under very specific conditions. These include a temperature of 20° to 22°C and for a period of 2 to 3 hours. Also, the proteolytic enzyme is maintained in an aqueous solution at its respective optimum pH range. For example, when pepsin is used, the pH of the digestion solution is maintained from 3 to 3.5, and preferably for 1 to 12 hours; while with pronase the pH of the digestion solution is maintained at 7.2 for 1 to 6 hours. The prepared film may be aged for a considerable period of time prior to such treatment. Preferably films which have aged 3 to 200 days are particularly susceptible to the treatment in order to greatly improve their capacity for enzyme complexation.

Finally, the thus-treated protein membrane may then be complexed with the particular enzyme which is desired. For these purposes, the swollen, washed membrane is soaked in an aqueous enzyme-containing solution, until complexing occurs. Usually, this requires 24 to 30 hours. The temperature range used during this time should be maintained from 4° to 10°C, depending upon the particular enzyme employed. Maximum enzyme uptake is measured by activity after washing and indicates when complexing is complete. The enzyme-protein complex membrane should then be carefully dried, preferably at room temperature or below, so as not to damage the bound enzyme. A wide variety of different types of enzymes can be complexed with natural proteins such as collagen and the like in this manner, depending upon the particular application intended.

Especially suitable, however, are lysozyme, invertase, lactase, urease and amylases. Lysozyme is widely used to hydrolyze microorganisms in pharmaceutical research, and in sewage treatment, either alone or in combination with other enzymes, and/or bacteria. One particularly important application for lysozyme-protein membrane complex is in the lysis of cells.

Example: Initially, a film approximately 0.05 mm thick was prepared from the protein collagen. Specifically, a dispersion of collagen in water, at a pH of 3 was prepared. The dispersion, maintained at room temperature, was prepared by high agitation of the collagen-containing suspension for a period of 1 to 5 minutes, and at room temperature, following which the solution was deaerated. The collagen dispersion was then cast into a film by spreading the collagen dispersion upon a flat surface, and allowing it to dry for 24 hours at room temperature.

The collagen film was then sliced into a number of small sections, and aged for various lengths of time. Both the unaged film, and the film aged for various periods of time was then divided into two portions, one of which was then contacted with the proteolytic enzyme pepsin. This contacting was carried out in an aqueous solution at 22°C, and for 3 hours, at a pH of 3.5. Both the treated and untreated collagen films were then impregnated with enzyme and tested for apparent specific enzyme activity by contacting with the substrate in the following conditions: e.g., an aqueous lactose solution, at a pH of 7, and temperature of 37°C for 10 hours.

Enzyme activity was then measured in terms of micromoles of glucose produced per minute per gram of complex employed. Also, the actual amount of enzyme bound to the collagen was then determined by measuring the tryptophan content of the complex. The results obtained are contained in the following table.

Effect of Pepsin Treatment on Enzyme Binding Capacity of Collagen Films

	Age of Collagen Film, days								
	0	7	14	21	28	49	58	113	153
Apparent specific enzyme activity (μmol glucose/min/g complex)									
Run 1									
Untreated	431.26	325.38	236.08	247.33	148.62	–	–	5.55	7.05
Pepsin-treated	490.22	438.65	431.25	431.62	457.01	–	–	439.94	468.40
Run 2									
Untreated	589.90	–	–	406.12	–	329.50	210.82	–	–
Pepsin-treated	766.77	–	–	767.47	–	727.50	733.60	–	–
Amount of enzyme* (mg enzyme/g complex)									
Run 1									
Untreated	–	–	–	–	–	–	–	–	12.5
Pepsin-treated	–	–	–	–	–	–	–	–	51.5
Run 2									
Untreated	–	–	–	–	–	–	31.5	–	–
Pepsin-treated	–	–	–	–	–	–	207.5	–	–

*Average of two determinations.

They demonstrate that for freshly cast film while the enzyme activity of a pepsin-treated collagen film was greater than that of an untreated film, that after aging the significant maintenance of enzyme activity with the treated film as compared to the untreated film was rather dramatic. Furthermore, the amount of enzyme taken up by the treated collagen film was significantly greater than that taken up by the untreated film, particularly after aging had taken place.

Granular Casein Coated with Saccharifying Enzyme Crosslinked to Albumen

S. Amotz, T.K. Nielsen, P.B.R. Poulsen and B.E. Norman; U.S. Patent 4,116,771; September 26, 1978; assigned to Novo Industri A/S, Denmark have produced an immobilized saccharifying enzyme by coating granular casein with a liquid-permeable proteinaceous layer comprising saccharifying enzyme and egg albumen, crosslinked by reaction with glutaraldehyde.

It was found that the physical and enzymatic properties of the immobilized-enzyme-coated products were greatly improved, provided that granular casein is used as core material and, furthermore, that glutaraldehyde crosslinking of the enzyme in the coating layer is effected in the presence of the water-soluble, nonenzymatic binder protein egg albumen, such materials evidently constituting a favorable combination for creating a sufficient number of interprotein enzyme-stabilizing crosslinks.

The weight ratio of saccharifying enzyme:casein is in the range of 1:200 to 1:2, preferably 1:25 to 1:5, the amount of amyloglucosidase and maltogenic alpha-amylase being calculated on the basis of products having activities of 10,000 and 3,000 units per gram, respectively. Likewise, a preferred range of the weight ratio of saccharifying enzyme:egg albumen is from 1:2 to 1:0.2.

In further detail, this process comprises the step of wetting a dry mixture of granular casein and egg albumen powder with an aqueous mixture consisting of the saccharifying enzyme and glutaraldehyde dissolved at pH 4-7, with vigorous stirring; followed by the step of maintaining the resulting wetted mixture in a quiescent state, usually at ambient temperature, to complete the combined carrier coating and protein crosslinking process. Wetting may be conducted either manually, for example by thorough mixing and kneading in a mortar, or mechanically, for example in a plough-type horizontal mixer or a similar type of industrial mixing apparatus. Optionally, the albumen may be dissolved together with the enzyme and the glutaraldehyde instead of being mixed in the dry state with the casein.

The weight ratio of glutaraldehyde (usually added as a 50%, w/v, aqueous solution) to coating layer proteins (enzyme plus albumen) may vary considerably, the preferred range being 0.1 to 1. The water content of the aqueous solution is usually adjusted to be 30 to 60%, w/w, of the total mixture.

The casein granules used in this process should possess sufficient physical stability to resist substantial deformation upon soaking under packed bed column conditions. In addition, the degree of swelling in water should be reasonably low, preferably not exceeding 200%. Exemplary of a product meeting such requirements is acid-precipitated granular casein, having particle diameters of 100 to 500 microns. Scanning electron microscope examination reveals that the surface of such particles has a very uneven and irregularly folded structure, bearing some resemblance to the macroscopic appearance of pumice. It is believed that a surface microstructure of this type would expose a large number of glutaraldehyde binding sites. It is essential that a substantial fraction of the albumen used as binder protein be almost instantly soluble in water. Hence, the preferred grade of albumen is a generally available spray-dried product.

Example 1: A relatively coarse grade granular hydrochloric-acid-precipitated casein (9 g) consisting of 100 to 500 micron particles, the 300 to 500 micron fraction being about 60%) was mixed with commercial spray-dried egg albumen (1.2 g). To this mixture was added a premixed solution of amyloglucosidase powder (6.5 ml of 18.5%, w/v, aqueous solution containing a total of 12,500 amyloglucosidase units) and glutaraldehyde (1.2 ml of 50%, w/w, aqueous solution). The amyloglucosidase powder was an ultrafiltrated, spray-dried product. The mixture was kneaded carefully in a mortar and then allowed to stand at room temperature for 1 hour. The resulting aggregate was disintegrated by mortaring to form a granular product which was allowed to dry at room temperature for one day. The dried granular product (11.4 g) had an activity of 425 units per gram, representing a recovery after immobilization of 38.8%.

Example 2: Granular acid-precipitated casein (1,500 g of type M60, consisting of 100 to 500 micron particles, the 150 to 350 micron fraction being about 70%) was mixed in the dry state with commercial spray-dried albumen (120 g) in a 20 liter plough-type horizontal mixer.

An ultrafiltrated concentrate of amyloglucosidase, (650 g of a solution having a dry matter content of about 25% and containing about 3,200 units of soluble amyloglucosidase per g of concentrate) was mixed thoroughly with glutaraldehyde (180 ml of a 50% w/w aqueous solution) at pH 4.9 and a temperature of 18°C, whereafter the resulting solution was poured into the mixing apparatus. Vigorous mixing was continued for an additional 0.5 to 1 minute, followed by removal of the wetted particulate product from the mixer. The moist product was left quiescent for 45 minutes to complete the crosslinking, whereby lumps of aggregated particles were formed. Granulation of the product was conducted on an oscillatory granulator provided with a 1.5 mm diameter screen. The granulate was washed with deionized water (10 liters) for 10 minutes, recovered by filtration and then subjected to fluid-bed drying. The dried product (1,600 g, 450 amyloglucosidase units per gram) was freed of fines by sieving.

Polypeptide Azides as Supports

In the process developed by *Y. Yugari and Y. Minamoto; U.S. Patent 3,985,617; October 12, 1976; assigned to Ajinomoto Co., Inc., Japan* polypeptides having repeating acidic amino acid groups with carboxyl groups as side chains are used to prepare immobilized enzymes. The carboxyl groups are converted to carbonyl azides which can then react with a biologically active protein such as an enzyme having an amino acid unit including an amino group in position α or ω. The carbonyl azide groups can also form crosslinks with unconverted carboxyl groups on the polypeptide molecules.

There are many methods for converting a carboxyl group to an acyl azide group, and most can be applied to the process. In one of the most suitable methods, a polypeptide containing repeating units of a lower alkyl ester of an acidic amino acid is reacted with hydrazine, and the hydrazide produced is reacted with nitrous acid until the hydrazide groups in a portion of the units are converted to azide groups.

A polypeptide azide may also be produced by reaction of a polyamino acid chloride with sodium azide. The biologically active proteins to which the process is applicable include enzymes, such as urease, uricase, urokinase, amino acid acylase, aspartase, amylase, lipase, glucose oxidase and protease, natural proteins of animal and plant origin, and natural peptides, such as antigens, antibodies and peptide hormones.

These biologically active proteins may easily be immobilized by immersing a polypeptide azide in solutions of the proteins buffered to a pH at which the activity of the active protein to be immobilized is not adversely affected, and whose temperature is kept below 40°C, preferably below 10°C, and by moderately stirring the mixture for 6 to 24 hours. The polypeptide on which a biologically active protein is immobilized may have any shape and be a membrane, tube, fiber, porous shaped body, bead or viscous liquid.

In order to enhance the mechanical strength of the composition, the polypeptide may be deposited on various carriers. When the coated carrier is immersed in a solution, the coating may peel off unless fastened to the surface of the carrier by an adhesive. The adhesive may also fix water-soluble constituents of the polypeptide. Since the original polypeptide is usually a linear polymer and is appreciably water-soluble, some constituents which are only sparingly crosslinked and whose carboxyl groups are not fully converted to carboalkoxy groups may be dissolved even when combined with biologically active proteins. Suitable adhesives include those capable of crosslinking, such as polyurethane resin, epoxy resin and polyester resin, and those incapable of crosslinking, such as polyvinyl chloride, and polyvinyl acetate.

The carriers capable of being coated with the polypeptide include beads of glass, synthetic resins, such as acrylic resin and vinyl chloride resin, or stainless steel, and the inner walls of glass and stainless steel tubes.

The adhesive is first applied to the surface of the carrier, a solution or a thin layer of the polypeptide is deposited on the coated surface, and the coating is dried. Alternatively, a solution of the polypeptide is first mixed with the adhesive, and the mixture is deposited on the surface of the carrier and dried. When a carrier is not used, the adhesive is still effective for fixing the water-soluble constituents of the composition and for increasing its mechanical strength.

Example: Small pieces of copolymer of L-methionine and L-glutamic acid γ-methyl ester (mol ratio 1:1) weighing 1 g were suspended in 10 ml methanol, 2 ml 80% hydrazine hydrate were added to the suspension, and the mixture was kept at 50°C for 2 hours. The hydrazide produced was filtered off and washed four times with 50% methanol. Thereafter, the washed product was immersed in 30 ml of a mixture of equal volumes of methanol and 0.5 N hydrochloric acid, allowed to stand for a few minutes, and filtered off. The immersing procedure was repeated once more, the hydrazide product was washed with methanol and dried. The hydrazide weighed 0.8 g and contained 1.1 meq/g hydrazino groups.

100 mg hydrazide was suspended in 20 ml methanol, cooled in an ice bath and 10 ml chilled 0.5 N hydrochloric acid, and then 2.0 ml chilled aqueous 3% sodium nitrite solution were added to the suspension. The mixture was stirred for 20 minutes at 0° to 4°C. The azide produced thereby was recovered by centrifuging at 3,000 rpm for 2 minutes and washed three times with chilled 50% methanol and subsequently twice with chilled water.

The recovered azide was immersed in 3.0 ml chilled 0.1 M phosphate buffer solution of pH 8.7 containing 2.55 mg/ml urease (Sigma Chemical Co., Type III, 2.2 U/ml) and 1 mM EDTA, and the mixture was stirred at 4°C for 48 hours. The resulting composition containing immobilized urease was centrifuged off, and washed successively once with 0.05 M phosphate buffer solution of pH 7.0, twice with the same buffer solution additionally containing 1 M sodium chloride, and twice with the same buffer solution without sodium chloride. The total volume of the filtrate was about 45 ml. The urease activities of the immobilized enzyme and of the filtrate were 8.6 U (86 U/g carrier) and 8.0 U, respectively. One unit of urease activity produces 1 mg nitrogen as ammonia from 0.25 M urea solution of pH 7.0 at 25°C in 5 minutes when the ammonia is colorimetrically determined by the indophenol method. In a comparison test, the same urease was immobilized on a carboxymethylcellulose in the manner described in *Nature,* 189, p. 576 (1961).

On the carboxymethylcellulose azide so produced, the urease was immobilized in the same manner as described above with reference to the copolymer of L-methionine and L-glutamic acid γ-methyl ester. An immobilized enzyme having 3.3 U (33 U/g carrier) of urease activity and a filtrate having 12.0 U of urease activity were obtained.

When the same urease was immobilized on polymethacrylic acid in the same manner as described with reference to CMC, the urease activities of the immobilized enzyme and of the washing solution were 1.4 U (14 U/g carrier) and 12.3 U, respectively.

BONDING ENZYMES TO CELLULAR MATERIALS

Covalent Linking of Isomerase to Cell Walls

Glucose isomerase is essentially an intracellular enzyme. That is, the enzyme is found primarily within the microorganism cell or tightly adherent thereto. Therefore, the enzyme usually must either be released from the cell by breaking the cellular structure or the entire cell must be used as the enzyme source. Although the former process can be carried out to produce an extracellular enzyme by disintegration of the cell with a sonic or supersonic disintegrater, this additional procedure is time-consuming and frequently difficult to perform on a practical scale.

The primary problem with using whole cells in the isomerization reaction is that the enzyme preparation tends to be unstable during multiple conversions and therefore cannot be used economically for recycling. Moreover, when live whole cells are used, undesirable side reactions occur during the isomerization due to other enzymatic material present, whereby products other than fructose are produced.

In the process developed by *G.J. Moskowitz; U.S. Patent 3,843,442; October 22, 1974; assigned to Baxter Laboratories Inc.* these disadvantages are overcome by covalently linking the enzyme to the cellular structures and other cell components within the cell itself.

The crosslinking reagent used in this process is a diazotized primary diamino compound which can contain, for example, a benzene, benzidine, pyridine, pyrimidine, purine, acridine, thiazole, oxazole, benzothiophene, cumarone, phenothiazine, or phenoxazine ring structure, and preferably, is a diazotized heterocyclic primary diamino compound having from one to three rings. The preferred crosslinking reagents are diazotized 2,6-diaminopyridine; 2,4-diamino-6-hydroxypyrimidine; 3,6-diaminoacridine; acriflavine; and other diazotized compounds containing a pyridine, pyrimidine, or acridine moiety.

This process can be carried out with wet living microbial cells freshly isolated from the fermentor, with solvent (e.g., acetone or methanol) dried cells or with a heterogeneous mixture of proteins such as occurs in the fermentation culture. The covalent linking of the enzyme to the cellular structure and other cell components is carried out by incubation of the whole cells with the crosslinking reagent, preferably at 0° to 25°C for 1 to 100 hours. Incubation at 4°C for 48 hours is optimum.

Although the glucose isomerase enzyme or enzyme preparation can be obtained from numerous microorganisms, for this process it is preferable to obtain the enzyme from streptomyces and lactobacillus species, for example, *Streptomyces phaeochromogenes* and *Lactobacillus brevis.*

Example: An immobilized whole cell preparation of glucose isomerase is produced as follows. All solutions are cooled to 0°C unless otherwise stated. 1.5 g of 3,6-diaminoacridine are dissolved in 235 ml of normal HCl. After stirring the solution for five minutes at 0°C, 56 ml of 0.5 molar $NaNO_2$ is added to effect the conversion of the primary amino groups to diazo groups. The mixture is stirred at 0°C under vacuum for one hour. The pH is then adjusted to 8.5 with 6 N NaOH and sufficient $CoCl_2$ is added to provide 0.001 M final concentration. Acetone-dried whole cells (3 g) are suspended in 100 ml of cold 0.1 M Na_2CO_3 solution containing 0.001 M $CoCl_2$. The pH is adjusted to 8.5 with HCl and then the cell mixture is added slowly with stirring to the above prepared diazoacridine and stirred at 4°C for 18 hours.

The resulting precipitate of immobilized whole cells is isolated by centrifugation with a yield of 18% of the initial activity (prior to immobilization), washed three times with 300 ml of water, and a portion used for glucose conversion by suspending in a reaction mixture containing 8 g of glucose, 0.03 M tris acetate [acetate of tris(hydroxymethyl)aminomethane], pH 8.0, 0.06 M $MgSO_4$ and 0.0006 M $CoCl_2$ in a final volume of 32 ml. The reaction mixture is incubated with shaking at 60°C for 24 hours. The immobilized enzyme is then isolated by centrifugation and resuspended in a fresh reaction mixture and the entire conversion procedure is repeated several times for a total of thirteen conversions.

In the foregoing conversion procedure, a sample containing 3 ml of 25% trichloroacetic acid is used as a zero time control. Untreated acetone-dried cells incubated in 0.1 M Na_2CO_3, pH 8.5, containing 0.001 M $CoCl_2$ are used as a control sample for comparison. In the conversion, conditions are selected which do not achieve equilibrium in order to detect small changes in enzyme activity during recycling of the enzyme. Therefore, an amount of enzyme is selected to produce 25 to 30% conversion of glucose to fructose in the indicated time. A mixture of 42 to 45% fructose and 58 to 55% glucose is the equilibrium mixture for the enzyme.

The assay for the conversion is carried out substantially according to the method of Tsumura and Sato, *Agr. Biol. Chem.,* Vol 29, p 1129 (1965) and the fructose is measured by the cysteine-carbazole method of Dische and Borenfreund, *J. Biol. Chem.,* Vol 192, p 583 (1951). In this assay, enzyme is incubated for 30 minutes at 60°C in a solution containing 1.0 M glucose, 0.1 M $MgSO_4$, and 0.001 M $CoCl_2$ in a final volume of 2.0 ml. A solution of 25% trichloroacetic acid (0.2 ml) is added to stop the reaction and the fructose produced is then measured. In this assay, one unit is defined as the amount of enzyme required to produce one micromol of fructose in the reaction mixture.

The table below sets forth the degree of conversion obtained with the crosslinked (immobilized) cells in comparison to the degree of conversion obtained with the untreated acetone-dried cells over thirteen conversion periods. Conversion number 1 is taken as the initial degree of conversion of the system and all subsequent conversions are expressed as a percentage thereof.

From the table results it can be seen that the immobilized cells showed a marked increase in stability as compared to that of the untreated cells. The untreated cells lost the greater part of their activity in the second conversion, whereas the immobilized cells retained their full activity throughout five conversions, and even after thirteen conversions their activity was still more than double that of the untreated cells after only two conversions.

	Conversion Number												
	1	2	3	4	5	6	7	8	9	10	11	12	13
Immobilized cells, degree of conversion per cycle	100	104	113	100	109	74	78	74	61	44	48	44	30
Untreated cells, degree of conversion per cycle	100	13*	-	-	-	-	-	-	-	-	-	-	-

*Discontinued after two conversions.

Glucose Isomerase Bonded to Homogenized Cell Mass

The process developed by *S. Amotz, T.K. Nielsen and N.O. Thiesen; U.S. Patent 3,980,521, September 14, 1976; assigned to Novo Industri A/S, Denmark* involves treatment of a bacterial cell concentrate, where not more than 75% of the cells are intact cells, with glutaraldehyde to convert the cells into a tough, dimensionally stable, controlled-particle-size enzyme product. The ruptured cells provide sufficient material reactive with glutaraldehyde to convert the concentrate into a coherent reaction product, which for example may be a gel. The coherent reaction product is dewatered, e.g, by drying, and shaped into a suitable size subdivided form greatly exceeding 10 microns in size.

The bacterial cells may be cultivated according to procedures known suitable for production of cells of high glucose isomerase activity. The cells are appropriately separated from the fermentation broth by filtration, centrifugation, etc., to form a concentrate containing from 3 to 30% weight by volume dry matter. In the process, the presence of autolyzed and disrupted cells and even free enzyme in the cell concentrate is crucial, therefore permitting concentration of the microbial cells by large scale commercial equipment such as self-cleaning sludge centrifuges. Relatively harsh handling conditions, even autolysis due to processing delays are acceptable.

Indeed, if a substantial degree of cell rupture, autolysis, etc., does not occur during the course of recovery and concentration, then rupture is deliberately caused to the point where the concentrate contains not more than about 75%, preferably less than 60% whole cells. Complete or 100% rupture of the cells is contemplated for this process.

Reaction with glutaraldehyde is carried out in an aqueous suspension of fragmented cells and likely some of the liberated cellular constituents, including glucose isomerase itself are in solution. Accordingly, cell disruption or fracturing should be carried out only after the cells have been concentrated beyond their usual (dilute) content in the fermentation medium. In practical terms this means that the microorganism is separated from its growth medium, e.g., centrifuged off, as a bacterial cell concentrate containing 3 to 20% dry matter. Then as incident to the concentration, or subsequent thereto, the desired disruption of the cells is carried out, as for example by autolysis or by homogenization.

The cell concentrate being immobilized and crosslinked by reaction with glutaraldehyde has then from 0 to 75% whole cells and a dry matter content of 3 to 30% by weight. Whatever glucose isomerase has been liberated in soluble form remains in the concentrate to become incorporated into the enzyme product. The dry matter is the residue left behind by drying at 105°C for 16 hours. Drying 24 hours at 60°C under vacuum will give an alternative dry matter content measurement.

The convenient range of ratios of the dry weight of the glutaraldehyde to the dry matter in the starting material is found to be preferably from 0.05 to 0.3. The relative proportion of glutaraldehyde to be employed may well vary from microorganism to microorganism, and may even require tailoring to fit different industrial scale glucose conversion installations.

In one preferred embodiment the starting material which may be either a concentrate or a homogenate is also treated with a flocculant so as to form aggregates and enhance the crosslinking reaction. In another preferred embodiment, the crosslinking mixture is frozen so as to form an inhomogeneous gel which leads to a more porous product.

Example 1: Preparation of a Concentrate and its Immobilization — Isomerase-containing cells of *Bacillus coagulans* NRRLB 5656 are cultivated, then the cells are recovered from the fermentation broth by centrifugation and the pH adjusted to 6.3. The concentrate is estimated to contain approximately 10% dry matter, and about 40% intact cells.

To 1 kg of the concentrate, 38 ml of commercial 50% glutaraldehyde was added with sufficient stirring to thoroughly mix glutaraldehyde and cell concentrate. Thereafter the reaction mass was left at ambient in a quiescent state. After an hour, the reaction mixture had gelled into a coherent mass approximately the consistency of cheese curd. The mass was broken up by mild stirring, and washed with two volumes of deionized water, and the water drained away.

The gel pieces were then transferred to a vacuum drum-dryer where the approximately 1 kg weight was dehydrated at 50°C to a weight of about 160 g. During the course of dehydration, the soft gel pieces became converted into a tough, dimensionally stable material. The dehydrated pieces were further comminuted into particles less than 1 mm diameter. The enzyme recovery will vary batch to batch from 50 to 60% of the initial activity. However, about 15% by weight of the product constitutes excessively fine material (of 1 to 70 microns).

Example 2: Immobilization with Freezing – The procedure of Example 1 was followed up to the preparation of the glutaraldehyde cell concentrate mixture and quiescent standing preparation of a gelled mass. Thereafter the gel (in its container) was transferred to a deep freeze and left overnight. The next day the frozen gel was thawed to ambient temperature, resulting in a watery mass. More water was added with stirring then the water drained off. The product was then dried in the rotary drier to a weight of about 160 grams. The ultimate product consisted of spongy particles, somewhat flaky in appearance. The recovery of apparent activity varied batch to batch between 60 to 70%. However, virtually no fines are produced.

Example 3: Immobilization with a Flocculant – 1,550 liters of a culture broth from a fermentation of glucose-isomerase-producing *Bacillus coagulans* NRRL B 5656 were concentrated by centrifuging at 10°C to give a sludge containing about 12 g dry weight (105°C) per 100 ml of concentrate.

11 kg of this concentrate (pH 7.9) was left at 20°C for 3 hours at room temperature with a mild stirring in order to let the *Bacillus coagulans* cells autolyze. The pH was adjusted to 6.5 with diluted acetic acid. To the sludge containing more than 70% of the activity in a soluble form, 330 ml of a 50% glutaraldehyde solution was added to give a concentration of about 1.4% of glutaraldehyde w/v in the reaction mixture. After one hour the partly crosslinked gel was agitated vigorously after addition of 20 liters of deionized water. To the suspension was added 80 ml of a 30% solution of Drewfloc EC 25 to make a clear solution.

The suspension was filtered to remove as much water as possible. The filtercake was dried in vacuum at 35°C. The dried cake was ground to a particle size of less than 300 μ. The activity was determined by isomerization in batch with a spray-dried powder made from the concentrate as a reference. The conditions were pH = 7.0, 65°C; $CoSO_4 \cdot 7H_2O$, 0.1 g/l; and $MgSO_4 \cdot 7H_2O$, 2.0 g/l; and 40% glucose w/v. The feed was flushed with nitrogen. The apparent activity of the immobilized enzyme was over 70% of the reference. After use the immobilized enzyme was removed by filtration and reused. This was done five times. After five successive uses the activity had not shown a significant drop.

A second portion of the autolyzed cell concentrate was treated with 20 ml of 30% Drewfloc EC 25 per kg of sludge before the reaction with 1.4% w/v glutaraldehyde. Essentially the same product resulted.

Diisocyanate-Modified Microbial Cells as Supports

A primary object of the disclosure by *Y. Takasaki; U.S. Patent 3,950,222; April 13, 1976; assigned to Agency of Industrial Science & Technology, Japan* is to provide a method of immobilizing of enzymes on microorganic cells such as molds and actinomycete cells. The process comprises the steps of treating microorganic cells or microorganic cells which have undergone various forms of treatments with a reagent having at least two functional groups and allowing an enzyme to react with resultant microorganic cells which have functional groups in a free state. This process is particularly useful for the immobilization of glucoamylase for conversion of starch to glucose. If a glucoamylase is immobilized by this method on the cells containing glucose isomerase, fructose can be directly produced from starch by use of the enzyme bonded cells.

Various other enzymes such as α-amylase, β-amylase, isoamylase, pullulanase, invertase, galactosidase, cellulase, protease, lipase, glucose-oxidase and catalase can likewise be immobilized. Not merely a single enzyme but two or more enzymes can be simultaneously immobilized on the cells. Such combinations are glucoamylase and glucose isomerase, β-amylase and isoamylase (or pullulanase), and α-amylase and glucoamylase, for example.

For the purpose of this immobilization, these enzymes are not required to be in a purified form but may be in a crude form containing extraneous proteins.

Molds and actinomycetes are particularly suitable microorganism host cells for immobilization of enzymes. This is because these microorganisms generally form relatively large pellets and, therefore, permit the preparation, recovery and other steps of handling immobilized enzymes to be carried out easily and economically. Other reasons are that these carriers are obtained very inexpensively, that they have appropriate degrees of specific gravity, that they are sufficiently dispersible in reaction liquids, and that they have large surface areas and therefore permit proportionately large quantities of enzymes to be bonded thereto.

Examples of the molds which are usable include those of genera *Rhizopus, Aspergillus and Penicillium.* Examples of the actinomycetes which are also usable include those of genera *Streptomyces* and *Actinomyces.* The host cells of these microorganisms may be those which occur as wastes from certain kinds of fermentation. In immobilizing glucoamylase, for example, if this enzyme is immobilized onto glucose-isomerase-containing microorganic cells, the immobilized enzyme can be employed for producing fructose directly from starch.

The cells to be used for the immobilization of an enzyme are either subjected to heat treatment at a temperature of 50°C or over, generally in the range of from 50° to 100°C, for a period of from several minutes to several hours or treated with a dilute acid or alkali so as to have the autolytic activity of cells destroyed or to be freed of an easily separable cell component. If occasion demands, however, microorganic cells in a form still retaining physiological activity may be used, and cell walls of microorganisms may also be used.

Examples of the crosslinking reagents having at least two functional groups which can be used for this process include tolylene diisocyanate, tolylene-2,4-diisocyanate, xylene diisocyanate, m-xylene diisocyanate, epichlorohydrin, glutaraldehyde, phenol-disulfonyl chloride, bis-diazobenzidine, toluene-2-isocyanate-4-isothiocyanate, nitrophenyl chloroacetate, 2-amino-4,6-dichloro-S-triazine and 2,4,6-trichloro-S-triazine. Particularly useful reagents are isocyanate compounds such as tolylene-2,4-diisocyanate.

Example: Streptomyces albus (ATCC 21132) was inoculated to a liquid medium (pH 7.2) containing 0.5% of xylene, 4% of corn steep liquor and 0.024% of $CoCl_2 \cdot 6H_2O$ and aerobically cultivated at 30°C for 24 hours. At the end of the culture, the microorganic cells were collected, washed with water and thereafter suspended in water and subjected to heat treatment at 65°C for 15 minutes. To 2 ml portions of the resultant cell suspension were respectively added tolylene diisocyanate, xylene diisocyanate, epichlorohydrin and glutaraldehyde, each in a volume of 0.2 ml.

The mixture was stirred at 30°C for 1 hour. Consequently, there were obtained cells of *Streptomyces albus* having thereon tolylene diisocyanate, xylene diisocyanate, epichlorohydrin and glutaraldehyde respectively. The cells thus produced were found to possess free functional groups capable of bonding various enzyme proteins in high yields. To these cells of the *Streptomyces* having the free functional groups retained thereon, 1 ml (9,200 units/ml) of glucoamylase produced by a genus *Rhizopus* was added and incubated at 30°C for 30 minutes. Thereafter, the cells were deprived of excess enzyme, washed thoroughly with water and analyzed to determine the quantity of glucoamylase bonded thereto. The results thus obtained are shown in the table below.

Functional Reagent	Glucoamylase Immobilized (units)
Tolylene diisocyanate	287.1
Xylene diisocyanate	188.8
Epichlorohydrin	165.5
Glutaraldehyde	20.4

Invertase and β-amylase were also coupled to the support of the above example in additional examples of the complete specification. Also *Aspergillus niger* and *Rhizopus oryzae* were used as support cells in other examples.

COVALENT BONDING TO CARBOHYDRATES

CYANOGEN-HALIDE-ACTIVATED CARBOHYDRATES

Cyanuric-Chloride-Modified Sheets or Membranes

M.D. Lilly, G. Kay, R.J.H. Wilson and A.K. Sharp; U.S. Patent 3,824,150; July 16, 1974; assigned to National Research Development Corporation, England have disclosed a process using a supported (insolubilized) enzyme and the preparation of the supported enzyme using cyanuric chloride. This process uses a support in the form of a permeable or impermeable sheet to produce a chemically modified stable intermediate containing reactive bridging groups and then reacting an enzyme with the reactive bridging groups.

The term sheet is used in the sense of a body whose breadth is large in comparison with its thickness. Thus, the insoluble support can be, for example, a pliable or rigid sheet or membrane made of natural or artificial material that is insoluble in the liquid medium in which the enzymatic reaction is carried out and for example, it can comprise a natural or synthetic polymeric material particularly a hydrophilic material, for example one having free hydroxyl groups such as cellulose, cellulosic material, crosslinked dextrans, starch, dextran, proteins such as wool, or polyvinyl alcohol.

The material may be woven, laid down as a nonwoven fabric, cast or extruded. It is particularly preferred that the sheet be permeable and preferably have pores which are large enough to allow solution molecules to pass through the sheet substantially without steric hindrance. A suitable average pore diameter is often greater than 10^{-5} cm, particularly between 10^{-5} and 10^{-2} cm such as for example about 10^{-3} cm.

Methods for treating an insoluble support so as to produce a stable intermediate containing reactive bridging groups include, for example, forming carboxymethyl groups on a polymer such as cellulose, esterifying the pendant carboxymethyl groups, and then reacting them with hydrazine and nitrous acid to form the azide derivative. This can then be reacted with the enzyme. A particularly preferred stable intermediate is one that is produced by reacting a triazinyl compound with a polymer so as to produce a stable intermediate having attached reactive bridging groups of the formula:

```
      N
     / \\
  -C     C-X
   ||    |
   N     N
    \   //
      C
      |
      Y
```

where X is a reactive radical, e.g., halogen, and Y is halogen or a nucleophilic substituent that is an amino group or an aliphatic or aromatic group.

Example: A twill filter cloth was washed thoroughly in water, detergent, sodium carbonate and sodium hydroxide (1 N) to remove noncellulosic material. After soaking in sodium hydroxide for 30 minutes the excess sodium hydroxide solution was removed. Cyanuric chloride (15 grams) was dissolved in 500 ml of acetone. The cloth was added to this solution followed by water (250 ml). After agitation for 5 minutes, the cloth was removed and washed in acetone-water (1:1 v/v) for 10 minutes. The washing was repeated until the smell of cyanuric chloride could no longer be detected. Excess moisture was removed by using absorbent paper.

Chymotrypsin (0.25 gram) was dissolved in water and this solution was poured over those parts of the cloth through which the substrate was to pass. 0.1 M phosphate buffer (pH 7, 25 ml) was then added. After about 5 minutes at 20° to 25°C, excess ammonia/ammonium chloride buffer (pH 8.6) was added. This mixture was left at 2°C overnight. The cloth was washed with large amounts of 1 M chloride. The cloth was fitted into a small filter press to give a cross-sectional area of about 100 cm^2. A low molecular weight substrate of acetyl-tyrosine ethyl ester, at a concentration of 10 millimolar, was pumped through the filter cloth at various flow rates. The results for the percentage of ester hydrolyzed at different flow rates are shown in Figure 8.1.

Figure 8.1: Flow Rate vs Percentage Ester Hydrolyzed

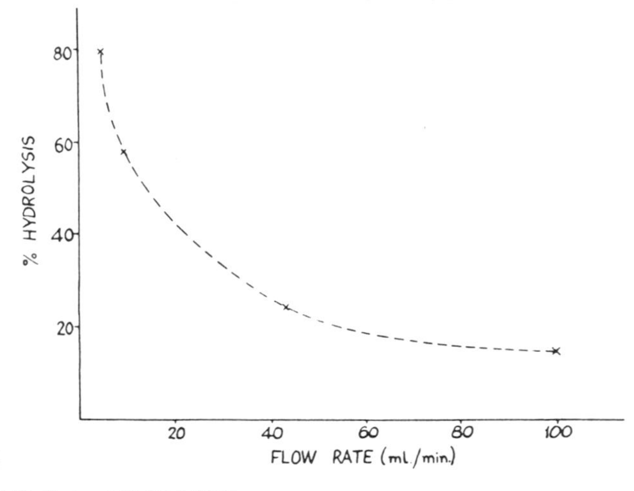

Source: U.S. Patent 3,824,150

Chymotrypsin, lactic dehydrogenase and penicillin amidase were also coupled to filter paper in other examples of the complete patent.

Modifying Dextran with Cyanogen Bromide in Nonaqueous Solvent

In the process developed by *M. Hanushewsky; U.S. Patent 3,876,501; April 8, 1975; assigned to Baxter Laboratories, Inc.* water-soluble carbohydrates can be activated using cyanogen bromide in a nonaqueous medium. The product has an improved level of binding reactivity toward enzymes and other proteins without crosslinking and consequent water-insolubility in the carbohydrates.

The steps of this process comprise combining the carbohydrate with a cyanogen halide in the presence of a nonreactive, nonaqueous solvent such as a lower alcohol of no more than 3 carbons, in the presence of an alkali of the formula MOR in a concentration sufficient to provide a pH of at least 10 if the same concentration of alkali were added to deionized water. (In MOR, M is an alkali metal and R is a lower alkyl radical or hydrogen.) The reaction should take place for a period of time sufficient to permit complete activation, which is relatively rapid. A period of time ranging from 10 to 40 minutes is usually sufficient to permit the activation reaction to go to completion. The cyanogen halide used is typically cyanogen bromide, but other cyanogen halides such as cyanogen iodide or cyanogen chloride can be used.

The nonreactive nonaqueous solvent used is most preferably absolute, i.e., essentially free of water as an impurity. The presence of water appears to promote undesirable side reactions which cause the insolubilization of the water-soluble carbohydrate and a reduction in the yield or activity of the product of the reaction. Lower alcohols such as methanol, ethanol, isopropanol, n-butanol, ethanediol, or propanediol can be used. The lowest molecular weight alcohols such as methanol are preferred for use in the reaction mixture at 20 to 80 ml of the nonaqueous solvent per gram of carbohydrate present. The alkali having the formula MOR can be an alkali metal alkoxide or hydroxide such as sodium methoxide, potassium ethoxide, potassium butoxide, lithium hydroxide, or sodium hydroxide.

The carbohydrates used can be dextran, soluble starch, soluble glycogen, or the like, and generally should have a molecular weight of at least 5,000. It is most preferable to use materials such as dextran of a molecular weight of 40,000 to 100,000, particularly Dextran 70 (MW 70,000), which is a nontoxic drug already used in various forms of therapy. However, soluble dextrans having molecular weights of 500,000 and above can be used.

Example 1: To 50 ml of absolute methanol was added 4 ml of a 25% solution of sodium methoxide in methanol. The beaker containing this mixture was packed in dry ice to cool it in an anhydrous manner, and then one gram of cyanogen bromide was added, followed by one gram of Dextran 70 (molecular weight 70,000). The reaction mixture was stirred for 20 minutes, at the reduced temperature of 0° to -10°C imparted by the dry ice. The product, a white solid, was recovered by filtration, followed by washing with methanol and drying for one hour in a vacuum in a container containing a drying agent (calcium chloride) separate from the activated dextran sample. The above reaction was repeated several times, as needed, to provide sufficient activated dextran for the binding reactions described below.

Portions of the activated dextran prepared above and samples of streptokinase enzyme were added to separate portions of 0.1 M sodium bicarbonate aqueous buffer, and thereafter combined with stirring at a temperature of about 4°C, to provide a reaction sample having a solution volume as indicated in the table below and concentrations of dextran and streptokinase as similarly indicated below. The weights of streptokinase present were calculated from a streptokinase activity of 80,000 streptokinase units per milligram of enzyme added. The reaction mixtures were maintained at 4°C for the number of hours indicated below.

After the reaction time, the dextran-streptokinase was isolated by column chromatography through a glass bead column, and the fractions containing the bound enzyme were lyophilized to give a solid product having the number of streptokinase units per milligram as indicated. The maximum theoretical yield of streptokinase units per milligram is also indicated, as is the percent of nitrogen found by analysis of some products.

Run	Total Volume of Reaction Mixture	Concentration, mg/ml		Binding Time, hr	Streptokinase . .Units/mg Product. .		Percent Nitrogen
		Dextran	Streptokinase		Found	Theoretical	
1	25	40	3.3	63	7,125	7,560	1.13
2	19	40	2.3	63	4,900	5,100	1.07
3*	12.5	80	4	17	3,750	4,100	1.18
4*	12.5	80	4	41	3,150	4,100	0.94
5	25	150	2	17	1,275	1,450	–
6	25	40	4	17	7,750	9,550	–
7	25	40	1	17	1,750	2,156	–

*These particular runs were prepared from a single reaction mixture and not by the separate dissolving of dextran and streptokinase in separate aliquots of buffer.

It can be seen that a very high percentage of the streptokinase used in this reaction mixture becomes chemically bonded to the dextran without loss in enzymatic activity. In fact, with respect to Run 1, the specific activity of the matricized product expressed in units of streptokinase activity per microgram of chemically bound nitrogen appears to be actually higher than pure streptokinase.

Example 2: About 7.8 mg (1,560 IU) of L-asparaginase were dialyzed in and against 0.1 M sodium phosphate buffer at a pH of 7.3. Following this, the asparaginase solution was diluted to 25 ml in the same buffer.

One gram of the activated dextran of Example 1 was added to this solution, with stirring, and allowed to react at 5°C for 41 hours. The resulting product was fractionated by a glass bead column, and the fractions yielding chemically bonded dextran-asparaginase were lyophilized to recover 1.19 grams of product having an enzyme activity of 0.717 IU of asparaginase activity per mg, about a 55% yield of bound enzyme.

Porous Cellulose Membranes Activated with Cyanogen Bromide

Known processes for preparing immobilized enzymes have not provided products with high uniformity and high activity. In prior processes, the FOM reported frequently is that corresonding to 3% loading and 3% activity or FOM equals 9×10^{-4}. The Figure of Merit (FOM) of an enzyme reactor is the weight or volume fraction of the reactor which is enzyme, times the ratio of the activity of the same native enzyme in solution at comparable pH and temperature levels. The FOM is proportional to the amount of mols of substrate converted per unit time for given weight or volume of reactor.

It is the purpose of the process developed by *H.P. Gregor; U.S. Patent 4,033,822; July 5, 1977* to use pressure-driven conditions to activate and couple enzymes to ultrafiltration membranes such as a homoporous cellulosic membrane having pores of about 15 to 200 Angstroms in diameter. The resultant enzyme-coupled ultrafiltration membranes can be used under pressure-driven conditions for enzymatic conversion of a substrate. The membranes which can be used can be of any known type, although cellulosic membranes have proven especially effective.

The matrix membrane or filter must have sufficient porosity so the substance or substances of biological activity can be introduced from an ambient solution or suspension phase. It should have sufficient mechanical strength (or be capable of being supported mechanically) so as to withstand the pressure gradients which are applied across it during processing and under conditions of use. In addition, it must have a multitude of small pores so as to have a large effective surface which, ultimately, will be lined with the enzymes. The pores advantageously range in size from 3 to 5 times the enzyme diameter. Most enzymes range in MW from 10,000 to 1,000,000 and in diameter from 5 to 75 A so membranes with pores of 15 to 200 A are preferred.

The pore surfaces of the matrix membrane or film can already carry active or reactive groups for attachment to enzymes or they can be treated in known manner to provide such groups. Such activation processes are advantageously carried out under pressure-driven conditions, especially where unstable substituents or by-products are involved, because the residence time of reactants in the membrane or filter can be controlled at will. For example, superior systems result when, after activating a regenerated cellulose membrane with cyanogen bromide at pH 11, excess reactant is rapidly washed away with water and the enzyme solution is immediately forced into the pores of the film.

Any superatmospheric pressure can be used to accelerate the passage of activating liquid and/or of enzyme through the membrane although a pressure of 30 to 120 psig gives particularly good results.

Example: A matrix membrane was cast from a 15% solution of cellulose acetate (39.4% acetyl, viscosity 45) in a DMF (dimethylformamide):acetone mixture (1:3.25 by volume) at room temperature onto a glass plate with a doctor blade of gate opening of 300 microns. The film was allowed to evaporate in the air for 3 minutes, then kept in a closed container for 1 hour to eliminate any appreciable "skin" formation and then coagulated by exposure to water vapor in the closed container for 15 minutes, followed by immersion in water at room temperature for 2 minutes, then in ice water for 1 hour. These were regenerated to cellulose by conventional hydrolysis in a buffer at pH 9.9 for 24 hours at 65°C.

The final films had a wet thickness of 30 microns, a water content of 80 to 87% and a hydraulic permeability (HP) of 0.001 to 0.002 cm/sec atm. These films were activated with a solution of 40 mg CNBr in 100 ml of water at pH 11 (with NaOH), forced through at 70 psig, and excess CNBr was removed by a one minute rinse with water, repeated with 0.1 M $NaHCO_3$, all at 70 psig, and immediately treated with a solution of 40 mg of chymotrypsin (CT)(3x crystallized from

4x crystallized CT, dialyzed salt-free and lyophilized) in 100 ml 0.1 M $NaHCO_3$ at 4°C, also at 70 psig, following which the membrane was incubated in the effluent for 24 hours at 4°C, then washed with distilled water. The final membrane contained 241 mg enzyme per gram of dried membrane, i.e., 24% loading. It retained its original HP, was now 33 microns thick with a water content of 73 to 80%. Membranes were stored at 4°C in distilled water. The activity of this membrane was determined by the method of Hummel using 0.00107 M BTEE (benzoyl tyrosine ethyl ester), wherein the substrate was forced through the membrane mounted in a Millipore cell at 25°C, with a pressure of 100 psi applied by nitrogen gas and the assay performed in the downstream effluent.

It was found that under these conditions the measured activity (as determined by the amount of substrate converted per unit time) was 75% of that of the same amount of native enzyme measured with the same substrate under conventional conditions. Accordingly, the FOM for this system was 0.24 x 0.75 or 0.18.

Activation with Cyanides plus Hypochlorous Acid

An inexpensive and simple means for preparing carriers for fixing or attaching enzymes has been developed by *M. Brenner; U.S. Patent 3,956,113; May 11, 1976.* This can be accomplished with the use of hydrocyanic acid and/or a water-soluble cyanide and a reagent which contains positive chlorine or bromine, namely hypochlorous acid or hypobromous acid and/or a water-soluble salt of hypochlorous acid or hypobromous acid or a chlorine compound or bromine compound which on hydrolysis yields one of the acids mentioned, or a mixture of such halogen derivatives.

Such materials are allowed to act on a water-soluble or swellable water-insoluble polyhydroxy compound in an alkaline medium, preferably in the cold. The reaction mixture is optionally neutralized and (a) in the case of a solution is freed of low molecular constituents by clarification and dialysis or gel filtration, (b) in the case of a suspension is freed of low molecular constituents by filtration or centrifuging and elution. The crosslinking and activation product present in solution or suspension or remaining as a filter residue or centrifuge residue is allowed to act, in the presence of water and optionally a buffer system, at a temperature below 50°C, preferably in the cold, on an optionally biologically active, water-soluble amino compound which is capable of substitution on at least one basic nitrogen atom (i.e., enzyme), and dialyzable material or material which has remained in a dissolved form is dialyzed or filtered or washed out of the reaction product by means of water or an aqueous salt solution.

The crosslinking and the evidently simultaneous introduction of amine fixing capacity can be formulated schematically somewhat as follows:

HO– –OH (soluble starch) $\xrightarrow{CN^-}$ HO– –O–C≡N (unstable intermediate) + O^-– –OH $\longrightarrow$ HO– –O–C(=NH)–O– –OH (insoluble crosslinking product)

Suitable polyhydroxy compounds include carbohydrates such as starch, starch paste, soluble starch, dextrins, dextrans, crosslinked dextrans, cellulose, especially mercerized cellulose, cellulose fibers, agarose, pectins and the like.

The degree of polymerization of the polyhydroxy compounds from the class of the carbohydrates which can be used as starting materials is optional, provided these compounds have the requisite solubility or swellability. Even saccharose gives an insoluble product having a very good enzyme fixing capacity. However, it is not essential to use carbohydrates. Water-soluble polyvinyl alcohols are also crosslinked, for example, in the system cyanide/hypochlorite, to give insoluble hydrophilic products of satisfactory amine fixing capacity.

The crosslinking and activation of the polyhydroxy compound take place between the solidification point of the solution and about 50°C; appropriately, they are carried out at between 0°C and room temperature.

If the polyhydroxy compound used is water-soluble and dialyzable such as, for example, sucrose (cane sugar), or if it is only water-soluble, such as soluble starch, dextrins and dextrans, they yield, within a very short time, a polycondensate, in practically quantitative yield, which precipitates in thick flocks. Substances of higher molecular weight, such as starch paste, lose their gel-like nature and become flocculent and filterable. Crosslinked dextrans, which are insoluble, undergo densification of their structure. Less distinct but nevertheless detectable changes are to be found in the case of mercerized cellulose and agarose. All these substances are distinguished by a pronounced capacity for amine fixing.

If the activated products are water-insoluble, they can remain in suspension and can, after the alkali has been neutralized, be reacted further directly with the amino compound. However, in some cases it is more advantageous to collect the insoluble material on a filter and to elute it, for which water, salt solutions and buffers, with or without addition of protein stabilizers (for example glycerine or complex-forming agents) can be used, and then to react the moist filter residue, again without delay because of the limited life of the active functional groups, with a water-soluble amino compound.

If they are water-soluble, the activated products can, if appropriate after neutralization of the alkali, be reacted further directly with the amino compound. However, in some cases it is more advantageous to remove low molecular weight material, if appropriate after neutralizing the alkali, by dialysis and only then to carry out the reaction with the amino compound.

The insoluble fixing products possess excellent filterability and because of their hydrodynamic properties are also very suitable for use in flowing reaction systems (continuous column operation).

Example 1: 2 grams of water-soluble starch are suspended in 100 ml of water and the mixture is heated on a water bath until dissolved and is allowed to cool. To 5 ml portions of this starch solution (each containing 100 mg of starch, corresponding to about 2 milliequivalents of -OH) were added, in each case, 100 mg (2 mmol) of sodium cyanide, diluted to 10, 25 or 50 ml with water, the mixture was cooled to -5° or 0° or 10°C and bleach corresponding to an amount of ½ or 1 or 2 or 4 mmol of sodium hypochlorite was added dropwise with good cooling. Flocculation commences after addition of 0.5 to 1 mmol of sodium hypochlorite, largely independently of the temperature and volume of the reaction mixture. The dry weight of the precipitates is about 100 mg. The same series of experiments with only 50 mg (1 mmol) of sodium cyanide gives approximately the same result.

If these experiments are carried out by adding the entire amount of hypochlorite at once to the starch/cyanide mixture, flocculation only takes place reliably if the cyanide is present in excess. If a less than equivalent amount of cyanide is present, the reaction appears to be restricted to the oxidation of the cyanide.

This is confirmed if the hypochlorite is taken first and the starch/cyanide mixture is then added. If this addition is made dropwise, no flocculation occurs even if there is an excess of cyanide. If, on the other hand, the addition is made all at once, product flocculation can be observed in the case of an excess of cyanide.

The pH-value of the above reaction solutions is about 12. If the starch/cyanide mixture is diluted with 0.1 N sodium hydroxide solution instead of water, no product flocculation takes place. A pH of 13 thus approximately represents the upper limit. Other experiments show that the lower limit of pH is about 10 to obtain good flocculation.

Example 2: A mixture of 25 mg of chymotrypsin and 25 mg of sodium sulfate is dissolved in 2 ml of water and lyophilized, and the residue is ground, with addition of 2 ml of 0.1 N sodium bicarbonate solution, with a filter cake obtained according to Example 1, at 0° to 5°C, from 200 mg of soluble starch, 200 mg of sodium cyanide, 20 ml of water and 4 mmol of sodium hypochlorite (bleach). The whole is left to stand for 30 hours at room temperature and is very thoroughly washed with 1.5 liters of water, and the residue is lyophilized (dry weight approximately 220 mg, containing 11.5% of water by Karl Fischer titration). A sample was hydrolyzed for 36 hours at 110°C in 20% strength hydrochloric acid and the hydrolysis product was analyzed for amino acids according to Stein and Moore. The expected amino acids were found in the expected ratio; their total amount corresponded to a fixing of 16.5 mg of chymotrypsin.

Activated Crosslinked Porous Cellulose Beads

The major disadvantage of using cellulose as a supporting material is that cellulose has a fibrous shape and lacks necessary mechanical strength. Reactors packed with cellulose have poor flow properties, develop severely high pressure drop and sometimes channeling. To overcome these problems, *G.T. Tsao and L.F. Chen; U.S. Patents 4,063,017; December 13, 1977 and 4,090,022; May 16, 1978; both assigned to Purdue Research Foundation* have prepared cellulose in a bead form which exhibited a better mechanical strength and provided enhanced flow properties compared to prior materials. This process involves the steps of:

(a) dissolving a cellulose derivative in an inert organic, water-miscible solvent to form a solution having a density greater than that of the precipitation solution used;

(b) distributing the solution in the form of droplets into a precipitation solution whereby the cellulose derivative is precipitated in the form of uniformly porous beads;

(c) separating the precipitated beads from the solution;

(d) washing the separated porous beads with water;

(e) hydrolyzing the washed beads to convert the beads to cellulose and to increase the active sites for attachment of enzymes and other biological agents; and

(f) washing the hydrolyzed beads to obtain porous cellulose beads.

The inert organic water-miscible solvent may be a combination of liquids which together with the cellulose derivative provides a solution which when mixed with the precipitating solution results in a phase inversion whereby the cellulose derivative is coagulated in the form of a porous bead. The inert organic solvent thus contains a component (a) which is characterized as a liquid which is capable of dissolving the cellulose derivative, such as cellulose acetate, and is soluble in the precipitation solution.

A second component (b) of the solvent system is a high density liquid which is soluble in component (a) and also in the precipitation solution and which is present in the solvent solution in an amount sufficient that the density of the final solvent solution (together with the cellulose derivative) is sufficiently higher than the density of the precipitation solution so that upon distributing the solvent solution in the form of droplets into the precipitation solution the cellulose will coagulate and precipitate out as a porous bead of desired size and porosity.

In addition to cellulose acetate, other cellulose derivatives may be used as a starting material for the preparation of the porous beads, for example, cellulose nitrate and methylcellulose. The terms "cellulose derivative" and "hydrolyzable cellulose derivative" are intended to include materials from which cellulose may be regenerated such as by means of, for example, hydrolysis or hydrogenation.

When using an aqueous precipitation solution, one may use as solvent component (a) acetone, formamide, a mixture of acetone and methanol or ethanol, methyl acetate, a mixture of methylene dichloride and methanol, methyl ethyl ketone and dimethyl sulfoxide. The solvent component (b) may thus be suitably chosen from dimethyl sulfoxide, formamide, methyl acetate, cyclohexanone, methylene dichloride, ethylene dichloride, a mixture of methylene dichloride and methanol, and a mixture of ethylene dichloride and methanol.

The table below sets forth a number of inert organic solvents for the precipitating of a cellulose derivative in an aqueous solution. The ratios set forth are the minimum needed in order to provide a solvent solution having a density greater than that of water. As can be seen, the greater the specific gravity of component (b), the less of that component is needed in order to achieve the minimum density.

Component (a)	Component (b)	Minimum Volume Ratio (a):(b)
Acetone	Dimethyl sulfoxide	70:30
Acetone	Ethylene dichloride	80:20
Acetone	Methylene dichloride	80:20
Acetone	Formamide	75:25
Acetone	Cyclohexanone	45:55
Acetone	Methyl acetate	35:65

After precipitation of the porous beads, cellulose is regenerated from the derivative by hydrolysis to create more active sites for enzyme attachment. In regenerating cellulose from its derivative after formation of the beads, one can remove the substituting groups (such as acetate from cellulose acetate) to regenerate all the hydroxyl groups normally present in the cellulose material. The higher the degree of regeneration, the more stability is to be found in the resulting beads. In some cases, where enzymes are to be immobilized on the cellulose bead carriers, it is desirable to convert the hydroxy or substituting groups into functional chemical groups, such as amino groups, which facilitate enzyme attachment.

Treatment of the porous cellulose beads with a crosslinking agent, either before or after hydrolysis of the beads, results in an increase of their physical strength. Attachment of enzymes onto the beads will also increase their physical strength. After treatment with, for example, a diisocyanate (e.g., tolylene-2,4-diisocyanate or hexamethylene diisocyanate), the beads in fact become quite rigid and strong. Crosslinking with epichlorohydrin also improves the physical properties of the porous cellulose beads.

The present process further provides for a method by which enzymes and other biological active agents may be immobilized by attachment onto the porous cellulose beads. For example, one may convert porous cellulose beads to diethylaminoethyl (DEAE) cellulose by reacting the beads with N,N-diethyl-2-chloroethylamine hydrochloride in a conventional manner. Cyanogen bromide, diisocyanates and dialdehydes may also be used to covalently bond the enzymes to the porous cellulose beads.

Example 1: 50 grams of cellulose acetate were dissolved in 400 ml of solvent (composed of acetone and dimethyl sulfoxide in a volume ratio of 6:4) to form a 12.5% (weight:volume) solution. With a spray gun, the cellulose solution was then sprayed at an air pressure of 20 psi as fine droplets into a water tank containing 40 gallons of water and four drops of common household detergent. Upon contacting the surface of the water, the cellulose acetate droplets coagulated into porous beads and sank to the bottom. The porous beads were then collected and washed. The washed beads were then deacetylated with about 0.15 N sodium hydroxide overnight at room temperature. The deacetylated beads were then washed and suction-dried, yielding a porous cellulose bead having a void space greater than 50% by volume ready for use in enzyme immobilization. The diameters of the beads ranged from 0.11 to over 1.68 mm with approximately 65% ranging in size from 0.30 to 0.84 mm.

Electron micrographs revealed that the beads were generally spherical, with the interior and surface thereof having the same structure. The pore sizes were quite uniform and the pores were distributed uniformly throughout the entire bead. The pore size of the beads was determined from scanning electron micrographs. The scanning micrographing requires dry samples and since the drying of the beads in air results in a size shrinkage, the beads were dried by the critical point technique with liquid carbon dioxide. The pore size was determined to be about 1000 A.

Example 2: 1 gram of porous cellulose beads, produced according to Example 1, was dispersed in 15 ml water which was adjusted to pH 11.5 with sodium hydroxide and kept at a constant temperature of 20°C. 1 gram of cyanogen bromide was added to this dispersion. The pH was maintained at 11.5 with 1 N NaOH. After 15 minutes, the beads were washed with a phosphate buffer (0.1 M) at pH = 7.0 and 0°C. 15 ml of glucoamylase solution (30 mg/ml) were then added to the beads. The mixture was left overnight. The beads so prepared contained 1,830 units of enzyme activity per gram dry weight of cellulose bead at 60°C using 5% maltose as substrate. One unit of enzyme activity is defined to be that which produces one micromol of product per minute.

Activated Glucan Beads

A matrix comprising a water-insoluble β-1,3-glucan gel in the shape of beads has been developed by *Y. Miyashiro, M. Ogawa, Y. Yamazaki and S. Igarasi; U.S. Patent 4,143,201; March 6, 1979; assigned to Takeda Chemical Industries, Ltd., Japan.* The term "matrix" means any of the carriers for immobilized enzymes, affinity chromatography, gel filtration, ion exchange and other applications.

Among the water-insoluble β-1,3-glucans which may be used are the polysaccharides elaborated by microorganisms of the genus *Alcaligenes* and of the genus *Agrobacterium.* The polysaccharide elaborated by the mutant strain, NTK-u (IFO 13140, ATCC 21680) of *Alcaligenes faecalis* var. *myxogenes* 10C3K (referred to as PS-1), the polysaccharide elaborated by *Agrobacterium radiobacter* (IFO 13127, ATCC 6466) and its mutant strain U-19 (IFO 13126, ATCC 21679) (referred to as PS-2) are particularly useful.

Preferred methods for producing beads of glucan gel comprise: (1) extruding, dripping or spraying a fluid containing a water-insoluble β-1,3-glucan into a heated oil bath and, thereby, causing the glucan to undergo gelation; (2) dissolving a water-insoluble β-1,3-glucan in an aqueous solution of sodium hydroxide and feeding the resultant solution through a drip nozzle into an aqueous solution of hydrogen chloride to neutralize it and cause the glucan to undergo gelation; and (3) adding an alkaline aqueous solution of a water-insoluble β-1,3-glucan dropwise to an organic solvent which is not freely miscible with water and neutralizing the resultant dispersion with an organic acid.

In the last method, examples of the alkaline aqueous solution able to dissolve water-insoluble β-1,3-glucans include aqueous solutions of sodium hydroxide, potassium hydroxide, barium hydroxide, calcium hydroxide, lithium hydroxide, ammonia and so forth. The pH range of the alkaline aqueous solution is preferably pH 11 to 13.

The concentration of the water-insoluble β-1,3-glucan has an influence upon the average diameter of the resultant beads. Normally, it is selected from 0.5 to 10%. Generally speaking, the lower the concentration of β-1,3-glucan is, the larger is the diameter of the resultant beads.

A method of encapsulating a core material of small diameter with the water-insoluble β-1,3-glucan has also been developed. The gel beads (microcapsules) consisting of a fine core material and encapsulated with the water-insoluble β-1,3-glucan may be made either to float on the water or to settle in water depending on the specific gravity of the particular core material selected. As core material, there may be used Sirasu (pumice ejected from volcanoes and the secondary deposit of such pumice) bead, pumice, alumina, silica, glass bead, hollow glass bead and so forth. Where the specific gravity of the core material is greater than 1, the microcapsules as packed into a column show flow rate properties even superior to those of water-insoluble β-1,3-glucan gel beads. Where the microcapsule has a specific gravity of less than 1, an enzyme or other active ingredient may be attached to the core to produce an immobilized enzyme preparation and, if this preparation is used in a batch reaction system where the reaction product separates out as a precipitate, the immobilized enzyme will float on the reaction mixture and, thus, facilitate separation of the reaction product.

Example 1: To 20 grams of PS-1 powder was added 660 ml of distilled water, followed by stirring to prepare a suspension. This suspension was added dropwise into 2 liters of corn salad oil heated to 80° to 85°C under agitation with a homomixer at 2,500 rpm, followed by stirring for 30 additional minutes. After the resultant PS-1 dispersion was cooled to room temperature, the corn salad oil was removed by decantation and then the sedimental gel was rinsed with 400 ml of toluene. This rinsing was repeated to completely remove the salad oil. The toluene and distilled water were then decanted, leaving 460 ml of PS-1 gel having the following diameters.

Diameter, μ	Percent
50-100	22.5
100-150	46
150-200	21
200-250	5
250-300	4
300-350	1.5

Example 2: 3 grams of PS-1 powder was dissolved by the addition of 144 ml of purified water and 16 ml of 1 N NaOH. To this solution was added 9 grams of Sirasu beads (40 to 80 mesh) followed by stirring. Under stirring at 360 rpm this mixture was added dropwise to a solution of 600 ml toluene and 2 grams of Emalex HC-30 (a polyoxyethylene-castor oil derivative). To this dispersion was added 15 ml of acetic acid to cause gelation. The reaction mixture was filtered through a nylon cloth and the cake was rinsed with water. A microscopic examination of the cake showed that there had been produced 20 to 40-mesh sperical microcapsules each consisting of Sirasu bead by way of core and PS-1 by way of shell.

Example 3: PS-1 gel beads prepared according to Example 1, were used as a carrier for immobilized enzyme. In a procedure for activating gel beads, 100 ml of distilled water was added to 115 ml of gel beads (corresponding to 5 grams of PS-1 powder) to obtain a suspension. On the other hand, 5 grams of BrCN was dissolved in 100 ml of distilled water. The suspension and solution were combined and the pH was gradually increased by the addition of 5 N NaOH until pH 11 was reached. The system was maintained at pH 11 for 15 minutes whereby an activated gel of PS-1 was obtained. The gel was filtered through a glass filter and the cake was rinsed with 500 ml of distilled water. 126 ml of the above activated PS-1 gel beads (corresponding to 5 grams of PS-1 powder) was combined with a solution in 50 ml distilled water of 50.9 mg of the α-amino acid ester hydrolase of *Xanthomonas citri* (IFO 3835) (CMC-treated, lyophilizate).

Following the addition of 50 ml of 0.2 M phosphate buffer (pH 8.0) and 24 ml of distilled water, the mixture was stirred at 5°C and pH 8 for 20 hours. The reaction mixture was filtered and the cake was washed with 0.2 M glycine solution, 0.5 M aqueous sodium chloride and distilled water in that order to prepare an immobilized enzyme preparation.

By the above procedure, 92.9% of the protein was immobilized. The enzyme activity of this preparation was determined based on the rate of hydrolysis of D-phenylglycine methyl ester. According to this assay, 90% of the activity was found to have been immobilized.

Stabilization of Immobilized Enzymes Using Quaternary Ammonium Compounds

In the presence of moisture, immobilized enzymes have a tendency to be slowly denatured or lose activity upon storage. This tendency can be somewhat lessened by drying the immobilized enzymes but since certain enzymes are extremely heat sensitive, such drying must be performed at relatively low temperatures, thereby requiring extended drying periods. Immobilized enzymes may also be stored frozen to lessen loss of activity upon storage. These methods, however, are not entirely satisfactory from a commercial viewpoint due to economics.

R.G. Dworschack, W.H. White and J.C. Chen; U.S. Patent 3,933,588; January 20, 1976; assigned to Standard Brands Incorporated have provided a convenient and economical method of treating immobilized enzymes in the presence of moisture to improve their storage stability. This stabilization may be obtained by treating an enzyme preparation consisting essentially of immobilized enzymes and water with a long chain quaternary ammonium compound, the amount of the compound being sufficient to increase the storage stability of the immobilized enzymes as compared to the storage stability of immobilized enzymes which have not been treated with the compound.

Immobilized enzymes which may be treated by this process are glucose isomerase, alpha-amylase, α-1,6-glucosidases, catalase, malt enzymes, glucose oxidase, cellulase and the like. The preferred immobilized enzymes which may be treated are glucose isomerase, alpha-amylase and pullulanase.

A relatively large number of long chain quaternary ammonium compounds may be used in the process. These quaternary ammonium compounds may be represented by the following formula:

$$R_4\text{—}\overset{\displaystyle R_1}{\underset{\displaystyle R_3}{\overset{|}{\underset{|}{N^+}}}}\text{—}R_2 \quad X^-$$

where R_1 contains at least 10 carbons, R_2 contains 1 to 20 carbons, and R_3 and R_4 contain not more than 6 carbons, the radicals being selected from alkyl, substituted alkyl, alkene, aryl, substituted aryl, aralkyl, and saturated and unsaturated cyclic, where the cyclic radical is formed by joining R_3 and R_4, and when the cyclic radical is unsaturated, there is no R_2. X may be any suitable inorganic or organic anion, such as a halide, nitrate, sulfate, acetate, etc.

Exemplary of quaternary ammonium groups represented by the above formula which may be used in the process are trimethyloctadecylammonium, dimethylbenzyldodecylammonium, N,N-diethylmorpholinium, cetylpyridinium, trihydroxyethyloctadecylammonium, diethyldioctadecylammonium, dioctadecylmorpholinium, dimethyldilaurylammonium, and 2-chloroethylbutyldistearylammonium.

Example: This example illustrates treating pullulanase which has been immobilized on a cellulose derivative with a long chain quaternary ammonium compound.

Aerobacter aerogenes NRRL 289 was cultivated in a suitable medium containing maltose and dextrose under submerged aerobic conditions at 30°C for 18 hours. The medium was filtered and the filter cake containing pullulanase was washed. The filter cake was slurried in water and the pullulanase extracted by the use of a nonionic surfactant. The slurry was filtered and into the filtrate containing the pullulanase was incorporated cyanogen-bromide-activated cellulose. The pullulanase was immobilized through covalent bonding to the cyanogen-bromide-activated cellulose.

Samples of the immobilized pullulanase were slurried in a 0.1 M NaCl solution, filtered, the filter cake washed with water and then with water containing 0.05 percent Variquat 415 (Northern Petrochemical). Another sample was prepared in the manner described above except it was not treated with Variquat 415. The initial activities of the samples were determined and the samples were then stored for a period and the activities of the samples again determined. The results of these determinations are shown in the table below. In the table, one IU is defined as that amount of material which will catalyze the hydrolysis of one micromol of α-1,6 bonds (based on release of maltotriose) per minute from a 0.5% pullulan solution at pH 5.0 and 45°C.

Sample Preparation	Initial Acitivity of Bound Pullulanase (IU/g)	Activity of Bound Pullulanase After 7 Days Storage
Treated with Variquat 415	53	52
No Variquat 415 treatment	53	36

Other examples in the patent show the process applied to immobilized intracellular glucose isomerase and to glucose isomerase adsorbed on Whatman DE-23 cellulose.

Stabilization of Cyanogen-Halide-Activated Polysaccharide Supports

Agarose activated by cyanogen bromide is known to couple different biological substances such as glucose oxidase. In this connection, the reaction is carried out in two stages, an activation of the agarose with the cyanogen bromide being carried out in the first stage and a coupling of the glucose oxidase to the cyanogen bromide activated agarose being carried out in the second stage. There is a need for cyanogen-bromide-activated agarose which does not need to be immediately coupled with the enzyme, but can be stored for a reasonable period of time and then used, without its activity having been reduced to an excessive extent and/or having lost its swellability in water.

In the process developed by *H.I. Johansson, H.K. Lundgren and M.K. Joustra; U.S. Patent 3,914,183; October 21, 1975; assigned to Pharmacia Fine Chemicals AB, Sweden* a stable activated support has been produced. This support comprises a composition in bead form of a water-insoluble cyanogen-halide-activated polysaccharide or a water-insoluble cyanogen-halide-activated derivative of polysaccharide, the polysaccharide or its derivative intended as a matrix for coupling amino-group-containing substances. The composition also contains: (1) a water-soluble acid capable of lowering the pH of its aqueous solution below 5; (2) at least one water-soluble polysaccharide; and (3) 0 to 4% of water. In the composition, the cyanogen-halide-activated polysaccharide or derivative thereof is stabilized against decrease in its reactivity and loss of swellability when stored.

According to the process, the water-insoluble cyanogen-halide-activated polysaccharide may be cyanogen-halide-activated agarose or a cyanogen-halide-activated derivative of agarose. An example of such a derivative is crosslinked agarose, epichlorohydrin having been used as crosslinking agent. A preferred composition may contain the following ingredients in the following contents:

	Percent
Water-insoluble cyanogen-halide-activated polysaccharide (or derivatives)	5-80
Water-soluble polysaccharide	25-95
Low molecular weight saccharide, if present	25-95
Water-soluble acid in a sufficient quantity to lower the pH of the composition after swelling in water to below 5	
Water	0-4

Other polysaccharides which may be used in this process include cellulose, dextran and their derivatives.

Example 1: A 500 ml reaction vessel was provided with an agitator, a thermometer and a glass calomel electrode coupled to a pH meter. The apparatus was placed in a ventilated cupboard. 3.3 grams of cyanogen bromide were dissolved with agitation in 300 ml of distilled water in the reaction vessel. 150 grams of sedimented agarose were weighed up and washed on a nutsch with approximately 2 liters of distilled water.

The agarose was then introduced into the reaction vessel and 2 N sodium hydroxide was then slowly added so that the pH rose to 11 whereupon the reaction began. Sodium hydroxide was consumed in the reaction which took place at about room temperature. The consumed sodium hydroxide was compensated for by adding 2 N sodium hydroxide to maintain the pH constant at 11. The supply of sodium hydroxide to the reaction vessel was stopped after 6 minutes. The reaction mixture was then transferred to a nutsch and washed with approximately 2 liters of cold, distilled water. The solution was finally washed with approximately 2 liters of 0.005 M citric acid. A solution containing 24 grams of dextran having the average molecular weight 40,000 in 120 ml of 0.005 M citric acid had been prepared previously.

The activated agarose washed with citric acid was then mixed with the solution of dextran while stirring for 1 hour. The mixture was then freeze-dried to 1 to 2% moisture content. Including the added substances, the yield was approximately 28 grams. The reactivity of the finished product against protein was tested by coupling α-chymotrypsinogen. In this way, 270 mg α-chymotrypsinogen were coupled per gram of dry matrix. The storage stability of the activated product was tested by storing the same at 40°C for 4 weeks. After this period, it was found that the coupling ability of the substance to protein had decreased by 19%.

Example 2: This example was carried out in the same manner as Example 1 with the exception that a solution containing 16.2 grams of dextran having an average molecular weight of 40,000 and 16.2 grams of lactose in 0.005 M citric acid was substituted for the solution containing only dextran. The yield was about 34 grams.

350 mg of α-chymotrypsinogen per gram of dry matrix was bound to the matrix by coupling. After storing at 40°C for 6 weeks, the coupling ability against protein was diminished by 15%.

COUPLING BY USE OF DIALDEHYDES OR ALDEHYDE GROUPS

Use of Glyoxal or Glutaraldehyde as Coupling Agents

Dialdehydes are used by *P.-C. Wirth and R. Tixier; U.S. Patent 3,836,433; September 17, 1974; assigned to Societe Generale De Recherches Et D'Applications, Scientifiques Sogeras, France* to fix nitrogenous materials, e.g., amino acids, proteins, enzymes, antigens, hormones, etc., to insoluble supports which form gels in aqueous media.

The support materials used in this process may be granular or may be spherical pearls or beads of varying dimensions. The support material may be any natural or synthetic polymer capable of forming a gel in aqueous media. Examples of such polymers are polysaccharides, for example agar-agar or agarose, or polyacrylamide. These polymers may be used alone or in admixture, for example a mixture of agarose and polyacrylamide. The dialdehydes which are used as bridge compounds may be represented by the formula OHC–R–CHO in which R is an aliphatic or aromatic chain of variable length and composition. Generally R will be a polymethylene chain of the formula $-(CH_2)_n-$ in which n is from 0 to 6.

In practice, the most convenient dialdehydes are glutaraldehyde and glyoxal. The dialdehyde compound reacts partly with the hydrophilic groups on the support gel (for example hydroxyl groups in the case of agarose or amide groups in the case of polyacrylamide) and partly with the amine function of the nitrogenous substance, such as protein, which it is desired to fix to the support.

The fixing or coupling of the nitrogenous substance to the support gel is carried out in two stages. In the first stage, a known quantity of the gel which may be hydrated or may be dry (a rehydratable lyophilized gel) is brought into contact with an excess of the dialdehyde (for example glutaraldehyde) in aqueous solution or in a suitable solvent medium except tris(hydroxymethyl)aminomethane which may react with the dialdehyde. The activated gel is then washed to remove excess aldehyde. In the second stage the nitrogenous substance is combined with the free aldehyde groups on the activated gel. The combination of nitrogenous substance and support is then washed to remove excess, nonfixed nitrogenous material.

This process is particularly suitable for use with enzymes. One of the major problems of the preparation of insoluble enzyme compositions is their loss of activity as a result of the chemical reactions used to fix or insolubilize them. It has been found that, in accordance with this process, not only is all the activity of the enzyme preserved but that in the majority of cases, the activity is increased, in particular when the enzymes are fixed to acrylamide-agarose gels, as shown in the table below.

Enzyme	- - Composition of Gel- - % Acrylamide	% Agarose	% Enzyme Fixed on Dehydrated Gel	Fixed/Stable Enzyme Activity %
Trypsin	4	2	4.6	116
	4	3	5.0	114
	6	4	6.5	156
Chymotrypsin	4	2	4.6	110
	4	4	6.5	140
	6	2	6.8	100
Pancreatic proteases	4	4	3.0	120
Papaya juice	4	2	3.2	120
protease	4	4	3.5	130
Alpha-amylase	3	4	2.5	80
Urease	3	2	4.2	100
Arginase	4	2	4.1	100
Lysozyme	5	1	2.5	
Cytochrome	4	4	6.0	

Note: For the proteases measurement was made on a hemoglobin. Better results were often observed on synthetic substrates.

Example 1: Fixation of Enzymatically Active Protein on a Hydrated Gel Using Glutaraldehyde — 10 ml of a 25% aqueous solution of glutaraldehyde are added to 20 ml of hydrated gel as measured in a graduated cylinder after sedimentation. The mixture is stirred in the cold for 5 hours and then the activated gel is recovered and washed several times with demineralized water to remove excess, noncombined glutaraldehyde.

The washed filtered gel is then added to a 1% solution of trypsin in a 0.02 molar calcium chloride solution (pH 7.8). The mixture is stirred for 60 minutes at 4°C and then the gel/trypsin composition is recovered by filtration or centrifuging. The gel is then washed several times with 30 ml lots of 0.02 molar calcium chloride solution, again recovered and then stored either in the hydrated or in the lyophilized state.

In two tests following the above procedure, when using agarose beads (4% agarose) as support, the resultant product contained 7% of fixed enzyme and when using an acrylamide/agarose gel (4% acrylamide, 2% agarose), the product contained 4.9% fixed enzyme, based on the dehydrated product.

Example 2: Fixation of Enzymatically Active Proteins Using a Glyoxal — 10 grams of a lyophilized acrylamide/agarose gel (4% acrylamide, 2% agarose) are suspended in 350 ml of a 40% solution of glyoxal and mechanically stirred for 5 to 6 hours at ambient temperature. The gel dehydrates slowly and, at the same time, reacts with the glyoxal. The gel is then recovered from the suspension and washed to remove all excess, uncombined glyoxal as determined by spectrophotometry at 260 millimicrons.

The washed and filtered gel is then suspended in 170 ml of a 1% solution of the enzyme in a 0.02 molar solution of calcium chloride (pH 7.8) and mechanically stirred for 2 hours at 4°C. The insoluble complex is recovered, washed with water to remove excess, unfixed enzyme and stored after a final filtration in the hydrated state at 4°C or is lyophilized.

When using trypsin as enzyme 5.5% of enzyme was fixed to the gel and when using chymotrypsin as enzyme 5.8% of the enzyme was fixed to the gel.

Bonding to Chitin by Use of Glutaraldehyde

Insolubilized enzymes which permit the percolation of liquids through the solid product have been developed by *W.L. Stanley and G.G. Watters; U.S. Patent 3,909,358; September 30, 1975; assigned to The U.S. Secretary of Agriculture.*

In accordance with this process, insolubilized but active enzymes are prepared from enzymes which are in a normal or native (soluble) state by reacting them with chitin and glutaraldehyde. The reaction is generally conducted in an aqueous medium, and preferably the chitin and glutaraldehyde are sequentially reacted with the starting enzyme added to the reaction mixture. For the sake of brevity, the insolubilized (immobilized) enzyme products are referred to by the term "enzyme-CG," and specifically as "lactase-CG," "amylase-CG," and the like.

Chitin is a polysaccharide where the primary repeating unit is 2-deoxy-2-(acetylamino)glucose. In general, about one out of every six units is not acetylated. Chitin is readily prepared by removing the impurities from shells of crab, shrimp, lobsters, crayfish, and the like, which are abundantly available from seafood processing plants.

For use in this process the chitin should be in granular form, obtainable by conventional comminuting procedures. It is generally desirable to apply a sieving operation to remove fine particles and over-size particles, retaining those having a mesh size of 10 to 50 mesh.

The enzyme to be insolubilized is dissolved in distilled water. Where necessary, the pH of the water is adjusted by conventional methods to a level at which the enzyme is soluble.

Following preparation of the aqueous solution of the enzyme, a mechanical separation step such as filtration or decantation can be applied to remove fillers, debris, or other undissolved material.

Next, the aqueous dispersion of the starting enzyme is added to moist chitin. Generally, about 1 to 10 parts of crude enzyme per 100 parts of chitin are used. The mixture is gently agitated by conventional means such as shaking, stirring, or the like while being held for 5 to 10 minutes at a temperature of 1° to 25°C in order to cause the enzyme to be adsorbed on the chitin.

Having adsorbed the enzyme on the chitin, an aqueous solution of glutaraldehyde is added to the suspension. The amount of glutaraldehyde is not critical. Usually, a large excess, e.g., 10 to 50 parts per part of enzyme, is used; the unreacted residue is removed in a subsequent washing step.

The resulting mixture is held for a period of time to ensure formation of the enzyme-CG product. Usually, the mixture is held for a short period, about 30 minutes, at ambient temperature. Subsequently, the mixture is held for a period of about 8 to 24 hours at 1° to 10°C. However, a somewhat shorter holding period can be realized if the holding is conducted at temperatures between 10° and 25°C. The product is then collected by filtration and washed several times with distilled water to remove excess reagents. The so-prepared enzyme-CG is then ready for use.

Usually, the starting enzyme contains inactive proteins and it is desirable to remove these from the final product. To this end, the enzyme-CG is washed with distilled water for a long period, e.g., 60 minutes. It is then soaked sequentially in (a) several volumes of 10 to 15% aqueous sodium chloride, (b) a potassium acetate buffer at pH 7, and finally, (c) a potassium acetate buffer at a pH whereat the enzyme-CG exhibits maximum activity. The purified enzyme-CG is collected by filtration and is ready for use. This process has wide versatility and can be applied to enzymes of all kinds.

One particular application of this process is preparing a lactase-CG for conversion of lactose in deproteinized whey to galactose and glucose.

Example 1: Preparation of Lactase-CG – To 5 grams of chitin particles (in a moist-condition, about 50% water) was added 50 mg of acid-tolerant lactase enzyme (a β-galactosidase) in 10 ml of water. The mixture was swirled and then allowed to stand for 5 minutes. An aqueous glutaraldehyde solution was added to the above mixture until the final concentration of glutaraldehyde in the mixture was 2%. This mixture was held at ambient temperature for 30 minutes, and then held at about 5°C for 8 hours.

The resulting lactase-CG was collected by filtration and while retained on a sintered glass funnel was backwashed with distilled water for 1 hour. The product was then serially soaked in several volumes of the following aqueous solutions tions: 2 N sodium chloride, 1 M potassium acetate (pH 7), and 0.1 M potassium acetate (pH 3).

The lactase-CG exhibited approximately 65% of the original enzyme activity. The pH at which the enzyme exhibited optimum activity was shifted from 4 in the case of the starting enzyme to between 3.0 and 3.5 for the lactase-CG. Two runs were made as described above except that the amount of starting enzyme was varied, in one case 100 mg, in the other 150 mg.

The activity of the products was measured in a shaker bath batch test at 40°C with a 0.4 M lactose solution in a 0.1 M potassium acetate buffer at a pH of 3. The production of glucose (μmol/min/g of lactase-CG) was measured. The results are tabulated below.

	........Lactase-CG.		
Run	Lactase (mg)	Chitin (g)	Activity
1	50	5	61
2	100	5	98
3	150	5	115

Example 2: Continuous Use of Lactase-CG Column – Lactase-CG prepared as described in Example 1, Run 2, was packed into a 9 mm x 2.2 cm jacketed column, and washed with 0.1 M potassium acetate solution (pH 3).

A lactose solution (4% in 0.1 M potassium acetate buffer at pH 3) was passed through the column at the rate of 60 ml per hour and at a temperature of 45°C. The column was so operated continuously for a period of 3 days. The percent hydrolysis of lactose was determined at the end of each day of operation. The results are given below.

Time (days)	Hydrolysis (%)
1	58.7
2	57.0
3	57.9

Enzymes Immobilized on Chitosan

Chitosan is a polyamino polysaccharide obtained by N-deacetylation of chitin with strong alkali and heat. Chitin is a polysaccharide wherein the primary repeating unit in the molecule is 2-deoxy-2-(acetylamino)glucose. In general, about one out of every six units in chitin is not acetylated, whereas in chitosan essentially all the repeating units are not acetylated. It should be noted that the extent of nonacetylation can be controlled by the severity of the deacetylation reaction.

In the process developed by *M.S. Masri, V.G. Randall and W.L. Stanley; U.S. Patent 4,089,746; May 16, 1978; assigned to The U.S. Secretary of Agriculture* insolubilized but active enzymes are prepared by mixing an aqueous solution of the enzyme with an aqueous solution of chitosan and then adding a polyfunctional crosslinking agent to form a gel. The gel is then reacted with a reducing agent to form a granular insolubilized enzyme, which has retained a substantial part of its original activity.

In the first step in the process, the enzyme to be insolubilized is dissolved in distilled water. Where necessary the pH of the water is adjusted by conventional methods such as adding an acid, buffer, etc., to a level at which the enzyme is soluble.

The aqueous solution then undergoes a mechanical separation step such as filtration or decantation to remove fillers, debris, or other undissolved material. Next, chitosan is dissolved in water. Generally, it is necessary to add a small amount of acid to the water in order to effect solution of the chitosan. The amount of acid is generally that necessary to adjust the pH of the dispersion to from 3 to 7. Usually, this amount is 4 to 5 milliequivalents of acid per gram of chitosan. As the acid, one may use hydrochloric acid, phosphoric acid, acetic acid, and the like. Furthermore, one may employ a buffer to attain the desired pH level and effect solubilization of the chitosan.

Next, the aqueous dispersion of starting enzyme is mixed with the aqueous dispersion of chitosan. Generally, about 10 to 100 milligrams of crude enzyme per gram of dry chitosan are used. The mixture is gently agitated by conventional means such as shaking, stirring, or the like while being held for approximately 5 to 20 minutes at a temperature of from about 10° to 25°C. It should be noted that the enzyme and the chitosan can be simultaneously dissolved in water to produce the mixture directly.

To the above mixture is added a polyfunctional crosslinking agent, i.e., one with more than one functional moiety, such as a di- or polyaldehyde, a di- or polyisocyanate, a di- or polyacid chloride, and the like. Usually, the polyfunctional crosslinking agent is dissolved in water and the resulting aqueous solution is added to the above mixture. The amount of polyfunctional crosslinking agent is not critical; 1 to 50 parts thereof per part of enzyme, may be used. The unreacted residue is removed in a subsequent washing step.

The resulting mixture is held for a period of time to ensure proper formation of a gel. Usually, the mixture is held for a short period, about 30 to 60 minutes, at ambient temperature. However, the holding can be conducted at temperatures between 10° to 25°C, the duration of the holding period decreasing as the temperature increases.

The gel is then mixed with a reducing agent to produce a granular material from the gel. It should be obvious that the reducing agent must be selected for its ability to form the gel into a granular product without interfering with the activity of the enzyme. Reducing agents that satisfy this limitation are certain borohydride reducing agents such as sodium borohydride, sodium cyanoborohydride, and the like. Contact between the gel and the reducing agent should be maintained for about 1 to 60 minutes at 4° to 20°C or until a suitable granular material has been formed. The concentration of reducing agent should be sufficient to produce a granular product from the gel; generally, the concentration of reducing agent is 2 to 10 parts per part of crude enzyme.

During the reduction reaction heat is generated. Consequently, the reaction mixture should be cooled to maintain the temperature between 4° to 20°C.

It is believed that formation of the products involves the following mechanism: The polyfunctional agent promotes crosslinking of the solubilized chitosan in the presence of an enzyme. The lysyl residues of the enzyme might participate in the crosslinking reaction and thereby become covalently fixed to the chitosan polymer. The crosslinked material has the texture of a gel and is a type of Schiff-base polymer. When the gel is treated with a reducing agent the Schiff-base polymer matrix is reduced and stabilized, thus yielding a granular product.

Example 1: Preparation of Insolubilized Lactase – Run 1: 7 grams of dry chitosan were dissolved in 250 ml of water containing 30 ml of 1 N HCl acid to effect solution. To this solution was added a solution of 0.5 gram of commercial grade lactase (a β-galactosidase) in 50 ml of water. The crude enzyme hydrolyzed about 10 micromols of lactose per milligram per minute at pH 4 and 40°C. The mixture was swirled for 5 minutes and added to 100 ml of aqueous solution containing 7 grams of dialdehyde starch (DAS) that had been solubilized with 2.5 millimols of sodium carbonate. The solubilized DAS solution had a pH of about 7.

The resulting mixture was held 30 minutes until a gel formed. Then, the mixture was cooled to and maintained at 10°C and stirred mechanically, and 2 grams of pelletized sodium borohydride (about 0.3 g each pellet) was added, over a period of 15 minutes. Stirring was continued for another 15 minutes. The insolubilized product took on a granular form, shortly after the start of addition of borohydride. The granular product was separated from the reaction mixture by filtration and washed with distilled water, followed by 0.5 M phosphate buffer (potassium dihydrogen phosphate and disodium hydrogen phosphate mixture) of pH 7.0, then with phosphate buffer of pH 5.6, and finally with distilled water.

Run 2: The procedure outlined above for Run 1 was followed with the following changes: (a) 1 gram of lactase was used; (b) 30 ml of a 30% aqueous solution of glyoxal (pH 7) was used instead of the DAS solution; and (c) 5 grams of sodium borohydride was employed as the reducing agent.

Run 3: The procedure outlined in Run 1 was used except that: (a) 2.4 grams of chitosan was dissolved in 5 ml of 2 M acetic acid, 115 ml of water, and 5 ml of 1 M sodium acetate buffer of pH 5.4; the final pH of the solution was 5.4; (b) 120 mg of crude lactase was used; (c) 375 mg of glutaraldehyde in 150 ml of water was used; and (d) 100 mg of sodium borohydride in 50 ml of water was used as the reducing agent.

The lactase activity of the products was measured in a shaker bath batch test at 40°C with 25 ml of 0.4 M lactose solution in a 0.1 M potassium acetate buffer, pH 4.0; 0.5 gram of moist product (containing 85 to 90% water) was used. The production of glucose (micromols per minute per gram of moist product) was measured. The results are tabulated below.

	. . . .Insolubilized Enzyme. . . .		
Run	**Enzyme**	**Crosslinker**	**Activity**
1	Lactase	DAS	21.3
2	Lactase	Glyoxal	43.0
3	Lactase	Glutaraldehyde	31.0

Example 2: Preparation of Insolubilized Invertase – The procedure outlined in Example 1, Run 3, was followed except that yeast invertase was used in place of lactase. The insolubilized product contained about 86% water.

The invertase activity of the product was measured in a shaker bath batch test at 40°C with 100 ml of 0.1 M sucrose solution in 0.1 M phosphate buffer at pH 5.0; using 1.0 gram of moist product. The rate of hydrolysis of sucrose was determined by measuring the micromols of glucose formed per minute per gram of moist product and was found to be 48.

A similar process has been developed by *J.-L. Leuba; U.S. Patent 4,094,743; June 13, 1978; assigned to Societe d'Assistance Technique pour Produits Nestle SA, Switzerland* except that the chitosan is first reacted with the dialdehyde and then with the enzyme.

The bifunctional dialdehydes may be, for example, a dialdehyde with a chain of 1 to 10 carbons, advantageously 3 to 6 carbons, more especially malonaldehyde, succinaldehyde, glutaraldehyde, adipaldehyde. It is normally preferred to use glutaraldehyde which has the advantage of being an inexpensive commercial product and which gives products with remarkable enzymatic activity.

Fixing of the enzyme is carried out in two stages. In a first stage, the so-called "activation" stage, chitosan is treated with the bifunctional reactant selected from dialdehydes, for example glutaraldehyde. The dialdehyde is fixed to the chitosan through one of its aldehyde groups, and an intermediate product is obtained, namely an "activated chitosan," whose surface is covered with aldehyde groups. In a second stage, the enzyme is fixed to the activated chitosan. These operations are carried out very simply by contact, for example, by immersion of the solid phase in the liquid phase under the following conditions:

(a) treatment of the chitosan solution of the dialdehyde with a pH value of from 6 to 8, preferably a buffer; temperature from 2° to 35°C, preferably ambient temperature; treatment time from 30 minutes to 24 hours; preferably treatment in the absence of air and elimination of the excess dialdehyde by washing with water

(b) fixing of the enzyme suspension of the activated chitosan in a solution with a pH-value of 2 to 11, preferably a buffer of pH 8.0 and containing $CaCl_2$ in a concentration of 0.02 M; addition of the enzyme in solution in an aqueous medium, temperature from 2°C to the temperature at which the enzyme is inactivated, preferably around 2°C; treatment time from 2 to 24 hours; elimination of the unfixed enzyme by washing with a saline solution, the buffer solution used for fixing, and then with water

In one preferred embodiment of the process, the enzymatically active product is stabilized by eliminating the residual aldehyde groups which have not reacted with the enzyme during fixing. For example, they may be reacted with a suitable chemical compound, for example, with a compound containing an amino group.

In this respect, it is favorable to use glycine, asparagine and, in particular, lysine, the enzymatically active product thus obtained showing a remarkable affinity for proteic substrates. It is also possible to reduce the residual aldehyde groups, for example, with sodium borohydride, which appreciably increases the stability of the enzymatically active product thus treated. These two treatments may, of course, be combined.

Example: 30 grams of chitosan are dissolved in dilute acetic acid (700 ml of water and 10 ml of glacial acetic acid) with vigorous stirring until the chitosan is completely dissolved. The viscous solution obtained is filtered through two layers of gauze and then injected in small quantities into 2 liters of water alkalized to pH 11.0 by the addition of sodium hydroxide. Under these conditions, the chitosan precipitates in the form of fine lightweight particles. Throughout this process, the pH of the solution is kept at its initial value by the addition of sodium hydroxide. After all the chitosan has precipitated, the pH-value of the solution is carefully lowered to 7.5 by the addition of dilute hydrochloric acid. The precipitate collected by centrifuging is washed twice with water and then freeze-dried giving a fine powder of chitosan.

2.5 grams of this chitosan are suspended in 50 ml of 0.05 M acetate buffer solution (pH 6.0) over a period of 24 hours. 16.5 ml of 20% glutaraldehyde (obtained by diluting 50% commercial-grade glutaraldehyde with the preceding buffer solution) are then added, followed by stirring for 30 minutes at ambient temperature. After washing with twice-distilled water, the "activated chitosan" is dispersed in 50 ml of 0.1 M borate buffer solution (pH 8.0) containing 0.02 mol/l of $CaCl_2$. This suspension is cooled to 2°C, followed by the addition of

250 mg of trypsin (Merck, crystallized) previously solubilized in 10 ml of 0.001 N HCl and stirring for 2 hours at 2°C. After centrifuging, the product is washed with 50 ml of the following solutions: 0.15 M NaCl solution, 0.1 M borate buffer solution (pH 8.0), and with twice-distilled water (3 times). It is then introduced into 50 ml of iced 0.1 M borate buffer solution (pH 8.0) to which 100 mg of solid sodium borohydride are added.

After 20 minutes at 2°C, the insoluble enzymatically active product is separated by centrifuging, rewashed with water and preserved in a 0.05 M borate buffer solution (pH 7.5) containing sodium azide as preservative.

The quantity of enzyme fixed to the chitosan by the glutaraldehyde is evaluated by the difference between the quantity added and the quantity present in the residual solution and the washing waters (determination based on UV absorption at 280 nm).

The amidase activity of the product is tested on a synthesis substrate, DL-BAPA (p-nitroanilide of N-benzoyl-DL-arginine), by the method of Erlanger et al, *Arch. Biochem. Biophys.,* 95, 271 (1961).

The results obtained are as follows: Of 100 mg of trypsin added to 1 gram of chitosan, 92.9 mg of trypsin are fixed. The specific amidase activity (BAPA), μmol of p-nitroanilide liberated/min/mg of enzyme, of the preparation is 49.16% and the specific caseinolytic activity 38.26%.

This product is freeze-dried in the presence of twice its weight of mannitol and the freeze-dried product is preserved for one month. Its enzymatic activity is then measured and found to be the same as that of the fresh product.

Use of Dialdehyde Cellulose

The dialdehyde groups used for immobilizing the enzymes have been produced as part of the cellulose molecule in the process disclosed by *B.S. Wildi and L.E. Weeks; U.S. Patent 4,013,514; March 22, 1977; assigned to Monsanto Company.*

The biologically-active conjugates of this process are prepared by reacting the dialdehyde cellulose, preferably in fibrous form, with an enzyme through a primary amino group of the enzyme not essential for biological activity. The reaction is carried out in an aqueous medium and at a substantially neutral or alkaline pH, i.e., at a pH of about 7 or higher.

Dialdehyde cellulose suitable for this process can be prepared by the well-known oxidation of cellulose with periodic acid.

The degree of oxidation of the cellulose to provide dialdehyde groups can vary. Normally about 5 to 80% of the glucoside units in the cellulose polymer are opened and oxidized to dialdehyde groups. A degree of oxidation higher than 80% is undesirable because the structural integrity of the dialdehyde cellulose matrix appears to be impaired. For these purposes about 10 to 50% of the glucoside units or rings are opened and oxidized to dialdehydes.

A wide variety of enzymes can be coupled directly to the dialdehyde cotton carrier. Many such enzymes are available commercially. Typical are the proteolytic enzymes such as the proteases, e.g., neutral and/or alkaline protease. In some instances another differently active enzyme such as amylase can be employed, alone or admixed with proteases so as to maximize the operative enzyme activity of the composite. Still other enzymes such as a lipase may be used instead of or in addition to amylase. Other suitable enzymes are esterase, nuclease, or other types of hydrolase. A hydrase or oxidoreductase may also be utilized, depending upon the ultimate activity and intended application.

Example 1: Oxidation of Cotton with Periodic Acid — Cotton roving (80 grams) cut in about 1 inch long pieces was washed in a 0.1% dodecylphenol-10 mol ethylene oxide adduct, plus 0.1% sodium carbonate solution (2,000 ml) for 5 to 10 minutes by manual stirring. The resulting slurry was then filtered over a Buchner funnel and the cotton fibers recovered.

The recovered cotton fibers were then rinsed five times with 4,000 ml aliquots of distilled water with intermittent stirring and filtering after each rinse, followed by rinsing three times with 2,000 ml aliquots of acetone to remove the water, again with intermittent stirring and filtering after each rinse. Ultimately the washed and rinsed cotton was air dried under a hood.

The dried cotton (50 grams) was placed in a 4 liter Erlenmeyer flask and aqueous, about 0.11 M, periodic acid solution (2.5 liters; pH 1.4) added. The flask was then stoppered using a polyethylene-coated stopper which resists oxidation and the outside of the flask was covered with aluminum foil so as to exclude light. Thereafter, the flask was placed on a platform shaker and shaken for 5 days at 25°C.

The cotton-periodic acid slurry was then filtered on a glass Buchner funnel, washed five times with 3- or 4-liter aliquots of water with intermittent stirring and with filtering after each wash, rinsed four times with 2-liter aliquots of acetone with intermittent stirring and with filtering after each rinse, and subsequently placed on a piece of aluminum foil and dried in air under a hood.

Yield: 49.5 grams. After drying in vacuo at 55°C the yield was 46.3 grams, i.e., a moisture loss of 6.5% was observed. A control sample (10 grams) after drying in vacuo at 55°C exhibited a moisture loss of 6.0%.

A 0.096 N sodium arsenite solution, standardized by 0.104 M potassium permanganate solution, was prepared and used to titrate aliquots of the periodic acid solution before and after cotton oxidation. From the titer data it was calculated that 57.8 atoms of oxygen per 100 glucoside units had been consumed, that is, about 57.8% of glucoside rings in the cotton were opened and oxidized to dialdehydes.

Example 2: Attachment of Invertase to Dialdehyde Cotton — A portion of fibrous dialdehyde cotton (2.5 grams) prepared in Example 1, above, was placed into a stainless steel pipe (0.625 inch i.d. by 1.5 inches long). Foamed polyethylene discs were placed at each end of the pipe to retain the fibers, and the pipe was capped at both ends with stainless steel caps provided with $1/16$" inlet and outlet fittings, respectively.

Invertase (0.500 gram; purified; derived from bakers' yeast, mellibiose free) was dissolved in water (40 ml) and pH of the resulting solution was adjusted to

8.0 by the addition of aqueous 0.1 N sodium hydroxide solution (0.4 ml). The invertase solution was then recirculated through the dialdehyde cotton reactor bed at a rate of 5 to 10 ml per minute for 1 hour. During recirculation pH of the solution was maintained at 8.0 by the addition of aqueous 0.1 N sodium hydroxide solution (2 ml).

After recirculation of the invertase solution through the reactor bed was completed, the reactor bed was washed by pumping therethrough 1 liter of water, 1 liter of aqueous 1 N sodium chloride solution, 1 liter of water, and ½ liter of aqueous 1 weight percent sucrose solution in aqueous 0.1 M sodium acetate at a pH of 4.5.

Example 3: Evaluation of Invertase Activity in Reactor — A substrate solution was prepared by dissolving sucrose (1 weight percent) in an aqueous 0.1 M sodium acetate solution at pH 4.5.

The substrate solution at 25°C was then pumped through the invertase reactor prepared in Example 2 and the obtained effluent mixed with 3,5-dinitrosalicylic acid solution and color developed at 95°C. Hydrolysis was monitored by measuring the absorbance of the effluent mixed with the 3,5-dinitrosalicylic acid solution at 540 nm. The 3,5-dinitrosalicylic acid solution was prepared by dissolving the acid (40 grams) in water (3 liters), adding aqueous 50 weight percent sodium hydroxide solution (128 milliliters), and potassium sodium tartrate (1 lb), and then adding sufficient water to make 1 gallon of reagent solution. The experimental data are compiled in the table below.

Time from Start (min)	Flow Rate (ml/min)	% Hydrolysis of Effluent
13	5	4.0
18	1	37.0
30	1	36.0
42	1	33.8
57	1	33.3
73	1	32.4
112	1	32.8

OTHER MODIFIED CARBOHYDRATES FOR BONDING ENZYMES

Oxirane Modified Polysaccharide Gels

The process developed by *J.O. Porath, N. Fornstedt, L. Sundberg, C. Eklund and R. Axen; U.S. Patent 3,853,708; December 10, 1974; assigned to Exploaterings AB T.B.F., Sweden* provides a method for chemically coupling biologically active substances, namely, enzymes and proteinaceous enzyme inhibitors, to an oxirane-containing polymer. The chemical coupling reaction is effected in a milieu where the biologically active substance does not lose its characteristic properties, e.g., enzyme activity, ability to specifically bind or inhibit other substances. The reaction is preferably carried out in a buffered aqueous solution. In this context, the term biologically active substances includes enzymes and proteinaceous enzyme inhibitors.

The oxirane-containing polymer can be prepared by treating a hydroxy-containing polymer with nonion building epoxides (oxiranes) containing at least two strongly reactive groups in the polymer. The chemically reactive product can then be transferred to a milieu suitable for further reaction with the biologically active substances. The course of reaction for a carbohydrate such as agarose, cellulose, etc., with epichlorohydrin in an alkaline milieu can be illustrated by the equation

$$ROH + Cl{-}CH_2{-}\underset{\diagdown O \diagup}{CH{-}CH_2} \longrightarrow R{-}O{-}CH_2{-}\underset{\diagdown O \diagup}{CH{-}CH_2} + HCl$$

where R represents a carbohydrate, e.g., cellulose, agarose, dextran, etc. The oxirane-containing polymer can also be prepared by the reaction between an amide-containing polymer, e.g., polyacrylamide or crosslinked polyacrylamide or a derivative thereof and a bifunctional epoxide, e.g., a bis-epoxide. If the activated polymer is an insoluble, granulated gel, it is washed, preferably on a filter. In a corresponding manner, a soluble polymer is freed from an excess of epichlorohydrin by dialysis.

The oxirane-containing polymer can also be prepared from polymers containing hydroxyl groups in 1,2-position, e.g., via tosylation and subsequent treatment with sodium methylate, whereafter coupling takes place, or an unsaturated alkylether, can also be oxidized with an organic peroxide or peracid.

Example 1: Activation of Sepharose 6B with Epichlorohydrin – Sepharose 6B was washed on a glass filter with distilled water. The withdrawn gel was then transferred to a round-bottomed flask and 1 N NaOH, as well as epichlorohydrin was added. After the activating reaction took place under extremely vigorous stirring for 1 hour at 60°C, the excess epichlorohydrin was washed away with distilled water to a neutral pH. The activated gel could be kept in this milieu at 5°C.

Example 2: Coupling of Soybean-Trypsin-Inhibitor (STI) – Activated Sepharose 6B buffered with bicarbonate pH 9.5 was reacted with soybean-trypsin-inhibitor (STI) dissolved in the same buffer. The coupling occurred at room temperature under mild stirring and with a reaction time of 20 hours. The product obtained was carefully washed with a coupling buffer, acetate buffer pH 3.0 containing 1 M NaCl, and finally with tris buffer pH 7.8.

Example 3: Coupling of Trypsin – Trypsin dissolved in bicarbonate buffer pH 9.0 was reacted with Sepharose 6B. Coupling and the washing of the product obtained occurred in the same way as in Example 2.

Other examples in the complete patent cover the (1) activation of cellulose tosyl ester with sodium methylate; (2) activation of agarose (sepharose) with epibromohydrin; and (3) activation of crosslinked polyacrylamide with bis-epoxide and their coupling with various enzymes and enzyme inhibitors.

Polysaccharides Modified with Cyclic Carbamate Groups

Modified polysaccharides with a minor portion of the hydroxyl groups substituted by an N-(2,5-dioxo-4-oxazolidinyl)alkyl carbamate or N-(2,5-dioxo-4-oxazolidinyl)aralkyl carbamate radical are disclosed by *M.H. Keys and F.E.*

Semersky; U.S. Patent 3,833,555; September 3, 1974; assigned to Owens-Illinois, Inc. These modified polysaccharides, useful for insolubilizing enzymes, can be represented by the formulas

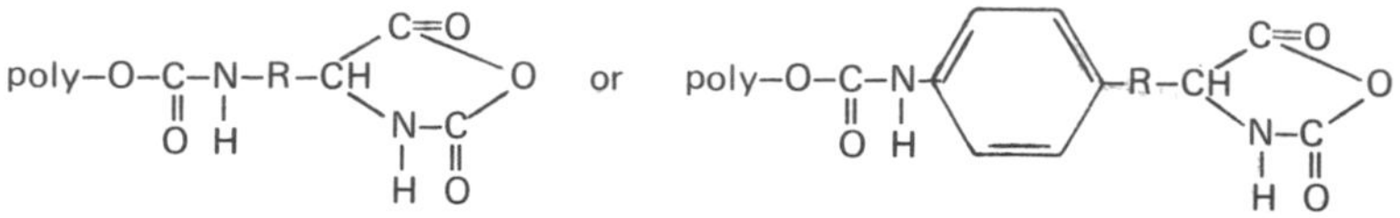

poly–O–C(=O)–N(H)–R–CH(C=O–O–C(=O)–NH) or poly–O–C(=O)–N(H)–C₆H₄–R–CH(C=O–O–C(=O)–NH)

(1) (2)

where poly is a polysaccharide such as dextran, agarose, carrageenan, cellulose, cotton or starch; R is an alkylene radical containing 1 to 8 carbons and in which less than 50% of the hydroxyl groups in the polysaccharide molecule are replaced by the radicals

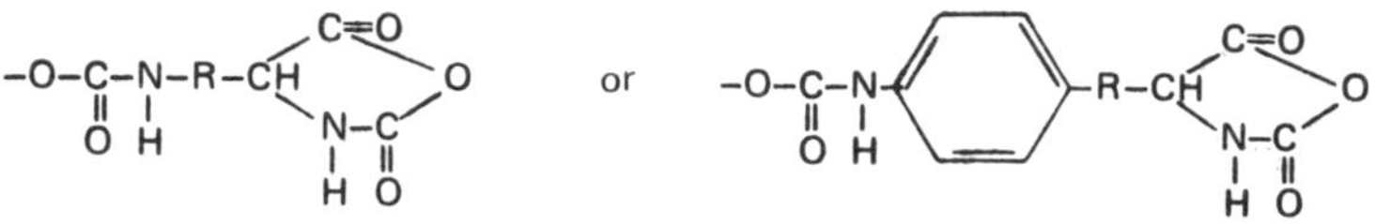

–O–C(=O)–N(H)–R–CH(C=O–O–C(=O)–NH) or –O–C(=O)–N(H)–C₆H₄–R–CH(C=O–O–C(=O)–NH)

These compositions can react with various water-insoluble enzymes such as chymotrypsin, ribonuclease, urease, peroxidase and the like to produce enzyme products which are water-insoluble and can be used repeatedly as catalysts in various reactions with minimal loss of enzymatic activity. These modified polysaccharides can be prepared by suspending the polysaccharide in water and treating the suspension with cyanogen bromide in the presence of an alkali metal hydroxide at room temperature. The activated polysaccharide is then condensed with an amino acid of the formula

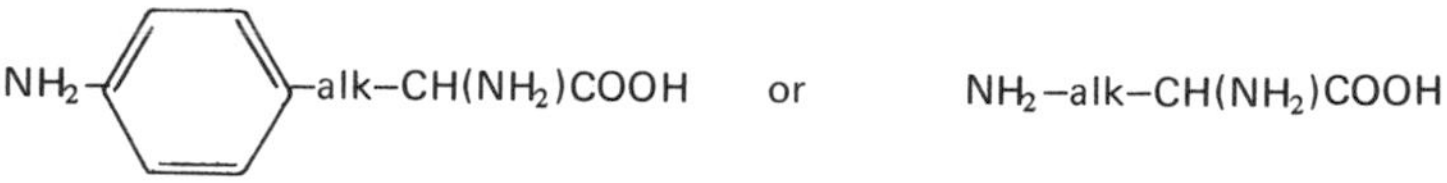

NH_2–C₆H₄–alk–CH(NH_2)COOH or NH_2–alk–CH(NH_2)COOH

where alk is an alkylene radical containing 1 to 8 carbons at a temperature of 0°C and the condensation product formed is reacted with phosgene at 35°C to effect cyclization.

Example 1: In formula (1) poly is agarose and R is tetramethylene. To 1 gram of agarose in 10 ml of water was added 1 gram of cyanogen bromide and sufficient 2.0 molar sodium hydroxide to maintain the pH at 10.5 at 25°C. The solid which formed was removed by filtration and washed with 500 ml of cold water. The activated agarose was then added to 10 ml of 0.1 molar sodium borate buffer containing 2 grams of lysine at a temperature of 0° to 5°C. The condensation product formed was washed with water, then 3 grams of solid product was mixed with 75 ml of dioxane and phosgene bubbled through the reaction mixture for 5 hours at 35° to 40°C followed by purging with dry nitrogen for 3 hours at 25°C. The off-white, cyclized, solid agarose product was separated by filtration and dried.

The condensation product composed of agarose and lysine but before cyclization with phosgene was quantitatively analyzed for amino acid side chains by adding 1 gram of the product to 3 ml of distilled water and mixing with 0.5 ml of 0.1 molar $CuCl_2$. After stirring for 30 minutes, the mixture was centrifuged and 1 ml of the supernatant was mixed with 3 ml of 0.05 molar ethylenediamine-tetraacetic acid. The optical absorbance of the resulting solution was measured at 740 nanometers and the concentration determined from a calibration curve.

If no cupric ions are detected, another aliquot of $CuCl_2$ solution is added as described above and the absorbance is again measured. The difference between the mols of $CuCl_2$ added to the condensation product and the mols of $CuCl_2$ present in the supernatant solution is equal to the number of mols of cupric ion complexed with the amino acid side chains in the product. The extent of lysine substitution was found to be 0.32 mmol/g of condensation product.

Example 2: A solution containing 0.5 mg of ribonuclease in 4.4 ml of water was saturated with 1.32 g of $KHCO_3$ and cooled to 0°C. To this solution was added with stirring 50 mg of the cyclized, agarose product prepared in Example 1. Thereafter, stirring was continued at 0°C for 3 hours. In this manner, the ribonuclease was rendered water-insoluble and after washing with water, it was stored at 3°C. This sample was assayed by packing in a chromatographic column and measuring the rate of change of cytidine-2',3'-cyclic phosphate to cytidine monophosphate by the method of Hammes and Walz as described in *Biochim Biophys Acta,* 198, 604 (1970). The activity indicated an equivalent of 1.1 mg of ribonuclease insolubilized per gram of cyclized agarose product.

Halogenated and Aminated Cellulose Products

J.-P. Mareschi, S. Sebesi and E. Braye; U.S. Patent 3,959,079; May 25, 1976; assigned to L'Air Liquide, Societe Anonyme pour l'Etude et l'Exploitation des Procedes Georges Claude; and Orsan, les Produits Organiques du Santerre, France have found a method of insolubilizing active proteins by the combination of chemical activation of the support and bonding the proteins to the support by intermolecular bridging agents.

The protein is bonded to a chemically-activated, water-insoluble polymeric support by means of an intermolecular bridging agent, which process comprises halogenating and aminating a polymer containing free hydroxyl or carboxyl groups in an organic solvent by means of a halogenating reagent and a bifunctional aminating reagent to obtain the chemically-activated, water-insoluble polymeric support; and bringing the resultant support into contact with the protein which is to be rendered insoluble and then adding the intermolecular bridging agent.

The water-insoluble, polymeric support can be selected from cellulose and its derivatives, particularly carboxymethylcellulose; polyacrylic acid; methyl methacrylic acid; the polyamino acids obtained from acidic or basic amino acids and polyvinyl alcohol. The halogenating agent used in the chemical activation of the support can be a thionyl halide, a sulfuryl halide, a phosphorus trihalide or pentahalide, a phosphorus oxyhalide and a halide of p-toluene sulfoxide.

In one embodiment of the process, the chemical activation of the support is effected under heat, at a temperature at most equal to 90°C. In this embodiment, the organic solvent in which the chemical activation reaction takes place is an aromatic hydrocarbon, which optionally may be substituted, for example, on the aromatic nucleus; suitable solvents are benzene, the halobenzenes, nitrobenzenes, benzonitriles, toluene and the xylenes.

In a further embodiment, the chemical activation reaction is effected in the cold, that is to say at a temperature between 0°C and ambient temperature. Pyridine is the solvent which is preferred when the chemical activation takes place in the cold. However, variants can be achieved by using mixtures with a high proportion of pyridine and completed either by an aromatic hydrocarbon such as those previously described or by dimethyl formamide, formamide and dimethyl sulfoxide.

The bifunctional aminating agent brought into contact and activated by the halogenation agent can be selected from compounds represented by the formulas: $NH_2–NH_2$, $NH_2–R_1–NH_2$ and $R_2–HN–R_1–NH–R_3$, in which R_1 is an alkyl radical containing at most 20 carbons optionally substituted by one or more phenyl nuclei and/or a hydroxyalkyl radical containing at most 20 carbons; and in which R_2 and R_3, which mutually may be the same or different, are lower alkyl radicals or phenyl nuclei.

The active proteins and the chemically activated support are crosslinked by intermolecular bridging agents. It is possible to use polyfunctional agents which are known as bridging agents, such as glutaraldehyde and bis-diazobenzidine. It was found that the bridging agent can also be a dihaloketone, preferably dichloroacetone. The process is believed to be applicable to any protein, examples being protease, amylase, lipase, and ribonuclease.

Example: Chemical Activation of the Support and Fixation of Amino Groups – 15 grams of cellulose or carboxymethylcellulose are suspended in 200 ml of benzene or toluene containing 3% of pyridine and 5 ml of thionyl chloride. The temperature is brought to 80°C for 2 hours and the suspension is left overnight at ambient temperature. The cellulose is then washed with the same solvent and is reintroduced into a further quantity of the solvent, to which is added 5% of ethylene diamine and 3% of pyridine.

The mixture is brought to its boiling point and is boiled for at least 2 hours. The mixture is then filtered and the residue washed with the same solvent as was used initially, then washed copiously with water and dried under vacuum. The resultant chemically-activated cellulose or cellulose derivative is colored to a greater or lesser degree, depending on the solvent used and the degree to which the polysaccharide has been attacked. Elementary analysis shows that the chloride content, Cl^-, can reach a maximum of 20%.

Crosslinking of the Proteases on the Support – 1 gram of the activated cellulose or carboxymethylcellulose, having been subjected to the treatment described above, is brought into contact with 2 ml of 0.02 N phosphate buffering agent, at pH 6.8, containing 10 mg/ml of a protease, subtilopeptidase. The mixture is left at 4°C under vacuum until evaporation of the water has occurred, then 2 ml of an approximately 3% glutaraldehyde solution is added. The crosslinked enzyme is thus anchored on the support. Copious washing is then carried out with a solution of sodium bicarbonate and sodium chloride. The residual proteolytic activity of the resultant polysaccharide-protein complex represents a fixation rate of about 0.5% by weight of pure enzyme based on the weight of the support.

Sulfochlorinated Lignin-Containing Cellulose Supports

The disclosure by *J.P. Bourdeau, J.L. Seris and R. Pornin; U.S. Patent 3,960,666; June 1, 1976; assigned to Societe Nationale des Petroles d'Aquitaine, France* arises from the discovery that celluloses which contain a certain proportion of lignin give enzyme supports which are much better than the pure cellulose. In the past, it was thought necessary to prepare a cellulose as pure as possible for serving as a support for enzymes, because it was assumed that the presence of lignin and other vegetable substances capable of accompanying the cellulose prevented the fixation of enzymes. The complex of enzyme fixed on the cellulose according to this process is characterized in that the cellulose support contains 5 to 25% of lignin by weight of the dry material, preferably 10 to 20%, the lignin being dispersed uniformly throughout the cellulose.

The cellulose with a moderate lignin content can be derived from any one of the well-known natural sources, as for example wood, of which it is possible to use the waste material, especially chips or sawdust, shavings or the like; paper pulp can likewise be used provided that it has a lignin content which is within the above limits. One source recommended for the quality of the cellulose as regards the retention of enzymes, is the stalks of cereals and particularly maize cobs.

The enzymes are fixed to the lignin-containing cellulose by a process involving treating cellulose pieces with an alkaline solution to remove lignin on the surface of the pieces, treating the resultant cellulose pieces with a sulfochlorinating or a chlorinating agent in an excess of pyridine until sulfur and/or chlorine is combined with the cellulose, washing the sulfur and/or chlorine-containing cellulose pieces and treating the pieces with an aqueous enzyme solution to attach the enzymes.

Example: Maize cobs containing about 80% of cellulose and 15% lignin, are reduced into grains of 1.5 to 2.5 mm. 50 grams of these grains are placed in 18% sodium hydroxide solution and this is stirred for 15 minutes at normal temperature. The grains are then separated and thoroughly washed with cold water, until the filtration has only a slight coloration. The washing is completed with 1 liter of methanol and then with 400 ml of pyridine.

The grains are then poured into 800 ml of pyridine, after which 100 ml of $SOCl_2$ are added dropwise to the agitated pyridine during 45 minutes. The temperature of the reaction mixture is raised to 55°C and the medium suddenly becomes intensely colored. One hour after completing the introduction of the thionyl chloride, the mixture is poured into ice-containing water, and the grains are then separated and washed with several liters of water and then with an HCl solution of pH 2, until there is complete elimination of the pyridine adsorbed on the grains. The washing is completed with 1 liter of methanol. The grains thus obtained are dried at 40°C. It is found that the treated cellulose contains about 3.5% of sulfur and only traces of chlorine.

Into 160 ml of a buffering solution of phosphates of pH 6.2 are introduced 16 grams of the grains prepared as indicated above and 40 ml of rennet which is known as Carlin. The whole is maintained at 40°C under stirring for 24 hours. The grains are then separated and there is carried out the desorption of that portion of the rennet which is not chemically fixed, by washing: first of all with

250 ml of the same buffering solution as above, then with 2 liters of distilled water and finally with 250 ml of acetate buffering solution of pH 4.4. The mother solutions from these treatments are used again. The grains thus obtained of the complex consisting of rennet and maize cob cellulose are kept at 4°C in 250 ml of acetate buffering solution.

Grains of the cellulose-rennet complex, prepared as above, were used for the continuous curdling of milk in an experimental production of cheese. Using this supported enzyme, 35 liters of milk were treated in 55½ hours with the single charge of 8 grams of cellulose complex containing only 0.04 gram of rennet. When the usual method of curdling milk by directly adding rennet is carried out, the amount of 0.04 gram of the enzyme can only make 0.4 liter of milk curdle. This means that the complex of this process leads to an 87-fold increase in efficiency of the enzyme, as it permits treating 35 liters instead of 0.4 liter.

Acid Polysaccharides Modified with Diimides

The process developed by *P.J. Mill, M.A. Cresswell and J.G. Feinberg; U.S. Patent 4,003,792; January 18, 1977; assigned to Miles Laboratories, Inc.* provides a conjugate consisting of a complex, biochemically active organic molecule (the active component) covalently linked to a polysaccharide containing acid groups (the acid polysaccharide) via a proportion of the acid groups such that the conjugate, like the polysaccharide, containing free acid groups itself, is capable of forming a water-soluble sodium salt and a water-insoluble calcium salt. Linkage is generally effected through basic amino or phenolic hydroxyl groups of the active component to the acid groups of the polysaccharide.

Examples of suitable acid polysaccharides are pectin, pectic acid, alginic acid, celluronic acid, and carrageenan. Pectin is a natural product isolated, e.g., from apples and citrus fruits, and consists of a polymer predominantly of galacturonic acid in which some of the carboxyl groups are esterified. Pectic acid is the free acid obtained by saponification of pectin. Alginic acid is isolated from algae and consists of a polymer predominantly of mannuronic acid with possibly some glucuronic acid. Celluronic acid is a polymer of glucuronic acid and glucose produced by controlled partial oxidation of cellulose with nitrogen dioxide.

These acid polysaccharides were used either as the sodium salts obtained by freeze-drying their solutions in vacuo, or as the free acids, obtained by treating the dried sodium salts repeatedly with excess ethanolic HCl, washing with ethanol, and drying in vacuo.

Carrageenan is obtained from certain algae and consists of a sulfated polymer of galactose and anhydrogalactose. Suitable material for conjugation may be obtained by adding a calcium salt to a solution of the carrageenan, collecting the precipitate of calcium salt which forms, dissolving it by the addition of ethylenediaminetetraacetic acid disodium salt and dialyzing the solution against water. The solution obtained is passed through a column of a strong cationic exchange resin in the hydrogen form and the resulting solution of the free acid carrageenan is treated with sufficient sodium hydroxide solution to half neutralize the acid groups. Finally, this solution is freeze-dried in vacuo.

The active component of the conjugates may be any complex organic material containing basic amino or phenolic hydroxyl groups. Examples of the active component of this process are extracts of allergenic or antigenic substances, extracts of microorganisms, microbial products, toxins, toxoids, hormones and enzymes.

The conjugates are produced by methods for the covalent bonding of complex organic compounds containing basic amino or phenolic hydroxyl groups to acids and in this regard it is important to choose a technique which will neither destroy the biochemical activity of the components nor introduce undesirable groups into the conjugate. These techniques generally utilize a coupling agent.

One method involves the use of a water-soluble diimide such as the methyl-p-toluene sulfonate of 1-cyclohexyl-3-morpholinylethylcarbodimide as the coupling agent. In this method, the acid polysaccharide and the antigen are mixed in aqueous solution with this reagent. The latter is converted into the corresponding urea, and the acid groups of the polysaccharide react with the basic amino or phenolic hydroxyl groups of the active component to form amido or ester linkages, respectively.

In a second suitable method, the acid polysaccharide is first reacted with ethyl chloroformate in the presence of triethylamine to produce a mixed anhydride, and the latter is then reacted with the active component to form amide or ester linkages between the constituents of the conjugation.

In a third method, the acid polysaccharide is reacted with N-hydroxypiperidine in the presence of N,N'-dicyclohexylcarbodiimide to produce the N-hydroxypiperidine ester of the acid polysaccharide. This latter ester is then reacted with the active component which becomes linked to the acid polysaccharide via amide or possibly ester linkages with liberation of N-hydroxypiperidine.

In a fourth method, a partially esterified acid polysaccharide is reacted with hydrazine hydrate with formation of the corresponding hydrazide. This hydrazide is then treated with nitrous acid to produce the corresponding azide, and the latter is reacted with the active component which becomes linked to the acid polysaccharide via amide or ester linkages.

In a fifth method, an amine or metal salt of the polysaccharide is first reacted with sulfur trioxide-N,N'-dimethylformamide complex to produce a mixed anhydride, which is subsequently reacted with the active component with the formation of amide or ester linkages. All these methods have the advantage that they can be carried out in aqueous solution at relatively low temperatures.

The conjugates of this process have been found useful in reagent systems using a soluble form of a biochemically active substance which can be or is insolubilized in situ, the conjugate retaining its activity throughout the process. Alternatively, the insoluble form of the conjugate can be prepared and used as such. An example of in situ insolubilization involves use of a soluble conjugated enzyme or other proteinaceous substance such as albumin to clarify fruit juices and then removing the conjugate by the addition of calcium ion. Another use resides in employing an insoluble antigen conjugate to effect a slow release of antigen in tissue. Such slow release in tissue is often called a depot effect. The conjugates can also be used for the preparation of antisera to toxins in animals, e.g., snake Naja Naja (Cobra venom). In the dry state, the conjugates have good storage stability. The following example illustrates the process.

Example: Alginic acid (10 grams, 0.046 equivalent) and 1-cyclohexyl-3-morpholinylethyl-carbodiimide methyl-p-toluene sulfonate (10 grams, 0.024 mol) were stirred with triethylamine (3.05 ml, 0.022 mol) and the sodium salt of ovalbumin (1 gram) in water (200 ml) for seven days. After this time, calcium acetate solution (10%, 100 ml) was added, and the resultant precipitated gel was filtered off and washed with distilled water (1 liter) for an hour and then centrifuged. This washing was repeated seven times, using on one occasion 2 liters of water and stirring overnight.

The washed gel finally obtained was then dissolved in an aqueous solution of sodium ethylenediaminetetraacetate and sodium carbonate, and the solution obtained was dialzyed against distilled water. The dialyzed solution was freeze-dried, and 9 grams of product was obtained. It was found to have a protein content of 5 to 6% by Kjeldahl analysis and was useful as a slow release antigenic substance for studying antigen-antibody response in mice over a prolonged period of time.

The following table lists enzyme conjugates which have been prepared, giving the polysaccharide component, the protein component, and indicating the coupling method (MA = mixed anhydride method using ethyl chloroformate; SO_3 = SO_3-complex anhydride methods; and azide = azide method using pectin hydrazide).

Polysaccharide	Active Component	Method
Alginic acid	Bakers' yeast extract	SO_3
Alginic acid	Trypsin	MA
Pectin	Trypsin	Azide
Alginic acid	Chymotrypsin	MA
Pectin	Chymotrypsin	Azide
Alginic acid	Ficin	MA

Polysaccharides Having Thiol Groups

Thiols are functional groups which engage in a multitude of reactions even in aqueous systems under mild conditions. Through different mechanisms the thiol group may take part in reactions involving substitution, addition and oxidation as well as complex and salt formation. The product resulting from such reactions would contain S–C, S–S, S–O, S–M, S-metal and S-metal-C bonds. It should be stressed at this point that the thiol group easily undergoes reactions involving radical mechanisms.

R.E.A.V. Axen, J.O. Porath and P.J.E. Carlsson; U.S. Patent 4,048,416; September 13, 1977; assigned to Exploaterings AB T.B.F., Sweden have developed polymeric products characterized in that they consist of a water-insoluble hydroxyl or amino group containing polymer substituted with organic side-chains each containing one or more thiosulfate groups or derivatives of such groups such as thiol or disulfide.

Beaded agarose gels are preferably used for preparation of this product but other hydrophilic, hydroxyl-containing polymers such as crosslinked dextran have also found certain use in this connection.

Basically, the preparative methods for forming these products from agarose comprise:

(1) preparation of thiol-agarose from epichlorohydrin and sodium hydrosulfide;

(2) prepartion of thiopolymer from epichlorohydrin and thiosulfate;

(3) preparation of thiol-agarose by means of 1,4-butanediol-diglycidylether, sodium thiosulfate and dithiothreitol (DTT) — the thiosulfate containing polymer can be recovered as a storage safe product which might then be readily converted into a thiol polymer or used to prepare a reactive polymer for fixation of thiol containing compounds;

(4) preparation of thiol-agarose from divinyl sulfone, sodium thiosulfate, and dithiothreitol (DTT); and

(5) prepartion of thiol-agarose from allylbromide, bromine, sodium thiosulfate, and dithiothreitol.

Enzymes often contain thiol groups which in many cases are directly associated with their activity. Most proteins however contain –S–S– bridges which are easily converted into thiols by treatment with suitable highly substituted thiol polymers of the kind described above.

Thiol polymers can, on the other hand, easily be transformed into polymers with disulfide groups of high reactivity in thiol-disulfide exchange reactions. If such a polymer is allowed to react with a thiol-containing protein, the resulting product is an –S–S– protein polymer. It is however possible, as will be shown below, to activate a thiol polymer even in other ways in order to make it readily available for fixation of thiol containing compounds.

Thiol polymers can be used either directly or after transformation of their thiol groups into suitable disulfides or other similar structures for chemisorption of natural products, for example, in protein separations.

Disulfide polymers which lead to a quick and complete reaction with a thiol component in solution are here referred to as "reactive" polymer disulfide structures. In principle such structures can be internal, in which case the enhanced reactivity could be explained by conformational strain, in particular ring tension. In the more important case of external structures such as polymer –S–S–L, the group derives its reactivity from the ligand, L.

Such reactive polymers include the disulfide with 2-thiopyridone or with a derivative of dithiocarbonic acid. Of special interest in this context would be the possibility for immobilizing enzymes via a disulfide bridge. The binding lends itself to reductive cleavage under suitable conditions. A column of immobilized enzyme, which after a certain period of use has lost its activity, can thus be regenerated in situ through reductive elimination of inactivated enzyme. After reactivation of the polymer (still in situ), the column reactor is then ready for renewed coupling of fresh enzyme.

Enzymes without a free thiol group can have their disulfide groups reduced by these thiol-containing polymers and be subsequently coupled. The following example shows the immobilization of enzymes through fixation of thiol agarose by means of thiol-disulfide exchange reactions.

Example: (a) Preparation of 2-Pyridine Disulfide Agarose — Beaded 2% agarose gel was crosslinked with epichlorohydrin; 180 grams of washed and suction-dried gel was suspended in 142 ml of 1 M sodium hydroxide solution followed by an addition of 5 ml epichlorohydrin. The reaction time was 1 hour at 60°C. The gel was then washed and suspended in 180 ml of 0.5 M phosphate buffer (20.5 grams $NaH_2PO_4 \cdot H_2O$, 14.0 grams $Na_2HPO_4 \cdot 2H_2O$ in 500 ml solution). 180 ml of a 2 M sodium thiosulfate solution was added to the suspension and the mixture was left to react at room temperature for 6 hours.

The resulting product was washed with water and reduced by 216 mg dithiothreitol dissolved in 180 ml of a 0.1 M sodium bicarbonate solution made 1 mM with respect to EDTA. The product was washed with 1 M sodium chloride solution, 1 mM with respect to EDTA, then with a 1 mM EDTA solution and finally with 50% aqueous acetone; 4 grams of pyridine disulfide was dissolved in 180 ml of 50% aqueous acetone, mixed with the gel, and reacted for 30 minutes at room temperature.

The resulting product was washed on a glass filter with 50% aqueous acetone and finally with 1 mM EDTA solution. Analyses showed the oxirane content after treatment with epichlorohydrin to be 80 μeq/g dry product; the content of 2-pyridine disulfide structures amounted to 50 μeq/g dry product as determined by N-analysis.

(b) Immobilization of Amyloglucosidase — Amyloglucosidase was reduced by a highly substituted thiol agarose. The reducing column was 11 cm in length and contained 17 ml of thiol agarose prepared from 2% agarose with epichlorohydrin, sodium thiosulfate and dithiothreitol. The degree of substitution was about 750 μeq/g dry product. In direct connection with the reducing column was installed a coupling column with an inner diameter of 2.1 cm containing 3.2 ml of 2-pyridine disulfide agarose prepared according to (a) above.

A solution of 41.4 mg amyloglucosidase in 6 ml of 0.1 M sodium bicarbonate solution, made 1 mM with respect to EDTA, was prepared, out of which a 2 ml portion was taken and pumped through the reduction column and also through the subsequent coupling column at a flow rate of 10.7 ml/hr. The coupling column was then washed at the same flow rate with the following solutions in the indicated order and durations:

Solutions	Time (hr)
0.1 M sodium bicarbonate solution/ 1 mM EDTA	2
0.1 M sodium acetate solution, pH 4.8	15
0.5 M sodium acetate solution, 1% with respect to starch	3
0.01 M sodium acetate solution, pH 4.8	3

The coupling column was shown to have adsorbed an activity corresponding to 0.32 enzymic units/mg conjugate. The amount of amylglucosidase activity pumped into the column was 80 units, corresponding to 1.25 units/mg dry gel.

The enzymic activity was determined from a standard curve based on glucose, and the enzymic unit was defined as the amount of glucose (mg) formed in a 3-minute period.

During a control experiment 2 ml of the amyloglucosidase solution was pumped directly into a column of pyridine disulfide agarose without previous passage through the reducing column. The washing procedure was carried out as described above. No activity was adsorbed onto the pyridine disulfide column.

COVALENT BONDING TO SYNTHETIC POLYMERS

POLYMERS HAVING ALDEHYDE GROUPS

Sulfited Acrolein Polymers Crosslinked with Dialdehydes

P.S. Forgione; U.S. Patent 3,847,743; November 12, 1974; assigned to American Cyanamid Company has found that a greater quantity of enzyme can be covalently bound to carbonyl polymers, such as aldehyde and ketone polymers, if the carbonyl polymer is crosslinked, i.e., rendered insoluble, with a dialdehyde rather than other crosslinking agents since the number of aldehyde groups available for enzyme binding is not decreased. The crosslinking reaction is an aldol condensation reaction and occurs between the reactive groups of the carbonyl polymer and the α-carbon atom of the dialdehyde. The reaction is carried out under typical aldol reaction conditions such as set forth in "The Aldol Condensation," Nielsen et al, *Organic Reactions,* Vol. 16, pages 1-85, John Wiley and Sons, NY, 1968. In general, the reaction is conducted at 0° to 80°C at a pH of over 7.0 by the addition of suitable materials, e.g., sodium hydroxide in ethanol.

The crosslinked polymer formed according to this process is insoluble and therefore precipitates from solution. These crosslinked polymers are preferably prepared for the subsequent binding of catalytically active enzymes. To this end, the carbonyl polymer must be first rendered susceptible to reaction with the enzyme. This is accomplished, in most aldehyde polymers, by reaction of the polymer with a suitable solubilizing agent such as a bisulfite, specifically an alkali metal or alkaline earth metal bisulfite. The reaction is conducted at a temperature ranging from 25° to 90°C, at atmospheric pressure, although superatmospheric or subatmospheric pressure can be used if desired, the process being more specifically disclosed in U.S. Patent 2,657,192. After the bisulfite treatment, the aldehyde polymers are then made hydrophilic by crosslinking.

For example, in a specific embodiment, polyacrolein, a water-insoluble polymer, is first contacted with a bisulfite such as sodium bisulfite to render it water-soluble. In such a condition, however, the polymer cannot be reacted with an enzyme because recovery of the product is relatively impossible. Crosslinking of the bisulfite-polymer product with a dialdehyde, however, renders it gel-like and effectively hydrophilic so as to allow reaction with the enzyme. The result of these two reactions is that the bisulfite breaks some of the heterocyclic rings of the polyacrolein creating more enzyme-reactive aldehyde groups in addition to a series of bisulfite groups. The dialdehyde reacts with some of these aldehyde and bisulfite groups with the formation of –CH=C(CHO)– linkages between two polymer molecules, thereby crosslinking the polyacrolein. Reaction of the enzyme, e.g., invertase, forms an adduct or covalent bond between the enzyme, the aldehyde groups of the crosslinking linkages and the other available aldehyde groups, and also through the bisulfite groups.

Examples of carbonyl polymers which may be used include those produced by known procedure from such aldehyde monomers as acrolein; α-alkyl acroleins, e.g., methacrolein, α-propylacrolein; crotonaldehyde; 2-methyl-2-butenal; 2,3-dimethyl-2-butenal; 2-ethyl-2-hexenal; 2-decenal; 2-dodecenal; 2-methyl-2-pentenal; 2-tetradecenal and the like, alone or in mixture with up to 95 wt %, based on the total weight of the copolymer, of each other and/or such other copolymerizable monomers known to react therewith.

Example 1: 4.0 parts of a 20% glutaraldehyde solution are slowly added with stirring to 40 parts of a 10% solution of polyacrolein-sodium bisulfite adduct (MW 80,000) previously adjusted to pH 10 with 10% sodium hydroxide solution. The mixture is stirred at 23°C for 1 hour and the off-white hydrophilic product which forms is then filtered and washed to neutrality.

The above wet hydrophilic adduct is suspended in 50 parts of water and reacted with 0.104 part of invertase (twice recrystallized) which has been first dissolved in 4 parts of water. The enzyme reaction mixture (pH 6.5) is gently stirred for 18 hours at 10°C and the resulting hydrophilic covalently bound enzyme polymer adduct is then washed free of unreacted enzyme. Assay of the hydrophilic enzyme-polymer adduct shows high activity with sucrose solution, in that 4 wt % of the isolated wet cake converts 100 ml of a 10% sucrose solution at pH 4.8 to 83.0% invert sugar in 15 minutes at 42°C.

Example 2: 5 parts of a copolymer of acrolein-styrene (93.8:6.2 weight ratio, respectively) are suspended in 40 parts of water containing 4.3 parts of sodium metabisulfite at pH 5.7. The mixture is stirred over nitrogen at 65°C for 5 hours, resulting in a completely water-soluble product. To this bisulfite addition product is slowly added, with stirring, 4.0 parts of a 20% solution of adipaldehyde and the pH of the reaction mixture is adjusted to pH 10.5 with 10% sodium hydroxide solution. The reaction mixture is stirred at 20°C for 2 hours. The resultant crosslinked, hydrophilic copolymer is filtered, washed to neutral pH, suspended in 20 parts of water and reacted with 4 ml of technical grade (k = 0.6) invertase at 18°C for 15 hours. The resulting covalently bound enzyme polymer adduct is washed free of unbound enzyme with distilled water and filtered (8.4 parts wet). Assay of 10 wt % of the wet product with 100 ml of 10% sucrose solution results in a 46.0% conversion to invert sugar.

Acrolein-Acrylic Acid Copolymers

E. van Leemputten; U.S. Patent 4,017,364; April 12, 1977; assigned to Societe d'Assistance Technique pour Produits Nestle S.A., Switzerland has provided enzymatically active products by reacting an enzyme with a polymer carrying both free aldehyde and free acid groups. The product obtained is characterized by its variable solubility in aqueous medium. This product combines the advantages of the corresponding free enzymes, i.e., accomplishment of the enzymatic reaction in homogeneous liquid medium, with the advantages of "insolubilized enzymes" because it is possible, by adjusting the pH, to recover the product on completion of the reaction by a simple mechanical separation and then reuse it.

A polymer containing free aldehyde and acid groups may be obtained by the polymerization of a suitable monomer or by the copolymerization of, for example, a mixture containing a monomer giving the free aldehyde groups and a monomer giving the free acid groups, or even by the creation of free aldehyde groups and/or free acid groups on a suitable polymer. It is essential to provide for the presence of a sufficient number of free acid groups to ensure the variable solubility required. It has been found that the presence of approximately 5% of free acid groups among all the free aldehyde and acid groups provides for this variable solubility. Obviously, a minimum number of free aldehyde groups is required for arresting the enzyme. It is desirable to have as many free aldehyde groups as possible to be able to fix as much enzyme as possible and hence to obtain, after fixation, a product with interesting enzymatic activity.

There are no crucial limits to the quantities and concentrations of enzyme and polymer to be used. Generally, any pH value between 1 and 11, providing it does not affect the integrity of the enzyme and the polymer, is suitable, although the fixing reaction is preferably carried out at a pH value of 7 to 8 at which the polymer is in solution. Similarly, although the fixing reaction is preferably carried out at ambient temperature, it may also be carried out at a higher temperature, but below the temperature at which the enzyme is inactivated. Finally, the reaction times may vary from a few minutes to several hours. However, the best results have been obtained with reaction times of 1 to 3 hours.

The product obtained has a variable solubility which, as already mentioned, is governed by the pH. It is pointed out that, in cases where the free acid groups are all carboxylic acid groups, the precipitation pH is also 4.5 which signifies that the presence of enzyme fixed to the polymer does not appreciably alter the solubility characteristics. Thus, at a high pH, the enzymatically active product will be soluble and will be extremely similar in its behavior to the free enzyme. By contrast, at a low pH, the product will precipitate and may readily be separated. After washing, this product may be reused in another cycle comprising dissolution followed by reprecipitation, and so on.

Example: (A) Preparation of the Acrolein/Acrylic Acid Copolymer — A mixture containing 450 ml of degassed oxygen-free water, 60 ml of freshly distilled acrolein, 9 ml of acrylic acid, 6.6 ml of a solution of water oxygenated to 30% and 12 ml of 2 N sulfuric acid, is prepared in a nitrogen atmosphere. 150 ml of a 2.75% solution of sodium nitrite in degassed water is then progressively added over a period of 40 minutes. On completion of the reaction, the pH value is between 3 and 4 and the copolymer is present in precipitated form.

After filtration, the precipitate is washed first with a solution of 0.002 M hydrochloric acid and then with acetone and dried. 7.75 g of the required copolymer are thus obtained. The presence of free aldehyde groups is detected by measuring the reducing power by the method of J.S. Thompson et al in *Anal. Biochem.* 22, 260 (1968). In addition, the presence of free acid groups is detected by titration. It is found that 14 wt % of the copolymer is in the form of free aldehyde groups and 17.6 wt % in the form of free acid groups.

(B) Fixation of Trypsin to the Copolymer – 600 mg of the above copolymer are dissolved at ambient temperature in 150 ml of a 0.2 M phosphate buffer with a pH value of 8. During dissolution and also subsequently during the fixation reaction, this pH value is kept at 8 by the addition of a 1 N soda solution. After complete dissolution of the polymer, 150 mg of crystallized trypsin are added and the mixture left to react for 1 hour at ambient temperature. The pH value is then lowered to 4 by the addition of hydrochloric acid. The precipitate formed is filtered and washed 3 times with water. 557 mg of enzymatically active product are obtained after drying. An adaptation to the determination of trypsin of the method of S. Blomberg et al described in *Eur. J. Biochem.* 15, 97 (1970) showed an enzyme activity of 80% for the bound enzyme.

(C) Solubilities of the Product – 93 mg of enzymatically active product, prepared in accordance with (B), are dissolved in 25 ml of a solution of a 0.2 M phosphate buffer of pH 8. The enzymatic activity of the solution is given the coefficient 100. 1 N hydrochloric acid is progressively added; the product precipitates gradually.

Samples are taken at regular intervals and centrifuged. The residual enzymatic activity of the supernatant phase is then measured. It is found that the enzymatic activity of the liquid suddently decreases around a pH of 8.4. The value measured at pH 6 is still 98%, whereas at pH 4.6 it has fallen to 2%. The residual activity is zero below pH 4.5. In other words, all the enzymatically active product has been converted into insoluble form. The insoluble product thus obtained is redissolved in a solution of pH 8 and the enzymatic activity of the solution is determined. It is found to have 98% of the enzymatic activity of the starting solution.

This enzymatically active product may thus be quantitatively separated from the enzymatic reaction medium by lowering the pH to a value below 4.5 and may be reused without significant losses by redissolution at a pH value of around 8 at which trypsin develops its maximum enzymatic activity.

Oxidized Polyethylene Glycol Monomethyl Ethers

There has been reported certain experimental work which involves attaching polyethylene glycol to enzymes via reaction of the enzyme with a polyethylene glycol derivatized triazine (F.F. Davis, *Abstract of the First Chemical Congress of North America,* December 1975). The product is suggested as being useful for enzyme therapy work since it may prevent immunogenicity, may lower susceptibility to proteolysis, and is harmless. For example, the polyethylene glycol derivatives of the enzymes asparaginase or phenylalanine-NH_3 lyase and uricase may be useful for the treatment of leukemia and gout, respectively.

G.P. Royer; U.S. Patent 4,002,531; January 11, 1977; assigned to Pierce Chemical Company provided a new method for preparing polyethylene glycol derivatives of enzymes. This method starts with the aldehyde derivative of a monoalkyl polyethylene glycol (formed by oxidation of the glycol) which is then used to reductively alkylate the enzyme in the presence of sodium borohydride. The method is considered to have the advantages of greater retention of enzymatic activity and greater stability of the glycol derivatized enzymatic product.

Example: Polyethylene glycol monomethyl ether (1,000 MW) is oxidized to the aldehyde with MnO_2; the isolated aldehyde is reacted in aqueous solution (borate buffer pH 8.5, 10 mM benzamidine) with trypsin for 10 min at 0°-4°C [Royer et al, *Biophys. Res. Commun.* 64, 478 (1975)], then treated with $NaBH_4$ to yield the polyethylene glycol enzyme derivative.

Oxidized Polyvinylene Glycols

A simple process for producing a bond between a substance containing primary amino groups and a synthetic polymer containing vicinal hydroxyl groups, preferably polyvinylene glycol is used by *A. Sieber; U.S. Patent 4,066,581; Jan. 3, 1978; assigned to Behringwerke AG, Germany.* The polymer is reacted with a medium to strong oxidant to form aldehyde groups and subsequently reacted with a compound containing primary amino groups. Suitable oxidants are those with a potential of approximately +1.0 to +1.5 V, e.g., compounds of hexavalent chromium such as $CrO_4^=$ or $Cr_2O_7^=$, MnO_2, or Cl_2, or even stronger oxidants can be used with a potential above +1.5 V, e.g., MnO_4^+, O_3 or F_2.

The degree of oxidation of the polymer depends, of course, on the concentration and the time of action, which should be considered when selecting the reaction conditions. The oxidants are generally used in a concentration of 0.01 to 1.0 M. Strong oxidants having a potential above +1.5 volts are preferably used in a concentration of 0.01 to 0.1 M, while oxidants having a potential of +1.0 to +1.5 volts are used in a concentration of 0.05 to 1.0 M. The time of action of the oxidant on the polymer depends on its efficiency; in the case of a relatively weak oxidant, for example, the system of dimethyl sulfoxide/glacial acetic acid, it is from 3 to 18 hours, for the system of dichromate/sulfuric acid it is from 5 to 60 minutes and for the system periodate/water 1 to 15 minutes. The time of action of other oxidants depends on their efficiency.

As the polymers used, for example, polyvinylene glycol, are chemically very stable compounds, the dispersion agent in which the reaction with the oxidant is to be performed is not very critical. Especially good results have been obtained by reacting, for example, polyvinylene glycol with periodate in water, with dichromate in about 0.1 N sulfuric acid and with dimethyl sulfoxide in glacial acetic acid. The reaction product of the oxidation can be directly reacted with biologically active compounds, e.g., proteins, containing primary amino groups. The oxidized polymer can be directly reacted with compounds of this type. The oxidation product of the polymer can be reacted not only with a biologically active compound, e.g., an enzyme, a plasma protein, or an amino acid, but also with a compound containing amino groups and capable of reacting with biologically active substances.

In the reaction of the aldehyde groups of the oxidized polymer with the primary amino groups of the participants in the reaction Schiff's bases are formed which are preferably stabilized by reduction. Suitable reducing agents to be used in known manner are complex boron hydrides, such as sodium boron hydride and cyano-boron hydride, complex metal hydrides, such as lithium aluminum hydride; Raney nickel is an example of a catalyst for the hydrogenation with hydrogen.

The process further relates to the compounds produced by binding compounds containing primary amino groups, preferably biologically active substances, to synthetic polymers containing vicinal hydroxyl groups, and to the use of the compounds as specific adsorbents or as selective reaction catalyst.

Example 1: Oxidative Activation of Polyvinylene Glycol (PVG) — 1 g of PVG was suspended in 50 ml of a 0.05 molar sodium metaperiodate solution and the suspension was stirred for 15 minutes at room temperature. Next, the suspension was filtered off with suction and the filter residue washed several times with distilled water.

Example 2: Direct Bond Between Protein (Immuno Globulin) and the Activated Carrier — The activated carrier of Example 1 was suspended in 59 ml of 0.67 molar phosphate buffer having a pH of 7.5 in which about 40 mg of protein had been dissolved and the suspension was stirred for 3 hours at room temperature. The mixture was then kept overnight at a temperature of 4°C. When the reaction had taken place, the suspension was centrifuged and the residue washed with 30 ml of NaCl. In this manner, about 18 mg of immuno globulin could be chemically bound to the carrier.

The oxidized carrier of Example 1 may also be reacted with various groups to provide active "spacers" which enlarge the distance of the biologically active grouping from the carrier molecule. Suitable spacers are, for example, hexamethylenediamine, vinyl sulfone derivatives containing diazotizable arylamino groups or sulfuric acid esters of β-hydroxy-ethylsulfones. By reacting the oxidized polymer with glutathione and then reacting the product obtained with 2,2'-dipyridyl disulfide and a compound containing free sulfhydryl groups, for example, a bond via a disulfide bridge can be produced. In this case, glutathione is the spacer.

POLYMERS WITH EPOXY GROUPS

Nitrile-Modified Epoxy Resins

High molecular weight glycidyl ethers of bisphenol A in which the hydroxyl groups have been reacted with acrylonitrile or methacrylonitrile are used by *J.S. Matthews; U.S. Patent 3,841,970; October 15, 1974; assigned to Gulf Research & Development Company* to immobilize enzymes such as glucose oxidase.

In reacting the solid polymer of diglycidyl ether of bisphenol A with the nitrile reactant, a stoichiometric excess of the nitrile reactant is used as a solvent for the polymer. The reaction is preferably carried out under reflux at the boiling point of the nitrile reactant, which is about 77°C for acrylonitrile and about 90°C for methacrylonitrile. If the polymer is not completely soluble in the nitrile reactant, a suitable liquid chlorinated hydrocarbon solvent can be used in conjunction with the nitrile reactant. The epoxy polymer and the nitrile reactant are heated under conditions of reflux, preferably in the presence of a tertiary amine base catalyst, such as triethylamine, until the reaction is completed.

The polymer product is then precipitated by pouring the solution into a liquid in which the polymer is insoluble such as a hydrocarbon solvent including hexane, heptane, and the like, a lower alkyl alcohol including methanol, butanol, water, or mixtures of water and alcohol, and the like.

The reacted polymer is then dried and finely powdered for reaction with the enzyme. This immobilization reaction is preferably carried out at a relatively low temperature, such as 0° to 10°C to avoid undue denaturation of the enzyme and with gentle agitation to insure good contact of the enzyme with the epoxy groups on the polymer. This enzyme immobilization reaction is carried out for a long enough time to bind a suitable amount of the enzyme on the polymer, generally for 1 to 8 hours.

In order to bind an enzyme to the epoxy containing polymer, the polymer is dispersed in a water solution of an enzyme and then the mixture is agitated. Generally the weight of the enzyme in the solution is no greater than about 1% of the weight of the polymer support which is used. The reaction binding each enzyme molecule to the polymer occurs through one or more of a large number of the epoxy groups available for reaction on the surface of the polymer particles and through one or more epoxy reactive groups in the enzyme.

Example: A 25 g portion of Epon 1010, a solid diglycidyl ether of bisphenol A having an average molecular weight of about 10,000, was introduced into a round-bottom 500 ml flask fitted with reflux condenser and heating mantle. This epoxy polymer was dissolved in 100 ml of chloroform by refluxing. After solution was obtained, 50 ml of acrylonitrile and 5 ml of triethylamine were added and the mixture was refluxed overnight.

The contents were then evaporated to dryness under a vacuum. The solid residue was dissolved in a solution containing 50 ml of dimethylformamide and 100 ml of acetone. The polymer was precipitated from solution by slowly pouring the polymer solution into two liters of a 1:1 volume ratio mixture of methanol and water. The solid precipitate was dried under vacuum yielding 22 g of a white powder. The powder was ground and sieved through 100 mesh. The epoxy equivalent value was found to be 31,000 and the nitrogen content was determined to be 1.13 wt %.

A 500 mg portion of this nitrile-modified epoxy resin was placed into a 30 ml screw cap bottle containing 15 ml of an aqueous solution of 1.725 mg of glucose oxidase having an activity of 183 IU. The mixture was shaken in a cold room at about 0°C for two hours. The solid product was filtered and by analysis was found to contain 1.664 mg of immobilized glucose oxidase (3.33 mg/g resin). The activity of this product was determined to be 54 IU (30% activity). The same procedure was repeated using unmodified Epon 1010 resin to give a product having 1.37 mg/g of enzyme on resin with 0.6% activity.

Glycidyl Ester Copolymers

J.S. Matthews; U.S. Patent 3,844,892; October 29, 1974; assigned to Gulf Research & Development Company has also immobilized enzymes with polymers made by copolymerizing an epoxy monomer containing a 1,2-epoxy group and terminal unsaturation with an olefin monomer such as acrylonitrile by free radical polymerization.

Suitable epoxy monomers include glycidyl acrylate, glycidyl methacrylate, allyl glycidyl ether, and the like. Other epoxy monomers which are useful include those which are prepared by the reaction of an epoxy resin with a difunctional olefin to form the epoxy monomer as defined.

The epoxy monomer is copolymerized with an olefin monomer by free radical polymerization to form the epoxy-containing polymer. In the final step this epoxy-containing polymer and the desired enzyme are reacted to immobilize the enzyme on the polymer support.

Example 1: A 27 g portion of methacrylonitrile and 250 ml of benzene were placed in a 500 ml resin kettle equipped with mechanical stirrer, condenser, nitrogen inlet and heating mantle. The conventional antioxidant-polymerization inhibitor had first been removed from the methacrylonitrile by passing it through a column of silica gel. To this solution were added 14.2 g of glycidyl methacrylate and one gram of azobisisobutyronitrile. The solution was heated at reflux with stirring for 18 hours. The cloudy solution was then slowly poured into two liters of rapidly stirred hexane to produce a flocculent white solid. The solid product was filtered, washed with hexane, dried under vacuum and then ground to a powder. This epoxy-containing product was determined to have an epoxy equivalent value of about 900 and was obtained in 32% (13 g) yield based on total reactants.

A 1 g sample of the polymer was placed in a 30 ml screw cap bottle along with 10.0 ml of a glucose oxidase solution having an activity of 100 IU. The bottle and contents were shaken overnight at a temperature of about 0°C. A colorimetric analysis of the polymer and solution demonstrated that the polymer had immobilized 0.037 mg of glucose oxidase per gram of polymer and was determined to have an activity of 10 IU.

Example 2: A 27 g (0.05 mol) portion of Epon 834 (a diglycidyl ether of bisphenol A having an epoxy equivalent value of 280) dissolved in 100 ml of toluene, 8.2 g (0.1 mol) of methacrylic acid and 1.0 ml of triethylamine were placed in the resin kettle described in Example 1. The reaction mixture was refluxed with stirring under a nitrogen atmosphere as 1 ml samples of the reaction mixture were periodically sampled and titrated with 0.1 N sodium hydroxide to analyze for methacrylic acid. When analysis indicated that about 95% of the methacrylic acid had been consumed, 50 ml of toluene were removed by distillation.

This was followed by the addition of 200 ml of benzene and 27 g (0.4 mol) of methacrylonitrile (purified as described in Example 1) and 1 g of azobisisobutyronitrile. The reaction mixture was stirred under reflux for 5 hours during which time the polymer product came out of solution. The polymer product was filtered, washed with benzene, ground and dried under vacuum yielding 61% based on total reactants (38 g) of a fluffy, white powdered polymer product having an epoxy equivalent value of 3,817.

A 0.5 g portion of this polymer was shaken for 4 hours with 5.0 ml of glucose oxidase solution having an activity of 105 IU. Colorimetric analysis revealed that the polymer had immobilized 0.117 mg of glucose oxidase per gram of polymer. This immobilized product had an activity of 27 IU.

Microbial Cells Coupled to Glycidyl Ester Copolymers

R.P. Nelson; U.S. Patent 3,957,580; May 18, 1976; assigned to Pfizer Inc. has found that microbial cells can be immobilized by chemical covalent bonding of the cells to water-insoluble particulate polymer matrix. The chemical bond can be formed either with the preformed polymer or with reactive monomer prior to polymerization. Further, treatment of the cells with a polyfunctional cross-linking agent either prior to, during or after bonding reduces enzyme loss from the cell.

In this process, intact microbial cells are chemically bonded to water-insoluble particulate polymer through reactive groups on the polymer. Such bonding will frequently occur by reaction of cell amino groups with the reactive polymer substituents. In addition, other groups such as hydroxyl and sulfhydryl are also available in the cells and can play a role in the bonding mechanism.

The bonding can occur either before or after the polymer formation. In the former case, the cells are contacted with a polymerizable ethylenically unsaturated monomer containing a reactive group, and the cell-carrying monomer is then polymerized or copolymerized in the presence of a crosslinking monomer and an initiator system. In the latter case, the cells are contacted directly with any particulate polymer containing a reactive group. This includes natural as well as synthetic organic polymers.

Monomers and polymers containing epoxide, halocarbonyl and halomethyl carbonyl groups are preferred for the bonding in this process. Especially desired are those polymers containing the reactive monomers 2,3-epoxypropyl methacrylate (glycidyl methacrylate), 2,3-epithiopropyl methacrylate, methacryloyl chloride and bromoacetylhydroxyethyl methacrylate. In the case of monomers and polymers containing nonreactive functional groups, the functional groups may be converted to reactive substituents by known methods.

The immobilization process is applicable to all microbial cells such as bacteria, yeasts, actinomyces and fungi. Preferred cells are those in which the primary enzyme systems are oxidoreductases, especially those such as glucose oxidase which do not require soluble cofactors; hydrolases such as α,β-amylases and peptidases, and especially penicillin acylase; lyases, preferably those concerned with breakage of carbon-oxygen bonds and especially those such as phenylalanine ammonia-lyase and aspartate ammonia-lyase which are concerned with breakage of carbon-nitrogen bonds; and isomerases such as racemases, epimerases, cis-trans isomerases, and especially glucose isomerase.

The covalent bonding of the cells to reactive polymer, or the bonding to reactive monomer followed by polymerization or copolymerization, is run in aqueous medium. The temperature for bonding to polymer or monomer, or for polymerization after bonding to monomer, is preferably from 10° to 30°C. The time required for bonding to polymer is 15 to 30 hours, while that for bonding to monomer is 1 to 5 hours. Polymerizations are normally completed in 1 to 6 hours. The weight ratio of polymer to cells (dry basis) can vary considerably, with the preferred ratio being about 0.5 to 4.

When cells are bonded to reactive monomer, the subsequent addition polymerization is run, with or without one or more comonomers, in the presence of difunctional crosslinking monomer and an initiator system. Any of those comonomers, crosslinking monomers and initiator systems commonly used in this type of polymerization and which do not destroy the cell enzymatic activity can therefore readily be used.

The polymerization and copolymerization reactions are initiated by free radicals with redox initiator systems being preferred. Representative of such systems are peroxy compounds such as ammonium, sodium, or potassium persulfate, benzoyl peroxide, and di(sec-butyl)peroxydicarbonate in combination with a reducing agent such as sodium thiosulfate, sodium metabisulfite, ferrous ammonium sulfate hexahydrate, dimethylaminopropionitrile or riboflavin. Preferred initiator system is dimethylaminopropionitrile-ammonium persulfate.

In certain cases, the cells are treated with a polyfunctional crosslinking reagent to reduce enzyme loss from the cells. While the exact mechanisms involved are not fully understood, it is believed that one class of the polyfunctional reagents crosslinks amino groups of the cell membrane, thus effectively reducing the porosity of the membrane; the other class accomplishes this by activating the carboxyl groups in the membrane which then react with amino groups through amide bond formation. Typical examples of the first class include pyruvic aldehyde, glyoxal, hydroxyadipaldehyde, cyanuric chloride, bis-diazotized benzidine, 1,3-difluoro-4,6-dinitrobenzene, toluene 2,4-diisocyanate, and especially glutaraldehyde, while the second class includes ethyl chloroformate and water-soluble carbodiimides.

Example 1: To 80 g glycidyl methacrylate plus 1.0 liter of water was added with stirring 30 g of N,N'-methylenebisacrylamide. After the mixture was stirred for 15 minutes at room temperature, 10 g of methyl methacrylate was added, the mixture was cooled to 5°C and nitrogen gas was bubbled into the cold mixture for 15 minutes. Then 20 ml of dimethylaminopropionitrile and 2 g of ammonium persulfate were added. The mixture under nitrogen was stirred at 10°C until polymerization began, then an additional hour at ambient temperature, and finally left to stand for 2 hours. About 750 ml of water was added to the resulting solid, and the mixture was stirred vigorously to break up the polymer. The solids were filtered, washed with water and air dried to give 120 g of white particulate polymer.

An *A. niger* (NRRL-3) fermentation broth, grown under submerged aerobic conditions at 33°C and pH 5.8 to 7.0 on glucose substrate, was filtered and the cells were washed with water and expressed to a semidry paste. The cells (250 g, 55 g dry) were placed in a solution containing 22 g of 25 wt % aqueous glutaraldehyde in 1,000 ml of water. The suspension was adjusted to pH 6.2 and stirred for 1¾ hours at room temperature. The solids were filtered and washed with water to give 104 g of wet treated cells possessing 76% of the original glucose oxidase activity.

A suspension of 32 g of wet treated cells and 16 g of the glycidyl methacrylate polymer in 240 ml of water was stirred for 30 hours at room temperature and then filtered. The solids were washed with water to give 126 g of wet immobilized cells having 58% of the glucose oxidase activity of the original untreated cells.

Example 2: The wet immobilized cells of Example 1 (47.3 g) were placed in 100 ml of an aqueous solution containing 10 g of glucose, and the mixture was stirred with aeration for 21 hours at room temperature. A 99% conversion to a mixture of gluconic acid and delta-gluconolactone was obtained, and 75% of the original enzyme activity was recovered.

Example 3: To 25 g of 2-hydroxyethyl methacrylate in 100 ml of water was added 12.5 g of cyanogen bromide dissolved in 100 ml of water. The mixture was immediately adjusted to pH 11 with 5 N NaOH, causing the temperature to rise to 46°C, and then maintained at that pH until no further base was required. The mixture was then cooled to room temperature, adjusted to pH 6.5 and treated with a slurry of 50 g of freeze-dried *Proteus rettgeri* cells (grown on lactic casein substrate at 28°C at pH 6.8 to 7.0 and isolated as a paste before drying) in 500 ml of water. The resulting suspension was stirred for 45 minutes.

Then 25 g of acrylamide and 2.5 g of N,N'-methylenebisacrylamide were added, and the suspension was stirred for 15 minutes, cooled to 4°C and treated with 4 ml of dimethylaminopropionitrile plus 425 mg of ammonium persulfate. The polymerization mixture was allowed to warm to room temperature and stand 1 hour. The mixture was then blended, using a Waring blender, and centrifuged. The gel-like centrifuge solids were resuspended in water, recentrifuged, washed with water and freeze-dried to give 112 g of dry immobilized cells containing 19% of the original penicillin acylase activity.

POLYMERS WITH CARBOXYL OR ACTIVE CARBONYL GROUPS

Crosslinked Maleic Anhydride Copolymers

Peptide materials such as enzymes and enzyme inhibitors are rendered water-insoluble by *F. Hüper, E. Rauenbusch, G. Schmidt-Kastner, B. Bomer and H. Bartl; U.S. Patent 3,910,825; October 7, 1975; assigned to Bayer AG, Germany* by binding the peptide materials to crosslinked copolymers.

These crosslinked copolymers comprise the following units: (A) 0.1 to 50 wt % of at least one α,β-olefinically unsaturated dicarboxylic acid anhydride having 4 to 9 carbons and (B) 99.9 to 50 wt % of at least one di- and/or polymethacrylate of a diol and/or a polyol, the copolymer having a bulk volume of 1.4 to 30 ml/g and a specific surface area of 1 to 500 m^2/g, and containing, after saponification of the anhydride groups, 0.02 to 10 milliequivalents of acid per gram. The copolymers preferably contain 2 to 20 wt % of units (A) and 80 to 98 wt % of units (B) and the preferred acrylate is a methacrylate.

Important examples of the α,β-monoolefinically unsaturated dicarboxylic acid anhydrides which provide the units (A) in the copolymers, are maleic anhydride, itaconic anhydride and citraconic anhydride, especially maleic anhydride. Mixtures of these anhydrides can also be used for the copolymerization. The diols and polyols from which the dimethacrylates, polymethacrylates, diacrylates and polyacrylates used to provide units (B) are derived comprise:

(1) Polyhydroxy compounds having at least two alcoholic or phenolic, preferably alcoholic, hydroxyl groups;
(2) The reaction products of the polydihydroxy compounds (1) with

alkylene oxides having preferably 2 to 4 carbons, or mixtures of such alkylene oxides, preferably 1 to 20, alkylene oxide units being added to 1 mol of the polyhydroxy compound (1). Examples of suitable alkylene oxides are ethylene oxide, propylene oxide, butylene oxide, trimethylene oxide, tetramethylene oxide, bischloromethyl-oxacyclobutane and styrene oxide, ethylene and propylene oxides being preferred.

(3) The reaction products of at least one of the alkylene oxides defined in (2) with a compound having at least two Zerewitinoff-active hydrogen atoms which is not an alcohol or a phenol.

The dimethacrylates of polyalkylene glycols with molecular weights of up to 1,000 (especially ethylene glycol, diethylene glycol, triethylene glycol and tetraethylene glycol), or their mixtures, are advantageous.

After production of the copolymer, it is combined with the peptide material in a second step. In this step the copolymer is directly introduced, generally at temperatures of 0° to 30°C, into the aqueous solution of the substance to be bound, preferably while the pH is kept constant.

If proteins are to be bound to the copolymers, the process is appropriately carried out with a pH-stat in a pH range of preferably 5.5 to 9.0. If penicillinacylase is to be bound, the process is approriately carried out between pH 5.7 and pH 6.8. During the binding reaction, a base should continuously be added to keep the pH range constant. For this it is possible to use both inorganic bases (for example, caustic alkali solutions) and organic bases (for example, tertiary organic amines). In contrast to experiences with other known resins, in which the reaction is carried out in buffered solutions, the yields of bound enzyme using the resins described were the better, the lower the salt content of the solutions.

Example 1: 80 g tetraethylene glycol dimethacrylate, 20 g maleic anhydride and 1 g azoisobutyronitrile are dissolved in 1 liter of benzene and the mixture is warmed to 60°C for 4 hours, while stirring. 1 g azoisobutyronitrile and 200 ml ligroin (BP 100° to 140°C) are then added and the polymerization is continued for 5 hours at 70°C. The pulverulent polymer is filtered off, suspended once in benzene and three times in petroleum ether (BP 30° to 50°C) and dried in vacuo. The yield was 94 g, the bulk volume 3.5 ml/g, the swelling volume in water 4.7 ml/g, the specific surface area 5 m^2/g and the acid content after saponification of the anhydride groups was 3.5 milliequivalents per gram.

Example 2: 1 g of carrier resin produced as described in Example 1 is suspended in 30 ml of an aqueous solution of 132 U of penicillinacylase (specific activity 1 U/mg of protein). The suspension is stirred for 20 hours at 25°C while keeping the pH value constant at 6.3 by adding 1 N sodium hydroxide solution with a pH-stat. Thereafter the resin is filtered off on a G3 glass frit and washed with 50 ml of 0.05 M phosphate buffer of pH 7.5, containing 1 M sodium chloride, and with 50 ml of the same buffer without sodium chloride. The activity of the resin is not changed further by additional washing.

Enzymatic activities were measured according to the NIPAB test as 132 U for the starting solution; 12 U for the supernatant liquid plus wash water; and 86 U (i.e., 65% of the starting activity) for the resin after reaction.

The enzymatic activity of the penicillinacylase was determined colorimetrically or titrimetrically with 0.002 M 6-nitro-3-(N-phenylacetyl)aminobenzoic acid (NIPAB) as the substrate at pH 7.5 and 25°C. The molar extinction coefficient of the resulting 6-nitro-3-aminobenzoic acid is 9090. One unit (U) corresponds to the conversion of 1 μmol of substrate per minute.

Crosslinked Maleic Anhydride Terpolymers

F. Hüper, E. Rauenbusch, G. Schmidt-Kastner, B. Bomer and H. Bartl; U.S. Patent 3,871,964; March 18, 1975; assigned to Bayer AG, Germany have also used crosslinked terpolymers similar to the process of U.S. Patent 3,910,825 for bonding enzymes. These terpolymers are prepared by copolymerizing (A) 0.1 to 30 wt % of at least one α,β-monoolefinically unsaturated dicarboxylic acid anhydride having 4 to 9 carbon atoms; (B) 35 to 90 wt % of at least one di- and/or polymethacrylate of a diol and/or polyol; and (C) about 5 to 60 wt % of at least one hydrophilic monomer.

Important examples of α,β-monoolefinically unsaturated dicarboxylic acid anhydrides which can provide units (A) are maleic, itaconic and citraconic anhydrides, especially maleic anhydride. Mixtures of these anhydrides can also be used for the copolymerization. For group (B), the dimethacrylates of ethylene glycol, diethylene glycol, triethylene glycol, tetraethylene glycol or higher polyalkylene glycols with molecular weights of up to 500, or their mixtures, are particularly advantageous. The hydrophilic monomers (C) preferably possess at least one carboxyl, aminocarbonyl, sulfo or sulfamoyl group, and the amino groups of the aminocarbonyl or sulfamoyl radical can optionally carry as substituents alkyl groups with 1 to 4 carbon atoms or alkoxymethyl groups with 1 to 4 carbon atoms in the alkoxy radical.

Examples of these hydrophilic monomers include acrylic acid, methacrylic acid, maleic acid half-esters with 1 to 8 carbon atoms in the alcohol radical, N-vinyllactams such as N-vinylpyrrolidone, methacrylamide, N-substituted-methacrylamides such as N-methyl- and N-methoxymethyl-methacrylamide and N-acryloyldimethyltaurine.

Example 1: 70 g of tetraethylene glycol dimethacrylate, 20 g of methacrylic acid, 10 g of maleic anhydride and 1 g of azoisobutyronitrile are dissolved in one liter of benzene and initially polymerized for 4 hours at 60°C. 1 g of azoisobutyronitrile and 200 ml of ligroin (BP 100° to 140°C) are then added and the polymerization is carried out for 2 hours at 70°C and 2 hours at 80°C. The pulverulent polymer is thoroughly washed with petroleum ether (BP 30° to 50°C) and dried in vacuo. The yield was 96 g, the bulk volume 8.8 ml/g, the swelling volume in water 12.4 ml/g, the specific surface area 8.6 m^2/g and the acid content after saponification of the anhydride groups was 3.85 milliequivalents per gram.

Example 2: 6 g of the carrier copolymer produced according to Example 1 are suspended in a solution of 610 U of penicillinacylase in 150 ml of water. The pH value is kept at 6.3 by adding 1 N NaOH, using a pH-stat, and the suspension is stirred for 20 hours at 25°C. The copolymer is then filtered off on a G3 glass frit and washed with 300 ml of 0.05 M phosphate buffer of pH 7.5, containing 1 M sodium chloride, and with 300 ml of the same buffer without sodium chloride. No further activity can be eluted by further washing. The supernatant liquid and the wash solutions are combined and their enzymatic activity is determined. The enzymatic activity of the moist copolymer is measured in aliquot amount. The enzymatic activities were determined as 610 U for the starting solution; 12 U for the supernatant liquid plus wash solution; and 561 U (92% of the starting activity) for the carrier copolymer after reaction. The NIPAB test as described in the previous U.S. Patent 3,910,825 was used.

Derivatives of Ethylene-Maleic Anhydride Copolymer

Reaction products of ethylene-maleic anhydride copolymers with diamines are used as catalyst supports by *L. Goldstein, E. Katzir, Y. Levin and S. Blumberg; U.S. Patent 4,013,511; March 22, 1977; assigned to Yeda Research & Development Co. Ltd., Israel*. These carriers include the reaction products of ethylene-maleic anhydride copolymer (EMA) and a suitable diamine, such as hydrazine (EMA hydrazide resin), p,p'-diaminodiphenylmethane (EMA-MDA resin) or a primary aliphatic diamine such as 2,6-diaminohexane (EMA-HMD resin). EMA-hydrazide and EMA-MDA resins are anionic. The process also relates to cationic resins prepared from these by reacting with N,N-dimethyl-1,3-propanediamine (DMPA) in the presence of an activating agent. The process relates also to the reaction products of the resins with biologically active proteins, and especially to carrier bound enzymes prepared by coupling such resins with suitable enzymes.

Example: Preparation of EMA-Hydrazide Resins – The EMA-hydrazide resins were prepared in two steps. EMA was first crosslinked by the addition of 30 mol percent hydrazine. The crosslinked insoluble EMA was then reacted with a twofold excess of hydrazine to achieve complete conversion of the anhydride residues into acylhydrazide groups.

The procedure was as follows. Oven-dried EMA (12.6 g; 0.1 base mol anhydride) was dissolved with stirring in redistilled dimethylsulfoxide, DMSO (100 ml). A solution of hydrazine hydrate in DMSO (1.5 g in 50 ml; 0.03 mol) was then added slowly with strong stirring. The reaction mixture solidified immediately into a glassy gel. The solid was broken with a glass rod and left at room temperature for 24 hours in a closed vessel, to allow the completion of the crosslinking reaction. The solid glassy material was covered with excess dry acetone, stirred for a few minutes and allowed to settle.

The liquid was decanted and the procedure repeated. The solid material was again suspended in acetone, ground in an homogenizer and allowed to settle. The grinding was repeated several times with fresh portions of acetone until the solid material acquired hard resinous texture as judged by its ease of filtration on a suction filter. After air drying for a few minutes on the filter to remove the acetone, the resin was suspended in 100 ml DMSO, reacted again with hydrazine (10 g hydrazine hydrate in 100 ml DMSO; 0.2 mol) and left stirring overnight at room temperature.

The reaction mixture was poured into acetone, the liquid decanted and the solid ground several times with acetone, washed with acetone on a suction filter and air dried. Traces of solvent were removed by leaving the material over P_2O_5 in a desiccator connected to a high vacuum pump for a few hours. The net weight of dry EMA-hydrazide resin was 10 to 12 grams.

Cationization of EMA-Hydrazide Resins — Anionic EMA-hydrazide resin (2 g; 1.3×10^{-2} base mols carboxyl) was suspended in a dimethylsulfoxide solution of dicyclohexylcarbodiimide (6 g in 60 ml; 3.0×10^{-2} mols). N,N-dimethyl-1,3-propanediamine, DMPA (10 ml; 0.1 mol) was slowly added to the magnetically stirred suspension. The reaction mixture was left stirring overnight at room temperature. The resin was separated by centrifugation, resuspended in dimethylsulfoxide, stirred magnetically for a few minutes and centrifuged down again. The procedure was repeated twice to ensure the removal of reagents.

The cationized resin was then suspended in acetone, stirred magnetically for a few minutes, and separated on a suction filter. After several washings with acetone the resin was air-dried and left overnight over P_2O_5 in a thoroughly evacuated desiccator, to remove traces of solvent. Other suitable aminoalkyl-N-dialkylamino derivatives can be used as well.

Activation of EMA-Hydrazide Resins — EMA-hydrazide resin (100 mg) was suspended in 50% acetic acid, 0.1 N in HCl (8 ml) and stirred for 30 minutes at room temperature. The stirred suspension was cooled to 4°C, aqueous sodium nitrite (40 mg in 1 ml of water) was added dropwise and the mixture stirred for 1 hour over ice. The activated resin was separated on a suction filter and exhaustively washed on the filter with cold water. It was then resuspended in water and brought to pH 9 by the dropwise addition of 0.5 N NaOH, crushed ice being added to keep the temperature down. The polymeric acyl-azide was separated by filtration, washed with 0.2 M borate buffer pH 8.4, resuspended and used directly in the coupling experiments.

Covalent Coupling of Anionic EMA-Hydrazide Resins to Proteins — A cold freshly prepared solution of an enzyme in 0.2 M borate buffer pH 8.4 (10 to 40 mg in 2 ml) was slowly added to the magnetically stirred suspension of EMA-hydrazide (100 mg in 5 ml of the same buffer). The reaction mixture was left stirring overnight at 4°C. The water-insoluble EMA-hydrazide enzyme derivative was separated by centrifugation and washed with water, 1 M KCl, again with water and then suspended in water.

Covalent Coupling of Cationic EMA-Hydrazide Resins to Proteins — This was performed in 0.2 M borate, pH 8. The subsequent washings of the insoluble enzyme derivative were carried out as described for the anionic EMA-hydrazide resin.

Water-Soluble Polymers Having Carboxyl or Amino Side Groups

The use of known water-soluble active enzymatic complexes has the drawback that an ultrafiltration of the reaction medium is required in order to separate the reaction product from the enzymatic complex and to recover the latter for its further reuse. Such ultrafiltration requires the use of costly equipment and is furthermore a cause of a decrease in the overall yield of the process.

M. Schneider; U.S. Patent 4,088,538; May 9, 1978; assigned to Battelle Memorial Institute, Switzerland is now able to combine the advantages of an insoluble active enzymatic complex with those of a soluble enzymatic complex, without having prior drawbacks. The process is characterized by the fact that the organic polymer is selected from polymers which form an active enzymatic complex which is reversibly precipitable and retains its enzymatic activity after redissolving, and that, after the desired degree of transformation of the substance to be transformed has been obtained, the enzymatic complex is precipitated and removed from the reaction medium.

The expression "reversibly precipitable" means that the enzymatic complex is capable of being redissolved after it has been precipitated, and that it can be again precipitated from the solution, which successive precipitating and redissolving operations can be repeated an indefinite number of times, without change in the physical-chemical and enzymatic properties of the complex, either in dissolved state or in the state of a precipitate.

As water-soluble organic polymer one can use, for example, a derivative of polyacrylic acid, such as polyacrylamide or a dextran, carboxymethylcellulose, polyethylene glycol, etc., the polymer having chemical groups which impart the property of reversibly precipitating or flocculating in aqueous medium as a result of a modification of at least one physicochemical parameter of the medium, such as the temperature, pH, concentration of solute, etc., or else by adding to such medium ions, such as bivalent or trivalent metallic ions, which are capable of bringing about the precipitation of the enzymatic complex, combined with the further use of at least one complexing agent for the ions in order to cause the redissolving of the active enzymatic complex.

In particular, as organic polymer, one can use a derivative of polyacrylamide bearing –COOH side groups which impart the property of precipitating quantitatively and reversibly by decrease of the pH of the medium below a value of the order of 4.5 to 5, while being soluble in this same medium at a pH higher than this value.

More particularly, one can use organic polymers having acid side groups of the benzoic or isophthalic type. Such polymers have the advantage that they are precipitable at a pH of less than 4.5 in a manner which is practically quantitative since the proportion of complex remaining in solution after the precipitation is generally less than 1 ppm referred to the initial quantity.

Example 1: Preparation of the Active Enzymatic Complex – Acrylic chloride and p-aminobenzoic acid are condensed and the acrylamide monomer obtained is polymerized in aqueous medium in a nitrogen atmosphere and in the presence of a small amount of N,N'-methylenebisacrylamide and of ammonium persulfate, the latter compound serving as polymerization catalyst. There is obtained a water-soluble polymer composed of interconnected macromolecular linear chains having the recurrent unit:

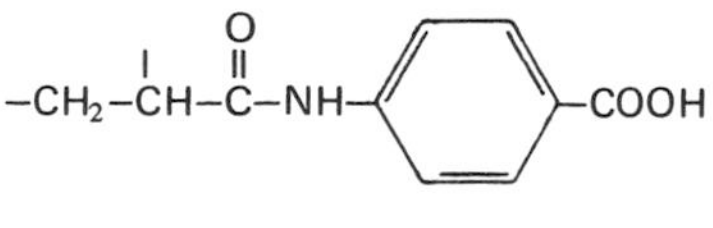

$$-CH_2-\underset{|}{C}H-\overset{O}{\overset{||}{C}}-NH-C_6H_4-COOH$$

An aqeuous solution of this polymer is filtered by passing it through a filter having pores of 0.22 micron, and the polymer is precipitated in the filtrate by lowering the pH to 4.5 by adding a dilute aqueous solution of citric acid. Thereupon the precipitate is collected and dried. The polymer precipitate obtained above is placed in suspension in dioxan (in a proportion of 10 ml of dioxan per g of precipitate) whereupon 1.1 ml of n-tributylamine per 10 ml of the suspension is added. The suspension is cooled to 0°C and 0.4 ml of ethyl chloroformate is then added.

The suspension is maintained at 0°C for 10 minutes whereupon 10 ml of aqueous glucose isomerase solution (20 g/l) is added for every 10 ml of the suspension and the mixture maintained for 15 minutes at 0°C, evaporated to dryness, and the solid residue obtained is dissolved in water. The enzymatic complex thus obtained is precipitated by lowering the pH to 4.5 by means of a dilute aqueous solution of citric acid. The precipitate is washed with an aqueous solution of sodium chloride (0.1 M) until no further enzymatic activity appears in the liquid, whereupon it is subjected to a final washing with distilled water and dried.

In this way one obtains an enzymatic complex which is soluble in water (at a pH of more than 4.5) and has an enzymatic activity corresponding to 1,500 enzyme units per g of complex (an enzyme unit being defined by the amount of fructose, expressed in micromols, produced in 30 minutes, at 50°C, from a 0.66 M glucose solution containing this enzymatic complex in solution).

Example 2: Use of the Enzymatic Complex to Effect the Transformation of Glucose into Fructose — Reaction medium: 250 ml of an aqueous glucose solution of 12 wt %. The reaction was kept at 50°C and pH 7.0. Two tests were carried out, one using 0.18 g of active enzymatic complex and the other using 1 g thereof (the complex having been previously dissolved in a small amount of water). The partial transformation of glucose into fructose (51 wt % fructose; 49 wt % glucose) is obtained at the end of a reaction time of 40 hours when using 0.18 g of enzymatic complex and 9 hours when using 1 g of enzymatic complex. After reaction, the enzymatic complex is precipitated by lowering the pH of the reaction medium to 4.5 by means of an aqueous solution of citric acid, whereupon this precipitate is separated from the reaction medium by settling.

This precipitate is then washed with an aqueous solution of citric acid having a pH of 4.5 and then dissolved in water; the resultant solution is filtered through a filter having pores of 0.22 micron and the enzymatic complex is precipitated by lowering the pH to 4.5, whereupon finally it is dried. The powdered enzymatic complex thus obtained is ready to be used again in the same manner as just described, with its enzymatic activity unchanged. Other reversibly precipitable polymers useful in such enzyme processes include the reaction product of carboxymethylcellulose hydrazide with o-aminobenzoic acid.

Water-Soluble Bead Copolymers with Active Carbonyl Groups

The improved carriers prepared by *D. Kräemer, K. Lehmann, H. Pennewiss, H. Plainer and R. Schweder; U.S. Patent 4,070,348; January 24, 1978; assigned to Rohm GmbH, Germany* have a well-defined head shape and are not accompanied by fine precipitation polymer. They are not enzymatically degradable and are able to bond, with covalent bonds, biologically active substances in high yield and with retention of their biological activity. Such carriers with bonded enzymes are capable of being separated easily and completely from substrate solutions and permit a high speed of flow through a column filled with them. The prerequisite for this is a high free volume between the individual particles in a pile of the particles.

The carrier is a bead-shaped, crosslinked, water-swellable copolymer which is formed of a mixture of the following comonomers:

(A) A radical-polymerizable compound which has a group that is capable of reacting with primary amino groups or hydroxyl groups of biologically active substances with the formation of a covalent bond, e.g., a compound having at least one α,β-ethylenic unsaturation and also a carboxylic acid anhydride group or a glycidyl group, or a compound of the formula

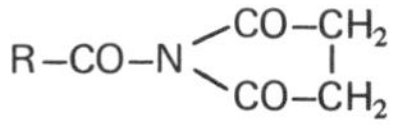

$$R-CO-N\begin{matrix} \diagup CO-CH_2 \\ \quad\quad | \\ \diagdown CO-CH_2 \end{matrix}$$

in which R is a hydrocarbon radical containing an α,β-ethylenically unsaturated group, preferably $CH_2=CH-$ or $CH_2=C(CH_3)-$;

(B) A crosslinking compound having at least two radical-polymerizable α,β-carbon double bonds, but otherwise free of the functional groups found in monomer (A);

(C) One or more radical-polymerizable monounsaturated water-soluble compounds, preferably containing a vinyl or vinylidene group and an amido, hydroxy, carboxylic acid, carboxylate, secondary or tertiary amino, quaternary ammonium or sulfonic acid group; and optionally

(D) Other water-insoluble radical-polymerizable monounsaturated compounds.

Comonomers (A), (B) and (C) constitute 50 to 100 mol percent. Comonomer (A) is present to the extent of at least 2 mol percent, comonomer (B) is present to the extent of 0.2 to 5 mol percent, and comonomer (C) is present to the extent of at least 10 mol percent, referred in each case to the total monomer mixture.

Due to their hydrophilic nature, these bead polymers swell rapidly and extensively in water, so that biologically active substances dissolved in swelling water can also penetrate simultaneously and be bonded by a covalent bond.

The active groups of comonomer (A) are groups which react at 0° to 40°C in aqueous solution with primary amino groups and hydroxyl groups and form covalent bonds with the oxygen or hydrogen atoms of the hydroxyl or amino groups, respectively. Since water is always present in considerable excess with respect to the hydroxyl and amino groups, groups which react spontaneously with water, such as, for instance, isocyanate groups, are less suitable. There are preferably present activated carboxyl groups such as known from peptide chemistry or N- or O-alkylating agents such as alkyl halide or epoxide groups. Examples of activated carboxyl groups which are used in peptide chemistry for the formation of peptide bonds are carbonyl chloride, carboxylic anhydride and carboxylic acid azide groups, as well as phenyl esters and the carboxylates of hydroxylamino derivatives.

Other compounds having activated carboxyl groups include, by way of example, acrylyl- and methacrylyl chloride, acrylic- and methacrylic anhydride, maleic anhydride, phenyl acrylate and methacrylate, glycidyl acrylate and methacrylate, 4-iodobutyl acrylate and methacrylate and 2-isopropenyl-4,4-dimethyl-oxazolone-5.

Upon reaction with enzymes, enzyme substrates, inhibitors, hormones, antibiotics, antibodies, antigens and peptides, the groups form covalent bonds to O- or N-atoms in aqueous solution at temperatures of 0° to 40°C. If the biologically active substances have a protein character, they have a terminal primary amino group as well as in many instances further OH or NH groups in amino acid radicals of ornithine, citrulline, arginine, lysine, serine, threonine or tyrosine.

Example 1: 167 g of n-heptane and 333 g of perchloroethylene which has been dried over molecular sieve 4 A were placed in a round-bottom flask (500 ml). 2 g of benzoyl peroxide and 0.02 g of a stabilizer (Bayer 4010 Na) were dissolved therein. At 20°C, a monomer solution was added consisting of 80 g dimethyl formamide, 60 g acrylamide, 60 g methacrylic anhydride, 1.5 g ethylene glycol dimethacrylate, and 0.1 g emulsifier (random copolymer of n-butyl methacrylate and methacryloyl choline chloride = 90/10). All monomers and solvents were free of water and furthermore both the receiver and the monomer solution were freed of oxygen by means of dry CO_2.

The monomer phase was distributed in the organic phase by constant agitation. The starting of the reaction was effected by addition of the second redox partner (1 g of dimethyl aniline). With a relatively short time of polymerization (30 min), the temperature was maintained substantially constant at 20° to 25°C by cooling with a coolant mixture. The resultant beads were freed from the organic phase by settling and short suction filtering. They were furthermore swollen slightly with dimethyl formamide and used in this condition for reaction with an enzyme.

Example 2: 3 g of the bead polymer produced in Example 1 were swollen for 5 hours at 90°C with anhydrous dimethyl formamide. A solution of 1 g of trypsin in 100 ml of 0.2 M phosphate buffer (pH 7.5) was then added with vigorous agitation and cooling by ice, the agitation was continued for 6 hours, and the batch was left overnight at 5°C.

The product was washed five times with sodium chloride solution, in which the beads were stirred for 30 minutes in water; solid sodium chloride was added up to a concentration of 6%, stirring was continued for an additional 30 minutes and filtration was effected. Finally, the beads were washed twice with, in each instance, 2 liters of 0.05 M phosphate buffer (pH 7.5, 0.01% sodium azide) and suction-filtered. Moist yield: 85 g. Enzyme content (determined by ultraviolet spectrometry after dissolving the beads in 0.5 N NaOH at 100°C): 22.8%, referred to dry weight.

The catalytic activity is as follows. Within 60 minutes, 100 mg of the trypsin beads (dry weight) split 220 μmol peptide bonds on casein. This corresponds to the degradation power of 5 mg of dissolved crystalline trypsin under the same test conditions (substrate concentration 3.3%, 37°C, automatic titration of the liberated carboxyl groups with 0.1 N NaOH at a pH of 8.1). The catalytic activity of the trypsin beads with respect to high-molecular substrate is about 20%, referred to the trypsin content thereof. After five repeated uses, no decrease in activity was observed.

Hydrolyzed Acrylate Esters Converted to Anhydrides

Prior processes have used only polymeric anhydrides prepared from corresponding monomers, namely from maleic anhydride, acrylic or methacrylic anhydrides. A high concentration of polymeric anhydrides may be prepared according to *J. Hradil, J. Coupek, M. Krivakova, J. Stamberg, A. Stoy and J. Turkova; U.S. Patent 3,964,973; June 22, 1976; assigned to Ceskoslovenska akademie ved, Czechoslovakia* in the hydrophilized superficial layer of polymers based on esters or nitriles of polyacrylic and polymethacrylic acid. The good mechanical properties of the starting hydrophobic polymers with hydrophilized surface and reactive groups enable not only successful bonding of the biologically active compounds, but also their easy and effective application.

Highly crosslinked polymers are best suited for the procedure. However, some completely nonswelling polymers such as polyacrylonitrile or poly(methyl methacrylate), may be used also in noncrosslinked or only lightly crosslinked state. The crosslinking is achieved not only by addition of crosslinking agents, but also by a chain transfer, especially for acrylonitrile.

The polymers containing various concentrations of anhydride groups can be prepared in a process whereby polymeric esters or nitriles are hydrolyzed up to the required degree by action of 2 to 9 M NaOH for 0.5 to 2 hours at 80° to 130°C. It is known that this reaction proceeds in microblocks, advantageously in regions with the isotactic structure. The places with a clean cut ability to close the preferred six-membered anhydride cycles are formed in the surface hydrophilized layer in this way. This cannot be attained by copolymerization of the monomeric acid, because the occurrence of carboxylic groups in the chains would be random and the probability of formation of the cyclic anhydrides would be suppressed.

According to the proposed procedure, the partial alkaline hydrolysis is followed by an acidic treatment, to transform carboxylic groups into the H^+ form, and eventually by anhydride formation through thermic dehydration at the temperature 200° to 250°C and the pressure below 0.1 torr or by treatment with known dehydrating agents such as, e.g., acetic anhydride, at 20° to 140°C or thionyl chloride at -10° to +20°C. Both reactions last mentioned may be advantageously catalyzed by pyridine or other acidobasic catalysts. The reactive anhydride groups formed can be proved by the infrared spectra (bands at 1,785 and 1,050 cm^{-1}).

Example 1: A low-hydrophilic macroporous gel (2.5 parts) consisting of 90 wt % of ethylene dimethacrylate and 10 wt % of 2-hydroxyethyl methacrylate and having the specific surface area 209 m^2/g, was hydrophilized by heating with 10 parts of 9 M NaOH to the temperature 125°C for 1 hour. The superficially hydrolyzed gel was filtered, transformed into the H^+ form with 200 parts of 2 M HCl, thoroughly washed with water and dried in a vacuum oven. Reactive anhydride groups were formed by heating of the superficially hydrophilized copolymer to 230°C for 4 hours under the pressure 0.1 torr. The dehydrated gel was swollen in 75 parts of 2 M phosphoric buffer solution of pH 7.5, the mixture was cooled and added under stirring into a solution of 1 part of chymotrypsin in 25 parts of 2 M phosphate buffer solution. The reaction mixture was stirred and cooled with ice to 4°C for 24 hours. The gel with chemically bound enzyme was then sucked off, washed with the buffer solution and water, and activated by washing with 400 parts of 0.1 M solution of sodium borate and 1 M sodium chloride (pH 8.5), further by 300 parts of 0.1 M sodium acetate (pH 4.1) and 300 parts of 0.01 M sodium acetate (pH 4.1). The concentration of the insolubilized enzyme prepared in this way was 3 mg related to 1 ml of the gel. The bound enzyme exhibited the esterase and proteolytic activities.

Example 2: A hydrophilic macroporous poly(ethylene diacrylate) 2,4 parts of the grain size 0.1 to 0.2 mm, was swollen in 10 parts of 9 M NaOH and hydrolyzed by heating to the temperature 130°C for 0.5 hour. Then it was transformed into the H^+ form and dehydrated by heating to 225°C for 4 hours at the pressure 0.1 torr. Chymotrypsin was linked to the gel by procedure described in Example 1 in the amount 1.5 mg/1 ml of the gel.

MODIFIED ACRYLONITRILE AND ACRYLAMIDE POLYMERS

Imidoesters of Acrylonitrile Polymers

A process has been disclosed by *O.R. Zaborsky; U.S. Patent 3,830,699; Aug. 20, 1974; assigned to Esso Research and Engineering Company* for preparing water-insoluble enzymes by covalently bonding enzymes to an insoluble polymer, by the reaction of imidoester functional groups on the polymer with enzyme amino groups.

The polymer is preferably an acrylonitrile-based polymer with at least a substantial portion of the surface acrylonitrile groups converted into imidoesters by contacting with an alcohol and a hydrogen halide (e.g., HCl) to form imidoester functional groups. The imidoester functional groups are then contacted with an aqueous enzyme-containing solution at reaction conditions whereby an insoluble enzyme composite is formed. The insoluble enzyme composite retains activity substantially equivalent to the enzyme in its native state and further shows increased resistance to degradation by heat as well as chemical denaturants.

Example 1: Preparation of the Imidoester of Polyacrylonitrile – To a cooled (–10° to +5°C) and magnetically stirred suspension of 10 g polyacrylonitrile powder in 100 ml dry methanol was added sufficient hydrogen chloride gas until complete saturation had occurred. The mixture, protected from moisture with a drying tube, was slowly allowed to come to room temperature, and then stirred for 2.5 days.

The modified polyacrylonitrile powder was then filtered, washed thoroughly with cool ether, and vacuum dried over a desiccant (potassium hydroxide flakes). Degree of imidoester formation was determined by infrared spectroscopy, alkaline titration or soluble chloride analysis. The above preparation gave 4 wt % imidoester groups.

Example 2: Immobilization of α-Chymotrypsin – 50 mg of the imidoester containing polymer of Example 1 was added to an aqueous solution of α-chymotrypsin (50 mg of enzyme in 5 ml water). The pH of the enzyme solution was adjusted at 10 by titration with 1 N NaOH in water, prior to the addition of the polymer. During the reaction of the enzyme and the polymer the pH was maintained at 10 by the addition of further aliquots of the sodium hydroxide solution. The reaction was continued for one hour at room temperature under a nitrogen atmosphere. The reaction mixture was then filtered through a Millipore filtration unit (RAWP filter; 1.2 microns; 25 mm) to separate out the insolubilized enzyme.

The insolubilized enzyme was washed with distilled water until the filtrate showed no more enzymatic activity. The insolubilized enzyme was evaluated for enzymatic activity and compared to the enzyme in its natural state, i.e., in aqueous solution. The insolubilized enzyme retained full enzymatic activity. Specific activity of insolubilized and free enzymes was 58 and 47 units/mg, respectively.

Example 3: Thermal Stability of the Insolubilized Enzymes – An appropriate amount of the imidoester-enzyme composite, suspended in buffer [0.05 molar tris(hydroxymethyl)aminomethane hydrochloride], was incubated in a test tube at the indicated temperature. Aliquots were withdrawn at appropriate time intervals and assayed for enzymatic activity. A similar procedure was used for the soluble native enzyme. It is apparent from the table that the enzyme composites of this process have increased thermal stability when compared to the soluble native enzyme.

α-Chymotrypsin

Incubation Time at 50°C (min)	Activity Remaining (percent) Insolubilized Enzyme	Activity Remaining (percent) Soluble, Native Enzyme
0	100.0	100.0
15	80.7	16.0
30	70.6	4.7
45	56.9	2.9
60	51.4	1.6
75	53.2	0.8
90	36.7	0
105	36.7	- -

Modified Polyacrylamides Having Alkylene-2,5-Dioxo-4-Oxazolidine Groups

Modified polyacrylamides in which one of the hydrogens in a minor portion of the amide groups is substituted by an alkylene-2,5-dioxo-4-oxazolidine are used by *M.H. Keyes and F.E. Semersky; U.S. Patent 3,839,309; October 1, 1974; assigned to Owens-Illinois, Inc.* for insolubilizing enzymes.

These polyacrylamide derivatives can be prepared as shown in the following sequence of reactions where R is hydrogen or lower alkyl containing 1 to 4 carbon atoms and alk is an alkylene group of 1 to 8 carbon atoms. In the first step of the reaction the polyacrylamide and hydrazine in water are heated at 45°C for 15 hours to form the corresponding hydrazide derivative of formula (2) which is reacted with nitrous acid at 0°C to form the azide of formula (3). This azide is further reacted at 0°C with a diaminoacid of the formula

$$NH_2–alk–CH(NH_2)COOH$$

to obtain the derivative of formula (4) which is cyclized with phosgene at about 35°C resulting in the compound represented by formula (5).

$$-CH_2-\underset{|}{\overset{R}{\overset{|}{C}}}-\overset{O}{\overset{\|}{C}}-NH_2 \xrightarrow{\text{hydrazine}} -CH_2-\underset{|}{\overset{R}{\overset{|}{C}}}-\overset{O}{\overset{\|}{C}}-NHNH_2 \xrightarrow{HNO_2} -CH_2-\underset{|}{\overset{R}{\overset{|}{C}}}-\overset{O}{\overset{\|}{C}}-N_3$$

(1) (2) (3)

$$\xrightarrow{NH_2-alk-CH(NH_2)COOH} -CH_2-\underset{|}{\overset{R}{\overset{|}{C}}}-\overset{O}{\overset{\|}{C}}-NH-alk-CH(NH_2)COOH \xrightarrow{COCl_2}$$

(4)

$$-CH_2-\underset{|}{\overset{R}{\overset{|}{C}}}-\overset{O}{\overset{\|}{C}}-NH-alk-CH\langle\begin{matrix} NH-C(=O) \\ \quad\quad\quad\; O \\ C(=O) \end{matrix}\rangle$$

(5)

Example 1: 5 g of dry polyacrylamide and 6 g (0.12 mol) of hydrazine hydrate in 200 ml of water were heated at 45°C for 15 hours. The resulting hydrazide was washed with 0.1 M sodium chloride solution and then suspended in 200 ml of 0.25 N hydrochloric acid. The suspension was cooled to 0°C and 25 ml of 1 M sodium nitrite solution was added. To the azide were added 18 g (0.12 mol) of lysine slowly over a period of 5 minutes with vigorous stirring.

After one hour, the lysine derivative formed was washed successively with 100 ml 0.1 M NaCl, 40 ml of 1 M NH_4OH plus 10 ml of 1 M NH_4Cl and distilled water, then dried at 110°C. In the final step, 3 g of the lysine derivatives were mixed with 75 ml of dioxane and phosgene was bubbled through the mixture with constant stirring for 5 hours at 40°C followed by purging with dry air for 15 hours at 25°C. The desired modified polyacrylamide was separated by filtration as a white solid and dried. Analysis showed the extent of lysine substitution to be 6 mol percent of which two-thirds (or 4%) was cyclized.

Example 2: 5 ml of a solution containing 1.5 mg of ribonuclease per ml of water was saturated with 0.1 M $NaHCO_3$ and cooled to 0°C. To this solution was added with stirring 120 mg of the modified polyacrylamide prepared in Example 1. Thereafter, stirring was continued at 0°C for 24 hours. In this manner, the ribonuclease was rendered water-insoluble and after washing with water, it was stored at 3°C. This sample was assayed initially and periodically by packing in a chromatographic column and measuring the rate of change of cytidine-2',3'-cyclic phosphate to cytidine monophosphate by the method of Hammes and Walz as described in *Biochim. Biophys. Acta*, 198, 604 (1970). The initial activitity was 0.283 mg of ribonuclease per gram of modified polyacrylamide. Eight months later, 0.071 mg of ribonuclease was present per gram of modified polyacrylamide.

By substituting the polyacrylamide of Example 1 with polymethacrylamide, polyethacrylamide, polypropylacrylamide and polybutacrylamide, the corresponding modified polymers are obtained which when reacted with a water-soluble hydrolytic enzyme such as papain or trypsin or alternatively, with a transferase enzyme such as glutamic-pyruvic transaminase will render the enzymes water-insoluble. Although lysine was used in Example 1, it will be apparent that any diaminoacid of the formula NH_2–alk–$CH(NH_2)COOH$ where alk is an alkylene group of from 1 to 8 carbon atoms can be substituted for lysine with equally effective results.

Polymers with N-Halogenoamide Groups

A stable immobilized enzyme is obtained by *R. Senju and H. Tanaka; U.S. Patent 4,073,689; February 14, 1978* by reacting an enzyme with a substance having an N-halogenoamide group in a weak alkaline aqueous solution. Studies on N-halogenation reaction of various amides have shown that an N-halogenoamide group has a low rate of conversion to the amino group under weak alkalinity and preferably reacts with groups containing active hydrogen, e.g., $-NH_2$, $-OH$, $-CONH_2$, etc., to form ureide, urethane, and acylurea, respectively. It has been found that the N-halogenoamide group is especially active upon the amino group and is very reactive even under mild reaction conditions such as pH 8 to 9 and reaction temperature of 30° to 40°C.

As the substance having an N-halogenoamide group for use in the process, water-soluble, low molecular weight, multivalent halogenoamide compounds (Substance A), water-soluble, high-molecular weight, multivalent halogenoamide compounds (Substance B) and solid carriers having an N-halogenoamide group (Substance C) are used. Substances A, B and C are prepared in the following manners, respectively.

Substance A is prepared by N-halogenating multivalent amide compounds such as polycarboxylic acid amide, e.g., adipic acid amide and succinic acid amide; carbamoylized polyhydric alcohols, carbamoylized sugars, etc., by known methods. As the polyhydric alcohol to be carbamoylized, glycerin, polyethylene glycol, polypropylene glycol, etc., may be used. As the sugar to be carbamoylized, glucose, fructose, sucrose, pentaerythritol, sorbitol, mannitol, etc., may be used.

Substance B is prepared by carbamoylizing water-soluble multivalent alcohols and then halogenating them according to the method of the abovedescribed halogenation reaction. As the multivalent alcohols, cellulose, starch, agar, alginic acid, Knojak (devil's-tongue jelly), polyvinyl alcohols, etc., may be used. Further, homopolymers and graft polymers of acrylamide or methacrylamide, or their copolymers with other polymerizable vinylpolymers, etc., are useful as the carbamoyl compounds.

The solid carrier (Substance C) having a halogenoamide group is prepared by carbamoylizing and halogenating a suitable solid material. Any material capable of being carbamoylized, irrespective of natural or synthetic origin, may be used. For example, cellulose, wood materials such as pulp, etc., wool, silk, cotton, polymeric materials such as nylon, polyvinyl alcohol, weakly basic anion exchange resin such as Dianion CR-20 may be used.

The solid material may be used in any desired form, such as powdery, granular, fibrous, fabric, membrane paper, plate form, etc. An especially preferable carrier is hydrophilic and porous, but not rigid, because the enzymatic reaction is generally carried out in an aqueous solution of the enzyme and the hydrophilic carrier does not impair diffusion of the substrate and mutual action of the substrate and the carrier. The reaction of immobilizing enzymes with the substances having an N-halogenoamide group is carried out at a temperature of -10° to 50°C, preferably, 15° to 40°C and at a pH of 8 to 11 in an aqueous solution of the enzyme.

Enzymes which may be immobilized in this process include purified or crude enzymes, enzyme mixtures and enzyme systems present in or isolated from animal, plant or microbial tissue. They are used in the form of whole cells, intact intracellular particles and crude extracts of these tissues.

For example, proteolytic enzymes such as trypsin, chymotrypsin and papain; hydrolases such as α-amylase invertase, β-galactosidase, ribonuclease, alkaline phosphatase, amyloglucosidase and dextranase; dehydrogenases such as creatine phosphokinase and pyruvate kinase; oxidases such as glycose oxidase and amidases such as amidase and penicillin amidases may be used. The concentration of the enzyme solution used in the method is not limited, but 0.01 to 10% enzyme solution is preferably used. In immobilizing an enzyme according to this process, the loss of the enzyme is very small since the immobilizing reaction is carried out in an aqueous solution of the enzyme under very mild reaction conditions.

Example 1: 140 ml of 1.0% aqueous solution of sodium hydroxide and 5.3 g (0.1 mol) of acrylonitrile (AN) are added to 21 g (0.1 mol) of commercially available hydroxyethylcellulose (HEC) and the mixture is allowed to react at 50°C for 60 minutes with stirring. Then, 23 g (0.2 mol) of 30% solution of hydrogen peroxide is added in two or three portions, and the solution is allowed to react at 20°C for 90 minutes. The reaction steps are illustrated by the following formula:

$$\text{cell–OH} \xrightarrow[\text{alkali}]{\text{AN}} \text{cell–O–CH}_2\text{–CH}_2\text{–CN} \xrightarrow{H_2O_2} \text{cell–O–CH}_2\text{–CH}_2\text{–CONH}_2 + \tfrac{1}{2}O_2\uparrow$$

The resulting aqueous solution of carbamoylethyl HEC is diluted to twice the volume with water. To the 10.0 g (0.0027 mol as the carbamoyl group) of dilute carbamoylethyl HEC solution is added 2.7 ml (0.0027 mol) of 1 M aqueous solution of sodium hypochlorite. The mixture is subjected to N-chlorination at 0° to 2°C for 60 minutes and then neutralized to pH 9 with 1 N acetic acid. To the resulting solution is added 2.0 ml of 250 mg/100 ml aqueous solution of α-amylase. Then, the mixture is sufficiently stirred, slowly heated up to 30°C and is allowed to stand for 20 minutes to obtain α-amylase immobilized in the gel of carbamoylethyl HEC. The enzyme immobilized in the gel is immobilized to the HEC chains not only by physical but also by chemical bond and therefore, will not wash out even by repeated washings.

Example 2: To 10.0 g (0.007 mol) of 5% aqueous solution of polyacrylamide is added 2.8 ml (0.0056 mol) of 2 M aqueous solution of sodium hypochlorite. The resulting mixture is subjected to N-chlorination at -5° to 0°C for 30 minutes and then neutralized to pH about 9 with 1 N acetic acid. Then, 2.0 ml of 250 mg/100 ml aqueous solution of α-amylase is added thereto. After sufficient stirring, the resulting mixture is slowly heated up to 30°C and is allowed to stand for 20 minutes to obtain α-amylase immobilized in the gel of polyacrylamide. The enzyme immobilized in the gel is immobilized to the polyacrylamide chains not only by physical but also by chemical bond and therefore, will not flow out even by repeated washings.

Example 3: To 1.44 g (10 mmol) of adipic acid amide are added 50 ml of water and 10 ml (20 mmol) of 2 M aqueous solution of sodium hypochlorite. The mixture is subjected to N-chlorination reaction for 60 minutes under ice cooling. To the reaction mixture is added 1.2 g (20 mmol) of acetone and then the resulting solution is adjusted to pH about 9.0 with acetic acid. To the resulting solution is added 30 ml of 1% enzyme solution (α-amylase, papain, trypsin or invertase). The solution is kept at 18°C for 50 minutes to precipitate the immobilized enzyme. Its activity is not lowered even by repeated washings.

POLYMERS REACTIVE WITH ENZYME THIOL GROUPS

Polymers with Thiol or Thiolactone Groups

S.A. Barker and C.J. Gray; U.S. Patent 3,846,306; November 5, 1974; assigned to Koch-Light Laboratories Limited, England disclose polymeric materials having thiol or thiolactone groups to which enzymes have been bonded. One can covalently bond enzymes to the hydrophilic water-insoluble polymers such that, when the polymer has thiol side groups, the functional groups with which it can react are other thiol groups and disulfide groups, and when the polymer has thiolactone side groups, the functional groups with which it can react are hydroxyl groups, phenolic groups and amino groups. The polymers also have repeating side groups such as primary amide groups and carboxylic acid groups which serve to make the polymers hydrophilic.

The primary amide side groupings can be derived from acrylamide. The cross-linking can be achieved with N,N'-methylenebisacrylamide linking units or other linking units. Repeating carboxylic acid and thiol groupings can be present as N-substituted cysteine residues derived from N-acryloyl S-benzyl-cysteine after polymerization, e.g., by treatment with sodium in liquid ammonia or by the more economic use of N-acryloyl thiazolidine carboxylic acid in the polymerization followed by treatment with hydrochloric acid.

The carboxylic acid groups can be present in the form of their N-ethylamide derivatives and such polymers may be prepared by the use of N-acryloyl S-benzyl cysteine N-ethylamide in the polymerization and subsequent removal of the S-benzyl groups with sodium in liquid ammonia. The carboxylic acid containing polymers may be converted to the thiolactone derivatives using dicyclohexylcarbodiimide in an anhydrous solvent.

Chemical coupling of biologically active molecules can be achieved by dissolution in aqueous buffer, adding the polymer, and allowing the resultant reaction to proceed over some hours with stirring. Alternatively, the same coupling can be achieved by passage of the biologically active molecules in aqueous buffer through a column packed with the polymer or a fluidized bed of the polymer particles. In all cases there may be an optimum pH for achieving maximum retention of the biologically active molecules which is a compromise between the optimum pH for coupling via the particular linkage used and the pH maximum stability of the free and bound organic species being recovered. In general, this optimum pH is slightly alkaline for the polythiol used as a disulfide derivative and oxygen is preferably excluded during the reaction process.

Example 1: Copolymer of Acrylamide and N-Acryloyl-S-Benzylcysteine (Copolymer A) — Stock solutions A and B were prepared. Solution A: 2-amino-2-(hydroxymethyl)propane-1,3-diol (36.6 g) and N,N,N',N'-tetramethylethylenediamine (0.46 cm^3 in 1 N hydrochloric acid, 45 cm^3), were mixed and diluted to 100 cm^3 with water. Solution B: Acrylamide (28 g) and N,N'-methylenebis-acrylamide (0.735 g) were mixed and dissolved in water (100 cm^3).

S-benzyl-N-acryloylcysteine (3.0 g) was dissolved in solution A (10 cm^3). N,N'-methylenebisacrylamide (1.0 g) was dissolved in solution B (20 cm^3). Then the two solutions were mixed together and ammonium persulfate solution was added (0.140 g in 100 cm^3, 40 cm^3). The solution was left to stand at room temperature, and after 30 minutes, set to a gel. After a further 12 hours the gel was broken up in a top drive macerator and washed with water. Then it was centrifuged down. The washing process using the macerator was repeated with water (3 x 150 cm^3), then with ethanol (3 x 150 cm^3). The solid was then filtered off and washed with ethanol (300 cm^3) using a vacuum filter, and dried in a desiccator over P_2O_5. The dried product was a white powder (7.8 g).

Example 2: Reduction of Copolymer A — Liquid ammonia (10 cm^3; dried by distillation from sodium metal) was added with stirring to copolymer A (1.0 g). Initially, after the addition of sodium metal in small pieces to this mixture, the blue color which was produced rapidly disappeared. After the addition of about 60 mg the blue color remained for 20 minutes. Then anhydrous ammonium acetate (2 g) was added. The ammonia was allowed to evaporate off at room temperature. Then 1 N hydrochloric acid (10 cm^3) was added to the residue. The solid was filtered off and washed with oxygen-free water (100 cm^3), cysteine hydrochloride solution (0.1 M, 100 cm^3), oxygen-free water (100 cm^3) and with ethanol (300 cm^3) and then dried under vacuum in a desiccator over P_2O_5. The dried product (Copolymer B) was a white powder (0.832 g).

Example 3: The Reaction of Trypsin with Copolymer B — (1) Conversion of Trypsin to Trypsin-Poly SH: N-acetylhomocysteine thiolactone (10 mg in carbonate buffer pH 10.6, 0.5 cm^3) was added to a solution of trypsin (100 mg in carbonate buffer pH 10.6, 10 cm^3). The mixture was allowed to stand at 4°C for 60 minutes and then it was immediately passed through a column of Sephadex G 25 (2 x 50 cm). The column was then eluted with carbonate buffer pH 10.6. Fractions (1 cm^3) were collected and scanned at 280 nm. Fractions representing the protein were combined and made up to 100 cm^3 with carbonate buffer pH 10.6 (this solution was called Trypsin SH I). Trypsin SH I (5.0 cm^3) was diluted to 100 cm^3 with carbonate buffer pH 10.6 (this solution was called Trypsin SH II).

(2) The Reaction of Trypsin SH I with Copolymer B: Crosslinking solution (1 cm^3) was added to a stirred suspension of copolymer B (50 mg) in Trypsin SH I (20 cm^3). The suspension was stirred for 16 hours at 4°C, then it was centrifuged and the solid washed with carbonate buffer pH 10.6 (7 x 3 m^3).

The supernatants and the washings were combined and made up to 100 cm^3 with the same buffer (this solution was called Trypsin SH III).

The solid was then washed with 0.1 M cysteine hydrochloride (3 x 2 cm^3) using a centrifuge. The supernatants were combined and made up to 10.0 cm^3 with 0.05 M phosphate buffer pH 7.6. This solution was supplied immediately to a column of Sephadex G 25 (1 x 45 cm). The column was eluted with 0.05 M phosphate buffer pH 7.6. Fractions (1 cm^3) were collected and scanned at 280 nm. The profile showed two peaks, the first one representing the protein and the second peak representing cysteine. Fractions corresponding to protein were combined and made up to 100 cm^3 in 0.05 M phosphate buffer pH 7.6 (this solution was called Trypsin SH IV). (The copolymer which had been washed with cysteine hydrochloride was called Free Solid.)

A calibration curve for protein was constructed using standard solutions of trypsin containing 50, 100, 150, 200 and 250 $\mu g/cm^3$. Trypsin SH II (1.0 cm^3), Trypsin SH III (1.0 cm^3) and Trypsin SH IV (1.9 cm^3) assayed in the same way. The data are presented in the following table. The amount of protein which had coupled with Copolymer B (50 mg) was calculated as 5.0 mg.

Determination of the Protein in Trypsin SH I and the Amount Which Coupled to Copolymer B

Sample	OD 500, nm
Trypsin, 50 $\mu g/1\ cm^3$	0.112
Trypsin, 100 $\mu g/1\ cm^3$	0.195
Trypsin, 150 $\mu g/1\ cm^3$	0.305
Trypsin, 200 $\mu g/1\ cm^3$	0.375
Trypsin, 250 $\mu g/1\ cm^3$	0.465
Trypsin SH II	0.115
Trypsin SH III	0.340
Trypsin SH IV*	0.110
Blank	0.042

*Represents 50 mg of trypsin attached to 50 mg of copolymer.

D.T. Cowling; U.S. Patent 3,884,761; May 20, 1975; assigned to Koch-Light Laboratories Limited, England has found an improved method of making water-insoluble hydrophilic polymers of U.S. Patent 3,846,306. An N-substituted thiazolidine is polymerized, possibly together with one or more copolymerizable monomers, in the presence of inert gas to give a polymer having thiazolidine side groups and those thiazolidines side groupings are then hydrolyzed to give a polymer having thiol side groupings.

The N-substituted thiazolidine which is used is preferably N-acryloyl-4-carbomethoxy-2,2 dimethylthiazolidine. A preferred polymer is obtained by copolymerizing N-acryloyl-4-carbomethoxy-2,2 dimethylthiazolidine with acrylamide and methylenebisacrylamide. This can be performed in an ethanolic solution in the presence of nitrogen gas, followed by hydrolyzing the resulting copolymer having pendant thiazolidine groups to form the thiol groups. The preferred hydrolyzing agent is hydrochloric acid, although other suitable agents can be used. The N-acryloyl-4-carbomethoxy-2,2-dimethylthiazolidine is prepared by reacting acryloyl chloride with 4-carbomethoxy-2,2-dimethylthiazolidine which had been prepared from cysteine methyl ester hydrochloride.

Example: A solution of α,α'-azobisisobutyronitrile (0.5 g) in ethanol (200 ml) was added to a stirred, filtered solution of acrylamide (55 g), methylenebis-acrylamide (8 g) and N-acryloyl-4-carbomethoxy-2,2-dimethylthiazolidine (48.5 g) in ethanol (500 ml) through which nitrogen was passed. After 15 minutes, the solution was heated to 75°C for 24 hours. The polymer started to separate after 15 minutes. The mixture was cooled to room temperature and the polymer collected by filtration, washing with ethanol and ether. The copolymer was dried over silica gel in a desiccator. Yield (92.5 g).

A suspension of the above-formed copolymer in 1 N hydrochloric acid (500 ml) was heated to 70°C for 3 hours, after which time it was cooled to room temperature. The solid was filtered off, washed with oxygen-free water (5 x 250 ml), cysteine hydrochloride solution (0.1 M in oxygen-free water 500 ml), oxygen-free water (5 x 250 ml) and suspended in acetone (1.0 liter). After 1 hour, the solid was filtered off, washed with ether and dried over silica gel in a desiccator. The thiol polymer was a white powder (16.2 g).

Polymers with Mercurated Aromatic Groups

It is known that in addition to other functional groups able to bind biopolymers characterized by the so-called "proteic character," through covalent bonding, aromatically bound atoms of bivalent mercury may be introduced into synthetic polymers which then are able to bind sulfhydryl groups covalently by reaction with a cysteine unit of bioprotein. A polymer carrier of this kind has been prepared, for instance, from polystyrene by the direct mercuration with mercuric perchlorate (L.H. Kent, J.H.R. Slade, *Biochem. J.*, 77, 12, 1960). The hydrophobic character of the polystyrene skeleton of this carrier is, however, not suitable for binding hydrophilic molecules.

This, therefore, has prompted a study of the transformations of some natural hydrophilic polymers. As a rule, however, these transformations are effected only by very intricate polymer analogous syntheses which require that several steps be carried out before derivatives containing bound aromatic mercury groups are formed. Polymers containing aromatically bound mercuric ions which overcome the above problems have been provided by *O. Wichterle, J. Coupek and M. Krivakova; U.S. Patent 4,087,598; May 2, 1978; assigned to Ceskoslovenska akademie ved, Czechoslovakia.* These polymers have the formula

$$\begin{array}{c} R^1 \\ \wr \\ R-\overset{|}{C}-Y-\underset{\underset{Hg^+}{|}}{Ar}-Z \\ \wr \end{array}$$

where R is H or CH_3, Y is $-CO(OCH_2{\cdot}CH_2)_x-O-$, where x = 0 to 20, or $-CONH-$, Ar is a bifunctional aromatic group selected from phenylene, naphthylene, tolylene, xylylene and biphenylene each of which has two free bonds; Z = H or $NH-R^2$, where R^2 is acyl, carboxy, C to alkyl or a sulfonyl group and R^1 is the chain of homopolymer of the formula

$$\begin{array}{c} \wr \\ R-\overset{|}{C}-Y-\underset{\underset{Hg^+}{|}}{Ar}-Z \\ \wr \end{array}$$

or the chain of a copolymer selected from the group consisting of methacrylic and acrylic esters of polyfunctional alcohols, which may be saturated, unsaturated, aromatic or aliphatic, straight or branched chain, substituted or unsubstituted. These polymers containing aromatically bound mercury ions and having a broad spectrum of hydrophilicity can be obtained by the homo- or copolymerization of a monomer selected from the group consisting of acrylanilide, methacrylanilide, and monomers of the formula

$$R^2{-}NH{-}Ar({-}OCH_2{-}CH_2)_x{-}O{-}CO\underset{\substack{|\\R^1}}{C}{=}CH_2$$

where Ar, x and R^1 are as above defined and R^2 is acyl, carboxyalkyl or sulfonyl. The polymer prepared is then treated with a mercuric salt, such as a mercuric acetate or perchlorate, nitrate sulfate, chloride, etc. The treatment is carried out at 10° to 30°C. The treatment is advantageously carried out in an aqueous medium, and preferably in water.

The aromatic nucleus is advantageously activated by an amide group for hydrogen substitution by a mercuric cation in the para position, so that the mercuration proceeds smoothly even at ambient temperature using dilute mercuric salt solutions. The degree of concentration is substantially noncritical and may vary from 0.1 to 50%, preferably 1 to 10%. Aminophenyl esters of acrylic and methacrylic acids require a moderately elevated temperature for this reaction. All monomers which are able to copolymerize with acrylanilide or methacrylanilide or with acetaminophenyl acrylate or methacrylate are suitable for use in the preparation of the copolymers.

The hydrophilic monomers of the methacrylic series, such as, acrylic acid, methacrylic acid, methylmethacrylate or methylacrylate, i.e., the monomers having a residue linked by an ester bond and which contain at least one hydroxy or amino group, are especially advantageous for use in the above synthesis. Most important from the standpoint of hydrolytic stability, a condition for versatile application of the polymer as a carrier, are copolymers of methacrylanilide with methacrylic hydrophilic monomers. Acrylic analogs of these polymers have a lower stability in strongly alkaline or strongly acid medium. Steric hindrance, caused by substituted or free amino group adjacent to the mercury atom on the aromatic ring, may hinder the application of the mercurated aminophenyl esters.

Example 1: A mixture consisting of 16 weight parts of methacrylanilide, 33 weight parts of 2-hydroxyethyl methacrylate, 33 weight parts of ethylene dimethacrylate, 98 weight parts of cyclohexanol, 10 weight parts of lauryl alcohol and 0.8 weight part of 2,2-azobisisobutyronitrile is stirred and heated to 70°C in 600 parts of water containing 6 weight parts of poly(vinylpyrrolidone) as a suspension stabilizer.

The suspension which is formed after 12 hours is cooled, filtered and washed with a tenfold volume of water and threefold volume of methanol. The particles are again stirred into a twofold volume of water; 35 weight parts of mercuric acetate and 3 parts of acetic acid in 100 parts of water are added and the resultant mixture is heated to 50°C for 2 hours. The resulting mercurated product is washed with water on a sintered-glass filter until the reaction of the bivalent mercury ceases and is then dried and analyzed for mercury content.

Example 2: A mixture consisting of 10 weight parts of acrylanilide, 83 weight parts of acrylamide, 7 weight parts of methylenebisacrylamide, and 1 weight part of potassium peroxodisulfate is allowed to react in 120 parts of water for 8 hours at 60°C between two glass plates separated by a spacer formed of a silicon rubber foil. The thus-obtained membrane is thoroughly washed with water and then mercurated as in Example 1.

MODIFIED LINEAR POLYAMIDES

Hydrolyzed and Modified Nylons

Side chains, useful for bonding enzymes, are introduced into hydrolyzed polyamides in the process disclosed by *M. Sokolovsky, A. Freeman and L. Goldstein; U.S. Patent 3,970,597; July 20, 1976*. A polyamide polymer such as any of the commercially available nylons, is subjected to controlled hydrolysis, so as to increase the number of free carboxyl and amino groups, especially on the surface of the polymer, by a factor of 2 to 5, as determined titrimetrically.

The polymer can be used in any convenient form, such as powder, fiber, film, membrane, sheet, web of fibers, etc. Most experiments were carried out with polymer powder, but experiments have shown that the process is applicable in a similar manner to all the other forms of the polymer. After this step of controlled hydrolysis, there is effected a four-component reaction involving the carboxyl and amino groups of the polymer with an aldehyde R_2CHO and with an isonitrile $R_3C{\equiv}N$ resulting in the formation of a side chain-substituted polymer, as shown by the following, where A and B are hydrocarbon-backbone residues of $-(CH_2)_n-$, with n an integer.

$$\text{CONHACONHBCONHACONHBCONHA} \xrightarrow[\text{hydrolysis}]{\text{controlled}}$$

$$\text{CONHACONHBCOOH} + \text{H}_2\text{NACONHBCONHA} \xrightarrow[\text{R}_2\text{CHO}]{\text{R}_3\text{CN}}$$

```
CONHACONHBCONACONHBCONHA
             |
             CHR2
             |
             CONHR3
```

In a preferred method, there is used a diisocyanoalkane, such as, e.g., 1,6-diisocyanohexane, resulting in the formation of a side chain having a terminal isocyano group, which is further reacted with a suitable aldehyde, carboxyl group and amino group to chemically bind a suitable group to the end of this side chain. By using nylon-6, acetaldehyde and 1,6-diisocyanohexane, a derivative of nylon designated as PIN-nylon is obtained, which is advantageously used for the chemical binding of various substituents, molecules, etc.

Example 1: Controlled Hydrolysis of Nylon Powders – Nylon-6 powder (10 g) was suspended in 3 N HCl (300 ml) and stirred at room temperature (20°C) for the desired amount of time. The powder was separated on a suction filter, washed exhaustively with water, ethanol, ether and air dried. Traces of solvent and moisture were removed in a vacuum desiccator over phosphorus pentoxide and the powders stored in closed vessels. The dependence of the carboxy content of nylon-6 powder on the time of hydrolysis is summarized in the table below, for a typical set of experiments. The nylon-6 samples routinely used were hydrolyzed for 4 hours.

Time of hydrolysis, hours	0	2	4	6	17	21	24
Carboxyl content, μmols/g	26	55	65	70	91	96	104

In a similar manner a partial hydrolysis was effected with nylon-6,6,in powder form and with nylon-11, the results being as follows. Nylon 6,6: unhydrolyzed, 30 μmols/g; 4 hour hydrolysis, 70 to 80 μmols/g. Nylon 11: unhydrolyzed, 10 μmols/g; 4 hour hydrolysis, 35 to 40 μmols/g; 7 hour hydrolysis, 60 to 70 μmols/g. Similar experiments were carried out with nylon fibers, with thin sheets of nylon-6, nylon netting, etc. The reaction was carried out under similar conditions and also in these cases an increase of titrimetrically determinable carboxyl groups by a factor of from about 2 to 5 as compared with untreated material, was attained.

Example 2: Preparation of Polyisonitrile Nylon (PIN-Nylon) — Partially hydrolyzed nylon-6 powder, 2 g (sample hydrolyzed for 4 hours; mean carboxyl content 60 to 65 μmols/g) was suspended in isopropanol (80 ml); acetaldehyde (20 ml) was then added, followed by 1,6-diisocyanohexane (8 ml) and the reaction allowed to proceed in a closed vessel for 24 hours with stirring at room temperature.

The acetaldehyde was pipetted with a precooled pipette, to prevent formation of bubbles. The PIN-nylon powder was separated on a suction filter, washed with isopropanol (50 ml) and then with ether (200 ml) and air dried. Traces of solvent were removed in a vacuum desiccator over phosphorus pentoxide. The PIN-nylon powder was stored at -5°C in a dark stoppered vial over silica gel. The coupling capacity of PIN-nylons was 25 to 50 μmols/g. See Example 3.

Example 3: Coupling of Trypsin to PIN-Nylon — PIN-nylon (50 mg) was suspended in 2 ml cold 0.1 M phosphate–0.5 M sodium acetate pH 8. A cold aqueous solution of trypsin (6 mg in 1 ml) was then added, followed by 0.1 ml acetaldehyde. The reaction mixture was left stirring overnight at 4°C. The insoluble enzyme derivative was separated on a filter, washed with water, 1 M KCl (0.1 M in $NaHCO_3$) and again with water, resuspended in water (4 ml) and stored at 4°C. The recovery of immobilized enzymic activity was 35%.

Example 4: Preparation of MDA-Nylon — p,p'-Diaminodiphenylmethane, MDA (2 g, 0.01 mol) was dissolved in 160 ml 100% propanol, cooled over ice and 4 ml acetaldehyde were then added. PIN-nylon powder (2 g) was suspended in 160 ml isopropanol, 2 ml glacial acetic acid (0.033 mol) were added, followed by the MDA-acetaldehyde solution. The reaction was allowed to proceed in a closed vessel for 24 hours with stirring at room temperature. The MDA-nylon was separated on a suction filter, washed with ethanol and then with ether and air dried. The coupling capacity of the MDA-nylon was 20 μmols/g.

Example 5: Coupling of Papain to MDA-Nylon — MDA-nylon (100 mg) was suspended in cold 0.2 N HCl (7 ml) and aqueous sodium nitrite (25 mg in 1 ml) added dropwise. The reaction mixture was stirred for 30 minutes over ice; the red brown diazotized MDA-nylon was separated on a suction filter, washed with cold water and finally with cold 0.1 M phosphate, pH 8, resuspended in the same buffer and used directly in the coupling experiments.

An aqueous solution of papain (10 ml, 1 to 1.3 mg protein per ml) was added to a magnetically stirred suspension of the above diazotized MDA-nylon (100 mg in 6 ml). The reaction mixture was left stirring overnight at 4°C. The insoluble MDA-nylon papain conjugate was separated by filtration, washed with water, 1 M KCl and again with water, resuspended in water (5 ml) and stored at 4°C. The recovery of immobilized enzymic activity was about 40%.

Modified Nylons Having Amidrazone Groups

W.E. Hornby and D.L. Morris; U.S. Patent 4,115,305; September 19, 1978; assigned to National Research Development Corporation have modified imidate esters of nylon for use as enzyme supports. The imidate esters may be prepared by reacting nylon with an alkylation reagent such as, e.g., a triethyloxonium salt.

The preferred supports may be produced by reaction of imidate groups in the polymer with compounds comprising an acid hydrazide group and in particular the alkyl and aryl hydrazides and dihydrazides, such as, e.g., succinic acid dihydrazide, adipic acid dihydrazide, phthalic acid dihydrazide, dipicolinic acid dihydrazide, oxalic acid dihydrazide and dihydrazides derived from polysaccharides. The product is then treated so as to provide the side chain with a suitably active functional group for linkage to a biologically active material, e.g., by modification of the free $-CONHNH_2$ group to an acid dihydrazide side chain. The reaction between a polymer comprising imidate groups and an acid hydrazide or dihydrazide may be carried out at ambient temperatures, in an inert organic solvent such as, e.g., formamide, which does not significantly degrade the polymer backbone.

Compounds comprising an acid hydrazide group are believed to react according to the following reaction scheme, as, for example, with the imidate ester of a nylon:

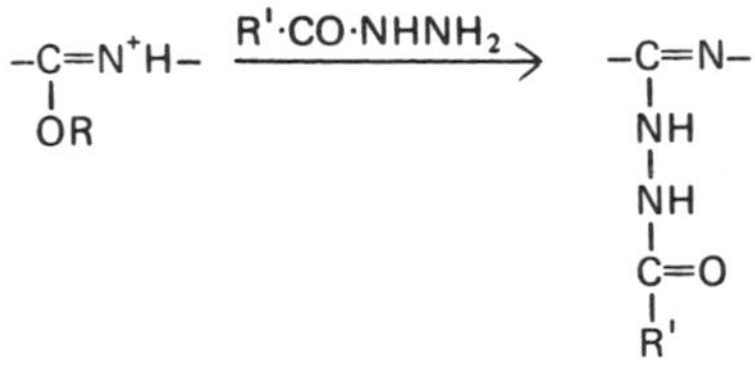

```
 -C=N⁺H-   R'·CO·NHNH₂     -C=N-
  |        ----------->     |
  OR                        NH
                            |
                            NH
                            |
                            C=O
                            |
                            R'
```

where R is an alkyl group and R' is a functional group capable of linking reaction with a biologically active material. It is believed that the amidrazone derivative shown above has a side chain which may tautomerize to produce a structure comprising double bonds in extended conjugation, i.e., that the structure 1 tautomerizes to structure 2:

```
      -C=N              -C-NH-
       |                 ||
       NH                N
       |                 |
(1)    NH         (2)    N
       |                 ||
       C=O               C-OH
       |                 |
```

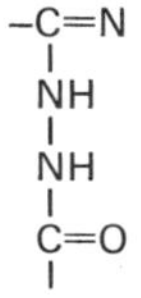

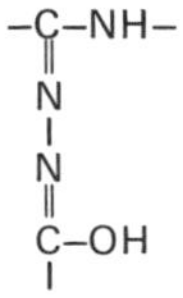

Structure 2 has carbon-nitrogen bonds in extended conjugation and is believed to protonate to a lower degree in aqueous media than structure 1. It is believed that the latter tautomerism contributes to reducing the buildup of localized positive charges on the support when in contact with aqueous media. A wide range of biologically active materials and particularly proteins may be linked to these organic polymers to produce biologically active matrices. These include enzymes present in or isolated from animal, plant or microbiological tissue.

The linking reaction in which the biologically active material is covalently bound to the organic polymer is preferably carried out under very mild conditions. Alkaline conditions are preferred, and the most favorable pH at which to bind the biologically active material to the organic polymer is the highest pH which the biologically active material in solution can tolerate without losing its activity. Usually the pH is 7.5 to 9. Binding is conveniently carried out in phosphate buffer but the choice of phosphate as a coupling buffer is not obligatory.

Buffers which do not contain free primary amino groups can be used, e.g., N-ethylmorpholine, imidazole, and ethylenediaminotetraacetic acid. Buffers such as Tris are not recommended, however, as the free amino groups in these compounds will react with the activated nylon. The amount of biologically active material bound to the organic polymer per unit weight of polymer, and the specific activity of the bound biologically active material can be maximized by controlling the pH of the reaction medium, the time of reaction, and the ratio of biologically active material to organic polymer in the reaction mixture. An adequate time of reaction is usually between 15 minutes and 1 hour.

Examples 1 through 4: A nylon tube is filled with a 10 wt % solution of triethyloxonium tetrafluoroborate in dichloromethane and the O-alkylation reaction allowed to proceed for 15 minutes to produce an imidate salt of the nylon tube. After washing with dichloromethane to remove any traces of unreacted alkylation reagent, the tube is cut into one-meter lengths which are filled with either (1) a 140 mM solution of adipic acid dihydrazide in formamide or (b) a 140 mM solution of succinic acid dihydrazide in formamide, and the reaction allowed to proceed for one hour at room temperature.

The tubes are then washed exhaustively with distilled water, and activated either (A) by filling the tube with a solution of diethyl adipimidate (30 mg/ml) in 20% (v/v) N-ethylmorpholine in methanol, incubating for 40 minutes at room temperature and then washing with methanol, or (B) by perfusing with a 5% (w/v) solution of glutaraldehyde in 0.2 M borate buffer, pH 8.5 for 15 minutes at room temperature and then washing free of excess glutaraldehyde by perfusion for 5 minutes with 0.1 M phosphate buffer pH 8.0.

Finally each tube is filled with 1 mg/ml aqueous solution of rabbit muscle lactate dehydrogenase in a neutral phosphate buffer and incubated for 2 to 3 hours at 40°C. The amounts of enzyme immobilized and the activities of the nylon tube supported enzymes are given in the table on the following page. Derivatives of glucose oxidase, catalase and lactate dehydrogenase are prepared in a similar fashion, and for comparison, derivatives of the same enzymes are preferred using 1,6-diaminohexane in place of the acid dihydrazide. It is found that when used in a continuous flow analyzer the acid dihydrazide derivatives have considerably improved carry-over characteristics by comparison with the diamine derivatives. For example, the lactate dehydrogenase dihydrazide derivative reveals considerably less carry-over due to pyridine nucleotide binding than the corresponding diamine derivative. It is found that dihydrazide derivatives can be used at a sample rate of 70 samples per hour with acceptable carry-over, whereas the diamine derivatives can only be used satisfactorily up to a sample rate of 40 samples per hour.

Ex. No.	Procedure	Nylon-Tube Derivative	Enzyme Immobilized (mg/M)	Tube Activity (mols/M-min)
1	A	Adipic acid dihydrazide-adipimidate-enzyme	0.56	0.37
2	A	Succinic acid dihydrazide-adipimidate-enzyme	0.56	0.19
3	B	Adipic acid dihydrazide-glutaraldehyde-enzyme	0.25	0.61
4	B	Succinic acid dihydrazide-glutaraldehyde-enzyme	0.22	0.52

Iminochlorides of Polyamides

Biologically active proteins are fixed onto substances which contain at least two secondary amino groups in the process disclosed by *J. Horn, H.-G. Batz and D. Jaworek; U.S. Patent 4,119,589; October 19, 1978; assigned to Boehringer Mannheim GmbH, Germany*. This process comprises reacting a compound containing at least two secondary amino groups at -80° to +50°C with a chlorination agent with the conversion of the secondary amino groups into imino-chloride groups of the formula

$$R-\underset{\displaystyle Cl}{\underset{|}{C}}=N-R_1$$

in which R and R_1, which can be the same or different, are aliphatic, aromatic or aliphatic-aromatic radicals, and reacting the carrier material product (1) directly with a biologically active protein, in an aqueous solvent, or (2) with a bifunctional compound containing at least one primary amino group, optionally in an aqueous or organic solvent, and reacting this product containing groups of the formula

$$R-\underset{\displaystyle NH-R_2-X}{\underset{|}{C}}=N-R_1$$

(in which R and R_1 are defined as above, R_2 is alkylene, α,ω-dioxyalkylene or a dicarboxylic acid amide group containing 2 to 8 carbon atoms and X is a primary amino group or a function capable of protein binding) either (a) directly with a protein or (b) first with at least one bifunctional or polyfunctional protein reagent and thereafter with a protein or with a reaction product of a protein with the protein reagent.

The imino-chlorination reaction is preferably carried out on liquid or solid polymers which contain secondary amino groups. Of these, the polyamides, such as nylon, Perlon, etc., are especially preferred. Polyurethane is another example of a polymer containing secondary amino groups. It is advisable to subject any solid polymers to the reaction in dissolved conventional solvents for the polymers in question. If the polymer containing the secondary amino groups is a solid, then it can be in the form of powder or of a solid body, e.g., a tube, etc. In the case of formed bodies with a relatively small surface area in relation to the mass, it is advisable to activate the surface area by deposition of a finely divided polymer from a solution in a finely divided, amorphous and thus more reactive form on the surface of a synthetic resin formed body.

A preferred chlorination agent for the preparation of the imino chloride is phosphorus pentachloride. Other chlorination agents which can be used include, e.g., phosphorus trichloride, thionyl chloride, phosgene, phosphorus oxychloride and sulfuryl chloride. The two latter agents are not so reactive as phosphorus pentachloride but have the advantage of being liquid so that they can be used without a solvent. However, in the case of phosphorus pentachloride, it is necessary to use a solvent, e.g., halogenated solvents, such as dichloroethane, methylene chloride, chloroform, carbon tetrachloride, etc. A preferred preparation of the imino-chloride is carried out using phosphorus pentachloride in dichloroethane at -38° to 0°C. These preferred conditions can be achieved by adding solid carbon dioxide to the solution. When the reaction is complete, excess chlorination agent should be removed before the reaction with the diamine.

The imino-chloride can be reacted directly in an aqueous solvent with the biologically active protein. The biologically active protein is preferably dissolved in an appropriate aqueous buffer, mixed with the solid imino-chloride and left to react. Such a method is recommended when the base structure of the imine is branched, e.g., in the case of polyamides which contain components with 3, 4 or more functions. Due to the steric relationships which arise, the direct reaction between the imino-chloride and the protein is favored. In the case of straight-chained polymers, on the other hand, the reaction with the proteins preferably takes place via an intermediate compound which contains at least one primary amino group (spacer).

The primary amino group can be connected directly with a carbon atom or can be present as a hydrazide group, i.e., is present bound to an acid amide group. If two primary amino groups are present, then the two amino groups are preferably separated from one another by an alkylene, α,ω-dioxyalkylene or dicarboxylic acid amide group containing 2 to 8 carbon atoms.

If, instead of a diamide, a monoamide is used, then this contains a further functional group which is preferably a group capable of forming a bond with a protein. This second functional group of the intermediate compound, regardless of whether it is a primary amino group or some other group, can be further reacted with a bifunctional protein reagent. Such compounds include, e.g., dialdehydes, such as glutardialdehyde, dihydroxy succinimide esters, diacetals, bismaleinimides, bifunctional imino esters, such as diethyl malonimidate, dimethyl adipinimidate, diepoxides, dicarboxylic acid chlorides, especially α,β-unsaturated dicarboxylic acid dichlorides, diisocyanates, diisothiocyanates, etc. They preferably contain 2 to 12 carbon atoms but can also have longer chains. Further details of the reaction is shown in the following example.

Example: 2-meter lengths of nylon-6 tubing with an internal diameter of 1 mm are rinsed with anhydrous methanol and subsequently with anhydrous methylene chloride to remove traces of moisture. The tubing is then precooled to -40°C and filled with a 5% solution of phosphorus pentachloride in carbon tetrachloride which has been cooled to -40°C. The precise maintenance of the temperature is not critical but temperatures below -20°C are preferred. After 5 hours, the phosphorus pentachloride solution is sucked out and excess reagent is rinsed out with dry carbon tetrachloride. Ethylenediamine or a 2% solution of oxalic acid dihydrazide in dimethyl sulfoxide is then introduced into the tubing. After standing for 2 hours with the amino or hydrazide solution, the tubing is rinsed overnight with water. The tubing is then rinsed with anhydrous dioxan in order to remove traces of water and a solution of 50 mg ethylene glycol/bis(propionic acid)/bis(hydroxysuccinimide) ester in 2.5 ml dioxan and 0.5 ml N-ethylmorpholine is introduced into the tubing. After 30 minutes, the solution is removed by suction and the tubing is rinsed with 0.1 M triethanolamine buffer (pH 8.5). The tubing is then filled with a solution of 2 mg glucose dehydrogenase in 1 ml triethanolamine buffer (pH 8.5) and left to stand for 3 hours at 4°C. It is thereafter rinsed with 5 liters 0.1 M triethanolamine buffer (pH 8.5) which is 1 M with regard to sodium chloride. The tubing subsequently displays an activity of 1.5 U/m (amine tubing) or of 1.6 U/m (oxalic acid hydrazide tubing).

In another embodiment, the biologically active protein is crosslinked with succinimide ester and then bound by glutardialdehyde, as bifunctional protein reagent, on to the free amino group of a diamine which, before reaction with the second primary amino group, has been reacted with the imino-chloride group. Since the succinimide ester is about 100 times more reactive than glutardialdehyde but, on the other hand, is also considerably more expensive, in this manner there can be achieved an increased activity yield but, at the same time, the bonding on to the polymer with the less expensive glutardialdehyde.

POLYURETHANES

Foamable Polyurethane Prepolymers Reacted with Aqueous Enzyme Solutions

L.L. Wood, F.J. Hartdegen and P.A. Hahn; U.S. Patent 3,929,574; December 30, 1975; assigned to W.R. Grace & Co. has disclosed a process for binding enzymes in active and reusable form on a polyurethane foam. The polyurethane is produced by the known reaction of di- and tri-isocyanates and other polyisocyanates with compounds containing active hydrogen, particularly glycols, polyglycols, polyester polyols and polyether polyols. This reaction makes an isocyanate-capped polyurethane. The enzyme preferably in aqueous media is bound by bringing it into contact with the polyurethane before the polyurethane is foamed. Since the polymerization reaction (e.g., of the polyol with the isocyanate) is exothermic, the temperature of the reaction mixture is maintained at below the temperature of thermal denaturation for the enzyme.

It is preferred that the polyurethane contain an average of two isocyanate groups per molecule. An even higher ratio can be used, e.g., 2 to 8 isocyanate groups per polyurethane molecule. Ratios higher than this are operable, but offer no advantage. In any case, all excess isocyanate groups left in the polyurethane foam (after binding of the enzyme) will be destroyed by hydrolysis upon the first contact of the foam with water, e.g., during the washing step preliminary to use of the bound enzyme.

In the binding step, the enzyme is maintained in a native conformation by the use of proper pH, ionic strength, presence of substrate, or necessary cations. The bound enzyme so produced is catalytically active. The binding reaction is a general one applicable to substantially all enzymes. For example, the following are suitable: urease, cellulase, pectinase, papain, bromelain, chymotrypsin, trypsin, ficin, lysozyme, glucose isomerase, lactase, and penicillin amidase.

Example 1: Ethylene glycol (100 g) and toluene diisocyanate (282 g) were mixed in a constant temperature bath at 65°C. After the material became water clear, it was cooled to 4°C and 100 ml of a fermentation broth containing cellulase activity was added with constant stirring at 4°C. After polyurethane foam formation was complete, about 15 minutes, the foam was washed with water and assayed for cellulase activity using carboxymethylcellulose as substrate and found to be active.

Example 2: The procedure of Example 1 was followed except that 100 ml of a 10% solution of pectinase in a 0.1 M phosphate buffer of pH 5 was used instead of broth. After polyurethane foam formation was complete, the foam was washed with water, cut into small pieces, and placed into a column. Freshly prepared apple juice containing 0.1% sodium benzoate was passed through the column at a rate of 1 liter per hour. The column clarified the apple juice as it passed through it, and the product had no changed flavor characteristics.

Two-Step Preparation of Polyurethane-Enzyme Foams

In the process detailed by *F.J. Hartdegen and W.E. Swann; U.S. Patent 4,098,645; July 4, 1978; assigned to W.R. Grace & Co.* an isocyanate-capped liquid polyurethane prepolymer acts as (a) a solvent to dissolve the protein which is to be bound; and (b) a reactant to react with the protein to bind it (the protein) to the poly(urea-urethane). The foam results when the resulting solution is mixed with water to form the polyurethane. The solidification temperature of the isocyanate-capped liquid polyurethane prepolymer used in this process will vary depending on the molecular weight of the prepolymer and on the structure of the backbone of the prepolymer.

The formation of the resulting solution is conducted in the absence of water and in the presence or absence of a diluent or a mixture of diluents. The diluents can be very soluble in water, e.g., acetone and the like; moderately soluble in water, e.g., methyl acetate, methyl ethyl ketone, and the like; or insoluble in water, e.g., benzene, etc.

The foaming step can also be conducted in the absence of a diluent or in the presence of one or more of such diluents. However, where conducting the foaming in the presence of a diluent care must be exercised to avoid the presence of so much diluent that the viscosity of the mixture comprising diluent and the resulting solution plus the water added to produce foaming is not reduced to an extent that carbon dioxide produced by the reaction of water and isocyanate groups fail to produce foaming or produces insufficient foaming to form a self-supporting poly(urea-urethane) foam comprising the immobilized protein.

Example 1: An isocyanate-capped liquid polyurethane prepolymer was prepared by reacting 31 g of glycerol (glycerine) with an amount of ethylene oxide to form 500 g of an intermediate compound (a hydroxyl-capped polyether) having an equivalent weight of about 500 (i.e., containing about 17 g of –OH group per 500 g of intermediate compound). This intermediate compound was reacted with commercial toluene diisocyanate using 1.05 mol of the toluenediisocyanate per 500 g of the intermediate compound. The resulting isocyanate-capped liquid polyurethane prepolymer was designated Prepolymer A.

Example 2: A mixture was prepared by mixing 1 g of the above Prepolymer A and 100 mg of invertase. The invertase dissolved and the resulting solution was stirred for 15 minutes in a dry atmosphere at about 25°C. Then a 2 g portion of water was added at 25°C while stirring. The resulting water-containing system began to foam; in 10 minutes foam formation was complete. The foam was thoroughly washed with water, which was collected and analyzed for invertase. 91% of the invertase was found bound to the foam. It was highly effective for inverting sucrose solutions. Its activity was retained even after 1,440 hours in which sucrose solution was continuously passed through a column packed with it.

Example 3: (Comparative Run) – When the invertase was mixed with the water and then added to Prepolymer A, only 68% was bound to the foam. The remainder was found in the water obtained by washing the polyurethane-enzyme foam with water.

Water-Dispersible Enzyme-Bonded Polyurethanes

In the process also developed by *F.J. Hartdegen and W.E. Swann; U.S. Patent 4,094,744; June 13, 1978; assigned to W.R. Grace & Co.* the isocyanate-capped liquid polyurethane prepolymer acts as (a) a solvent to dissolve the protein which is to be bound; and (b) a reactant to react with the protein to bind it (the protein) to the polymer. The formation of the protein/prepolymer solution is conducted in the absence of water. The solution can be formed in the presence or absence of a diluent or in the presence of a mixture of diluents. The diluents can be very soluble in water, e.g., acetone, etc.; moderately soluble in water, e.g., methyl acetate, methyl ethyl ketone, etc.; or insoluble in water, e.g., benzene, etc.

Following the formation of the protein/prepolymer solution certain proteins will cause the prepolymer to solidify if the protein is present in sufficiently large amounts. An example of this phenomenon is penicillin amidase. Where the amidase level exceeds about 10 wt % based on the weight of the prepolymer, the prepolymer solution exhibits increased viscosity and cannot be stirred after about 60 minutes. At concentrations below about 5 wt %, the solution can still be stirred and dispersed into water.

It is believed that other proteins will react with the prepolymer to form biologically active solid polymers, assuming the proteins are present in the prepolymer solution in large enough amounts. There is no way to determine in advance which protein will react with the prepolymer to form solids or at what levels the proteins must be employed in the solution. However, using the process, protein/prepolymer solutions can be formed without destroying the biological activity of the protein, for any protein, solid formation can be determined simply by dissolving the protein at successively larger levels in different batches of prepolymer.

Once it is decided to immobilize a particular protein in active form by binding it to a polyurethane matrix, it is a simple matter to form solutions using increasingly large levels of protein as solute and determine if solid formation occurs. Conversely, if solid formation is to be avoided the above test can also be performed to determine the maximum binding level of proteins.

Example: Lysozyme Bound to Polyurethane – A water-dispersible lysozyme/polyurethane reaction product was prepared by mixing 0.1 g lysozyme and 1.0 g prepolymer (obtained by mixing polyethylene glycol, MW 1,000, with trimethylolpropane and capping with toluene diisocyanate; their respective molar ratio was

2:0.66:6.3). These materials were allowed to react for 1 hour in a desiccator at 70°F. The enzyme/polyurethane composition was slowly added to 500 ml deionized water containing 3 drops poly(oxyethylene/oxypropylene) surfactant (Pluronic L-61), while the water was agitated rapidly. It was stirred 35 minutes more. The material separated into two phases. The soluble phase was separated by filtration through a Whatman #40 filter, then a Millipore 0.22 μ filter. From activity against *Micrococcus lysodeikticus,* it was found that 50% of the enzyme activity was present in the soluble phase. A control sample was prepared by curing and filtering 1.0 g of prepolymer (without enzyme) as above.

Sephadex G-50 (fine) was swollen in 0.05 M ammonium acetate buffer (pH 7.0), and used to fill a 2.5 x 45 cm glass chromatography column for a finished column of 2.5 x 41 cm. Samples were eluted with 0.05 M ammonium acetate buffer (pH 7.0) at 1.35 ml min^{-1} at ambient temperature and collected in 4.5 ml fractions. Absorbance at λ = 280 nm was read on each fraction. For each sample the absorbance was measured serially for 60 fractions eluted from the column. The samples were applied to the column as follows.

A 2.0 ml sample of soluble polymer-bound lysozyme was applied to the column and eluted (as described). The soluble polymer-bound lysozyme peaked at fraction 37 with an absorbance of 0.1. The free polyurethane used as a control peaked at 41 and an absorbance of 0.12. The shape of the absorbance curves for the polymer-bound enzyme and free polymer were similar and did not give sharp peaks as did the dextran and free enzyme. The relatively broad distribution and lack of a sharp peak is believed to reflect the presence of a broad distribution of polymer chain lengths in the bound enzyme and free polymer samples. 2.0 ml sample containing 1.5 mg free dextran (MW about 2,000) and 4 mg free lysozyme was applied to the column and eluted with buffer. The dextran peaked at fraction 13 (absorbance 0.71) and free lysozyme at fraction 30 (absorbance 0.32).

OTHER SYNTHETIC POLYMERS

Derivatives of p-Chloromethylstyrene Polymers

The process developed by *M.H. Keyes and F.E. Semersky; U.S. Patent 3,860,486; January 14, 1975; assigned to Owens-Illinois, Inc.* is based upon the discovery that water-soluble enzymes can be made water-insoluble by reaction with a modified polychloromethylstyrene in which less than 50% of the chloro groups are substituted by the 2,5-dioxo-4-oxazolidine group to form units having the following structure.

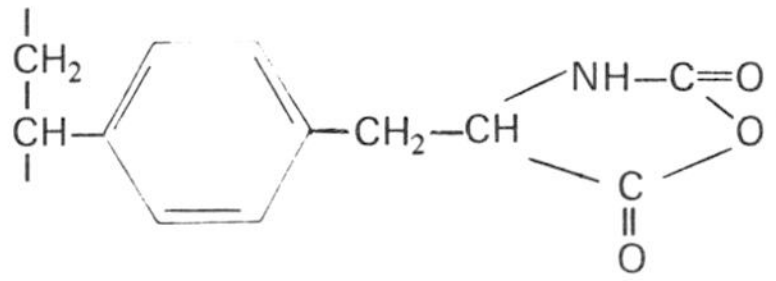

```
  |
  CH2                      NH—C=O
  |                       /      \
  CH—C6H4—CH2—CH                  O
  |                       \      /
                            C
                            ||
                            O
```

The enzyme binding reaction is carried out in an aqueous medium at about 0°C and in the presence of an alkali metal bicarbonate (preferably sodium bicarbonate). When the reaction is complete, the solid, water-insoluble, enzyme product formed is separated by filtration, washed repeatedly with water and assayed for activity.

The compounds of the above formula can be prepared by various methods and more specifically by refluxing 4 mols of ethyl acetamidocyanoacetate and 1 mol of poly(chloromethylstyrene) in acetone in the presence of sodium iodide and sodium ethylate for 24 hours to form a condensate which is separated by filtration and purified by extraction with alcohol and water.

This purified product is then slurried with hydrobromic acid at the reflux temperature for about 24 hours to form the acid salt of the alanine moiety formed which is separated and neutralized with triethylamine at room temperature. The polystyryl alanine thus formed, is then separated, dissolved in dioxane and gradually heated to 40°C while phosgene gas is passed through the mixture to cyclize the polystyryl alanine and form the compound of the above formula which is removed by filtration and dried.

Example 1: 2½ ml of dimethylformamide and 2.5 ml of 0.2 M aqueous sodium bicarbonate containing 3.4 mg of ribonuclease per ml were mixed, cooled to 0°C and 200 mg of the compound shown in the formula was added with stirring. The temperature and stirring were maintained for 72 hours after which time the water-insoluble ribonuclease formed was separated by filtration and washed repeatedly with distilled water.

The sample thus prepared was assayed initially and periodically by measuring the rate of change of cytidine-2',3'-cyclic phosphate to cytidine monophosphate by the method of Hammes and Walz as described in *Biochim. Biophys. Acta,* 198, 604 (1970). The initial activity was equivalent to 0.2 mg of soluble ribonuclease per gram of modified polystyrene which diminished slightly after 10 days of storage at 0°C.

Example 2: 5 mg of chymotrypsin is dissolved in 10 ml of water and 4 ml of the solution is mixed with 25 mg of the compound shown in the formula. The resulting mixture is saturated with $NaHCO_3$ and stirred for 3 hours at 0°C. The activity of the water-insoluble chymotrypsin produced is measured by the method of Hummel as described in *Can. J. Biochem. Physiol.,* 37, 1,393 (1959) using benzoyl tyrosine ethyl ester as a substrate. If desired, other enzymes such as peroxidase, papain, trypsin, xanthine oxidase, glutamic-pyruvic transaminase, etc., can be rendered water-insoluble by reaction with the compound of the formula as described above.

Modified p-Nitrostyrene-Grafted Polypropylene

Derivatives of graft polymers are used by *R.S. Kenyon, J.L. Garnett and M.J. Liddy; U.S. Patent 3,981,775; September 21, 1976* to immobilize enzymes. This process comprises the steps of (1) grafting p-nitrostyrene or p-nitrophenylacrylate onto the surface of an inert polymeric carrier, (2) converting the nitro group of the grafted molecule to an amino, diazo or isothiocyanato group, and (3) covalently attaching the enzyme to the amino, diazo or isothiocyanato group to produce an enzymatically active conjugate.

The inert polymeric carrier may be any suitable natural or synthetic trunk polymer to which p-nitrostyrene or p-nitrophenylacrylate can be grafted. Such carriers include polyvinyl chloride, polyethylene, polypropylene, polystyrene, polyacrylamide cellulose, wool and long chain carbohydrates such as starch and modified dextrose such as Sephadex. The preferred carrier is polyvinyl chloride dispersed in a liquid medium in which it is substantially insoluble.

The carrier may be in the form of beads, pellets, powder, plates, tubes or any other suitable configuration. If desired, the polymeric carrier may itself be a shell graft polymer having a core of one polymer, e.g., polypropylene and a surface graft of a monomer such as styrene.

The grafting of the p-nitrostyrene or p-nitrophenylacrylate to the trunk polymer is preferably induced by UV or ionizing radiation. The radiation may be derived from any suitable source such as an x-ray or γ-ray source, an electron beam facility or the like.

Suitable conditions for the grafting of the monomers to a given trunk polymer can be determined by a person skilled in the art; however, the following points are relevant:

(1) The radiation dose is preferably given at a high dose rate and for a short time in order to maximize the surface graft as opposed to graft within the body of the carrier.
(2) The solvent should preferably not swell the carrier as swelling of the carrier will increase the amount of nonsurface grafting.
(3) The reaction should be carried out so as to minimize homopolymerization of the monomer.
(4) The reaction is preferably carried out in a solvent in which the carrier is insoluble but in which the monomer is soluble.

In a typical example a dose of from 1 to 10 Mrad was given at a dose rate of from 10 krad/hr to 1 Mrad/sec most preferably from 50 krad/hr to 1 Mrad/min in air the carrier being polypropylene or polyvinyl chloride powder suspended in a 1 to 40 wt % monomer solution in a suitable organic solvent such as dimethylformamide or methanol.

The nitro group of the monomer may be reduced to an amino, diazo or isothiocyanato group by any suitable reaction. If the nitro group is reduced to an amino group the enzyme may be covalently bonded to the amino group using a crosslinking agent such as N,N-dicyclohexyl-carbodiimide or N-ethyl-5-phenylisoxazolium-3-sulfonate. Alternatively, the amino group may be converted to a diazo or an isothiocyanato group to which the enzyme may be attached directly.

Example: (A) Preparation of a Surface-Grafted, Inert-Core Copolymer Poly-(p-Nitrostyrene)-γ-Polypropylene — Polypropylene powder is suspended in a solution of p-nitrostyrene in N,N-dimethylformamide or methanol (30%) and the suspension irradiated by γ-radiation to a total dose of up to 10 Mrad at a dose rate of 200 to 400 krad/hr. Nitrostyrene monomer and homopolymer are removed by extraction of the product with a mixture of chloroform:benzene (3:1) in a Soxhlet extractor for 75 hours.

The extracted product is further extracted with methanol in a Soxhlet extractor for 24 hours to remove chloroform and benzene. It is then dried in a vacuum oven at 70°C to remove methanol. Elemental analysis indicated a graft of 4 wt %.

(B) Reduction to Poly(p-Aminostyrene)-γ-Polypropylene – The copolymer (10 g) was added to a 500 ml 2-neck flask in about 100 ml of N,N-dimethylformamide. Heat to 100°C in an oil bath while stirring and add a solution of stannous chloride $SnCl_2{\cdot}2H_2O$ (100 g), in N,N-dimethylformamide (25 ml) and continue to stir for 6 hours at 100°C. Add hydrochloric acid (10 M, 100 ml) and stir for a further 6 hours at 100°C. Filter the cooled reaction mixture on a sintered-glass Buchner funnel and wash the resin with N,N-dimethylformamide (6 x 50 ml) and hydrochloric acid (6 N, 6 x 50 ml) alternately, then finally, with N,N-dimethylformamide (50 ml).

(C) Conversion to Poly(p-Isothiocyanatostyrene-γ-Polypropylene – The resin is treated with a mixture of N,N-dimethylformamide (90 ml) and triethylamine (5 ml). After stirring for 10 minutes to ensure complete neutralization, the resin is filtered and washed with N,N-dimethylformamide. The resin is then transferred to a 3-neck flask (500 ml) in N,N-dimethylformamide (100 ml) cooled to 4°C. Then, with rapid mechanical stirring, carbon disulfide (25 ml) and triethylamine (50 ml) are added dropwise and simultaneously. Sufficient N,N-dimethylformamide is added to the reaction mixture to nearly fill the flask and it is warmed to ambient temperature over a 4-hour period with continued stirring.

The flask is again cooled to 4°C and ethylchloroformate (45 ml) added dropwise with stirring; then, with slow stirring, the reaction mixture is allowed to warm to ambient temperature over a 12-hour period. The resin is filtered and washed with chloroform until all visible triethylammonium chloride has been removed. Finally it is washed with pyridine (50 ml) and then with methanol (5 x 50 ml) and dried in a vacuum oven at 50°C.

(D) Covalent Bonding of Trypsin – The resin (500 mg) is suspended in bicarbonate buffer (ph 9.6, 0.1 M, 10 ml) and cooled to 5°C with gentle shaking and a trypsin solution in bicarbonate buffer (10 mg/ml, 5 ml) is added. After shaking at 5°C for 16 hours, the suspension is centrifuged and the supernatant decanted. The copolymer-trypsin conjugate is washed with water and centrifuged to remove unbound enzyme. The conjugate is then dried by lyophilization.

Aliphatic Dialdehyde-Aromatic Polyamine Condensates

The manufacture of a condensation resin having a large bonding capacity for proteins, by the reaction of 1,3-phenylenediamine and formaldehyde, as well as diazotization of the reaction product, is known. Support resins having higher activity than known resins are prepared by *V. Krasnobajew and R. Böeniger; U.S. Patent 4,066,504; January 3, 1978; assigned to Givaudan Corporation* by condensing a carbocyclic, aromatic polyamine other than 1,3-phenylenediamine with an aliphatic dialdehyde or acrolein.

Carbocyclic, aromatic polyamines which can be used in the process include primary diamines or triamines, i.e., mononuclear compounds such as phenylenediamines (e.g., p-phenylenediamine) and phenylenetriamines (e.g., 2,4,6-triaminophenol, 2,4,6-triaminotoluene), polynuclear, e.g., dinuclear compounds such as biphenyldiamines (e.g., benzidine) or polynuclear compounds with aliphatic bridges, e.g., with up to 2 methylene bridges such as diaminodiphenylalkanes (e.g., diaminodiphenylmethanes) and their mixtures. The aromatic nuclei can also carry other substituents such as halogen atoms (e.g., chlorine), lower alkyl groups (e.g., methyl, ethyl), lower alkoxy groups (e.g., methoxy, ethoxy) or hydroxy, carboxy, mercapto, lower alkylthio or sulfonyl groups.

The preferred polyamine reaction components are benzidine and 2,4,6-triaminotoluene. Examples of aliphatic dialdehydes which can be used in the process include dialdehydes with up to 6 carbon atoms, e.g., glutardialdehyde, succinic acid dialdehyde, glyoxal, etc. Glutardialdehyde is preferably used. Proteins which can be bonded to the condensation products include polypeptides, antigens, antibodies, protein inhibitors and, especially, enzymes.

The molar ratio of the polyamine and aldehyde reaction partners is between 1:1 and 1:10, a ratio of 1:3 being preferred. The reaction can be carried out in a known manner, e.g., in aqueous solution, preferably with the addition of an acid such as mineral acid (e.g., hydrochloric acid). In a particular embodiment, the reaction is carried out in the presence of an inert, fine-grained, preferably inorganic, especially silicate-containing, adjuvant.

Inert, water-immiscible organic solvents (e.g., dichloromethane, chloroform, carbon tetrachloride, benzene, toluene, ethyl acetate, dioxane, carbon disulfide) are suitable as the second phase. Chloroform is preferably used as the inert, water-immiscible solvent, but acetone is also suitable for this purpose. The reaction temperature is not critical; it can lie, e.g., at between 0° and 50°C, preferably at room temperature (i.e., at about 18° to 22°C).

In order to fix the protein to the carrier, the latter is treated with an aqueous solution, preferably a buffered solution, of the protein (1 to 50 mg/ml) at a temperature of 4°C to room temperature. This treatment can be carried out while stirring on shaking. Because of the high activity of the carrier material, the fixing of the proteins can also be advantageously carried out by simply filtering the protein solution through a carrier layer, preferably a column filled with the carrier material, as is common, e.g., in column chromatography.

Example 1: To a solution or suspension of 5 g of the polyamine in 200 ml of chloroform, there are added portionwise with vigorous stirring, 20 g of aldehyde (80 ml of a 25% solution in water) and, after a further 5 minutes, 20 ml of 7 N hydrochloric acid. The reaction mixture solidifies. It is treated with 300 ml of water and shaken for 5 minutes until it again becomes liquid. The polymeric particles are left to stand for 1 hour with occasional stirring. After vacuum filtration of the mixture over a Buchner funnel, the polymeric particles are washed with three 200 ml portions of acetone and then with 0.1 N sodium hydroxide solution. The residual chloroform is removed by washing with acetone. The thus-obtained carrier particles are spherical, homogeneous and capable of good filtration; they are stored in water. They can be employed for enzyme fixing, optionally after diazotization.

Example 2: A chromatography column (1.5 x 15 cm) is filled with 1 g of carrier material and washed with the buffer solution used in the following enzyme fixing. The buffered enzyme solution (1 to 100 mg of protein per ml of buffer) is passed through the column with a flow rate of 20 ml/hr. The coupling capacity is exhausted when protein can be detected in the eluate, which can be detected, e.g., by measurement of the absorption spectrum at 280 mμ. The nonfixed enzyme is removed by washing the column with 1 M potassium chloride and the buffer solution. There then follows an activity determination of an aliquot of the enzyme-fixed material with the respective substrate under given reaction conditions. Then an aliquot of the enzyme solution is washed with water, dried and the dried material determined analogously. The number of units of enzyme activity per gram of carrier material can be determined in this manner.

The condensation products manufactured according to Example 1 and their corresponding activities determined according to Example 2 are compiled in the following table. The amount of fixed enzyme (mg/g of carrier) is about 50 to 200.

Condensation Product		Amyloglucosidase Activity units/g of Carrier Material*
Polyamine	Aldehyde	
1,4-phenylenediamine	glutardialdehyde	800
2,4,6-triaminophenol	glutardialdehyde	800
2,4,6-triaminotoluene	glutardialdehyde	1,400
2,4-diaminotoluene	glutardialdehyde	700
2,4-diaminoanisole	glutardialdehyde	4,000
3-amino-4-chloroaniline	glutardialdehyde	820
3-methylmercaptoaniline	glutardialdehyde	500
benzidine	glutardialdehyde	600
4,4'-diaminodiphenylmethane	glutardialdehyde	200
4,4'-diaminodiphenylsulfone	glutardialdehyde	250
benzidine:1,3-phenylenediamine (1:3)	glutardialdehyde	500**
1,3-phenylenediamine	glyoxal	600
1,3-phenylenediamine	succindialdehyde	800
1,3-phenylenediamine	acrolein	600

*Analogous results are obtained for catalase, β-galactosidase, subtilopeptidase, naringinase and α-amylase.
**Glucose oxidase

Oxidized Polymers Having p-Phenylenediamine Side Groups

M. Bleha, E. Votavová and Z. Plichta; U.S. Patent 4,085,005; April 18, 1978; assigned to Ceskoslovenska akademie ved, Czechoslovakia have prepared insoluble biologically active compounds by bonding soluble biologically active compounds to a polymer containing in its side chains the p-phenylenediamine skelton as the active bonding site. The system can be represented by the following formula:

$$R_1{-}CO{-}(O{-}CH_2{-}CH_2)_x{-}N(R_2){-}C_6H_4{-}NH_2$$

where R_1 is a polymer residue, R_2 is a lower alkyl, and x = 1 to 4. Oxidation of this aromatic system yields a reactive form capable of reacting with the amino group of the biologically active compound with formation of a strong covalent bond. The oxidations are carried out in an aqueous medium in the presence of the biologically active compound and under suitable conditions of pH values. Various oxidizing agents can be used which do not interfere with the biologically active compounds present in the reaction mixture, such as, e.g., potassium ferricyanide, Cu^{2+} ions in an alkaline medium, or in other words, complex forms of the redox ions, and also, e.g., the arsenate of an alkali metal, or a compound possessing a similar redox potential.

It is also possible to work under biologically tolerable conditions in a suitable medium, with oxygen (also air oxygen) being the oxidation agent in the presence of ions suitable for redox systems, such as, e.g., Cu^{2+} ions in an alkaline medium or a cupric ammonium complex. Organic redox systems, such as quinone-hydroquinone, can also be used. The temperature of the oxidation reactions is defined by the thermal stability of the biologically active compounds and varies advantageously from 0° to 5°C.

The above biologically active compounds can be prepared using hydrophilic copolymers in various forms, such as copolymers soluble in water or crosslinked in the form of blocks, plates, tubes, etc. For uses of the biologically active compounds in various applications, crosslinked copolymers having a porous character in the form of spherical particles have proved to be of advantage. Such materials possess a high specific surface area and thus, after bonding of the biologically active compounds, also a rather high effectivity. They are therefore used in various application processes, such as various forms of affinity chromatography, separation and purification processes, and the like.

Example 1: The polymer carrier was prepared by heterogeneous suspension copolymerization from a mixture of monomers consisting of glycol monomethacrylate, glycol dimethacrylate and N-ethyl-N-(2-methacroylethyl)-N'-acetyl-p-phenylenediamine so that the final product contained 14% of the functional monomer and had a porous character with a specific surface area of 65.6 m^2/g. After acid hydrolysis, 0.5 g of the material was suspended in 10 ml of a 0.05 mol borax solution containing 40 mg of trypsin. 2 ml of a 0.01 mol solution of potassium ferricyanide in a 0.05 mol borax solution were added to the suspension during 1 hour.

The gel material with bonded enzyme was then washed by standard procedure with 1 N NaCl containing 0.1 N acetate buffer solution, pH = 4.7, and eventually with 0.01 N acetate buffer solution. The activity of the bonded enzyme was determined, (6.7 milliunits per milligram) and the bond strength was checked by determining bonded amino acids after washing of the enzyme with a solution of 6 N guanidine hydrochloride.

Example 2: The polymer carrier prepared as in Example 1 was used for bonding trypsin. 0.5 g of the polymer was mixed with a solution of 50 mg trypsin in 5 ml of distilled water. 5 ml of ammonium buffer solution containing 22 mg of hydroquinone was added, and stirring was continued at 5°C for 3 hours under access of air. The polymer material was washed, and the activity of the bonded trypsin was determined (5.06 milliunits per milligram).

Example 3: Trypsin was bonded onto the polymer carrier prepared in Example 1 by oxidation reaction with potassium arsenate in a 0.05 mol borax solution. On washing by the prescribed procedure the enzymatic activity of bonded trypsin was determined (6.2 milliunits per milligram).

IMMOBILIZED ENZYMES AND COENZYMES FOR AFFINITY CHROMATOGRAPHY

The ability of biopolymers to recognize and bind specific molecular groups or structures has been termed affinity binding, bioaffinity or bioselectivity. When such selectivity is used in chromatographic techniques, it is termed affinity chromatography. In this technique, the insoluble factor, the ligand, is covalently bonded to a solid support to form a sorbent useful as a chromatographic adsorbent. The ligate, the matching biopolymer is adsorbed during the chromatography.

A solution containing the ligate plus other materials is applied to the column and a ligand-ligate bond is formed. Nonadsorbed materials are washed from the column and finally a solution capable of displacing the ligate is used to desorb the ligate.

Enzymes are useful in affinity chromatography both covalently immobilized as the ligand or reversibly immobilized as the ligate. Materials which show such affinity with the enzymes include coenzymes or cofactors, enzyme inhibitors and substrates.

IMMOBILIZED ENZYMES

Enzymes Coupled to Neutralized Polyamphoteric Resins

A process is provided by *E. Werle and H. Fritz; U.S. Patent 3,834,990; Sept. 10, 1974; assigned to Farbenfabriken Bayer AG, Germany* for the enrichment of polypeptides such as enzymes or enzyme inhibitors by linking a polypeptide in a covalent linkage to a carrier to form an addition product of resinous nature into which basic groups are introduced to render the product electrostatically neutral.

The resulting addition product in water is contacted with a solution of the substance to be enriched and capable of complexing with the fixed polypeptide and the complex is then washed, caused to dissociate and the complex-forming substance isolated. Prior investigations by these workers have used resins without the addition of basic groups (termed A-resins). However, the use of neutralized resins (termed N-resins) provides the advantage of higher binder capacity, use of milder isolation conditions and use with enzymes or enzyme inhibitors of low isoelectric point.

The N-resins are prepared by adding to the system of carrier and enzyme, or of carrier and enzyme inhibitor, at the suitable time, aliphatic and/or aromatic polyamines in which only one amino group is free while all other amino groups are protected from acylation. Examples of such amines are N,N-dimethyl-ethylenediamine, agmatine, N-methyl-N'-(3-aminopropyl)-piperazine, 4-amino-pyridine, 2-amino-pyridine, 3-amino-1-cyclohexyl-aminopropane, etc.

Examples of enzymes which can be enriched, purified and isolated from their more or less impure solutions with the aid of this process are primarily trypsin, chymotrypsin, kallikrein, plasmin, pepsin, renin, ribonuclease, thrombin, amylase, papain, hyaluronidase, carbopeptidase A and B, pancreatopeptidase E, penicillinase and cholinesterases.

Inhibitors which can be enriched, purified and isolated according to this process are primarily kallikrein-trypsin inhibitors; the specific trypsin inhibitor obtained from pig pancreas and bovine pancreas, which exclusively inhibits trypsin; the trypsin inhibitor obtained from spermatoceles; and the trypsin inhibitors from plant seeds and potatoes.

Example 1: Preparation of the Water-Insoluble Trypsin-N-Resin – 200 mg of a copolymer of ethylene and maleic acid anhydride with an average molecular weight of about 30,000 (EMA-31-resin) were added with stirring to a mixture of 20 ml of 0.1% hexamethylenediamine and 75 ml of a 0.2 M salt-buffer solution (0.1 M triethanolamine, 0.1 M NaCl), pH 7.8, which had been cooled to 0° to 4°C.

The suspension was briefly homogenized with cooling and mixed, 2 minutes after the addition of the resin, with a cooled (0° to 4°C) solution of 1 g of trypsin in 75 ml of the salt-buffer solution. After stirring this reaction mixture for 4 minutes, there was added an aqueous solution of 4 g of N,N-dimethylethylenediamine (mixed with water in a molar ratio of 1:1 and adjusted to pH 7.8 with 2 N HCl), the cooled reaction mixture was stirred for 2 hours and subsequently centrifuged.

The supernatant solution still contained 320 mg of trypsin, i.e., 69% of the trypsin used was linked to the water-insoluble resin. The water-insoluble trypsin N-resin was mixed with twice its volume of formylated or ethanolized cellulose powder, poured into a column cooled to 0° to 4°C, and then washed with a salt-buffer solution (0.1 M triethanolamine, 0.1 M NaCl, 0.01 M $CaCl_2$), pH 7.8, for a sufficient period of time (overnight) so that no more tryptic acitivity could be detected in the flow. Similar yields of trypsin N-resins are obtained when the N,N-dimethylethylenediamine is replaced with the other amines mentioned above.

Example 2: Isolation of Inhibitors with the Aid of the Trypsin-N-Resin – The extract from pig pancreas which has been freed from high-molecular proteins by treatment with 3% perchloric acid (cf H. Fritz, G. Hartwich and E. Werle, *Z. Physiol. Chem.*, 348, 150, 1966) contained the inhibitor with a specific activity of 0.014 ImU/μg of biuret-protein (for definition of ImU see the above literature reference; 1 ImU for trypsin inhibits approximately 1 μg Trypure Novo).

The trypsin N-resin, prepared from 10 g of trypsin, was stirred into the inhibitor-containing neutral solution (2.14 liters; 745,000 ImU for trypsin) while cooling to 0° to 4°C. After stirring for 2½ hours (0° to 4°C) the mixture was centrifuged and the supernatant solution was discarded. The insoluble inhibitor-trypsin resin complex was subsequently stirred 3 times (0° to 4°C) with a 0.2 M salt-buffer solution (0.1 M triethanolamine; 0.1 M NaCl; 0.01 M $CaCl_2$), and each time the solution supernatant after centrifuging was discarded. The resin complex was then slurried with a 0.2 M KCl solution, the pH of the suspension adjusted to 2.0 by the addition of 2 N HCl, and the suspension was then stirred for 1½ hours while cooling.

The mixture was subsequently centrifuged and the residue (enzyme resin plus resin complex) again treated with a 0.2 M KCl/HCl solution, pH 2.0, as described above. The two combined supernatant solutions contained a total amount of inhibitor of 458,600 ImU for trypsin, i.e., 68.5% of the amount used for isolation. The specific activity of the inhibitor preparation amounted to 0.32 ImU per microgram biuret-protein and after desalting through a Sephadex-G-25 column to 2.73 ImU/μg biuret-protein. An inhibitor-(Trasylol)-N-resin inhibitor can also be used to isolate trypsin and chymotrypsin enzymes.

Isolation and Purification of Coenzyme A

I. Chibata, T. Tosa and Y. Matuo; U.S. Patent 3,935,072; January 27, 1976; assigned to Tanabe Seiyaku Co., Ltd., Japan have found that a biologically specific interaction between Coenzyme A (CoA) and its associated enzymes can be utilized advantageously for the purification of CoA. In other words, enzymes or proteins, which exhibit a special and unique affinity to CoA, are inherently contained in living cells of a CoA-producing microorganism. Such enzyme or protein (i.e., CoA-affinity-substance) can be advantageously used as a liquid which is coupled to a solid support for affinity chromatography. The coupling of the ligand can be readily achieved by immobilizing the cell-free extract of a CoA-producing microorganism with a cyanogen halide activated water-insoluble polysaccharide.

According to the process, a purified CoA solution is obtained by the steps of

(a) immobilizing the cell-free extract of a CoA-producing microorganism with a cyanogen halide activated water-insoluble polysaccharide to prepare an immobilized cell-free extract involving CoA-affinity-substance,

(b) washing the immobilized cell-free extract,

(c) contacting a crude CoA solution with the immobilized cell-free extract to have CoA adsorbed thereon,

(d) washing the immobilized cell-free extract, and

(e) eluting CoA from the immobilized cell-free extract.

Alternatively, CoA-affinity-substance may be separated from the cell-free extract of a CoA-producing microorganism prior to the immobilization step mentioned above. In this alternative method, the separation of CoA-affinity-substance is carried out by the steps of (a) immobilizing CoA with a cyanogen halide activated water-insoluble polysaccharide, (b) contacting the resultant immobilized CoA with the cell-free extract of a CoA-producing microorganism to have CoA-affinity-substance absorbed on the immobilized CoA, and (c) eluting CoA-affinity-substance from the immobilized CoA. CoA-affinity-substance thus obtained is then immobilized with a cyanogen halide activated water-insoluble polysaccharide. A purified CoA solution is obtained by contacting a crude CoA solution with the resultant immobilized CoA-affinity-substance to have CoA adsorbed thereon, washing the immobilized CoA-affinity-substance, and then eluting CoA from the immobilized CoA-affinity-substance.

Examples of the CoA-producing microorganisms include *Sarcina lutea* IAM 1099, *Sarcina aurantiaca* IAM 1059, *Sarcina aurantiaca* IFO 3064, *Micrococcus rubens* IFO 3768, *Mircrobacterium flavum* IAM 1642, *Brevibacterium ammoniagenes* IAM 1641, *Corynebacterium alkanophilum* nov. sp. ATCC 21071 and *Pseudomonas alkanolytica* nov. sp. ATCC 21034.

The cell-free extract of each one of the above microorganisms can be prepared by known methods such as, for example, homogenization, ultrasonic destruction or lysozyme treatment. The cyanogen halide activated water-insoluble polysaccharide of the process can be prepared by known methods such as that disclosed in *Nature,* 214, 1302-1304 (1967).

The aqueous nutrient media used in the example below contained glucose (10 w/v %), corn steep liquor (2.2 w/v %), peptone (1.35 w/v %), monopotassium phosphate (0.5 w/v %), dipotassium phosphate (0.5 w/v %), magnesium sulfate heptahydrate (0.2 w/v %), ammonium acetate (1.0 w/v %) and biotin (0.05 μg/ml).

Example: (A) Preparation of the Cell-Free Extract of a CoA-Producing Microorganism — 10 liters of the aqeuous nutrient medium was charged into a jar fermentor, adjusted to pH 7.0 and then sterilized. *Sarcina lutea* IAM 1099 is inoculated and cultivated at 30°C for 72 hours under aeration (3 liters/minute) and agitation (350 rotations/minute). The fermentation broth is centrifuged, and the cells thus collected are lyophilized, whereby 500 g of the lyophilized cells of *Sarcina lutea* IAM 1099 are obtained. 100 g of the lyophilized cells are suspended in 3.5 liters of a 0.05 M phosphate buffer solution (pH 7.0). The suspension is subjected to ultrasonic treatment (20 kilocycles/hour) under ice-cooling and then to centrifugation. 3 liters of the cell-free extract of *Sarcina lutea* IAM 1099 are obtained.

(B) Preparation of a Cyanogen Halide Activated Water-Insoluble Polysaccharide — 700 g of agarose (wet form), sold under the trade name Sepharose, are suspended in 28 liters of an aqueous (25 mg/ml) cyanogen bromide solution, and the suspension is stirred at 25°C for 8 minutes. The suspension is kept at pH 11 with an aqueous 5 N-sodium hydroxide solution during the reaction. After the reaction, the resultant precipitate is collected by filtration. The precipitate thus collected is washed with water and an aqueous 0.1 M sodium bicarbonate solution. 650 g of cyanogen bromide activated agarose (wet form) are obtained.

(C) Preparation of an Immobilized Cell-Free Extract — 320 g of cyanogen bromide activated agarose (wet form) are added to 3 liters of the cell-free extract obtained in (A) above. After the mixture is stirred at 10°C for 17 hours, 320 g of cyanogen bromide activated agarose (wet form) are again added. The mixture

is then adjusted to pH 6.0 and is further stirred at 10°C for 6 hours. The resultant precipitate is collected by filtration. The precipitate thus collected is washed with water and an aqueous 0.01 M sodium acetate solution (pH 7.0). 640 g of an immobilized cell-free extract (wet form) are obtained.

(D) Purification of CoA – A seed culture is prepared by cultivating *Sarcina lutea* IAM 1099 for 20 hours in the aqueous nutrient. 20 ml of this nutrient are charged into a 500 ml shaking flask, and 0.5 ml of the seed culture is added. The nutrient medium is cultivated at 30°C for 48 hours under shaking (140 rotations per minute). 20 mg of potassium pantothenate, 20 mg of adenine and 20 mg of cysteine hydrochloride are dissolved in the nutrient medium, and the medium is diluted with water to bring its volume to 25 ml. Then, the medium is further cultivated at 30°C for 16 hours under shaking (140 rotations per minute). The fermentation broth thus obtained is heated at 100°C for 10 minutes and then is centrifuged. 340 ml of water are added to the supernatant solution, whereby 510 ml of a crude CoA solution (pH 6.5; ionic strength, 0.04) are obtained. The solution contains 94 mg of CoA (purity, 0.7%).

640 g of the immobilized cell-free extract obtained in (C) are charged into a 5 cm x 50 cm column, and the column is washed with an aqueous 0.01 M sodium acetate solution (pH 6.5). 510 ml of the crude CoA solution thus obtained are passed through the column at a flow rate of 18 ml/hr. The column is then washed with an aqueous 0.01 M acetate buffer solution (pH 6.5) containing 0.03 M sodium chloride until contaminants are removed from the column. Then, an aqueous 0.01 M sodium acetate buffer solution (pH 6.5) containing 0.5 mol sodium chloride is passed through the column to elute CoA therefrom. The eluate thus obtained contains 59 mg of CoA (purity, 71%).

IMMOBILIZED COENZYMES

Bonding Thio Derivatives of Coenzymes to Supports

A coenzyme-support matrix having improved bonding capacity has been developed by *P.D.G. Dean and D.B. Craven; U.S. Patent 3,904,478; September 9, 1975.* This process comprises reacting a thio-derivative of the coenzyme in which an amino group on the coenzyme has been replaced by a thio group, with

(a) a bifunctional organic compound having a nucleophilic group so as to couple the bifunctional organic compound to the coenzyme and then coupling the resultant coenzyme reaction product to a water-insoluble support material through the unreacted functional group of the bifunctional organic compound or

(b) a water-insoluble support material having a plurality of pendant nucleophilic groups so as to couple the coenzyme to the support material.

The process may be applied to a variety of naturally occurring coenzymes, and also to fragments, analogues and derivatives of naturally occurring coenzymes having coenzyme activity. Among the naturally occurring coenzymes which may be used are, for example, nicotinamide adenine dinucleotide (NAD), nicotinamide adenine dinucleotide phosphate (NADP) and their reduced forms, nicotinamide mononucleotide (NMN), adenosine diphosphate ribose (ADP-ribose), adenosine triphosphate (ATP), adenosine diphosphate (ADP), an adenosine monophosphate (AMP), pyridoxamine phosphate, a pterin, or a nucleoside phosphate. Fragments of coenzymes having coenzyme activity which may be used include, nucleosides containing a purine or pyrimidine ring, and phosphate-containing organic molecules such as pyridoxamine phosphate, acetyl phosphate, creatine phosphate, sugar phosphates such as glucose-6-phosphate, phospho-amino acids, alcohol phosphates such as 6-aminohexan-1-ol phosphate, nucleoside phosphates, phospholipids, and pyrophosphates.

Thio-derivatives of the coenzyme which may be used in the process include thiol, thioether, thioester, disulfide, sulfonyl halides, sulfonic acids, sulfonic esters and alkylsulfonyl and sulfone derivatives. These may be prepared by standard methods, for example, the thio-derivative of adenosine monophosphate with P_2S_5 after protecting the –OH groups with acetonide or an acetate.

The bifunctional organic compound which is coupled to the coenzyme in procedure (a) contains a nucleophilic group which can react with the thio group on the coenzyme to form a covalent bond, for example, an amino group. The bifunctional organic compound also contains a group through which the coenzyme can be coupled to a support material, and this may be an amino group, or a hydroxyl, ester, or other suitable functional group. Preferably, the bifunctional organic compound is an aliphatic diamine, having a chain of 2 to 12 carbons between the amino groups, for example, 1,6-hexanediamine, 1,8-octanediamine and 1,12-dodecanediamine. The bifunctional compound may also be an α-, β-, γ- or ω-amino acid, for example, ϵ-aminocaproic acid. The reaction of the bifunctional organic compound with the thio-derivative of the coenzyme may be carried out in aqueous or alcoholic solution, preferably at 70° to 90°C.

The insoluble support may be an organic or inorganic material. Thus, the support may be a natural or synthetic polymeric material, particularly a hydrophilic material, for example, a polymer having free hydroxyl groups such as cellulose, a cellulose derivative such as carboxymethylcellulose; starch; a dextran or crosslinked dextran such as Sephadex, an agarose such as Sepharose; agar and agar derived polymers; a polyvinyl alcohol; a polyacrylamide; a nylon; a polyester such as polyethyleneterephthalate; cellulose acetate; and substituted crosslinked polystyrenes such as chloromethylated polystyrene. Inorganic supports which may be used include, for example, glasses and aluminum silicates.

The insoluble support material may take the form of beads, or a sheet of fabric, for example, a woven fabric or any other convenient cast or extruded shape. Preferably, it is produced in the form of a permeable material which is adapted to make up a packed column. The reaction product of the coenzyme and the bifunctional compound may be attached to the polymeric material by a variety of chemical techniques. In a preferred technique, cellulose or other polymeric materials containing a plurality of vicinal diol groups are treated with cyanogen bromide and then reacted with the modified coenzyme.

Example: This example describes the production of insolubilized adenosine monophosphate (AMP). 6-Mercaptopurineriboside-5'-phosphate is prepared by reacting inosine monophosphate with P_2S_5 after protecting the –OH groups with acetate. 6-Mercaptopurineriboside-5'-phosphate barium salt (7 mg) and 7 ml of a 40% aqueous solution of 1,6-hexanediamine are introduced into a thick Pyrex glass ampoule. The ampoule is sealed and then heated to 70° to 80°C for 16 hours. The ampoule is then opened, the solution frozen and then freeze dried. The resulting residue is redissolved in approximately 20 ml H_2O and freeze dried again. This process was repeated once more to give a residue almost free of unreacted diamine.

The product is further purified by preparative TLC on cellulose (binder-free) using 95% ethanol:1 M ammonium acetate, pH 7.5 in the ratio 7:3 as the solvent system. The UV (254 cm^{-1}) absorbing band is scraped off, the derivative eluted with water and recovered by freeze drying. This is coupled to 2 g Sepharose activated with 0.5 g of cyanogen bromide, by reaction in pH 10.0, 0.1 M sodium bicarbonate buffer overnight at 4°C. 90% of the derivative is bound in this manner.

The insolubilized AMP preparation is an admirable affinity chromatography support for the purification of crude dehydrogenase extracts. This is shown by the following results obtained for pure enzyme heavily adulterated with bovine serum alumin, the concentration of salt required to elute the bovine serum albumin being 10 mM phosphate, pH 7.5.

Enzyme	Concentration of Salt Required to Elute Enzyme
Lactate DH (muscle)	More than 1 M potassium chloride
Lactate DH (heart)	5 mM NADH will elute enzyme in presence of 1 M potassium chloride
Glucose-6-phosphate DH (yeast)	10 mM phosphate, pH 7.5
Alcohol DH (yeast)	405 mM potassium chloride

Coupling of Nicotinamide Adenine Dinucleotide to Supports

H.H. Weetall; U.S. Patent 3,957,748; May 18, 1976; assigned to Corning Glass Works has insolubilized the coenzyme NAD by chemically coupling it to an essentially water-insoluble carrier in such a manner that the coenzyme retains its redox properties. The NAD is coupled to a variety of diazotized organic and inorganic carriers by means of an azo linkage to form an insoluble, biochemically active composite. The composites can be easily removed from a reaction and used repeatedly in either the oxidized or reduced form.

Generally, the only requirements for the carriers are that, if organic, they can be modified for the coupling procedures disclosed below, and if inorganic, they should have available oxide or hydroxyl groups for coupling with the silicon portion of the silane coupling agents. A preferred carrier is glass, either in bead or porous form. The porous glass may be either particulate or in the form of an integral piece such as a disc.

Example: Porous glass particles of 40 mesh containing an average pore diameter of 550 A were used as a carrier to which NAD was coupled through a silane coupling agent. The glass particles were initially cleaned and etched by brief exposure to a concentrated HF solution and further incubation in hot 10 N NaOH for one hour. The porous glass particles were then thoroughly washed and dried.

A 20 g portion of the porous glass particles was mixed with 250 ml of dry toluene containing 10 wt % α-aminopropyltriethoxysilane (A-1100) and the slurry was refluxed for 24 hours. Then the porous glass was washed with toluene and acetone and dried in a vacuum oven at 60°C. The α-amino group was acylated by refluxing a slurry of the silanized glass particles in 250 ml chloroform containing 10 wt % p-nitrobenzoyl chloride and 10 wt % triethylamine for 24 hours. The resulting acylated glass particles were washed thoroughly with chloroform and dried at 60°C in a vacuum oven. The aryl nitro group was reduced to an aryl amine group by refluxing a slurry of the glass particles in 250 ml aqueous solution of 20 wt % sodium dithionite for 1 hour.

The treated glass particles were diazotized with 100 ml of 2 N HCl containing 2.5 g sodium nitrite at 0°C in a filter flask which was continually aspirated to remove evolved NO_2. After 20 minutes, the reaction slurry was filtered. The diazotized glass particles were then washed with 500 ml of cold, aqueous 1 wt % sulfamic acid to quench the reaction and remove traces of sodium nitrite. The particles were then immediately slurried in 50 ml of a buffered solution (0.10 M Tris/HCl, pH 8.5) of 10^{-2} M NAD. The coupling reaction was carried out at room temperature. The reaction was complete in 20 minutes and was monitored by withdrawing aliquots of the supernatant and measuring the decrease in OD of the diluted samples at 260 mμ.

Evidence of significant coupling of NAD to the porous glass was indicated by the intense orange color of the composite. Since the composite was thought to hold both chemically coupled and adsorbed NAD, the composite was treated to remove all adsorbed NAD by thorough washing with 0.01 M, pH 8.5, phosphate buffer.

Reversible Bonding of Enzymes to Immobilized Coenzymes

In a series of seven disclosures *P.D.G. Dean and C.R. Lowe; U.S. Patents 4,011,205; March 8, 1977; 4,011,377; March 8, 1977; 4,012,568; March 15, 1977; 4,012,569; March 15, 1977; 4,012,570; March 15, 1977; 4,012,571; March 15, 1977* and *4,012,572; March 15, 1977* have attached coenzymes to insoluble supports which can immobilize enzymes in a reversible process.

It has been found that those enzymes which require a coenzyme for their reactivity can be readily separated from other organic materials and from one another, by taking advantage of their affinity for the coenzyme and attaching the coenzyme to an insoluble support.

The insoluble polymeric support material can comprise a natural or synthetic material, particularly a hydrophilic material such as, for example, a polyacrylamide or a polymer having free hydroxyl groups such as cellulose, cellulose derivatives, starch dextran and crosslinked dextrans, for example, Sephadex, Sepharose, proteins such as wool, and polyvinyl alcohol. Other polymeric materials which can be used include nylon, polyesters such as polyethylene terephthalate, cellulose acetate, and substituted polystyrenes.

The polymeric material may take the form of beads, or a sheet of fabric, for example, a woven fabric or any other convenient cast or extruded shape. Preferably it is produced in the form of a permeable material which is adapted to make up a packed column.

The coenzyme that is chemically attached to the polymeric material can be, for example, nicotinamide adenine dinucleotide (NAD), nicotinamide adenine dinucleotide phosphate (NADP) and their reduced forms, adenosine diphosphate ribose (ADP-ribose), adenosine triphosphate (ATP), adenosine diphosphate (ADP), adenosine monophosphate (AMP), pyridoxamine phosphate, pyridoxal phosphate, and pterins.

The coenzymes may be attached to the polymeric material by a variety of chemical techniques. Azide groups or chloro s-triazinyl groups may be formed on the polymeric matrix which will react directly with the coenzyme without affecting its ability to attach itself to an enzyme molecule. In a particularly preferred technique, cellulose, a cellulose derivative, or another polymeric material containing a plurality of vicinal diol groups is treated with cyanogen bromide and then with the coenzyme according to the following reaction scheme.

$$\left[\begin{matrix}-OH \\ -OH\end{matrix}\right. + CNBr \longrightarrow \left[\begin{matrix}-O \\ -O\end{matrix}\right\rangle C{=}NH \xrightarrow{RNH_2} \left[\begin{matrix}-O \\ -O\end{matrix}\right\rangle C{=}N{-}R + NH_3$$

In the above reaction scheme RNH_2 represents the coenzyme. Preferably, the coenzyme is linked to the backbone of the polymer chain by an intermediate carbon chain containing either 2 to 20 carbon atoms or oligopeptides or polyaromatic systems such as benzidine bridges.

The process is preferably carried out by flowing a solution of the enzyme mixture through a packed column of the reactive matrix. After washing the column to remove unwanted materials the enzymes attached thereto may be removed sequentially by elution, preferably with buffered solutions of increasing ionic strength or buffered solutions of cofactors for the enzymes. It is found that different enzymes are removed from the column at different and characteristic ionic strengths thus affording a method of separating and purifying the enzymes. Inorganic phosphate buffer solutions and potassium chloride solutions have been found to be eminently suitable as means for increasing the ionic strength.

Where the mixture contains only two enzymes an alternative and simpler procedure may be adopted. In the first stage, the enzymes are attached to the reactive matrix as previously described and then washed with a solution of a substrate or effector for one of the enzymes. It is found that this technique often binds more tightly to the support the enzyme whose substrate or effector has been flowed through the column and removes the other enzyme in the eluent. The enzyme remaining on the support may be removed by washing with a buffer solution of higher ionic strength or by removal of the effector or by addition of a competitor for this effector.

Example 1: This example describes the separation of a mixture of L-malate dehydrogenase and L-lactate dehydrogenase on NAD-cellulose. 10 g of cellulose powder are activated by the cyanogen bromide technique of Axen, Porath and Emback, *Nature* 214, p. 1302 (1967) and reacted with 100 mg of nicotinamide adenine dinucleotide in 0.1 molar sodium carbonate buffer pH 9 for 12 hours. A synthetic mixture is made up of L-malate dehydrogenase (MDH), L-lactate dehydrogenase (LDH) and bovine serum albumin (BSA) and dialyzed overnight against 10 mM phosphate buffer. The NAD-cellulose is packed into a column 5 mm in diameter and 20 mm in height and equilibrated with 10 mM phosphate buffer pH 7.5. A 50 μl sample containing 0.335 unit of MDH, 1.85 units of LDH and 0.8 mg of BSA is applied to the column and nonadsorbed protein washed off with 2 ml of phosphate buffer. The column is then eluted with a potassium chloride gradient in 10 mM phosphate buffer, the gradient being from 0 to 0.5 M KCl (20 ml total); MDH, LDH and BSA are assayed in the effluent from the column, and the initial and final specific activities, recoveries and purification factors are given in the following table.

Ingredient	Amount Added	Initial Specific Activity (units/mg)	Units Recovered	Final Specific Activity (units/mg)	Purification (fold)
MDH	0.335 U	0.42	0.3	20	47.6
LDH	1.85 U	2.30	1.3	26	11.3
BSA	0.8 mg	–	–	–	–

It is found that MDH is eluted at 0.1 M KCl, BSA at 0.175 M KCl and LDH at 0.3 M KCl.

Example 2: This example demonstrates the chromatography of D-glucose 6-phosphate dehydrogenase from a crude yeast extract on ε-amino-caproyl-NADP-agarose prepared as follows. ε-Amino-caproyl-agarose was prepared by coupling ε-amino-caproic acid (1 g) to Sepharose 4B (10 g) by the method of Axen, Porath and Emback (1967). To 10 g of this polymer were added dicyclohexyl carbodiimide (5 g) in pyridine (12 ml) and NADP (100 mg) in water (3 ml). This mixture was rotated slowly for 10 days at room temperature and the gel washed successively with 80% aqueous pyridine, ethanol, butanol, ethanol, water, 2 M KCl and finally water.

A sample (1 ml) of the crude yeast extract was applied to a 1 cm x 5 cm column of ε-amino-caproyl-NADP-agarose equilibrated with 0.05 M triethanolamine-NaOH buffer, pH 7.6. Nonabsorbed protein was washed off with 50 ml of the same buffer and enzyme eluted with a potassium chloride gradient, 0 to 0.5 M, 200 ml total volume, at a flow rate of 12 ml/hr. D-glucose 6-phosphate dehydrogenase and other enzymically-active protein were assayed in the effluent. Enzyme was eluted at 0.15 M KCl, its specific activity increasing from 0.29 μmol NADPH/min/mg in the initial extract to 6.50 μmol NADPH/min/mg in the purified product. Recovery of enzyme activity was quantitative.

Example 3: This example summarizes in table form some results obtained with differing coenzymes, matrices, enzymes and extension arms. The results are expressed in terms of the concentration of potassium chloride required to elute the enzyme or bovine serum albumin in synthetically prepared mixtures. The procedure described in Example 1 was used in each case.

Matrix	Coenzyme	Extension Arm	Enzyme	Salt Concentration to Elute: Enzyme	BSA
Cellulose	ADP	none	L-lactate dehydrogenase	280	155
Cellulose	ADP-ribose	none	L-lactate dehydrogenase	295	140
Cellulose	ADP-ribose	none	L-glutamate dehydrogenase	460	155
Sephadex G-200	NAD	none	L-lactate dehydrogenase	120	0
Sephadex G-100	NAD	none	L-malate dehydrogenase	90	0

(continued)

Matrix	Coenzyme	Extension Arm	Enzyme	Salt Concentration to Elute Enzyme	BSA
Sepharose	AMP	ε-amino-caproyl	L-malate dehydrogenase	90	0
Sepharose	AMP	ε-amino-caproyl	isocitrate dehydrogenase	190	0
Sepharose	NAD	ε-amino-caproyl	glyceraldehyde 3-phosphate dehydrogenase*	>1,000	0
Sepharose	NAD	triglycyl	L-lactate dehydrogenase**	85	0
Sepharose	NAD	ε-amino-caproyl	L-lactate dehydrogenase***	>1,000	0
Sepharose	NADP	ε-amino-caproyl	D-glucose 6-phosphate dehydrogenase	125	0

*Enzyme eluted with a 200 μl pulse of 5 mM NADH.
**Enzyme eluted with 50 mM NAD.
***Column equilibrated with 5 mM Na_2SO_3 in all buffers. Enzyme eluted with 5 mM NADH.

IMMOBILIZED ENZYME INHIBITORS

Trypsin Inhibitors Supported on Macroporous Polyacrylates

Hydrophilic macroporous copolymers are used by *J. Coupek, J. Truková, O. Hubálková, M. Kriváková and V. Mansfeld; U.S. Patent 3,983,001; Sept. 28, 1976; assigned to Ceskoslovenska akademie ved, Czechoslovakia* as supports for use in affinity chromatography.

This process for the isolation of biologically active compounds comprises forming a sorption complex between a solvent soluble biologically active compound to be isolated and a biologically active compound linked to a solid carrier by a covalent bond, the solid carrier being a hydrophilic macroporous copolymer derived from hydrophilic monomers selected from hydroxyalkyl acrylates, hydroxyalkyl methacrylates; oligo- and polyglycol acrylates, oligo- and polyglycol methacrylates; acrylonitrile, methacrylonitrile; aminoalkyl acrylates, aminoalkyl methacrylates; acrylic acid, methacrylic acid or methylolacrylamide; crosslinked by copolymerization with divinyl or polyvinyl monomers selected from alkylene diacrylates, alkylene dimethacrylates, oligo- and polyglycol diacrylates, oligo- and polyglycol dimethacrylates, alkyl tri- and polyacrylates, alkyl tri- and polymethacrylates, glycol tri- and polyacrylates, glycol tri- and polymethacrylates, alkylenebisacrylamides, alkylenebismethacrylamides and divinylbenzene.

Preparation of the hydrogel copolymers useful as the carrier is carried out in accordance with known polymerization procedures as is the formation of the physical form thereof ultimately used.

It is to be understood that throughout the patent, the method is described either by using a biologically active compound bound in the gel, which compound corresponds to that sought to be separated, or by using a different biologically active compound. It is not necessary that both compounds be the same; however, the ability of one biologically active compound to separate either a like compound or a dissimilar one is dependent upon the following: The isolation makes use of the ability of biologically active substances to form specific reversible complexes with other active substances, e.g., enzymes form specific complexes with their inhibitors, antibodies with antigens, toxins with antitoxins, receptors with hormones, etc. If one component of the specific complex is bound to a solid carrier, it is possible to adsorb the other component from the solution.

Example 1: Isolation of highly active chymotrypsin by affinity chromatography on a hydroxyethyl methacrylate gel with covalently bound trypsin inhibitor was carried out as follows: 0.02 part by weight of crystalline chymotrypsin was dissolved in 1 part by weight of 0.05 M Tris-HCl buffer solution of pH 8.0. The solution was supplied to a column (10 mm x 80 mm) containing the hydroxyethyl methacrylate gel, which had been prepared by a suspension copolymerization of 2-hydroxyethyl methacrylate with ethylene dimethacrylate in the presence of an inert solvent; the resulting copolymer had a molecular weight exclusion limit of 300,000. Pancreatic trypsin inhibitor was covalently linked to the gel which was then equilibrated with 0.05 M Tris-HCl buffer solution having a pH of 8.0.

After the sample had soaked into the column, the column was eluted with the same buffer solution using a flow rate of 300 ml/hr and fractions thereby obtained were collected at ten-minute intervals. As soon as the fractions did not contain any compound absorbing in the ultraviolet region of the spectrum (at 280 nm), the elution with 0.05 M Tris-HCl buffer solution was stoppered and the gel column was further eluted with about 0.1 M acetic acid solution having a pH of 3. The dissociation of the sorption complex occurred at the change of pH of the elution agent and chymotrypsin was eluted from the column which exhibited high activity after lyophilization.

Its proteolytic activity to hemoglobin and esterase activity to acetyltyrosine ethyl ester were measured at a pH of 8; both were related to 1 mg of chymotrypsin. The chymotrypsin concentration was determined photometrically from the absorbance of chymotrypsin solution in 1 mM HCl at 280 nm.

Example 2: Chymotrypsin inhibitor was isolated from potatoes by means of affinity chromatography on a hydroxyethyl methacrylate gel which gel was crosslinked by ethylene dimethyacrylate and carried covalently linked chymotrypsin. Potatoes (1500 parts by weight) were homogenized in a high-speed blender and extracted with 200 parts by weight of a mixture of 0.9% NaCl and 0.03% Na_2SO_3 solutions mixed in a 1:1 ratio. The homogenized mixture was kept for 2 hours in a refrigerator at a temperature of +4°C and then was centrifuged. The supernatant was filtered through a 0.5 cm thick layer of kieselguhr and lyophilized. The crude light brown extract was obtained in the yield of 33 parts by weight. This crude lyophilized extract (from potatoes, 12 parts by weight) was dissolved in 75 parts by weight of 0.2 M Tris-HCl buffer solution of pH 8.0 and after the whole amount had been dissolved the solution was adjusted with 1 M NaOH to a pH of 8.

The sample was again filtered through the kieselguhr layer and introduced into a column (14 mm x 455 mm) packed with a hydroxyethyl methacrylate gel carrying covalently linked chymotrypsin and equilibrated with 0.2 M Tris-HCl buffer to pH 8. After the sample had soaked into the column, the column was eluted with 0.2 M Tris-HCl at pH 8.0 by a 3.5 ml/hr flow rate and fractions were collected each hour. When the fractions indicated that they did not any longer contain any protein, the column was eluted with 0.2 M KCl-HCl solution of pH 2.0. By changing pH in the column the inhibitor was eluted in a narrow zone. The inhibition activity of the isolated inhibitor increased 13 times by the single operation.

Trypsin Inhibitors Supported on Activated Polysaccharides

Enzymes such as trypsin, kallikrein, plasmin, chymotrypsin and the like, which take part in the kinin production and consumption in a living organism play an important role in regulating the circulatory system. The inhibitors against such enzymes derived from the blood and organ origins also play an important role in the maintenance of the normal functioning of the living body together in cooperation with these enzymes.

S. Fujii, Y. Kajita, S. Hiraku, H. Kira, H. Aishita, K. Muryobayashi, H. Terashima, A. Akimoto, K. Taniguchi and M. Wada; U.S. Patent 4,030,977; June 21, 1977; assigned to Ono Pharmaceutical Co., Ltd., Japan provide a method for isolating and purifying enzymes which take part in the production and consumption of kinin, and their naturally occurring inhibitors using an insoluble enzyme-enzyme inhibitor system or an enzyme-insoluble enzyme inhibitor system. This method comprises either converting the enzymes into an insoluble type, specifically combining the corresponding enzyme inhibitors and releasing of the inhibitors using an eluent having a specific composition, or converting the inhibitors into an insoluble type, specifically combining the corresponding enzymes and releasing of the enzymes using an eluent having a specific composition.

Depending upon the purpose, it is also possible to simultaneously obtain the related enzyme and its inhibitor from organs or body fluids if a series of the above treatments is carried out in parallel using a combination of an insoluble type of the enzyme and the corresponding enzyme inhibitor.

The proteolytic enzymes useful in this process are those capable of producing kinin from kininogen, e.g., trypsin, kallikrein or plasmin, enzymes capable of consuming kinin, i.e., chymotrypsin, and enzymes which can be inhibited by an inhibitor capable of inhibiting the above enzymes, i.e., thrombin. The naturally occurring proteolytic enzyme inhibitors used in this process are inhibitors which are proteinous in nature and capable of reversibly inhibiting the activity of the above enzyme. The term "physiologically active substance" includes both the above proteolytic enzymes and an inhibitor which is proteinous in nature and inhibits the activity of the above proteolytic enzyme.

The source material which can be used includes organs such as the heart, lung, pancreas, parotid, kidney, liver, spleen, etc., of mammals, for example, a horse, bovine, swine, dog, rabbit, sheep, guinea pig, rat and the like and body fluids such as blood, lymph, urine and the like of humans and mammals enumerated above.

The inhibitors can be chemically coupled to insoluble carriers which are inactive, porous and also hydrophilic, such as agarose, dextran, polyacrylamides and the like. Sepharose, Sephadex and Biogel can be used as the agarose, dextran and polyacrylamide, respectively. When a polysaccharide such as agarose or dextran is used as an insoluble carrier, the polysaccharide is first activated with cyanogen bromide in a known manner. Various other known modifiers or linkages may also be used to activate the carriers.

Example 1: Purification of Trypsin Inhibitor from Bovine Lung – To 1 kg of bovine lung which had been frozen and cut into pieces was added 1 liter of a

0.1 M triethanolamine buffer (containing 0.3 M sodium chloride and 0.01 M calcium chloride, pH 7.8). The resulting mixture was ice-cooled and then homogenized using a Waring blender for 10 minutes. The homogenate was stirred for an additional one hour and 100 g of Celite 545 (a diatomaceous silica) was added. The resulting suspension was centrifuged. The precipitate was suspended in the same volume of the above buffer and centrifuged again.

To the combined supernatant solution was added 50% trichloroacetic acid to a final concentration of 2.5%, and the resulting solution was allowed to stand for 30 minutes and then centrifuged. The supernatant solution was adjusted to a pH of 7.8 with 3.5 N sodium hydroxide and filtered to obtain 1950 ml of a yellow transparent solution containing a physiologically active substance.

On the other hand, 1 g of trypsin reacted overnight at a pH of 9.2 with 100 ml of Sepharose 4B, which had previously been activated with 10 g of cyanogen bromide, to form trypsin-Sepharose 4B was transferred to a column (2.2 cm x 20 cm), and 1950 ml of the above obtained solution (1,053,000 trypsin-kallikrein inhibitor units) was applied on the column. The column was washed with a 0.1 M triethanolamine buffer (pH 7.8) until the eluent was entirely protein-free. The column was then eluted with 0.25 M KCl-HCl (pH 1.8) to obtain bovine lung trypsin inhibitor in a yield of 80%. The inhibitor thus obtained was found to have a specific activity increase of 53 to 130 times compared with the solution as initially applied. The product thus obtained was subsequently desalted by passing it through a column of Sephadex G-25.

Example 2: Fractionation of Pancreas Proteolytic Enzyme — A mixture of 1 g of pancreas powder and 50 ml of cold water was stirred for a period of 45 minutes, and adjusted to a pH of 4.1 with 2 M acetic acid. The mixture was further stirred for an additional 1.5 hours, followed by centrifuging to obtain 48 ml of the supernatant. A 2 ml portion of the resulting supernatant was diluted with 10 ml of a 0.1 M trisaminomethane-hydrochloric acid buffer, pH 8.0 (containing 10 mM $CaCl_2$) and the solution was applied to a 4 ml column packed with bovine lung trypsin inhibitor-Sepharose 4B which had previously been equilibrated with the same buffer as used above. The unabsorbed substance showing an absorption at a wavelength of 280 mμ was washed out, and the column was then eluted successively with a 0.1 M acetate buffer (pH 4.0), 0.1 N acetic acid and 0.02 M hydrochloric acid to fractionally obtain chymotrypsin, kallikrein and trypsin, respectively.

SUPPORTS FOR AFFINITY CHROMATOGRAPHY

Supports Having Aminoalkylbenzoic Acid Groups

A process for the separation and purification of D-amino acid oxidase by affinity chromatography has been disclosed by *H. Mazarguil, F. Meiller and P. Monsan; U.S. Patent 3,941,657; March 2, 1976; assigned to Rhone-Poulenc Industries, France.*

The process comprises contacting a solution of mixture of proteins, containing the D-amino acid oxidase, with a carrier onto which the enzyme is fixed, then separating the D-amino acid oxidase from the carrier, and is characterized in that the carrier is a mineral carrier bearing haloalkylsilane grafts, upon which are fixed residues having the formula $-NH(CH_2)_3NH(CH_2)_2C_6H_4COOH$. The purified D-amino acid oxidase is then separated from the carrier in a solution of ethylene glycol.

The grafted carrier comprises oxides, hydroxides or other porous, insoluble mineral compounds on which are grafted, substituted or unsubstituted halo-alkylsilane groups whose alkyl residue comprises from 3 to 11 carbon atoms. The grafted carrier is modified by replacing the halogen of the graft by a residue having the above formula. The process for modifying the graft comprises:

(a) reacting the grafted carrier with the compound having the formula: $H_2N(CH_2)_3NH(CH_2)_2C_6H_4COOC_2H_5$ in accordance with any known processes, and in particular in suspension at at boiling temperature; then

(b) converting the ester function into an acid function by the action of any concentrated acid, in particular hydrochloric acid.

When purifying the D-amino acid oxidase, a column for chromatography is filled with the modified carrier, balanced at the pH-value which is compatible with the enzyme, then the mixture of proteins containing the D-amino acid oxidase in buffer solution is circulated in the column. The carrier is then washed with water to remove the absorbed inactive proteins.

The D-amino acid oxidase is then separated from the carrier by dissolution in a solution of ethylene glycol in a concentration of less than 50% by volume. The D-amino acid oxidase produced is pure and virtually all of the activity of the initial mixture is to be found again in the purified enzyme.

Example 1: Preparation of the Grafted Carrier — An aqueous solution of sodium silicate, corresponding to 220 g/l of SiO_2, is added dropwise to 230 ml of a 120 g/l aqueous solution of H_2SO_4, which is agitated. When the pH is 3.8, the addition of sodium silicate is stopped and the resulting sol, together with 2 drops of a sodium alkyl sulfonate, are poured into 8 liters of vigorously agitated trichloroethylene. The hydrogel balls formed precipitate. 1 liter of ammoniated water pH 9 is added, followed by filtration.

The balls are washed 3 times with N/10 HCl, then with water. The hydrogel produced contains 80% of water. 120 g of the hydrogel, 15 g of triethoxyiodopropylsilane and 200 ml of benzene are then heated at boiling temperature. In 3 hours, 95 ml of water is separated by azeotropic distillation. After cooling, the product formed is drained, washed with acetone and dried. The result is a silica grafted by iodopropylsilane groups, containing 2.1% by weight of iodine, whose grain size is less than 200 μ, specific surface area is 425 m^2/g and pore volume is 1.1 ml/g.

Example 2: Modification of the Grafted Silica — 11 g of the grafted silica is introduced into 25 ml of a benzene solution containing 1.5 g of the compound $H_2N(CH_2)_3NH(CH_2)_2C_6H_4COOC_2H_5$ and the dispersion is heated under reflux for 24 hours. After cooling, the silica is filtered and washed with ethanol at 95°C to dissolve any unreacted compound. This compound being colored, the washing operation is continued until the solvent is colorless. The silica is then washed with water.

The resulting product and 50 ml of 4 N HCl are then heated at boiling temperature for 24 hours, in order to convert the ester function into an acid function. After cooling, the balls are filtered and washed with distilled water, and then the pH is balanced at 8.6 with a pyrophosphate buffer solution. The amount of compound fixed is approximately 1 g.

Example 3: Purification of the D-Amino Acid Oxidase — 11 g of the modified grafted silica produced is introduced into a column for chromatography, then 5 ml of a pH 8.5 pyrophosphate buffer solution, containing 150 mg of the mixture of proteins, comprising D-amino acid oxidase and inactive proteins, are circulated at a speed of 9 ml/hr.

Enzymatic activity is determined on the solution issuing from the column, by the addition of D-alanine in 0.2 M pH 8.3 pyrophosphate buffer, then dinitrophenylhydrazine, and the absorbence is measured at 440 nm. The solution has no enzymatic activity, which shows that the D-amino acid oxidase has been retained on the carrier.

Water is then circulated in the column, at the same speed, until metering by spectrometry at 280 nm, carried out on the solution issuing from the column, confirms that there is no longer any inactive protein absorbed on the carrier. The inactive proteins represent 90% of all of the proteins introduced. 20 ml of a pH 8.5 pyrophosphate buffer solution, containing 30% by volume of ethylene glycol, is then introduced into the column at a rate of 9 ml/hr. Enzymatic activity is measured in the solution issuing from the column, as described above. The activity represents 90% of the activity of the initial mixture. The enrichment factor is approximately 100.

Modified Polysaccharides as Supports

While nearly all of the earlier methods for coupling ligands or proteins to agarose depend on the initial modification of the gel with CNBr, an alternative method exists which is rapid, simple, and safe. This method depends on the oxidation of cis-vicinal hydroxyl groups of agarose (or cellulose) by sodium metaperiodate ($NaIO_4$) to generate aldehyde functions. These aldehydic functions react at pH 4 to 6 with primary amines to form Schiff bases (aldimines) which are reduced with sodium borohydride ($NaBH_4$) to form stable secondary amines.

P. Cuatrecasas and I. Parikh; U.S. Patent 3,947,352; March 30, 1976 have determined that the reductive stabilization of the intermediate Schiff base is best achieved with sodium cyanoborohydride ($NaBH_3CN$) since this reagent, at a pH between 6 and 6.5, preferentially reduces Schiff bases without reducing the aldehyde functions of the agarose. Since sodium cyanoborohydride selectively reduces only the Schiff bases, it drives the overall reaction to completion.

Moreover, useful derivatives can be prepared by allowing the periodate-oxidized agarose to react with a bifunctional, symmetrical dihydrazide, more advantageously succinic dihydrazide (SDH), since carbonyl groups react with hydrazides more completely than with primary amines. The matrix is thus converted to a hydrazido/hydrazone form and can be reduced with $NaBH_4$ or preferably, $NaBH_3CN$, to produce unsymmetric hydrazides while the unreacted hydrazido groups are unaffected. The stable hydrazido-agarose derivatives can be stored and used at will for coupling proteins or ligands that contain carboxyl or amino groups with carbodiimide reagents according to known procedures or by conversion of the acyl hydrazide to the acyl azide form, also according to known procedures.

Other solid supports, such as crosslinked dextrans and cellulose, unlike agarose contain very large numbers of cis-vicinal hydroxyl groups and are thus very well suited for activation and coupling by the periodate oxidation procedures. Cellulose, because of its stability to the conditions of oxidation, can be substituted to a very high degree with the periodate method (see table below). The more porous Sephadex gels (G-75 and G-200) shatter after periodate treatment (5 mM, 2 hours) and therefore are not recommended.

Incorporation of Radioactive Ligands Into Activated Gels*

Matrix	SDH	Alanine	Albumin
	. . . . .(μmol/ml)		(mg/ml)
Agarose	1.2	0.25	0.20
Particulate cellulose	210	1.1	0.21
Beaded, porous cellulose	500	1.1	0.16

*Gels were oxidized with 0.5 M sodium metaperiodate for 2 hours and allowed to react with excess succinic dihydrazide (SDH). SDH incorporation was calculated from elemental analysis.

Example 1: Sodium Periodate Oxidation of Agarose – To a suspension of 100 ml of agarose in 80 ml of water was added 20 ml of 1 M $NaIO_4$. The suspension, placed in a 500 ml tightly closed polyethylene bottle, was gently shaken on a mechanical shaker for 2 hours at room temperature. The oxidized agarose was filtered and washed on a coarse sintered-glass funnel with 2 liters of water.

Example 2: Reductive Amination of Oxidized Agarose with Sodium Borohydride – The washed, oxidized agarose was allowed to react with a solution of ligand containing an amino group. For example, an ω-aminoalkyl agarose derivative was prepared by adding the oxidized gel, 100 ml, to 100 ml of 2 M aqueous diaminodipropylamine at pH 5.0. After 6 to 10 hours of gentle shaking at room temperature; the pH was raised to 9 with solid Na_2CO_3. 10 ml of freshly prepared 5 M $NaBH_4$ were added in small aliquots to the magnetically stirred suspension and kept at 4°C over a 12-hour period with precautions to avoid excessive foaming. The reduced agarose derivative was washed in a 250 ml sintered-glass funnel with 2 liters of 1 M NaCl without suction over a 4-hour period.

After stopping the outflow (with a rubber stopper), the agarose derivative was incubated in an equal volume of 1 M NaCl for 15 hours while on the funnel, and washed with an additional 2 liters of 1 M NaCl over a 2-hour period. The filtrates of the wash were occasionally checked for the presence of free diamine by the TNBS test or with ninhydrin. The extent of substitution of diamine, as determined by reaction with ^{14}C-labeled acetic anhydride, was 2 to 3 μmol/ml of agarose.

The TBNS (2,4,6-trinitrobenzene sulfonic acid) test is as follows: 2 to 3 drops of 1.5% ethanolic solution of TNBS are added to a mixture of 0.5 ml of filtrate and 0.5 ml of saturated sodium borate. The presence of diamine is indicated by formation of intense yellow color (420 nm) and of hydrazide by brick red color (500 nm).

Example 3: Reductive Amination with Sodium Cyanoborohydride – The reaction product of $NaIO_4$-oxidized agarose with the amino ligand was also subjected to reductive amination with sodium cyanoborohydride ($NaBH_3CN$). Although sodium borohydride is efficient, the use of $NaBH_3CN$ can be most advantageous, e.g., when the quantity of the amino ligand is limited. Since the latter reducing agent drives the reaction toward completion, relatively more efficient use of the amino ligand results. This method is also preferable when the ligand to be coupled is sensitive to the higher pH values (pH 9 to 10) necessary with $NaBH_4$.

A between 1 to 50 mM solution of the amino ligand, a designation which includes proteins as well as smaller molecules, in 0.5 M phosphate buffer at pH 6, containing 0.5 mM sodium cyanoborohydride was prepared at room temperature and centrifuged at 3,000 *g* for 10 minutes to remove insoluble material. The pH of the solution was adjusted to 6. The solution was then added to an equal volume of periodate-oxidized agarose which had previously been washed with 1 to 2 volumes of 0.5 M phosphate buffer at pH 6. The suspension was gently shaken for 3 days in a closed, capped polyethylene bottle at room temperature with a mechanical shaker. The gel was extensively washed as described above.

The unreacted aldehyde groups on the agarose matrix were then reduced with a solution of 1 M $NaBH_4$ (1 ml per each milliliter of agarose gel) for 15 hours at 4°C. The substituted agarose was washed extensively again. The substitution of ligand was about 2 μmol/ml of agarose when 50 mM diaminodipropylamine was used.

Modified Agarose with Affinity for Trypsin

An insoluble affinity matrix or carrier material which is capable of biospecifically binding trypsin and trypsin-like enzymes has been provided by *A.H. Nishikawa and H.F. Hixson, Jr.; U.S. Patent 4,020,268; April 26, 1977; assigned to Xerox Corporation.* This affinity matrix contains chemical compounds called ligands which have affinity for the desired enzymes and satisfy the following structure:

$$X{-}A{-}R$$

where X is

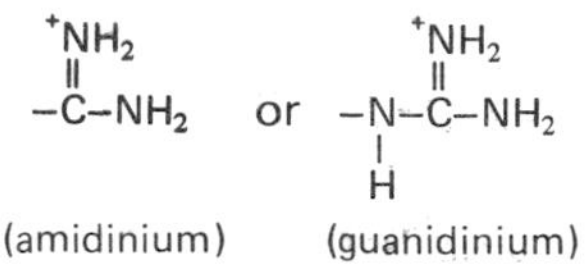

$$\underset{\text{(amidinium)}}{-\overset{\overset{+}{N}H_2}{\overset{\|}{C}}-NH_2} \quad \text{or} \quad \underset{\text{(guanidinium)}}{-\underset{H}{\underset{|}{N}}-\overset{\overset{+}{N}H_2}{\overset{\|}{C}}-NH_2}$$

where R is $-(CH_2)_yZ$, $-O(CH_2)_yZ$, $-S(CH_2)_yZ$, or $-SS(CH_2)_yZ$; Z is $-NH_2$, COOH, I, Br, Cl, COCl, NCS, NCO, SO_2Cl; y is 0-3; and A is preferably a phenyl group satisfying the formula

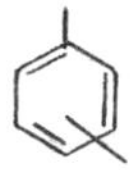

[phenyl ring structure]

or any suitable aliphatic or aryl aliphatic compound including benzoyl, norvalyl, cyclohexyl, hexyl and pentyl.

The ligand molecules are covalently attached to the insoluble carrier or matrix via a linear molecule called a leash.

Typical matrix or carrier materials include: cellulose and all of its refined forms; aminoethylcellulose; phosphocellulose; diethylaminoethylcellulose; carboxymethylcellulose; ECTEOLA-cellulose; p-aminobenzylcellulose; polyethyleneiminocellulose; triethylaminocellulose; sulfoethylcellulose; guanidoethylcellulose; agaroses such as, e.g., Sepharose or Bio-Gel A; crosslinked dextrans such as, e.g., Sephadex; crosslinked polyacrylamides such as, e.g., Bio-Gel P agaropectin; and crosslinked or noncrosslinked collagen. These materials are characteristically insoluble in aqueous solutions and preferably have low hydrophobicity and a nonionic character. A matrix material having an ionic character or having high hydrophobicity may prove to be an impediment in the affinity process.

Suitable leash molecules should be linear and hydrophilic, as well as free of strong ionic groups. Typical leash molecules include: ε-aminocaproic acid; β-alanine; 1,6-diaminohexane; bis(3-aminopropyl)amine; glycine; glycylglycine; glycylglycylglycine; succinamidoethylamine; succinamidobutylamine; succinamidohexamethylamine; aminoethanol; 2,2'-diaminodiethyl ether and its higher homologs. The molecular length of the leash compound used should be long enough to provide the proper relationship of ligand to matrix material so that effective binding of the desired enzyme may be accomplished. In some cases it is found that no leash is required. In most cases the compound used is preferably 7 A or longer.

Coupling reactions suitable for attaching leash and/or ligand molecules to the matrix or ligand molecules to leash compounds are generally limited by the solvent medium in which they can be carried out. The choice of the solvent medium is predicated by the requirements of the carrier material. Polysaccharide and polyacrylamide carriers are usually used in water. When necessary, water-miscible organic solvents in appropriate ratios with water can be used, e.g., agarose gels may be mixed with either 50% ethylene glycol or 50% dimethylformamide.

Although any suitable reaction temperature may be used, preferably from 1° to 30°C is conveniently used in an aqueous buffer solution of appropriate pH which may or may not include water miscible solvents. The process may be practiced by using any suitable coupling reaction. A typical coupling reaction includes the coupling of carboxylic groups with amino groups or alcoholic groups which may be accomplished, e.g., by the addition of carbodiimides (water-soluble type preferred), or via acyl azides, acyl halides, and acyl anhydrides by the addition of appropriate reagents. A ligand (or leash) containing an amino group may be coupled to a leash (or ligand) containing, e.g., isocyanate, isothiocyanate, sulfonyl chloride, alkyl halide, and methyl or ethyl amidate.

The affinity matrices may be used in clinical and/or veterinary therapeutic procedures, e.g., (1) in an in vivo binding agent or trapping device for trypsin or trypsin-like enzymes; (2) an in vivo agent or device for the administration of trypsin or trypsin-like enzymes for therapeutic purposes; (3) an extracorporeal device such as in a vascular shunt system for removing trypsin or trypsin-like enzymes from the blood; and (4) an injection or similar administrative device for introducing into the patient trypsin or trypsin-like enzymes for therapeutic purposes.

Example 1: To a suspension of agarose gel is added cyanogen bromide at 100 mg/ml of the gel and then titrated to a pH of 11.0 with 4 M sodium hydroxide. After alkali uptake ceases, the activated gel is collected on a sintered glass funnel and washed with cold borate buffer having a concentration of 0.1 M and a pH of 9.5. Then the gel is resuspended in cold borate buffer and mixed in equal volumes with a solution of 1,6-diaminohexane at 100 mg/ml of gel and the pH is adjusted to 9.5 The mixture is gently stirred overnight at 4°C. The gel is then washed by pouring into a coarse sintered glass funnel and slowly percolating distilled water through it. The aminohexamethyleneagarose obtained is suspended in an equal volume of distilled water into which is mixed 0.2 g of succinic anhydride/ml of gel. The pH is adjusted to 6.0 and held there by addition of 4 M NaOH.

When base uptake ceases, the mixture is stirred gently for 3 hours at 4°C and then allowed to stand at that temperature overnight. The succinamidohexamethylagarose (SHA) so obtained is washed on a coarse sintered glass funnel as previously described.

The ligand p-aminobenzamidine monohydrochloride in the amount of 37 mg/ml gel is coupled to the succinyl groups in the SHA gel by reaction in a 0.1 M water-soluble carbodiimide solution employing the technique of Hoare and Kishland [*J. Biol. Chem.* 242: 2447 (1967)]. The water-soluble carbodiimide used is 1-ethyl-3-(3'-dimethylaminopropyl)carbodiimide monohydrochloride (EDC). The coupling reaction is carried out in an automatic pH-stat at pH 4.75 using 1.0 M hydrochloric acid. When proton uptake is observed to cease, the gel slurry is washed on a sintered glass funnel using distilled water applied at 5 times the volume of the gel. The affinity matrix material obtained is referred to as pBz-SHA gel having a ligand capacity of 0.21 meq/ml of gel. The specific affinity of this matrix is illustrated in the following example.

Example 2: A standard 0.05 M tris(hydroxymethyl)aminomethane/0.5 M potassium chloride (having a pH of 8.0) buffer (Tris/KCl buffer) is used to equilibrate the affinity column, to dissolve enzyme samples, and to wash the enzyme loaded affinity column. A small chromatographic column tube (0.7 cm x 20 cm) is filled with 7 to 8 ml of pBz-SHA gel. After equilibration the column is loaded with 40 mg of the purest commercially available trypsin dissolved in about 2 ml of Tris/KCl buffer. When 0.01 M benzamidine hydrochloride (Bz) in the Tris/KCl buffer is percolated through the trypsin-bound column, the trypsin is desorbed by the soluble inhibitor and is specifically washed out in a highly purified state devoid of any degraded trypsin.

The column is subsequently eluted with glycine/KCl buffer (0.05 M glycine and 0.5 M KCl, pH 3.0) resulting in the emergence of little or no protein. The active trypsin so obtained is better than 95% pure as determined by the active-site titration method of Chase and Shaw, *Biochem. Biophys. Res. Comm.* 29:508 (1967). The yield of 65% obtained indicates that the purest commercially available trypsin is only 65% pure. The specificity of the affinity material is verified by loading 40 mg of α-chymotrypsin onto a preequilibrated column, then washing with the Tris/KCl buffer containing Bz. Typically, α-chymotrypsin is found to pass immediately through the affinity gel. Further, verification of affinity specificity is made by eluting the column with glycine/potassium chloride (0.05M/0.5M, pH 3.0) buffer. The column was found to elute protein indicating a properly functioning column.

IMMOBILIZED ENZYMES FOR ANALYTICAL PROCESSES

DETERMINATION OF ENZYME SUBSTRATES

Measuring Conductivity Change in a Glucose-Glucose Oxidase Reaction

R.A. Messing; U.S. Patent 3,839,154; October 1, 1974; assigned to Corning Glass Works has developed an apparatus and method for continuously monitoring conductivity changes attributable to a glucose-glucose oxidase reaction in which the glucose oxidase is immobilized within porous inorganic carriers and kept separate from the reaction product solution. An illustrative diagram of the apparatus is shown in Figure 11.1.

Figure 11.1: Apparatus for Measuring Conductivity Change

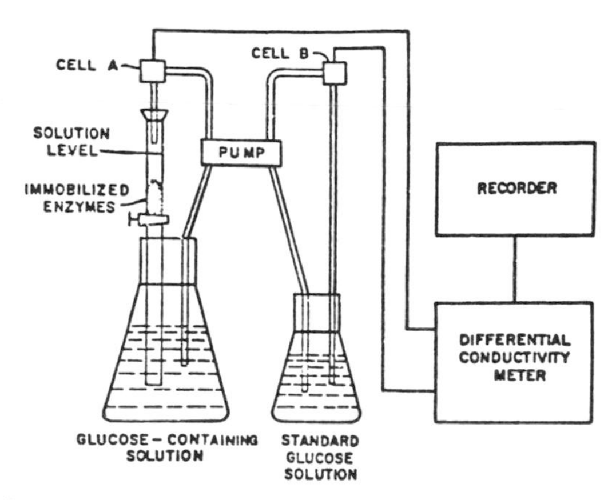

Source: U.S. Patent 3,839,154

In this figure a glucose-containing solution can be continuously pumped from a flask and through a glass plug flow-through column containing the immobilized enzyme composite consisting of glucose oxidase and catalase immobilized within the pores of porous inorganic carrier materials. The conductivity of the solution after passage through the column is measured by cell A which is a simple flow-through container having two electrodes electrically connected to a differential conductivity meter. Also connected to the differential conductivity meter is a second cell, B, also having two electrodes, which measures the conductivity of a standard glucose solution which is also continuously pumped through the conductivity cell. The differential conductivity meter indicates the conductivity change in the glucose-containing solution (also containing reaction products) relative to the conductivity of the standard solution and this change can be continuously recorded on a conventional recorder as illustrated. These recordings can be used to determine the concentration of the glucose-containing solution or the enzyme activity via standard curves prepared beforehand.

The immobilization of the enzymes was performed in a 25 ml Erlenmeyer flask. A volume of dialyzed glucose oxidase-catalase solution (containing 4,500 GOU) was added to the 300 mg of preconditioned porous titania carrier having an average pore diameter of 350 A. The vessel was then placed in a shaking water bath at 35°C and reacted with shaking for about 2 hours and 20 minutes. The vessel was then removed from the bath and the adsorption and diffusion was allowed to continue overnight (approximately 15 hours) at room temperature without shaking. The enzyme solution was decanted and the immobilized enzyme system was washed several times as follows. As the first wash, 9 ml of distilled water was added to the vessel. The vessel was stoppered and inverted several times to mix thoroughly.

The wash was then decanted. The second wash, performed as above, contained 9 ml of 0.5 M sodium chloride; the third wash was a 9 ml volume of 0.2 M acetate buffer, pH 6.1, and the final wash was 9 ml of distilled water. The immobilized enzyme system was then transferred with distilled water to a stopcock column fitted with a fiber glass retainer just above the stopcock fittings. The immobilized enzymes were stored between measurements at room temperature in the columns filled with distilled water and stoppered with a cork stopper.

The column conductivity measurements were performed at 22°C using a 125 ml Erlenmeyer flask containing the 100 ml of glucose substrate solution with a magnetic stirring bar. This flask was mounted on a magnetic stirrer and stirring was commenced. The inlet (the column outlet) and outlet tubes were inserted below the surface of the substrate solution and circulation was initiated by turning on the pump. The meter and recorder were then balanced and, after a stable base line was achieved, the outlet tubing from the flow-through cell A was inserted into the top of the column containing the immobilized enzymes.

The stopcock on the flow-through column was adjusted to maintain a three-fourths inch head of glucose-containing solution above the immobilized enzyme as the substrate solution circulated through the column and back into the reaction flask. This circulation was maintained throughout the assay. The results were obtained from the initial slopes of the differential conductivity recording which were then multiplied by the cell constant 80, corrected for dilution by multiplying by four (100 ml of substrate was used in place of the 25 ml of standard solution of 6% glucose containing 0.004% H_2O_2 at pH 5.7 to 6.4), correlated to GOU by dividing by 2.65 and finally reduced to a per-gram basis by multiplying by 3.33.

The apparent activity of the enzyme composite was affected by the flow rate through the column as shown by the data below.

Flow Rate (ml/hr)	Apparent Activity (GOU/g)
125	30.2
182	46.3
235	52.4
280	60.3
390	66.4

It should be noted from the above table that increases in flow rate result in increases in the apparent activity of the immobilized glucose oxidase-catalase. Since the diffusion of the glucose substrate molecule possibly plays a part in the apparent activity of the immobilized enzymes, then anything that would increase the diffusion rate of the glucose should increase the apparent activity of the preparation. Increasing flow rates of the substrate solution not only removes the reaction products from the pores but renews the concentration of glucose at the surface of the carrier, thus increasing the diffusion of the glucose.

The effect of flow rates upon the enzymes immobilized in carriers having other pore sizes was observed. In all systems studied, flow rate increases above 350 ml per hour have little or no effect on increasing the apparent activity.

Determination of Urease by Conductivity Measurements

R.A. Messing; U.S. Patent 3,915,804; October 28, 1975; assigned to Corning Glass Works has also used the apparatus as described in U.S. Patent 3,839,154 to measure minute changes in the electrical conductivity of a solution containing the products of urea-urease reaction without the necessity of having the enzyme present in the reaction products. The two flasks of Figure 11.1 of U.S. Patent 3,839,154 would contain urea in place of glucose.

Urease has a molecular weight of about 480,000 and a largest dimension of about 125 A when the enzyme exists as a monomer. If urease exists as a dimer, the largest dimension is about 250 A. Inasmuch as the substrate for urease, urea is extremely small relative to the enzyme, consideration need not be given to the substrate size in making an initial determination of suitable average pore size range for the carrier. In the examples below, urease was bonded to porous alumina and titania carriers having the following characteristics.

Table 1: Porous Carriers

	Al_2O_3	Al_2O_3-TiO_2	.... TiO_2		
Average pore diameter, A	175	220	350	420	855
Minimum pore diameter, A	140	140	220	300	725
Maximum pore diameter, A	220	300	400	590	985
Pore volume, cc/g	0.6	0.5	0.45	0.4	0.22
Surface area, m^2/g	100	77	48	35	9
Particle mesh size	25-60	25-60	25-60	30-80	25-80

The activities of the immobilized enzymes described above were determined by circulating 40 ml of 1 M (6%) urea solution at 390 ml/hr through the columns containing the immobilized enzymes. The slope that developed was multiplied by 4 before utilizing the conversion factor because 40 ml of substrate solution was used while only 10 ml of the standard urea was used to determine the conversion factor. The dilution-corrected slope was then divided by 77.7 to determine the number of Sumner Units. The assays were performed at room temperature and the immobilized enzymes were stored in water at room temperature between assays over the half-life determination period. All immobilized enzyme samples were repetitively assayed. Results are shown in Table 2.

Table 2: Assays of Immobilized Enzymes Having Carriers of Varying Average Pore Diameter

	Al_2O_3	Al_2O_3-TiO_2	.... TiO_2	....	
	175 A	220 A	350 A	420 A	855 A
Days	....	....	(Sumner Units)	....	....
0	0.35	0.81	1.18	4.36	2.46
0	0.22	–	–	–	–
1	0.11	–	–	–	–
4	–	0.59	0.63	2.32	1.77
5	0.05	0.45	0.43	1.98	0.95
6	–	0.06	0.33	1.98	0.61
7	–	0.05	0.29	1.56	0.36
11	–	0.03	0.16	1.11	0.19
18	–	0.03	0.11	0.73	0.09
27-28	–	0.01	0.07	0.47	0.03
32	–	–	0.05	0.34	–

From the above table, it can be seen that the optimum average pore size for the porous ceramic for urease is about 420 A, while a smaller amount of enzymatic activity is retained with composites using carriers having an average pore of as low as 175 A or as high as 855 A. This would appear to indicate that the form of urease immobilized in these carriers is probably predominantly the dimer form which has a maximum dimension of approximately 250 A.

Immobilized Enzymes Encapsulating Measuring Electrode

A hydrophobic membrane having the capability of immobilizing a protein has been developed by *J.A. Janata and J. Janata; U.S. Patent 3,966,580; June 29, 1976; assigned to The University of Utah.* The membrane comprises a thin substrate of a hydrophobic polymer capable of being swollen by solvent action and in which an aliphatic compound having a reactive terminal group can be absorbed from a solvent system. The hydrophobic membrane has pendant therefrom an essentially hydrocarbon chain which contains a protein reactive group, such as an oxirane group (epoxide) or other protein reactive group.

The hydrophobic membrane having protein-immobilizing ability is used to encapsulate a highly conductive electrode, such as platinum. A protein, e.g., an antibody, having a selectivity for reaction with a particular protein (antigen) is reacted at the protein-reactive site. Immersion of such a coated electrode in association with a reference electrode into an aqueous solution provides an electrically sensitive system capable of measuring the change in electrical charge of the solution-polymer interface caused by the capture of a particular protein (antigen) by the electrode with an immunoreactive antibody.

Polymers useful for the membrane are hydrophobic polymers which contain no pendant polar groups. Typical polymers include thermoplastic polymers such as polyvinyl chloride, polystyrene, polyethylene, polypropylene, silicone rubber, polyurethane, polycarbonate, polytetrafluoroethylene and the like. Thermosetting polymers such as epoxy resins and crosslinked polyesters may also be used.

Preferred polymers are those which may be coated upon the electrode by dip-casting or shrink-fitting. The polymeric membrane is then treated with a solvent system capable of swelling the membrane. The solvent system contains, besides an appropriate solvent, an aliphatic compound having a reactive site, preferably at one end of the hydrocarbon chain. The solvents used to swell the polymeric membrane are preferably those which may be readily removed by drying of the polymer.

A typical solvent mixture comprises petroleum ether of a 30° to 60°C boiling range and toluene. A typical aliphatic compound having a reactive site is n-decanol. Other useful aliphatic compounds are n-hexanol, n-decylamine, n-hexylamine, n-decanoic acid and like compounds having a labile hydrogen. The soaked membrane is then dried in a vacuum oven at a temperature preselected to remove the solvent without removing substantial quantities of the aliphatic compound having a reactive site. Driving off the solvent results in a membrane having a low concentration of pendant aliphatic groups having a reactive group such as hydroxyl, amine or carboxyl.

After the polymeric membrane is dried, it is immersed in a solution containing a compound having a protein-reactive site and another reactive site reactive with the reactive group of the aliphatic compound now attached to the membrane. For example, if the reactive group of the aliphatic compound is a hydroxyl group, then a typical protein-reactive compound is epichlorohydrin wherein an epoxide ether is formed. Other reactive compounds which may be included in the solvent system in place of an aliphatic alcohol are such compounds as aliphatic amines, carboxyl-containing aliphatic compounds and the like. These may be reactive with epichlorohydrin or with a bis-epoxide wherein a pair of oxirane rings are present for reaction.

The protein-reactive compound is then placed in a solution containing the protein to be immobilized. A preferred reaction temperature is room temperature and it is generally preferred to allow one or two days for the reaction to proceed. The reaction is generally conducted in a slightly basic medium. After the protein is attached, it is treated to wash off all residues of unreacted materials and further reacted with a compound to neutralize any unreacted protein-reactive, i.e., immobilizing groups which remain after reaction with the protein molecule. Unreacted protein-reactive groups tend to be polar in nature and are also undesirable in that they react nonspecifically with proteins and could give erroneous results in a protein detection device.

The protein or other component of an immunochemically reactive pair, such as an enzyme, is a very large molecule and it is likely that after immobilization of such a large molecule upon the reactive membrane, that other relatively large molecules, e.g. haptens, cannot penetrate through the barrier of large molecules to reach the membrane surface, although small molecules such as water may.

The membranes are particularly useful as they can be used in a device for electrically detecting the presence of a particular compound, both qualitatively and quantitatively in a given solution. Typical electrodes are glass encased platinum electrodes. A sheath of the hydrophobic polymeric membrane containing a reactive group is formed on the measuring electrode in a thickness of 10 to 50 microns, with a thickness of 20 to 40 microns being preferred. A protein or other compound of an immunochemical pair is immobilized in the membrane. The measuring electrode is used in conjunction with a reference electrode. The two electrodes are immersed in a solution which contains a protein or other compound of the type sought to be identified. The measuring electrode and the reference electrode are electrically connected to a meter sensitive to very slight changes in electrical potential. As the particular protein is captured by the measuring electrode, the electrical potential at the polymer-solution interface changes. The slight change is detected by the meter, which has a high impedence in electrode circuitry, thus indicating the presence of the compound. By calibration, the meter may be used to determine quantitatively the amount of compound present in the solution.

Electrode-Enzyme Assembly for Amperometric Analysis of Glucose

In prior enzyme electrodes, the enzyme has been included in the electrode structure either in the form of a solution or physically bound on a porous film having spaces large enough to hold the enzyme molecules or on a polymeric gel matrix. Such electrodes have suffered from the limitation of a rather quick loss in activity with time due to the relative instability of the enzyme, requiring frequent replacement of the enzyme portion. Accordingly, it is a primary object of the present process to provide an improved enzyme electrode for rapid and accurate amperometric analysis.

G.G. Guilbault and G.J. Lubrano; U.S. Patent 3,948,745; April 6, 1976; assigned to U.S. Department of Health, Education and Welfare have provided an enzyme electrode having improved enzyme stability and which retains its useful activity for longer periods of time without replacement of the enzyme portion. Figure 11.2 is an elevational sectional view which illustrates a preferred embodiment of the enzyme electrode in accordance with the present process.

Figure 11.2: Enzyme Electrode Assembly

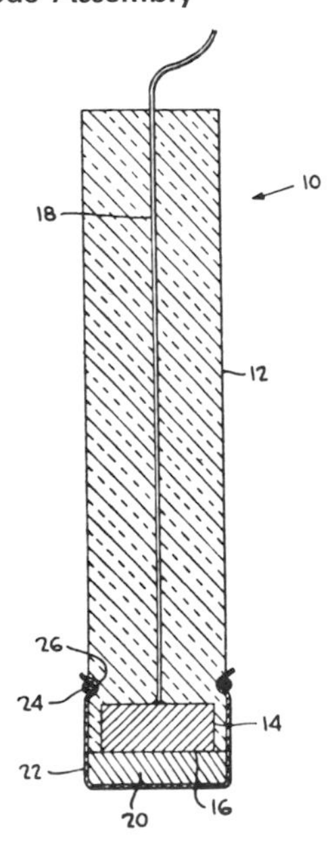

Source: U.S. Patent 3,948,745

Figure 11.2 shows an electrode assembly **10** which includes an electrically insulating support body **12** of plastic or glass which is preferably cylindrical and which supports a platinum electrode **14**, the latter including an active or exposed face **16**, and a conductor **18** attached to the electrode **14** and which passes through the support body **12**. Covering the exposed face **16** of the electrode **14** is a layer **20** of solid organic material having an enzyme chemically bound thereto. The layer **20** is held in contact with the exposed face **16** of the electrode **14** by means of a liquid-permeable membrane **22** of cellophane or the like which is held in position on the support body **12** by an O-ring **24** or the like which fits around an annular groove or detent **26** formed on the lower end of the support body **12**.

The particular enzyme used with the electrode assembly will depend upon the substance to be detected and will be one which is capable of reacting with the substance to produce hydrogen peroxide or other amperometrically active material to which the electrode is responsive. Likewise, the particular solid organic material used will be one which is insoluble in the substance being analyzed and capable of reacting with and chemically binding the enzyme, and will generally be a derivative of a polymeric material, such as polyacrylamide, polyacrylic acid, nylon, cellulose or the like, containing enzyme-reactive groups.

For example, in an analysis for glucose content, the enzyme used would be glucose oxidase. A suitable solid organic material to chemically bind the glucose oxidase would be a polyacrylamide derivative formed by converting a hydrazide derivative of polyacrylamide to an acyl azide intermediate that couples to amine groups on the enzyme. Another solid organic material which is suitable for use with glucose oxidase is a diazo derivative of polyacrylic acid. Its preparation and coupling to the glucose oxidase enzyme are described below.

Acrylic acid is polymerized by heating with a few mg of ammonium persulfate at 80° to 110°C for several hours. The viscous polyacrylic acid formed is then saturated with p-nitroaniline by stirring overnight with an excess. A portion of the N-p-nitrophenyl amide is then diluted with an equal amount of water and the nitro group on the N-p-nitrophenyl amide is reduced by adding titanium trichloride (26%) dropwise with vigorous stirring. When reduction is complete the yellow-orange color of the amide which is soluble is completely changed to a blue-black precipitate, the p-aminophenyl amide.

The precipitate is then washed several times with water and then cold (0°C) 2 M nitrous acid is added slowly with vigorous stirring until the blue-black polymer completely turns white. The diazonium derivative formed is quickly washed several times with cold 0.1 M phosphate or acetate buffer (pH 6.0). A cold glucose oxidase solution is then added and the mixture is stirred for one to eight hours in an ice bath, after which the precipitate is washed several times with cold buffer solution.

The mode of operation of the electrode assembly **10** may be illustrated in regard to analysis for glucose content, wherein the layer **20** comprises glucose oxidase chemically bound, for example, to either the polyacrylamide derivative or the polyacrylic acid derivative described above. The electrode assembly is placed in a solution of glucose and a potential of 0.6 volt versus SCE is impressed across the electrode. Glucose diffuses from the solution through the cellophane membrane **22** and into the layer **20** where it undergoes hydrolysis to form gluconic acid plus hydrogen peroxide. The hydrogen peroxide formed diffuses both out of the layer **20** and towards the face **16** of the platinum electrode **14** where it is oxidized. A current is produced that is proportional to the concentration of hydrogen peroxide and thus to the initial concentration of glucose. Both the initial rate of formation of hydrogen peroxide (within 4 to 12 seconds reaction time) and steady state current (at 1 minute reaction time) can be used as a measure of the glucose concentration.

When interferences are present that give a significant background current, the rate method would be the method of choice, because background current has no effect on the measurement of a rate. The steady state current method can also be used successfully with background current if another electrode similar to the enzyme electrode, but without the enzyme is used in parallel with the enzyme electrode and its response subtracted from the glucose electrode response by means of a differential amplifier or some other subtracting technique.

The output can be automated by using operational amplifier circuitry to take the derivative of the current signal as a function of time and using further transistorized circuitry to detect and display the maximum value of the rate within the first fifteen seconds of reaction time. In the case of steady state values, the circuitry can be set to display the value of the current at one minute of reaction time.

Determination of Blood Alcohol with Stabilized Enzyme Systems

The use of enzymes in chemical analysis for the determination of various components in biological fluids is gaining rapidly as a practical analytical technique. One specific reaction which has been found very useful is the detection of ethyl alcohol in biological fluids using the enzyme alcohol dehydrogenase (ADH). It has been found that ADH catalyzes the reaction between certain alcohols, for example, ethyl alcohol and nicotinamide adenine dinucleotide (NAD) to produce the corresponding aldehyde plus $NADH_2$ The above reaction is normally conducted with a slight excess of NAD to assure that all the ethyl alcohol is oxidized. The amount of $NADH_2$ can then be measured spectrophotometrically. This has been done, for example, by measuring the $NADH_2$ at 340 nm using a conventional UV spectrophotometer.

R.B. Koch and J.D. Skogen; U.S. Patent 3,941,659; March 2, 1976; assigned to Honeywell Inc. have used stabilized immobilized enzymes for the determination of alcohol in blood by the use of fluorescence spectroscopy. In this process, a measured sample is combined with an amount of a buffered carrier solution and introduced into a flow-through reaction chamber which may be a packed-column, for example, containing an amount of an immobilized alcohol dehydrogenase, where the desired reaction is catalyzed and takes place.

After traversing the packed column, the solution, in which the desired reaction is substantially complete, then continues to flow, while remaining in a closed system, through a transparent cell where the fluorescence of the reacted solution is measured utilizing a conventional fluorescence spectrophotometer which can, by conventional electronic means, produce a recorded or other output indicating the concentration of alcohol in the sample. The effluent containing the reacted sample is then discarded via a drain. The flow-through system of the present process may be one utilizing a constant-head type pump or may be operated by gravity. The flow-rate may be conveniently controlled by selection of size and length of the column and conduit components of the flow system such that the reaction is essentially completed by the time the reacted solution reaches the measurement cell. The activity of the immobilized alcohol dehydrogenase (ADH) is really extended in life span and increased in amount by the provision of a specially buffered carrier solution, gelling and storage techniques. This involves the use of certain combinations of ADH stabilizers for gel preparation, storage and use. These stabilizers are normally selected from a group including ammonium sulfate (0.1 to 1.0 M), dithiothreitol (DTT)(approximately 0.001 M), glycine (approximately 1%), sodium pyrophosphate (approximately 3%), zinc chloride (approximately 0.01 M). In the carrier solution of the preferred embodiment pyrophosphate or Tris buffer is employed in combination with ammonium sulfate and DTT.

If desired, a second column measuring cell system may be utilized wherein the packed-column does not contain any immobilized alcohol dehydrogenase but is identical to the first column in other respects. Samples may then be run through both systems and the difference in the resulting fluorescence spectrophotometer measurements utilized to derive the concentration of alcohol in the sample. This approach is most useful wherein the determination of baseline fluorescence of samples such as blood may change from sample to sample.

The gel used for entrapping the enzyme is a specially prepared crosslinked acrylamide gel in which the crosslinking agent is normally N,N'-methylenebisacrylamide and where the mol ratio of the acrylamide monomer to the crosslinking agent is normally from 45:1 to 55:1 and the solids content is about 21%. The high ratio gel used yields a much improved enzyme lattice in which leaching out of ADH molecules is greatly reduced. An important aspect of the process in regard to the immobilization and longer-term use of the ADH lies in the particular technique of keeping the ADH activity at a high level for a relatively long period of time.

It has been found that ADH, in contrast to many other successfully immobilized enzymes, required chemical stabilizers in gelling, storage and use in order to keep its activity at a sufficiently high level for more than a few hours. The stabilizers which have been effectively used include ammonium sulfate (0.1 to 1.0 M), dithiothreitol (DTT)(approximately 0.001 M), glycine (approximately 1%), sodium pyrophosphate (approximately 3%), and zinc chloride (approximately 0.01 M). Tests on supernatant and gel activity show that the above stabilizers indeed keep the ADH from being inactivated during the gelling process. Experimentation indicates that use of all of the above stabilizers during gelling produces the most active immobilized ADH.

After successful preparation of active immobilized ADH was achieved, much experimentation was required to determine the best method of retaining this activity both in storage of the immobilized ADH and its use in the analytical column environment. Some of the more important results of the experimental work in varying the preparation and storage conditions for the immobilized ADH are shown in the following table. The following conditions apply to the table: Gel 1 was prepared with all stabilizers (ammonium sulfate 0.1 to 1.0 M, dithiothreitol ~ 0.001 M, glycine ~ 1%, sodium pyrophosphate ~ 3%, and zinc chloride ~ 0.01 M) including 1.0 M $(NH_4)_2SO_4$. Gel 2 was prepared with all stabilizers including 0.1 M $(NH_4)_2SO_4$. Each gel was divided into three equal portions

(A, B and C) and soaked in 10 ml of each of the following solutions: Solution A: 0.1 M phosphate buffer, pH 7.4, $(NH_4)_2SO_4$ (gel conc.), 10^{-3} M DTT; Solution B: $(NH_4)_2SO_4$ (gel conc.), 10^{-3} M DTT; Solution C: $(NH_4)_2SO_4$ (gel conc.), 10^{-3} M DTT, then dried in N_2 stream with rehydration prior to tests.

Comparative Activity of Two Gels Under Different Storage Conditions

	Same Gels Tested in Column Repeatedly					
	Gel 1 in Solution			Gel 2 in Solution		
Days	A	B	C	A	B	C
2	–	–	99.3	–	–	124.6
3	–	–	79.5	–	–	97.4
6	–	–	82.5	–	–	77.4
7	–	–	88.0	–	–	84.9
8	14.8	–	–	0	–	–
9	–	71.3	–	–	69.9	–
10	–	49.4	–	–	–	–
13	–	32.7	–	–	21.8	–
14	–	–	68.8	–	–	69.6
42	–	–	11.3	–	–	9.5
	Stored Gel Portions Previously Unused					
2	–	–	99.3	–	–	124.6
8	14.8	–	–	0	–	–
9	–	71.3	–	–	69.9	–
10	11.5	–	–	–	–	–
15	–	–	95.3	–	–	114.8
28	–	–	54.7	–	–	–
35	–	75.0	–	–	109.0	–
36	1.0	–	–	0	–	–
44	–	58.5	–	–	–	–
48	–	–	–	–	–	119.2

Note: Activity values are expressed in % and correspond to an arbitrary $NADH_2$ reference concentration of 0.14 μM set at 100%. This is an accurate gauge of the ability of the enzyme to catalyze the ETOH + NAD $\rightarrow$ CH_3CHO + $NADH_2$ reaction.

The above table shows activity of the same gel column on different days. Between testings, gels were returned to the indicated (i.e., A, B, or C) storage conditions. The results obtained from corresponding portions of Gels 1 and 2 are similar indicating no significant difference due to the indicated levels of $(NH_4)_2SO_4$ concentration used in gel formation. However, the gels soaked in the presence of 0.1 M phosphate (A portions) even initially had very low or no enzyme activity, indicating phosphate buffer should not be used to store the gel.

The gels soaked in $(NH_4)_2SO_4$ and DTT (B portions) had more activity than those in the A portions. However, the gel activity did drop substantially with time over a period of only 3 days. The gels soaked in $(NH_4)_2SO_4$ and DTT, which were dried for storage and rehydrated for each test (C portions) maintained activity quite well over a period of about 2 weeks.

IMMUNOASSAY PROCESSES USING COVALENTLY BOUND ENZYMES

The technique of enzyme immunoassay is gaining increased acceptance, especially with regard to the measurement of therapeutic drugs. Instead of coupling a radioactive label to a specific antigen, an enzyme (typically glucose-6-phosphate dehydrogenase) is used as a label. When the enzyme-labeled antigen is bound to the antibody, the enzyme's activity is modified. Competition between the labeled and unlabeled antigens changes the proportion of labeled antigen bound to the antibody, thereby altering the measured enzyme activity. From determined enzyme activity, the concentration of the drug in the serum can be calculated. The technique is simple, rapid, and readily adaptable to many different mechanized instruments although it lacks the precision of gas chromatography. Enzyme-linked immunosorbent assays (ELISA), in which enzyme-labeled antibodies are used, have had some applications in the measurement of specific proteins.

Determination of Steroidal and Other Haptens

A.H.W.M. Schuurs and B.K. Van Weemen; U.S. Patent 3,850,752; November 26, 1974; assigned to Akzona Incorporated have developed a process for determining the amount of an antigen or hapten (the bindable substance) using an enzyme conjugate. This process comprises the steps of: (a) providing a given quantity of the coupling product of low molecular weight organic compound with an enzyme; (b) providing a corresponding given quantity of an insolubilized antibody against low molecular weight organic compound; (c) contacting a sample of a fluid containing the low molecular weight organic compound to be determined with components (a) and (b) to form a reaction mixture; and (d) determining the enzyme activity of the liquid or the solid phase of the resulting reaction mixture, which activity is a measure of the quantity of low molecular weight organic compound to be determined.

The enzyme activity of a fraction of the reaction mixture is demonstrated or measured by incubating that fraction with a substrate and other substances required for processing the relative enzyme reaction. For preference a reaction is used in which a colored compound is formed or removed whose absorption can also be measured quantitatively in an easy manner. Low molecular substances which are eligible for demonstration by this method and have a molecular weight of up to about 1,500 are for example: steroids, vitamin B_{12}, folinic acid, thyroxine and triiodothyronine, releasing factors, histamine, serotonin and other biogenic amines, digoxine, digitoxin, prostaglandins, adrenalin, noradrenalin, vegetable hormones such as auxin, kinetin and gibberellic acid, and antibiotics such as penicillin.

The choice of the enzyme that is to be a component of the conjugate (low molecular substance-enzyme) depends on properties such as the specific activity (a high conversion rate enhances the sensitivity of the test system) and the simplicity of the determination of the enzyme. The determination of an enzyme which catalyzes a conversion involving colored reaction components is simple. Such colorimetric determinations can be automatized in a simple manner.

According to the process it is also possible to use enzymes which catalyze conversions involving reaction components that can be determined spectrophotometrically or fluorimetrically. These determinations can also be automatized. For the preparation of the conjugates, enzymes such as catalase, peroxidase, β-glucuronidase, β-D-glucosidase, β-D-galactosidase, urease, glucose-oxidase and galactose-oxidase are preferred, particularly the group of the oxido-reductases. The insoluble specific binding protein or the insoluble low molecular compound to be used in the present determination can be prepared by a known method, for example, by crosslinking with chloroformic acid ethyl ester, by covalent binding with insoluble carriers such as agarose, crosslinked dextran or filter paper, or by physical coupling to insoluble carriers such as plastic objects.

Example: (A) Preparation of Testosterone-3-HRP for Determination of Testosterone – 100 mg of testosterone-3-(O-carboxymethyl)-oxime and 0.143 ml of tri-n-butylamine were dissolved in 5 ml of dioxan, cooled to 2°C and then 0.03 ml of isobutylchlorocarbonate were added. After 30 minutes the solution was added to 100 mg of HRP (horseradish peroxidase) in a mixture of 9 ml of water and 6 ml of dioxan, and adjusted to pH 9 with 0.1 N NaOH. This solution was stirred for 4 hours at 2°C and dialyzed overnight. The precipitate obtained after the dialysate had been adjusted to pH 4.6 was centrifuged after having been left to stand overnight, suspended in 10 ml of water and dissolved by means of caustic soda solution. The material was precipitated three times with 15 ml of acetone at pH 4.5, dissolved in 15 ml of water which had been adjusted to pH 7.8 with caustic soda solution, dialyzed and finally lyophilized.

(B) Preparation of Testosterone-3-BSA – This conjugate was prepared in the same manner as the testosterone-3-HRP, but the starting materials were 50 mg of testosterone-3-(O-carboxymethyl)-oxime and 150 mg of BSA (bovine serum albumin).

(C) Preparation of Antibodies Against Testosterone-3-BSA – Five rabbits were injected intramuscularly with increasing dosages of testosterone-3-BSA in complete Freund's adjuvant (0.5, 1 and 2 mg) at intervals of 3 weeks. Two weeks after the last injection the animals were injected intravenously with 2 mg of antigen dissolved in physiological salt. One week after that, blood was taken from the animals. The antibodies formed against BSA were removed by treating the serum portionwise with BSA-m-aminobenzyloxymethyl cellulose.

(D) Preparation of Antitestosterone Cellulose – Aminocellulose (350 mg) treated with ammoniacal copper solution was suspended in distilled water (50 ml). The suspension was cooled down to 0°C. Ten ml of a 36% hydrochloric acid were added and after that dropwise 10 ml of a 10% $NaNO_2$ solution. The suspension was centrifuged, washed with cold distilled water and then with 0.05 M sodium borate of pH 8.6. The cellulose was suspended in 43 ml of 0.05 M sodium borate of pH 8.6. To this suspension were added 7 ml of γ-globulin solution isolated from the antibodies against testosterone-BSA. The mixture was stirred for 26 hours at 4°C, centrifuged and washed with 0.02 M phosphate buffer of pH 6.0. From the antiserum of each of the 5 immunized rabbits a cellulose suspension was prepared.

(E) Determination of Testosterone by Means of Testosterone-3-HRP and Antitestosterone Cellulose – The following test system was built up: (1) Immunoreaction: 0.5 ml of sample containing testosterone, 0.2 ml of testosterone-3-HRP (100 ng/ml) and 0.3 ml of an antitestosterone-cellulose suspension were rotated at room temperature for 2 hours and then centrifuged for 5 minutes at 1,000 *g*. The immunoreaction took place in 0.02 M phosphate buffer of pH 6.0 and containing 2% sheep serum. (2) Enzyme reaction: 0.5 ml of the supernatant liquid was incubated at room temperature with 1.5 ml of substrate for 30 minutes. The extinction was measured at 460 nm.

Figure 11.3 shows the results of incubation of a testosterone dilution series with testosterone-3-HRP and four different concentrations of antitestosterone-cellulose: 1 mg/ml (I), 2 mg/ml (II), 4 mg/ml (III) and 16 mg/ml (IV). It is obvious that with this system a quantity of about 10 ng of testosterone can be demonstrated.

Figure 11.3: Results of Incubation of a Testosterone Dilution Series

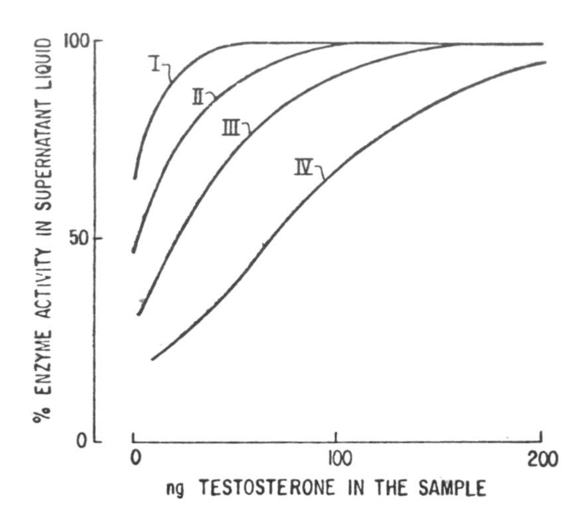

Source: U.S. Patent 3,850,752

Pregnancy Tests Using Enzyme Conjugates

A.H.W.M. Schuurs and B.K. Van Weemen; U.S. Patent 3,839,153; October 1, 1974; assigned to Akzona Incorporated have also used the techniques of U.S. Patent 3,850,752 for a pregnancy test. Chorionic gonadotropin and other haptens are determined by this means.

The coupling product of antigen, hapten or bindable substance with an enzyme, can be produced in a known manner. The choice of the enzyme which is taken up in the coupling product, is determined by a number of properties of that enzyme. It is, of course, essential that the enzyme should be resistant to the coupling with another molecule, i.e., modification of one or more amino acid side chains. Also of great importance is the specific activity of the enzyme. As less enzyme conjugate needs to be added to reach a measurable enzyme effect, the test system grows more sensitive. Further those enzymes are to be preferred, of which the determination of the activity can be made in a simple manner. In the first place those enzymes are considered that can be determined colorimetrically, spectrophotometrically or fluorimetrically. This kind of determination is also suitable for automation.

Colorimetrically those enzymes can be determined that catalyze a reaction in which a colored substance appears or disappears, either in the primary or in the secondary reaction. As enzymes considered to act as enzymatically active component in conjugates, are mentioned catalase, peroxidase, β-glucuronidase, β-D-glucosidase, β-D-galactosidase, urease, glucose oxidase, galactose oxidase, and alkaline phosphatase.

At the end of the reaction between the components and the reagents, the enzyme activity of the liquid or solid phase of the reaction mixture or of the two phases can be determined. Most simple is, however, to determine the enzyme activity of the liquid phase. A preferably applied method for the determination of the enzyme activity consists in contacting an indicator-paper impregnated with enzyme reagents, e.g., in case use is made of a peroxidase, a H_2O_2-supplier like urea-H_2O_2, and a color-reagent like o-tolidine. The antibodies made insoluble against the specific binding proteins, which are also an essential reagent for the process, can also be prepared in a known way. The antibodies can be prepared by taking a purified preparation of the specific binding protein, or of proteins which have at least partly the same antigen properties as the specific binding protein, and injecting this in a known way into another animal species than from which it was obtained.

The serum of the treated animal, or the gamma-globulin fraction thereof, can be made insoluble by crosslinking with compounds such as glutaric aldehyde and chloroformic acid ethyl ester, or by binding to solid carrier particles either physically by adsorption, or chemically by the formation of covalent bonds. As solid carriers can be considered materials such as cellulose (modified or not), agarose, crosslinked dextran, polystyrene and the like. Covalent binding of the antibodies to these materials can be effected with the help of substances such as carbodiimides, di- and tri-chloro-s-triazines, glutaric aldehyde, cyanogen bromide, and, e.g., by diazotization.

An important embodiment of this process is the determination of gonadotropic hormones, and particularly for the determination of HCG (Human Chorionic Gonadotropin) as a means to diagnose pregnancy in a very early stage.

Example 1: Determination of Human Chorionic Gonadotropin (HCG) – (a) Preparation of HCG-HRP: 5 mg HCG and 20 mg horseradish peroxidase (HRP) were dissolved in 2 ml 0.05 M phosphate buffer of pH 6.2. After addition of 40 μl 25% glutaric aldehyde solution the mixture was shaken for 2 hours at room temperature. After 5 minutes centrifugation at 250 *g*, the liquid was fractionated over Sephadex G-200 in 0.05 M phosphate buffer of pH 6.2. The fractions of which the highest percentage enzyme activity was bound by antibodies against HCG were used in the test system.

(b) Preparation of Antibodies Against HCG: Antibodies against HCG were induced in rabbits as described by Schuurs et al, *Acta Endocr. (Kbh.)* 59, 120 (1968).

(c) Preparation of Antibodies Against Rabbit-γ-Globulin: Rabbit-γ-globulin was isolated from normal rabbit serum by precipitation with 18% w/v solid sodium sulfate. Antobodies against this were prepared by immunizing a sheep according to the following schedule. On day 70 the sheep was bled.

Days	Amount (mg)	Freund's Adjuvants	Injection Manner
0	0.5	+	intramuscular
14	0.5	+	intramuscular
28	1	+	intramuscular
42	1	-	intravenous
56	1	-	intravenous

(d) Preparation of the Immunoadsorbent [Sheep-Anti-(Rabbit-γ-Globulin)] Cellulose: The γ-globulin fraction of the sheep serum described under (c) was prepared by precipitation with 16% w/v solid sodium sulfate. After washing, the precipitate was taken up in enough 0.05 M borate (pH 8.6) to give a 10 mg/ml protein concentration.

350 mg m-aminobenzyloxymethylcellulose was suspended in 50 ml distilled water, and diazotized by adding 10 ml 36% hydrochloric acid and, dropwise, 10 ml 10% $NaNO_2$ solution at 0°C. The suspension was centrifuged, washed and the precipitate resuspended in 43 ml 0.05 M sodium borate of pH 8.6. Then 7 ml of the prepared γ-globulin solution was added. The mixture was stirred for 26 hours at 4°C, then centrifuged and washed with 0.02 M phosphate buffer of pH 6.0.

(e) Determination of HCG: A dilution series (32-16-8-4-2-1-0.5-0 IU/ml) of HCG in 0.02 M phosphate buffer of pH 6.0 was prepared, which contained 2% v/v normal sheep serum. 0.5 ml of each of the HCG-containing samples was incubated with 0.1 ml rabbit-(anti-HCG) serum and 0.1 ml HCG-HRP conjugate, both in suitable dilution, for half an hour at room temperature. Then 0.3 ml of the immunoadsorbent (10 mg/ml) prepared according to (d) was added, and the resulting mixture was rotated at room temperature for 1 hour. After centrifugation the enzyme activity in the supernatant was measured by mixing 0.5 ml of this liquid with 1.5 ml substrate (10 μl 30% H_2O_2 and 20 mg 5-aminosalicylic acid in 150 ml 0.02 M phosphate buffer of pH 6.0) and after 30 minutes at 25°C measuring the extinction at 460 nm. In this way it was proved possible to detect a HCG concentration from 0.5 to 1 IU/ml in the sample.

With this method also urine samples could be tested; the test is therefore suitable for a pregnancy check. The correlation with an existing method of test, a haemagglutination inhibiting test, was good. It proved possible to raise the sensitivity of the system by the application of a preincubation. Here, first the sample only was incubated with the antiserum, and then the HCG-HRP conjugate was added.

Example 2: In a bottle the following reagents were subsequently lyophilized in separate layers. (1) 0.3 ml of the immunoadsorbent suspension (10 mg/ml) as described in Example 1(d). (2) 0.1 ml of a 1% mannitol solution. (3) 0.1 ml of HCG-HRP as described in Example 1(a). (4) A second layer of 0.1 ml of a 1% mannitol solution. (5) 0.1 ml of rabbit (anti-HCG) serum as described in Example 1(b).

To this lyophilized mixture 0.5 ml of a urine sample and subsequently 0.5 ml distilled water were added. After 10 minutes the enzyme activity of the supernatant was measured by means of a slip of paper impregnated with urea-hydrogen peroxide and o-tolidine. If the urine sample was that from a pregnant woman ($>$2 IU HCG/ml) a blue color appeared within 5 minutes, whereas in case of urine of a nonpregnant woman no discoloration took place within the same period.

Drug Conjugates with Glucose 6-Phosphate Dehydrogenase

A wide variety of haptenic compounds, particularly drugs of abuse, and drugs used in repetitive therapeutic applications, and steroids are conjugated to glucose-6-phosphate dehydrogenase by *K.E. Rubenstein and E.F. Ullman; U.S. Patent 3,875,011; April 1, 1975; assigned to Syva Company* for use in immunoassays. The resulting product has a higher turnover rate, so as to provide a high multiplication factor when employed in a homogeneous enzyme immunoassay. Haptenic conjugated glucose-6-phosphate dehydrogenase is provided having from 2 to 12 ligands, normally the majority of all of the ligands being bonded to amino groups, particularly of lysine. The haptens or ligands will normally have molecular weights of at least 125 and generally not exceeding 600 molecular weight. The ligands will have at least one heteroatom and may have two or more heteroatoms, which will normally be oxygen, nitrogen and sulfur, although halogens, particularly chlorine and iodine may also be present. The ligands for the most part will be naturally occurring, physiologically active compounds and synthetic drugs, which will be modified to the extent necessary for conjugation to the glucose-6-phosphate dehydrogenase. Such ligands include derivatives of morphine, phenobarbital, estradiol, diphenylhydantoin, and related compounds.

Example: A survey was carried out using the isobutyl chloroformate mixed anhydride of O^3-carboxymethylmorphine and methyl O^3-morphinoxyacetimidate. The conjugations were carried out as follows. The enzyme was obtained as a 4.8 mg/ml solution in 30% glycerol. Specific activity was 561 IU/mg protein for NAD (cofactor) reduction at 30°C. The glycerol solution was dialyzed against 0.05 M sodium phosphate, pH 7.5 and diluted with that buffer to a concentration of 2 mg/ml. The pH was adjusted to 7.0 with 1 M HCl.

To one ml of this cooled (4°C) stirred enzyme solution was added in five portions during 5 minutes, 37.5 μl of a 0.2 M solution of the mixed anhydride (N-methyl-^{14}C) in dimethylformamide. After each addition, the pH rose slightly and was readjusted to 7 with 1 M HCl. The solution was then maintained for 5 hours at 4°C and dialyzed exhaustively against 0.055 M Tris-HCl, pH 7.9. The resulting solution was diluted to 2 ml with dialysis buffer. Scintillation counting was then employed to determine the number of ligands conjugated to the enzyme on the average.

To determine the activity of the enzyme and its inhibition, immunoassays were carried out. The assay mixture had a total volume of 1 ml and was prepared from 20 μl of 0.1 M NAD in water (pH 5 to 6), 50 μl of 0.066 M glucose-6-phosphate in assay buffer, and enzyme solution. The remaining volume was made up by the assay buffer which was 0.055 M Tris-HCl, pH 7.9. The mixture was incubated for 60 seconds in a spectrophotometer flow cell at 30°C, and the increase in absorbance at 350 nm was then read over a 1 minute interval. 10 μl of the above enzyme solution diluted 1:100 in assay buffer containing 0.1% RSA (rabbit serum albumin) gave a rate of 0.160 optical density units per minute (OD/min).

This corresponded to 52% of the activity of the native enzyme. Addition of a large excess of an antiopiate gamma-globulin preparation (5 μl of a solution that was 8 x 10^{-5} M in binding sites) prior to addition of the enzyme solution to the assay mixture reduced the activity by 78% (0.035 OD/min). When 50 μl of 10^{-4} M morphine and water was added to the substrates prior to addition of the antibody and enzyme, the total enzyme activity was recovered.

A number of preparations were carried out using differing buffers, pH, and mol ratios. Also, in some instances the enzyme substrates were included to determine their effect on deactivation. It was found that above 14 ligands per enzyme molecule on the average, substantial deactivation of the enzyme had occurred. The table on the following page indicates the results of the above.

Reagent*	Buffer**	pH	G-6-PDH (mols x 10^{-8})	Reagent (mols x 10^{-6})	Deactivation (%)***	Inhibition† (%)	G-6-PDH Ligands††
1	P-C	9	3.85	1.25	56	52	7.5
2	P	8.5	3.85	5.0	41	48	4.7
2	P	8.5	3.85	15	66	80	10.5
2†††	T-M	8.5	3.85	15	45	75	9.5
2§	T-M	8.5	3.85	15	48	82	11.5
2§§	T-M	8.5	3.85	15	63	87	12.0
2	P	6.0	1.92	7.5	16	9	1.3
2	P	7.0	1.92	7.5	52	78	13

*1 = O^3-carboxymethylmorphine; 2 = methyl O^3-morphinoxyacetimidate.
**P = 0.5 M sodium phosphate; C = 0.5 M sodium carbonate; T = 0.055 M Tris-HCl; and M = 0.003 M magnesium chloride.
***Percent of original enzyme activity after conjugation and dialysis.
†Maximum inhibition by excess antimorphine.
††Both ligands contain ^{14}C. Determined by liquid scintillation counting of aliquots of product.
†††Prior to conjugation added G-6-P to 50 mM and NAD to 40 mM. pH readjusted to 8.5.
§Prior to conjugation added G-6-P to 50 mM and NAD to 40 mM.
§§G-6-PDH had activity of 561 IU/mg, while other G-6-PDH was 460 IU/mg.

Enzyme Bonded Morphine-Type Compounds

Morphine and related compounds are linked to enzymes by *K.E. Rubenstein and E.F. Ullman; U.S. Patent 3,852,157; December 3, 1974; assigned to Syva Corp.* Primarily the hydroxy group on the 3-position of the morphine is used as the site for introduction of linking groups. The preferred compounds having the basic morphine structure will have the formula:

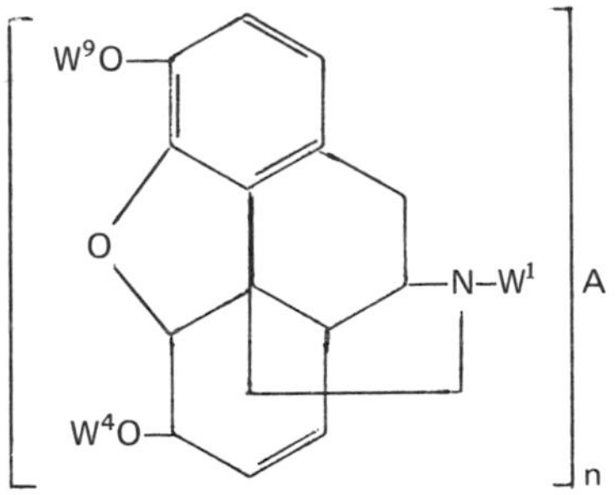

W^9O—, O, —N—W^1, W^4O—, A, n

where one of W^1 and W^9 is X; when other than X, W^1 is methyl; and W^9 is hydrogen, methyl, acetyl or glucuronyl; W^4 is hydrogen or acetyl, usually hydrogen; X is:

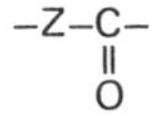

$$-Z-\underset{\underset{O}{\|}}{C}-$$

where Z is hydrocarbylene of from 1 to 7 carbon atoms, preferably aliphatic, having from 0 to 1 site of ethylenic unsaturation; and A is an enzyme, either specifically labelled, with n equal to 1 to 2 ligands or randomly (random as to 1 or more particular available reactive functionalities) labelled with n equal to 2 to 30, more usually 2 to 20, the enzyme retaining a substantial proportion of its activity. The enzyme will be of 10,000 to 150,000 molecular weight and is preferably an oxidoreductase, e.g., malate dehydrogenase, lactate dehydrogenase, glyoxylate reductase, or glucose 6-phosphate dehydrogenase, or a glycosidase, e.g., lysozyme or amylase. Illustrative opiates which can be bound to an enzyme include morphine, heroin, hydromorphone, oxymorphone, metopon, codeine, hydrocodone, dihydrocodeine, dihydrohydroxycodeinone, pholcodine, dextromethorphan, phenazocine, and Dionin and their metabolites.

Example 1: Morphine (900 mg) was dried for 4 hours at 50°C, 0.01 mm Hg. The dried morphine was dissolved in 18 ml of abs ethanol and 125 mg sodium hydroxide was added, followed by the addition of 350 mg dry sodium chloroacetate. After purging with nitrogen, the solution was stirred and refluxed for 4 hours. The hot solution was treated with 3.8 ml ethanolic hydrogen chloride (0.85 M) and then filtered while still warm. On cooling overnight, a precipitate (272 mg) formed which was collected and recrystallized from ethanol/water. On addition of ether to the original filtrate an additional precipitate was obtained which was also recrystallized from ethanol/water. Total yield was 600 mg (55%) On heating this product to 75°C in vacuo there was a weight loss corresponding to 0.48 molecule of ethanol or 1.15 molecules of water. The dried compound decomposes at 190° to 220°C (depends on rate of heating).

Example 2: Carboxymethylmorphine (240 mg) suspended in 8 ml dry dimethyl formamide (DMF) was cooled to -15°C and treated with 84 μl isobutyl chloroformate. The solid dissolved while stirring for 30 minutes at -15°C.

Example 3: To a solution of 100 mg of amylase in 15 ml of water containing 600 mg of sodium bicarbonate was added 4.6×10^{-2} M of the mixed anhydride of morphine (prepared in Example 2) in 1 ml DMF at 0° to 5°C. The solution was stirred at 0°C for 18 hours, transferred to a dialysis bag and dialyzed against water for 2 days at 0°C. The residue was lyophilized to a white solid.

The activity of the morphine modified amylase was determined as follows. Morphine modified amylase was mixed with a suspension of dyed amylose at 37°C. After an arbitrary time, the reaction was quenched, centrifuged and the supernatant liquid transferred to a cuvette and the optical density measured. This was repeated a number of times so that the optical density could be plotted versus time to give a linear graph. The slope of the line indicated the enzyme activity. It was found that the modification of the amylase with morphine had little, if any, effect on the enzymatic activity of the starting amylase.

A measured amount of morphine antiserum (concentration of active sites equals 2×10^{-7} M per liter) was added to a 6.4×10^{-8} M carboxymethylmorphine modified amylase solution. The amylase activity test was then carried out at 37°C for a total time of 20 minutes. The following table indicates the results.

Volume of Carboxymethylmorphine Modified Amylase (μl)	Volume of Antibody Solution (ml)	Optical Density
100	0.0	0.910
100	0.020	0.720
100	0.050	0.560
100	0.100	0.450
100	0.150	0.390

The above table shows that with increasing concentration of antibody, there is decreasing activity of the carboxymethylmorphine modified amylase. The effect of addition of codeine was determined by running one sample containing antibody and carboxymethylmorphine in the absence of codeine and one sample in the presence of codeine.

With 0.2 ml of a solution containing antibody-carboxymethylmorphine modified enzyme complex employed in the amylase determination, the resulting optical density was 0.450. However, when 0.030 ml of a 10^{-7} M of codeine solution was added to the same amount of antibody-enzyme complex, the resulting optical density in the amylase assay was 0.580. In this manner, a solution containing 10^{-7} M codeine could be assayed where only 30 microliters were available. This is not intended to indicate, however, that this is the minimum concentration required, but rather only that it can be successfully employed. The above method could also be used with morphine, morphine glucuronide, as well as other close structural analogs of morphine and codeine.

Cardiac Glycoside-Enzyme Conjugates

E.F. Ullman and K.E. Rubenstein; U.S. Patent 4,039,385; August 2, 1977; assigned to Syva Company also provide ligand-enzyme compositions for use in the determination of cardiac glycosides and aglycones. For the most part, these enzyme-bound-ligands will have the following formula:

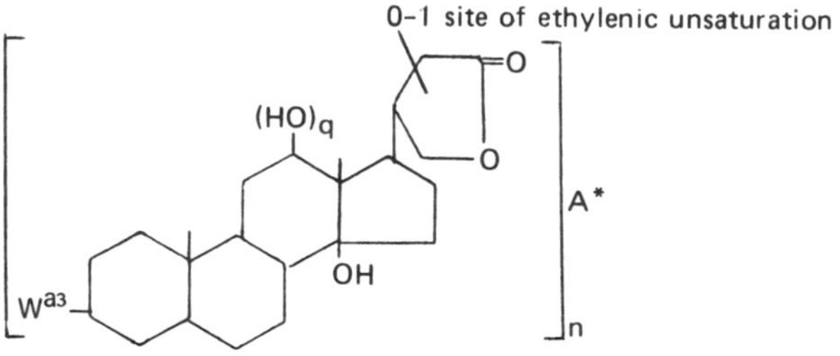

0–1 site of ethylenic unsaturation
$(HO)_q$
OH
W^{a3}
A*
n

where n is preferably 2 to 10; q is 0 or 1; A* is an enzyme, preferably an oxidoreductase, and particularly preferred a dehydrogenase; and W^{a3} is a linking group which may be singly or doubly bonded to the annular carbon atom to which W^{a3} is attached. For the most part, the compounds will have one site of ethylenic unsaturation in the E ring, particularly in the alpha, beta position.

The nature of the linking group will vary, depending on whether the glycoside is involved or the aglycone. Where the glycoside is involved, preferentially the terminal sugar will be cleaved to a dialdehyde, which may be conjugated directly to the enzyme by reductive amination or the dialdehyde may be derivatized to provide a carboxylic acid group, for example, with glycine, and the carboxy functionality employed to provide the covalent bond with the enzyme.

When the aglycone is used, the linking group may be varied widely and will usually be of 1 to 10 atoms other than hydrogen and normally including from 1 to 4 heteroatoms which are chalcogen (oxygen and sulfur) or nitrogen. The linking group will normally be aliphatic, and may be aliphatically saturated or unsaturated, usually having 0 to 1 site of ethylenic unsaturation as the only aliphatic unsaturation. Chalcogen will be present as nonoxo-carbonyl or oxy, e.g., bonded solely to carbon and hydrogen, particularly carbon, and nitrogen will be bonded solely to carbon, with the proviso that when neutral or when imino, nitrogen may be bonded to from 0 to 1 hydrogen or when present as oximino or hydroxylamino, singly bonded to oxygen. The following example uses a derivative of digoxigenin prepared via formation of the 3-keto derivative which is then converted to the O-carboxymethyl oxime.

Example: Conjugation of the O-Carboxymethyl Digoxigenin-3-Keto Oxime to G-6-PDH – To a dry flask, fitted with serum stopper and drying tube was introduced 23.05 mg (0.05 mmol) of the oxime and 250 μl of DMF (dried over 4 A molecular sieves) and 7.1 μl (0.052 mmol) of dry triethylamine added through the serum stopper with a syringe with stirring at room temperature.

After cooling the mixture to -14°C, 9.3 μl (0.05 mmol) of carbitol chloroformate was added below the surface of the solution and the mixture stirred for 30 minutes. In a separate flask, to 2 ml of glucose-6-phosphate dehydrogenase (G-6-PDH) at a concentration of 1 to 2 mg/ml in 0.055 M tris buffer, pH 8.1 with stirring is added 20 mg of glucose-6-phosphate disodium salt and 40 mg NADH. (During the reaction aliquots are taken and the enzyme activity is determined by diluting a 5 μl aliquot of the enzyme solution to 5 ml, and taking a 50 μl aliquot of the diluted enzyme solution and diluting with 1 ml buffer and 50 μl substrate, introducing the solution into a 1.5 ml sample cup and employing a flow cell, reading the enzyme activity over a 60 second interval in a Gilford spectrophotometer.)

The mixture is cooled to 0°C and with stirring 1.08 ml carbitol added slowly with a syringe below the surface of the solution. After standing for 30 minutes, any precipitate is removed by centrifuging for 4 minutes. The supernatant is adjusted to a pH of 9.0 with 1 N NaOH. The enzyme activity is checked at this time.

To a stirring solution of the enzyme, 1 μl aliquots of the mixed anhydride are added to the enzyme at a rate of about 1 μl per minute. After the addition of 10 μl of the mixed anhydride, the percent inhibition and the percent deactivation are determined.

Percent inhibition is determined by employing approximately 5 μl of full strength antidigoxin in the above assay. About 35 to 45 μl of the mixed anhydride are added to obtain an inhibition of about 50% and a deactivation of about 36%. When desired inhibition and deactivation are obtained, the enzyme conjugate is purified by dialysis against 0.055 M tris-HCl buffer, pH 8.1 containing 0.05% NaN_3 and 0.005% thimerosal.

Following the above procedure, in a first reaction, an enzyme conjugate was obtained having 5 digoxins conjugated to the enzyme, which was 36% deactivated and was 64% inhibited, while in a second reaction sequence, an enzyme conjugate was obtained having 9.4 digoxins, which was 26% deactivated and 48% inhibited.

Benzdiazocycloheptane-Linked Enzymes

Tranquilizers having the benzdiazocycloheptane structure are also bound to enzymes by *E.F. Ullman and K.E. Rubenstein; U.S. Patent 4,046,636; September 6, 1977; assigned to Syva Company*. Examples of such tranquilizers are Librium, Valium, diazepam or oxazepam. For the most part, the enzyme-bound benzdiazocycloheptane will have the following formula:

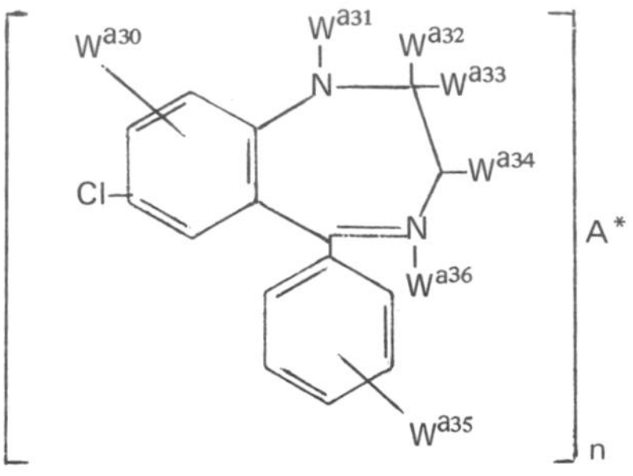

where any of the W groups other than W^{a36} can be X*; X* is a bond or linking group bonded to the enzyme at other than its active site; A* is an enzyme, preferably a hydrolase, e.g., lysozyme, or an oxidoreductase, particularly a dehydrogenase, e.g., malate dehydrogenase and glucose-6-phosphate dehydrogenase; n is usually 2 to 12, there being a sufficient number of benzdiazocycloheptane groups to provide a significant reduction in activity upon binding of receptor to one or more conjugated benzdiazocycloheptane groups; when other than X*: W^{a30} and W^{a35} are hydrogen; W^{a31} is hydrogen, lower alkyl of one to three carbons, e.g., methyl, or may be taken together with W^{a32} to form a double bond between the carbon and the nitrogen; W^{a33} is amino or lower alkylamino of 1 to 3 carbons, e.g., methylamino, or may be taken together with W^{a32} to form a carbonyl; W^{a34} is hydrogen or hydroxyl; and W^{a36} is oxy or an unshared pair of electrons.

By active site is intended those amino acid groups which are directly involved in the binding of substrate or occupy the cleft or area to which substrate is bound. In effect, attachment at the particular group will result in substantially total inhibition.

Illustrative groups for X* include carboxypropylenecarbonyl, oxypropyl-, oxybutylcarbonyl, propylenecarbonyl, carboxyethyleneoxyethylenecarbonyl, oxyethyleneimido, succindioyl, and N-methyl 3-azaglutardioyl. In forming the various amide products which find use in this process, the carboxylic acid will normally be activated. This can be achieved in a number of ways. Two ways of particular interest are the reaction with a carbodiimide, usually a water soluble dialiphatic or dicycloaliphatic carbodiimide in an inert polar solvent, e.g., dimethylformamide, acetonitrile and hexamethylphosphoramide. The reaction is carried out by bringing the various reagents together under mild conditions and allowing sufficient time for the reaction to occur.

A second method is to form a mixed anhydride employing an alkyl chloroformate, e.g., isobutyl chloroformate. The mixed anhydride is formed by combining the carboxy substituted benzdiazocycloheptane, the alkyl chloroformate and a tertiary amine. The temperature is normally below ambient temperature.

Example 1: Oxazepam Hemisuccinate – Oxazepam (2 g, 7 mmol) and succinic anhydride (1.2 g, 11.2 mmol) in pyridine (40 ml, dried over barium oxide) were heated under a nitrogen atmosphere at 95°C for 7 hours. The mixture was cooled, and the pyridine removed at reduced pressure. The residue was taken up in ethyl acetate and extracted into aqueous potassium carbonate, pH 13. After neutralizing the basic extracts with aqueous acid, the hemisuccinate was extracted into ethyl acetate, and the extracts washed with saturated brine. They were then dried, filtered, and concentrated in vacuo to give 2.1 g (78%) of the crystalline hemisuccinate, which was recrystallized from ethyl acetate-cyclohexane: MP 204° to 206°C (lit. 204.5° to 205.5°C).

Example 2: Oxazepam Hemisuccinate Conjugated to Lysozyme – Into a reaction flask was introduced 20.4 mg (5.04×10^{-2} mmol) of oxazepam hemisuccinate, 1 ml dimethylformamide and 15 μl triethylamine and the mixture cooled to -15°C. Isobutyl chloroformate (6.94 μl, 5.3×10^{-2} mmol) is added, the mixture stirred for 45 minutes, while the temperature is allowed to rise to -5°C.

This mixture is then added to a solution of 120 mg (0.84×10^{-2} mmol) of lysozyme in 10 ml water, pH 8.7, at 4°C, the pH being adjusted with 0.05 N sodium hydroxide. During the addition the pH is maintained at 8.7 and the reaction allowed to continue until the pH is constant for about 30 minutes. The pH is then adjusted to 7.0, the product centrifuged and dialyzed against pH 6.0, 0.025 M Tris-maleate buffer.

Lactam Conjugated with Enzymes

The enzyme conjugates described by *K.E. Rubenstein and E.F. Ullman; U.S. Patent 3,905,871; September 16, 1975; assigned to Syva Company* are lactams having from 5 to 6 annular members. The lactams include amide, imide, and urea functionalities, including compounds based on glutethimide, barbiturates, primidone and diphenyl hydantoin. The conjugates of these compounds to enzymes will for the most part have the following formula:

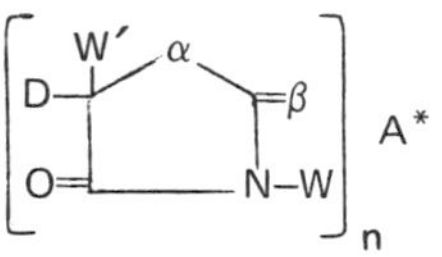

where one of the W groups is X*, or a hydrogen of one of the W groups is replaced by X*; D is hydrogen or hydrocarbyl of 1 to 8 carbons, usually having 0 to 1 site of ethylenic unsaturation, where hydrocarbyl includes alkyl, alkenyl, cycloalkyl, cycloalkenyl, or aryl hydrocarbon, preferably D is 1 to 6 carbons and will usually be phenyl, alkyl of 2 to 6 carbons, alkenyl of 3 to 4 carbons, e.g., allyl, cyclohexyl or cyclohexenyl, the alkyl group being either straight chain or branched, usually branched when over 2 carbons; β is H_2 or chalcogen (O, S) when α is amido and is otherwise oxygen; α is ethylene, amido, or imino.

When other than X*, W is hydrogen and W′ is hydrocarbyl of 1 to 8 carbons, usually 2 to 6 carbon atoms having 0 to 1 site of ethylenic unsaturation, including alkyl, alkenyl, cycloalkyl, cycloalkenyl or aryl hydrocarbon; A* is an enzyme bonded at other than its reactive site, having a number (n) of ligands in the range of 1 to the molecular weight of A* divided by 2,000, usually of 2 to 30 and more frequently 6 to 20; and X* is a bond or linking group of 1 to 10 atoms other than hydrogen, usually of 2 to 8 atoms other than hydrogen, which are usually carbon, oxygen, and nitrogen, wherein the carbon is present as aliphatic, alicyclic or aromatic hydrocarbon, usually aliphatic or aromatic hydrocarbon, generally having 0 to 1 site of ethylenic unsaturation as the only aliphatic unsaturation, the oxygen being present as carbonyl, both nonoxo and oxo, or oxy particularly ethereal, and the nitrogen being present as amino, particularly tertiary amino, or amido, there generally being 0 to 4 heteroatoms, there usually being 0 to 2 heteroatoms in the chain; normally, there are 1 to 4, more usually 1 to 2 heterofunctionalities either in or along the chain.

When α is imino, D is normally phenyl. When α is ethylene, D is phenyl or ethyl. When β is hydrogen, α is amido. The preferred enzymes are the hydrolases and the oxidoreductases, particularly the hydrolases that act on glycosyl compounds, and oxidoreductases that act on the CH-OH group of donors, more particularly with NAD or NADP as acceptor. These enzymes are illustrated by lysozyme, malate dehydrogenase, and glucose 6-phosphate dehydrogenase. Within the above formula, are a number of different subgenera, e.g., barbiturates, based on barbituric acid, disubstituted at the 5-position.

Example: Sodium phenobarbital (5.08 g, 0.02 mol), methyl chloracetate (2.16 g, 0.02 mol), methanol (14 ml) and a catalytic amount of DMF (1 ml) were refluxed for 2 hours. A white precipitate separated out during this period. The reaction mixture was cooled to room temperature and filtered. The methanolic filtrate was evaporated to dryness to yield about 5 g of a gummy material which solidified on standing. (The precipitate from the above filtration partially dissolved when rewashed with distilled water. The water-insoluble material, about 50 mg, proved to be the dialkylated product.)

The solidified material was stirred with 20 ml of 1 N NaOH solution for 15 minutes and then filtered. This separated the alkali-insoluble derivatives, the monoalkylated product and unreacted phenobarbital. The alkaline filtrate was acidified with concentrated HCl to a pH 2 and the white gummy precipitate which formed was taken up in methylene chloride.

Drying ($MgSO_4$) and evaporation of the organic solvent yielded 4 g of gummy material. This was dissolved in benzene and chromatographed over a column of silica gel (40 g). Elution was with chloroform and 100 ml fractions were collected. (The progress of the chromatography was followed by TLC, since the dialkylated product has a R_f 0.9, the monoalkylated material R_f 0.6 and phenobarbital R_f 0.1 with chloroform/methanol 95:5.) Fractions 2 to 5 combined yielded on evaporation 1.6 g of a gum which solidified on standing. Trituration with petroleum ether and filtration yielded 1.5 g of a white powder which was shown by NMR to be the required monoalkylated derivative, N-methoxycarbonylmethyl phenobarbital. Further elution with chloroform (500 ml) yielded 1.5 g of a white solid which was shown to be unreacted phenobarbital.

The monoester prepared above (1 g) was refluxed with 10 ml of 20% HCl solution for 3.5 hours. The cooled reaction mixture was diluted with water (20 ml) and extracted with ether. Evaporation of the ether extract yielded 0.98 g of a colorless gum which very slowly solidified on standing. NMR and TLC showed that complete hydrolysis had occurred to the acid. A pure sample of the acid was prepared by preparative TLC for UV analysis, with chloroform/methanol (5:1) as eluent.

To a cold (0°C) solution of 29.6 mg N-carboxymethyl phenobarbital (0.1 mmol) and 14.3 μl triethylamine (0.1 mmol) in 1.0 ml dry dimethylformamide was added 13.1 μl isobutyl chloroformate (0.1 mmol). The solution was stirred at 4°C for 1 hour before use.

The cold solution of mixed anhydride was added dropwise with stirring to a cold (4°C) solution of 0.100 g lysozyme (6.9 mmol) and 0.100 g sodium bicarbonate in 10 ml water. The resulting heterogeneous solution was stored at 4°C for 48 hours before being dialyzed against water for 48 hours. (The water was changed 3 times daily.) The dialysate was then chromatographed on Bio-Rex 70 employing a 0.05 to 0.20 M pH 7.15 phosphate buffer gradient for elution.

The assay employing the phenobarbital conjugate had an enzyme concentration in the enzyme conjugate stock solution of 1.71 X 10^{-5} M, an antibody concentration based on binding sites in the stock solution of 1.66 X 10^{-5} M and a binding constant for the antibody of 5.94 X 10^{7}. The assay solution had a total volume of 0.800 ml employed a urine volume of 0.080 ml, had an enzyme concentration of 2.14 X 10^{-7} M and an antibody concentration based on binding sites of 2.08 X 10^{-7} M. The assay was carried out for 40 seconds and the sensitivity was found to be 0.3 μg/ml, the minimum detectable amount.

Methadone-Enzyme Conjugates

Methadone derivatives are provided by *D. Wagner and E.F. Ullman; U.S. Patent 3,843,696; October 22, 1974; assigned to Syva Company* for use in preparing reagents for immunoassays and for preparing antibodies to methadone.

2,2-Diphenyl-4-dimethylaminopentanoyl substituted aliphatic carboxylic acids are used in the preparation of these derivatives for preparing antibodies and for joining with detector systems to be used in immunoassays for methadone. Detecting systems of special interest are enzymes and stable free radicals. Various enzymes may be used, such as peptidases, esterases, amidases, phosphorylases, carbohydrases, oxidases, and the like. Of particular interest are such enzymes as lysozyme, peroxidase, α-amylase, dehydrogenases, particularly malate dehydrogenase, alkaline phosphatase, β-glucuronidase, cellulase and phospholipase, particularly phospholipase C. The substituted proteins will for the most part have the following formula: $(M{-}R^2{-}CO{-})_nY$, where Y is a polypeptide residue (e.g., enzyme), M has the formula:

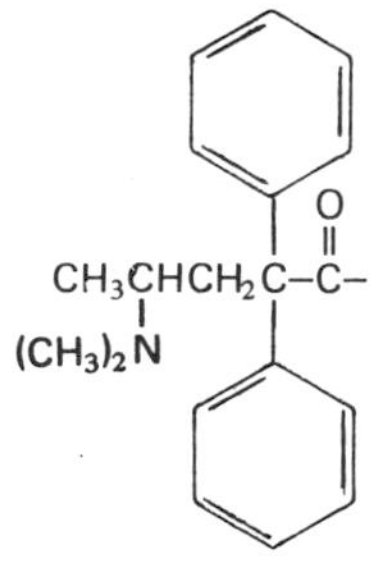

O
‖
$CH_3CHCH_2C{-}C{-}$
$(CH_3)_2N$

R^2 is a divalent aliphatic group having 0 to 1 heteroatom (O,N) of 4 to 10 carbon atoms, more usually allylene of 4 to 6 carbon atoms, having 0 to 1 site of aliphatic unsaturation, usually ethylenic, and may be straight chain or branched chain, preferably straight chain, there being at least 4 carbon atoms between the two carbonyl groups and n will usually be 2 to 35 when Y is an enzyme residue. The mixed anhydride which may be used to form the conjugate will have the following formula: $M-R^2-CO_2CO_2R^3$ where R^3 is alkyl of 1 to 6 carbon atoms, more usually of 2 to 4 carbon atoms, e.g., ethyl, isopropyl, butyl and hexyl and M and R^2 are defined above.

Example 1: Preparation of 7,7-Diphenyl-6-Keto-9-Dimethylaminodecanoic Acid Hydrochloride – A solution of tetramethylene bromide (32.4 g, 150 mmol) in dry ether (150 ml) was added to magnesium (10.9 g, 450 mmol) in ether (80 ml) at such a rate that the ether refluxed. The reaction was carried out under argon. After the addition was completed, the reaction mixture was boiled for one hour. A solution of 2,2-diphenyl-4-dimethylaminopentanonitrile [prepared according to J.W. Cusic, *J. Am. Chem. Soc.,* 71, 3,546 (1949)] (8.4 g, 30 mmol) in dry xylene (100 ml) was added during 30 minutes at room temperature, and the mixture was stirred at 55°C for 1 hour.

The reaction mixture was cooled in ice-water bath and CO_2 was passed through it with fast stirring for 4 hours. Water (200 ml) and concentrated HCl (100 ml) were added, the magnesium was filtered off, and the filtrate was boiled under reflux for 2 hours. The cooled, clear solution was washed with ether (3 x 150 ml) and extracted with dichloromethane (3 x 140 ml). This extract was evaporated to dryness, and the residue dissolved in 0.5 liter of 0.5 N sodium hydroxide. The solution was washed with ether (3 x 100 ml), made acidic with concentrated HCl (150 ml), saturated with sodium chloride and extracted with dichloromethane (3 x 200 ml). Evaporation of the solvent left an oil (7.55 g, 60%) which ran as a single spot on TLC.

Example 2: Conjugation of Methadone-Carboxylic Acid and Lysozyme – The acid of Example 1, (21.0 mg, 50 μmol) was dissolved in 1 ml dry DMF, two drops of triethylamine were added, and the chilled solution was treated with 6.5 μl of isobutylchloroformate to form the mixed anhydride. Lysozyme, (120 mg) (50 μmol of lysine) was dissolved in 12 ml of water. The pH was adjusted to 10.0 with 0.05 N NaOH and maintained there during the dropwise addition of the mixed anhydride solution. After 30 minutes additional stirring, the mixture was centrifuged.

The pellet (containing only a small fraction of the enzyme) was dissolved in 4 ml of 8 M urea. It remained soluble during dialysis against water. The resulting solution proved quite dilute and the ΔOD/time in the assay was grossly nonlinear. The supernatant fraction stayed homogeneous through dialysis against water, and proved to be a very high quality product as judged from its usefulness in the assay.

Enzyme Conjugates of Amphetamine and Analogs

Acyl derivatives of p-(2-aminopropyl-1)phenol are used by *R.S. Schneider and D. Wagner; U.S. Patent 3,878,187; April 15, 1975; assigned to Syva Company* to prepare enzyme conjugates of amphetamine (2-aminopropylbenzene). The acyl group will normally have 12 to 16 carbon atoms. The only heteroatoms will be the heteroatoms of the acyl group, the oxy group and the amine, which will be oxygen, nitrogen and sulfur. The aliphatic divalent group joining the phenoxy group to the nonoxo carbonyl group may be branched or straight chain, usually from 0 to 2 branches of one carbon atom, i.e., methyl, aliphatically saturated or unsaturated, usually from 0 to 1 site of ethylenic unsaturation, and except for methylene, will usually have its free valences other than geminal.

Of particular interest are compounds where the nonoxo carbonyl substituted amphetamine (including N-methyl) is bonded to an amino group which is part of a polypeptide structure. Included are derivatives of polypeptides such are enzymes which are used as the detector in an immunoassay system.

Various enzymes may be used, such as peptidases, esterases, amidases, phosphorylases, carbohydrases, oxidases, and the like. Of particular interest are such enzymes as lysozyme, peroxidase, amylase, dehydrogenase, particularly malate dehydrogenase and mannitol 1-phosphate dehydrogenase, β-glucuronidase, cellulase, and phospholipase, particularly phospholipase C. Usually there will be at lease one amphetamine per 50,000 molecular weight, and more usually at least one amphetamine per 30,000 molecular weight.

The substituted polypeptides will for the most part have the following formula: $(\text{Amphetaminyl–O–R–CO–})_a Y'$ where Y' is a polypeptide residue, e.g., enzyme. R is hydrocarbon, usually allylene, having from 0 to 1 site of ethylenic unsaturation, and of from 1 to 6 carbon atoms, usually of 1 to 4 carbon atoms and a is a number of at least one.

The carboxylic acid employed in this process can be prepared from the p-(2-aminopropyl-1)phenol by first protecting the amino group with a removable acyl group and then substituting the halogen atom of a halo-substituted carboxylic acid with the phenolic oxygen. The reaction is readily carried out in an inert polar solvent, such as a ketone or ether, e.g., acetone. The protective group may then be removed, if desired, or retained during preparation of the mixed anhydride. The protective group should be one which can be readily removed despite the presence of other amide functionalities. Trifluoroacetyl is an illustrative group.

The mixed anhydrides may then be used for reacting with various amino groups, or in the alternative, the carboxylic acid can be activated with carbodiimide. The formation of amides employing either carbodiimide or mixed carbonate anhydride follows conventional procedures. In the following examples, the p-(2′-aminopropyl-1′)phenoxyacetic acid was prepared from p-hydroxyamphetamine by first forming the N-formyl derivative which is condensed with methyl chloroacetate followed by removal of the ester and formyl moieties.

Example 1: N-Trifluoroacetyl p-(2′-Aminopropyl-1′)Phenoxyacetic Acid – p-(2′-aminopropyl-1′)phenoxyacetic acid (0.5 g, 2.4 mmol) was suspended in trifluoroacetic anhydride (3.5 ml) and the mixture warmed slowly to 80°C (at 70°C the amino acid was completely dissolved). The clear solution was cooled to room temperature and cold water (30 ml) added. The crystalline trifluoroacetylamino acid (TFA) was filtered and washed with a little water and dried. Yield: 6.14 mg, 84%, MP 173° to 174°C. An analytical sample was recrystallized from aqueous methanol, MP 174° to 175°C.

Example 2: Conjugation of p-(2′-Aminopropyl-1′)Phenoxyacetic Acid – To a solution of the N-TFA of Example 1 (475 mg, 1.5 mmol) in dry dimethylformamide (DMF, 4ml), cooled in an ice-salt bath (~-10°C), dry triethylamine (0.21 ml, 1.5 mmol) was added; the mixture stirred for two minutes, and isobutyl chloroformate (0.19 ml, 1.5 mmol) added. The reaction mixture was stirred for 15 minutes at -10° to -6°C, and then for 15 minutes at 0°C. The mixed anhydride thus obtained could be added to a solution of lysozyme in water, with the pH initially adjusted to 9.5. The mixed anhydride was at a ratio of 2.5 mols of mixed anhydride per lysine amino group present. There are six lysine groups per lysozyme molecule. The pH was maintained at 9.5 by the addition of 0.1 N sodium hydroxide as required. The final lysozyme conjugate will have about four molecules of the amphetamine group per molecule of lysozyme. The same technique may be used with a wide variety of other enzymes, such as horseradish peroxidase, malate dehydrogenase, alkaline phosphatase, α-amylase, phospholipase, and cellulase. These enzymes are only illustrative of the wide variety of enzymes which may be employed.

Cocaine Linked to Enzymes via Isocyanate or Thiourea Groups

Isothiocyanate-modified benzoyl ecgonine compounds (e.g., p-isothiocyanatobenzoyl ecgonine) are provided by *M.J. Soffer and R.S. Schneider; U.S. Patent 3,917,582; November 4, 1975; assigned to Syva Corporation* for conjugation to an enzyme. The enzyme conjugate is useful in immunoassays for the determination of benzoyl ecgonine or cocaine.

The isothiocyanate compound is prepared by esterifying the appropriate ecgonine derivative with an appropriate nitrobenzoic acid. The ester may then be reduced to the aminobenzoate, conveniently employing catalytic hydrogenation. The amino compound may then be derivatized to the isothiocyanate using thiophosgene. The amino compound is combined with the thiophosgene under mild conditions. Conjugation of the isothiocyanate to a poly(amino acid), e.g., enzyme, is carried out by combining the appropriate ratio of the isothiocyanate to the poly(amino acid) under mild conditions and at a constant mildly basic pH, normally in the range of 8 to 9. These conjugates will, for the most part, have the following formula:

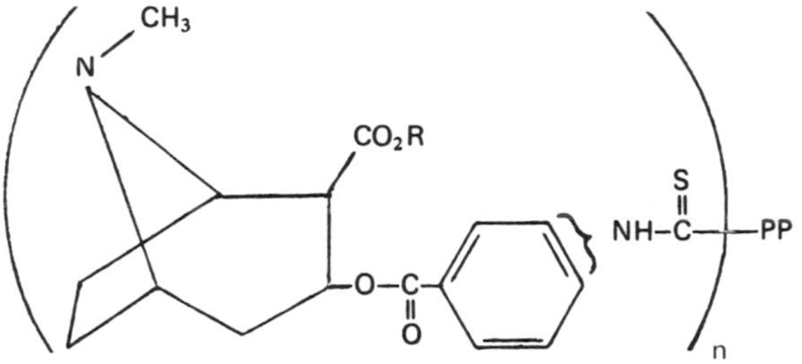

where R is hydrogen or methyl, usually hydrogen; the thiocarbamoyl group is either meta or para, usually para; PP is a poly(amino acid), e.g., an enzyme; and n is the number of benzoyl ecgonine groups bonded to the poly(amino acid), which will generally be from 1 to 30, preferably 3 to 16. The p-aminobenzoyl ecgonine used in the following example was prepared by forming the methyl ester of ecgonine (3-hydroxy-2-tropane carboxylic acid), then reacting the 3-hydroxy group with p-nitrobenzoyl chloride to form p-nitrococaine, reducing the nitro group to form p-aminococaine and removing the ester group to form p-aminobenzoyl ecgonine.

Example 1: Preparation of p-Isothiocyanatobenzoyl Ecgonine – Into two ml of 2 N HCl under nitrogen was introduced 100 mg (0.33 mmol) of p-aminobenzoyl ecgonine. To the solution was added 31 μl (46 mg) of thiophosgene and the heterogeneous mixture stirred vigorously under nitrogen at room temperature. After 10 minutes, the thiophosgene could no longer be observed. The product crystallized. The mixture was cooled in ice, filtered, and the filtrate washed with water. After drying the solid over phosphorus pentoxide and potassium hydroxide, 62 mg was isolated, MP 257°C (dec). To the mother-liquor was added approximately 3 ml of water, the solution cooled in ice and the precipitate collected. The second crop yielded 66 mg, MP 257°C.

Example 2: Conjugation of p-Isothiocyanatobenzoyl Ecgonine to Lysozyme – A solution of 60 mg (25 μmol) lysozyme in 5.0 ml water was cooled to 4°C and adjusted to pH 9.0 with 0.05 M NaOH. A total of 99.5 mg (25 μmol) p-isothiocyanatobenzoyl ecgonine was added in one portion to the alkaline protein solution. This conjugation was run on the pH-Stat at 4°C with the machine maintaining the pH at 9.0 with 0.05 M NaOH. (The reaction can also be performed by the manual addition of base using a pH meter to follow the course of the reaction.) After 3¾ hours, the clear solution was adjusted to pH 9.5. Since no precipitation occurred, the pH was lowered to 7.0 with dilute HCl. The clear solution was dialyzed against water for 48 hours. The dialysate was immediately suitable for the assay of benzoyl ecgonine.

The immunoassay using an enzyme as the detector is carried out as follows. With lysozyme, a bacterial suspension of *M luteus* is used, dissolving 0.2 ml of a suspension of 300 mg of the bacteria in 400 ml of 0.025 M, pH 6, Tris-maleate buffer. First the bacterial suspension is introduced into the assay vessel. When testing a sample, 50 μl of the sample is then introduced. This is followed by 50 μl of antibody solution in 0.025 M, pH 6, Tris-maleate buffer and the transfer made quantitative by washing with 325 μl of the same buffer solution. The benzoyl ecgonine conjugate to lysozyme (50 μl) is then added to give a binding site to benzoyl ecgonine ratio of about 1:1.5 and 325 μl of buffer used to insure quantitative transfer.

The results are then read by observing the decrease in optical density at 436 nm for 40 seconds at 36°C. The results are reported in OD/min. In the subject assay, the antibody employed was obtained in response to a conjugate of p-diazabenzoyl ecgonine with bovine serum albumin. The binding constant was about $1 \times 10^{8} M^{-1}$. The concentration of antibody was 6.9×10^{-6} M based on binding sites as determined employing a free radical assay technique with a cocaine spin label.

Conjugates with Anesthetics such as Lidocaine

Derivatives of anesthetics involving anilides, e.g., lidocaine, having an annular amino group are used by *P. Singh; U.S. Patent 4,069,105; January 17, 1978; assigned to Syva Company* to prepare conjugates for enzyme immunoassay. A difunctional linking group is provided, which provides a link to the annular amino group and the enzyme. A compound of particular interest is the derivative of lidocaine which has the following formula:

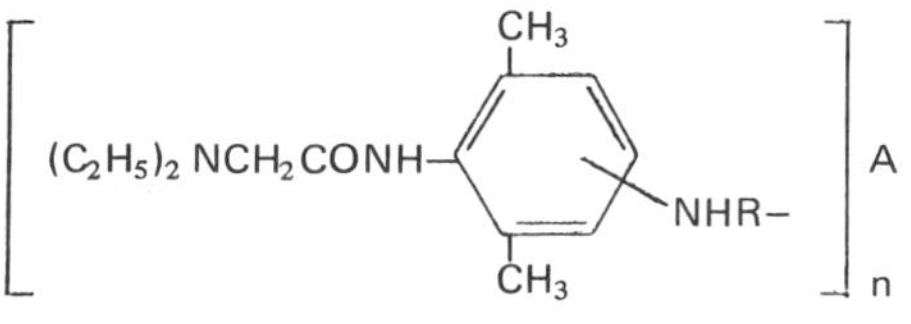

CH_3

$(C_2H_5)_2NCH_2CONH$– [benzene ring] –NHR– $\Big]_n$ A

CH_3

where R is a linking group of 1 to 8 carbon atoms having 1 to 4 heteroatoms, which are oxygen, nitrogen and sulfur, where the oxygen is present as oxy or nonoxocarbonyl, particularly the latter; nitrogen is present as tertiary amino, amido or imino and sulfur is present as thioether or thiono; and having 0 to 1 site of ethylenic unsaturation; A is a poly(amino acid); n is preferably 2 to 12.

Various enzymes may be employed, such as oxidoreductases, hydrolases, lyases, and the like. Groups of particular interest include those enzymes which employ nicotinamide adenine dinucleotide (NAD) or its phosphate (NADP) as an acceptor, such as the dehydrogenases, including glucose-6-phosphate dehydrogenase, malate dehydrogenase, alcohol dehydrogenase, lactate dehydrogenase, and the like; peroxidases; oxidases; glycoside hydrolases; and the like. In preparing the p-aminolidocaine, an appropriate aniline derivative with the amino group protected (e.g., by tosyl group) is nitrated, so as to provide the meta- or para-nitroaniline. The protective group may then be removed from the amino group. The amino group is then acylated to provide the desired anilide, followed by reduction of the nitro to amino to provide the amino functionality for linking.

Example 1: Preparation of the Hemisuccinamide of p-Aminolidocaine – A solution of 1.54 g (5.1 mmol) p-aminolidocaine and 520 mg (5.2 mmol) of succinic anhydride in 50 ml of dry THF (freshly distilled from LAH) was stirred overnight. TLC (alumina, chloroform) showed starting material remaining. Small portions of the anhydride were added until TLC showed no starting material remained. The solvent was removed in vacuo to yield a yellowish foam. The product was dissolved in water and 6 N hydrochloric acid was added to pH 2, then the solution was evaporated to dryness. The residue was dissolved in a small amount of water and acetone was added, with heating, to the cloud point. The hydrochloride crystallized as small hexagonal crystals, MP 152° to 154°C. The material is very hygroscopic.

Example 2: Glucose-6-Phosphate Dehydrogenase Conjugated to the Hemisuccinamide of p-Aminolidocaine – A stock solution of glucose-6-phosphate dehydrogenase (G6PDH) was prepared by reconstituting a dry powder of G6PDH with 0.055 M Tris-HCl buffer, pH 8.1 (room temperature) to provide a concentration of about 2 to 3 mg/ml. To a small flask was introduced 1 ml of the enzyme solution and the solution cooled to 0°C with stirring. To the solution was then added 20 mg glucose-6-phosphate and 20 mg NADH, followed by the slow addition of 300 μl carbitol and the pH adjusted to 9 with 2 N NaOH.

A solution was prepared of dry lidocaine hemisuccinamide HCl (0.075 mmol) in 375 μl of dry DMF, the mixture stirred and 21 μl of triethylamine added under the surface of the liquid with stirring, while maintaining the mixture at -10°C. To the solution was slowly added 14 μl of carbitol chloroformate under the surface of the stirred solution. After 1.5 hours, the mixed anhydride was ready for use.

Over an interval of approximately 15 seconds, 25 μl of the mixed anhydride was added under conditions to minimize local concentrations. After approximately a 15 minute reaction time, an aliquot of the enzyme was removed and checked for activity and the addition of the mixed anhydride repeated as described above until the desired activity and inhibitability were obtained. The pH was maintained above 8.5 during the course of the reaction by the addition of 2 N sodium hydroxide as required. The following table indicates the results obtained.

Run No.	Content*	pH	Corrected ΔOD	Conjugate Volume (ml)	I** (%)	D*** (%)	Total MA† Added
1	Enzyme + carbitol	9.04	745	1.31	–	–	–
2	Run 1 + 25 μl MA	9.00	725	1.335	7.9	3.2	25
3	Run 2 + 50 μl MA	8.97	681	1.385	16.1	9.0	75
4	Run 3 + 50 μl MA	8.94	626	1.435	27.9	16.5	125
5	Run 4 + 100 μl MA	8.90	553	1.535	57.5	23.1	225
6	Run 5 + 50 μl MA	8.89	495	1.585	69.6	34.0	275
7	Run 6 + 50 μl MA	8.88	432	1.635	81.4	42.4	325
8	Run 7 + 50 μl MA	8.87	310	1.685	93.5	58.6	375

*Reaction mixture with each run having additional amount of MA added to the mixture and a 5 μl aliquot taken for testing.

**I = inhibitability and refers to reduction in rate in presence of excess anti(lidocaine).

***D = Deactivation and is the loss of enzyme activity due to conjugation of the lidocaine derivative.

†MA = mixed anhydride.

Use of Conditioners in the Preparation of Conjugates

The disclosure of *J.K. Weltman and M.B. Rotman; U.S. Patent 4,002,532; January 11, 1977* provides a method for coupling enzymes to macromolecules in the presence of conditioners. Conditioners are chemical compounds which, if used during conjugation reactions of enzymes and macromolecules with polyfunctional reagents, have the following effects.

There is a substantial elimination of precipitation of enzyme conjugates; a reduction of nonspecific adsorption of enzyme conjugates to solid phase immunosorbents; the preservation of enzyme activity of the conjugates; the preservation of the specific binding properties of the conjugated partner or partners and the resistance of the enzyme activity of the conjugates to denaturants, such as detergents, urea and salts. The conjugates are the products of enzymes and other biological macromolecules coupled with a polyfunctional coupling agent in the presence of a polyamine conditioner.

The biological macromolecules can be nucleic acids, proteins, glycoproteins or lipoproteins, all of which contain amino groups. Examples of such macromolecules are deoxyribo- and ribonucleic acids, viral proteins, allergens, immunoglobulins, blood group substances, transplantation antigens, carcino embryonic antigen, α-fetoprotein and other tumor-specific antigens, growth hormone and other polypeptide hormones. Any enzyme can be used to form the conjugates. Particularly useful are enzymes which can be detected either chromogenically or fluorogenically with great sensitivity. Among the useful enzymes are acid and alkaline phosphatases, alcohol dehydrogenase catalase, glucose and galactose oxidases, α and β-galactosidases, lactate dehydrogenase, lysozyme, luciferase, peroxidases, ribonuclease and esterases.

Coupling agents for use in forming conjugates are polyfunctional, most often bifunctional, reagents with active hydrogen, halogen, aldehyde, sulfonate or ester groups capable of reacting with one or more types of substituent groups in the macromolecule reactants such as $-NH_2$, $-COOH$, $-SH$, phenolate or imidazol. Organic polyamines of either high or low molecular weight are used as conditioners. For example, macromolecules with multiple amino groups, i.e., 2, 3, 4, etc. amino groups, such as synthetic polypeptides or natural proteins are conditioners. The conditioner must be different from either of the coupling partners used in the formation of the conjugate so that it will not have an adverse effect on the assay of the conjugate. Diaminodipropylamine is an example of a low molecular weight conditioner. Other low molecular weight conditioners may be aromatic amines such as phenylenediamine, 4,4'-diaminobenzophenone, p,p'-diaminodiphenyl methane naphthalenediamine, benzidine, 2,6-diaminopurine, or the series of aliphatic diamines, such as, ethylenediamine, propylenediamine or hexamethylenediamine.

The conditioner should be free of groups which may react with the amine such as reactive acyl, including carboxy, keto, acylhalide, aldehyde groups, sulfonic acid groups and, in general, any other groups which react in accordance with known principles with the amino groups of the conditioner to block the amino groups. However, any other substituents may be present on the conditioner molecule, e.g., alkyl, aryl, alkoxy, nitro, mercapto, and peptidyl. The use of conditioners enabled the preparation of useful antibody conjugates of α-D-galactosidase, a bacterial enzyme which had been thought previously to be unsuitable for use in enzyme-antibody conjugates [Hermann and Morse, *Immunochem.* (1974) 11, 79-82].

Conjugates of α-D-galactosidase are desirable because of the characteristics of this enzyme, namely, sensitivity of assay and availability of stable chromogenic and fluorogenic substrates. Assays for this enzyme have been developed which are more sensitive than those for any other enzyme, i.e., a single molecule of β-D-galactosidase can be detected. The use of conditioners to improve the properties of conjugates should find application in many tests involving enzyme conjugates. For example, conditioners may be used to produce improved reagents for detection of macromolecules such as allergens, serum proteins, hormones, antigens characteristic of bacteria, viruses, fungi, protozoa and helminths and for histocompatibility typing.

Example: Immunospecific Detection of Insolubilized Sheep Immunoglobulin – Rabbit antibodies against sheep immunoglobulin were conjugated to β-D-galactosidase (Z) in the presence of a conditioner. To this end, rabbits were immunized by subcutaneous injections of one mg purified sheep immunoglobulin dispersed in complete Freund's adjuvant. After the immunizing schedule, immunospecifically purified antibodies against sheep immunoglobulin (RaShIg) were isolated from the rabbit serum by affinity chromatography. The purified RaShIg was conjugated to Z.

In order to achieve the conjugation, a mixture was prepared in phosphate buffered saline (PBS) pH 7.6 containing 6×10^{-6} M of Z, 2×10^{-2} M glutaraldehyde and the conditioner, 3,3'-diaminodipropylamine, at 1.3×10^{-2} M. This mixture was permitted to react for 30 minutes at room temperature, following which the RaShIg was added to a final concentration of 9×10^{-6} M, and the reaction was allowed to proceed for an additional 45 minutes; $NaHSO_3$ at a final concentration 2×10^{-2} M was then added to stop the reaction. The conjugation mixture was diluted 14 fold with PBS and then dialyzed against PBS. The dialyzed solution was clarified by filtration through a 0.45 μ Millipore membrane. This procedure produced significantly less insoluble protein than similar reactions in which conditioners were omitted.

For instance, without conditioner only 2% of the initial enzymatic activity remained in solution, the other 98% being either insolubilized or inactivated. The conjugates made by the procedure described above (RaShIg-Z) contained 97% of the initial Z activity in soluble form. The prepared RaShIg-Z conjugates were diluted in buffer E so that the solution contained 10,000 units of enzyme activity per ml. Buffer E was 0.14 M NaCl, 0.075 M sodium phosphate, 1% Tween 20, 1% BSA and 0.05% NaN_3. Various amounts of this RaShIg-Z solution were incubated overnight with cellulose disks onto which either sheep immunoglobulin (ShIg) or rabbit immunoglobulin (RaIg) had been insolubilized. The disks were washed five times with buffer E without BSA and then placed in a solution of chromogenic substrate for the assay of Z activity. The enzyme was quantified from the rate of hydrolysis of the substrate, o-nitrophenyl-β-D-galactopyranoside, signaled by the appearance of yellow color in the solution.

The amount of RaShIg-Z-activity immunospecifically bound to ShIg and nonspecifically bound to RaIg is given in the following table. Between 65 and 174 times more Z activity was bound to ShIg, the specific antigen, than to the irrelevant protein, RaIg. Thus, conjugates prepared as described with conditioners exhibited high degree of specificity as shown by the S/N ratios in the following table. Conjugates prepared without conditioners were found to be significantly less specific, with S/N ratios of almost half of those shown in the table below.

Antibody Specificity of a Rabbit Antisheep Immunoglobulin Conjugate of β-D-Galactosidase

Enzyme Units.			
	Bound to		
Total Added	Sheep Immunoglobulin Disk (S)	Rabbit Immunoglobulin Disk (N)	(S/N)
45	1.5	0.017	89
220	5.2	0.030	174
450	9.9	0.069	144
2,200	15.0	0.230	65

RADIOIMMUNOASSAY PROCESS

In radioimmunoassay or competitive protein binding radioassay, the compound to be measured, generally the antigen, is allowed to compete with a similar or a chemically related radioactive compound for a limited number of binding sites on the antibodies or on the specific binding proteins. The antibody-bound radioactive labeled compound is then separated from the free labeled compounds and measured. Separation methods previously used, for example, electrophoresis, gel filtration, precipitation of free antigen with charcoal and precipitation of bound antigen with salt or another antibody, are often time consuming, complicated and do not give clear cut separation. Fixation of antibodies to the wall of a test tube enables the separation of the bound and free antigens by simple decantation.

M.J. Barrett; U.S. Patent 4,001,583; January 4, 1977; assigned to Smith Kline Instrument, Inc. has developed means of covalently linking the "bindable protein" to plastic materials for such assays. Antibodies, specific binding proteins or other protein materials such as enzymes are covalently linked through their amino and other reactive groups to active aldehyde groups of aliphatic dialdehydes such as glutaraldehyde, that have been previously polymerized on the inside surface of a plastic material, such as a plastic test tube, for example, a polypropylene or a polyethylene test tube.

The polymerized glutaraldehyde may in turn be attached to aliphatic primary amines of the formula $CH_3-(CH_2)_n-NH_2$ where n is an integer of 5 to 20, preferably 18, or aliphatic diamines of the formula $CH_3-(CH_2)_n-NH-(CH_2)_m-NH_2$

where n and m are integers of 3 to 20, preferably where n is 18 and m is 3. In this embodiment, prior to reaction with glutaraldehyde the plastic material is heated in a solution of the amine or diamine at above 50°C, preferably at 90°C. The excess amine or diamine is washed away and the plastic is treated with a solution of glutaraldehyde at room temperature, or a slightly elevated temperature, for example, at 56°C, for from 1 to 2 hours to 1 to 2 days. Thus, glutaraldehyde is polymerized directly on the inside surface of a plastic material such as a plastic test tube with or without prior treatment with an aliphatic amine or diamine.

In order to carry out a radioassay, a mixture containing buffer, labeled antigen or analogs and unlabeled antigen or analogs from a biological fluid, such as serum, or from a standard solution is incubated in a plastic test tube with antibody or other specific binding protein covalently bound to the inside surface of the test tube. After a suitable incubation period the free antigen is removed rapidly and simply by decantation, leaving no time for readjustment of the established equilibrium. The radioactive antigen bound to the test tube is counted after the tube has been rinsed with buffer. The entire procedure is simple, rapid and no special skill is required.

It is believed that there is a self-polymerization of the aldehyde material on the surface of the plastic followed by a Schiff base-type coupling of the protein to the active aldehyde group.

The term "biological substance" used here denotes a material of biological origin such as an antigen, antibody or enzyme, all of which being capable of chemically reacting with an aldehyde group. Unreacted amino groups present in the protein chains of most biological substances provide such chemically reactive groups.

An aqueous solution of glutaraldehyde is dispensed into a plastic test tube, for example a polypropylene or polyethylene test tube, and is allowed to remain in contact with the inner surface of the test tube at room temperature or slightly elevated temperature, for example 56°C, for a period of time such as 1 to 2 hours to 1 to 2 days. The glutaraldehyde polymerizes on the surface of the plastic and forms a thin layer of polymer on the inner surface of the plastic test tube with a large number of active aldehyde groups that can react covalently with primary amino groups of antibodies or proteins.

The polymerization of glutaraldehyde occurs at a pH of 3 to 10. Also, the glutaraldehyde solution polymerizes on plastic surfaces over a wide range of aldehyde concentration, namely 0.1, 0.2, 0.5, 1 or 2% glutaraldehyde. The amount of glutaraldehyde polymer and consequently the number of active aldehyde groups on the plastic surface can be increased or decreased by varying the concentration of the aldehyde solution, incubation time and temperature. Excess active aldehyde groups on the plastic surface that are not used up in subsequent protein coupling can be blocked by reacting with compounds having primary amino groups such as monoethanolamine or lysine.

After the glutaraldehyde is polymerized on the surface of the test tube, the aldehyde solution is removed by aspiration and the tube is washed thoroughly with deionized water. Tubes thus prepared are ready for use in a radioassay comprising protein or antibody coupling.

The aldehyde treated tubes are very stable and retain their ability to couple protein even after washings with concentrated salt solutions and detergents.

Antibodies and proteins are coupled to the active aldehyde groups on the glutaraldehyde treated test tube surface through their primary amino groups. The rate and the amount of protein coupled to the test tubes is directly proportional to the concentration of the protein solution used. However, the maximum amount of protein that can be coupled to the surface is governed by the size ot the protein molecule and the area of the plastic surface. Using I^{125}-gamma globulin (γG), molecular weight 170,000, as a model protein, experimental data indicates that approximately 1.2 μg of γG is the maximum amount that can be coupled to 1 cm^2 of a glutaraldehyde treated plastic surface. The following example shows the binding of gamma globulin to the tube.

Example: 2 ml of 0.1% glutaraldehyde in 0.1 M carbonate buffer, pH 9.0 is dispensed into each of 50 polypropylene tubes measuring 1.1 X 5.5 cm. The tubes are incubated at 56°C for 3 hours, cooled at room temperature, and the aldehyde solution removed by aspiration. The tubes are washed 10 times with deionized water.

An I^{125} labeled human gamma globulin solution at 15 μg protein per ml of 0.1 M phosphate buffer pH 7.0 is prepared and 2 ml of the human gamma globulin (HGG) solution is dispensed into each glutaraldehyde treated tube. The amount of I^{125} HGG dispensed into each tube is counted (79,398 $\pm$ 514 cpm). The I^{125} HGG solution is left inside the tube at 4°C for 15 hours. After the incubation period, the I^{125} HGG solution is removed by aspiration. The I^{125} HGG bound to the wall is counted after the tubes have been thoroughly rinsed.

The amount of I^{125} HGG coupled to the wall has a mean of 3.013$\pm$0.067 μg (represented by 7975$\pm$178 cpm) a coefficient of variance of 2.24. The I^{125} HGG coupled to the aldehyde treated surface cannot be washed away with 1% sodium lauryl sulfate or dodecyl sodium sulfate.

IMMOBILIZED ENZYMES IN SUGAR PRODUCTION

HYDROLYSIS OF STARCH TO DEXTROSE (GLUCOSE)

Processes for hydrolyzing starch to dextrose (also termed glucose in the literature) can be grouped into two broad categories. These are the acid-enzyme and the enzyme-enzyme conversion processes. In the acid-enzyme process, generally starch is first partially hydrolyzed or liquefied using an acid such as hydrochloric acid. The suspension is then heated to relatively high temperatures to partially hydrolyze the starch and then cooled and treated with a glucoamylase preparation under suitable conditions to enzymatically convert the partially hydrolyzed starch to dextrose. Glucoamylase has been referred to as glucamylase glucogenic enzyme, starch glucogenase and γ-amylase. Glucoamylase is an exo-amylolytic enzyme which catalyzes the sequential hydrolysis of glucose moieties from the nonreducing ends of starch or amylodextrin molecules.

In the enzyme-enzyme conversion process, generally a starch slurry is formed and a starch liquefying enzyme, for instance, bacterial α-amylase, is added and the starch slurry heated to 80° to 90°C to partially hydrolyze the starch. The partially hydrolyzed starch, which generally has a DE of 10 to 20, is then treated with glucoamylase. α-Amylase is an endo-amylolytic enzyme capable of promoting almost random cleavage of α-1,4-glucosidic bonds within the starch molecule.

There has been a great deal of interest shown in the use of starch debranching enzymes for dextrose production. The use of such enzymes increases the amount of dextrose formed since they can readily act upon bonds or linkages in the starch molecules which are not acted upon by α-amylase or which are only slowly acted upon by glucoamylase. Debranching enzymes are generally referred to as α-1,6-glucosidases. A number of enzymes having considerably different specificities have been identified in the art as being capable of hydrolyzing α-1,6-glucosidic linkages. Of these, probably the two most important from the commercial standpoint are pullulanase and isoamylase. The major difference in regard to the specificity of these enzymes is that pullulanase will degrade the linear polysaccharide pullulan whereas isoamylase will not to any significant degree.

Using Immobilized Glucoamylase and α-Amylase

Mixed immobilized enzymes have been used by *K.N. Thompson, N.E. Lloyd, and R.A. Johnson; U.S. Patents 4,102,745; July 25, 1978; and 4,011,137; March 8, 1977; both assigned to Standard Brands Incorporated* to provide a substantially complete conversion of starch to dextrose. A partially hydrolyzed starch solution (10% hydrolyzed) is contacted with immobilized glucoamylase and α-amylase selected from soluble α-amylase, immobilized α-amylase or their mixtures.

It has been found that, when a glucoamylase preparation is subjected to immobilization, the resulting immobilized glucoamylase does not convert partially hydrolyzed starch so rapidly nor so completely as the soluble glucoamylase preparation from which the immobilized glucoamylase is prepared. During the immobilization of glucoamylase, the α-amylase, which is inherently present, is rendered substantially inactive or inert regardless of the method of immobilization used. Apparently, the small amount of α-amylase, which is inherently present in soluble glucoamylase preparations, has a beneficial effect on the overall conversion of starch to dextrose with glucoamylase. Thus, to obtain maximum use of immobilized glucoamylase in the conversion of partially hydrolyzed starch to dextrose, there must also be present soluble and/or immobilized α-amylase.

Surprisingly, this finding is true even when the partially hydrolyzed starch has been prepared by treatment of unmodified starch with α-amylase and, therefore, would be assumed to be rendered readily susceptible to conversion with glucoamylase by such treatment. Moreover, it has been discovered that α-amylase added to immobilized glucoamylase is effective for increasing the conversion of partially hydrolyzed starch to dextrose even during the latter stages of the conversion. Apparently, branched dextrins are formed during the initial stages of the hydrolysis reaction which are not readily hydrolyzed by the immobilized glucoamylase but which are readily hydrolyzed by α-amylase and, thus, the overall conversion of the starch hydrolysate is enhanced.

This process may be performed by a number of techniques. For instance, soluble or immobilized α-amylase and immobilized glucoamylase may be used concurrently or sequentially. It is preferred that they be used concurrently as, for example, when partially hydrolyzed starch is contacted with a mixture of immobilized glucoamylase and immobilized α-amylase. Of course, it will be realized that the α-amylase and glucoamylase may be immobilized on or within the same carrier and results will be obtained which are substantially equivalent to those given by mixtures of α-amylase and glucoamylase immobilized on separate carriers.

In the case where the enzymes are used sequentially, the conversion process will comprise at least three steps in the following sequence: (1) contacting the partial hydrolysate with immobilized glucoamylase, (2) contacting the resulting hydrolysate with a soluble or immobilized α-amylase, and (3) contacting the resulting hydrolysate with immobilized glucoamylase. The last two steps of the sequence may be repeated a number of times depending on the conditions under which the reactions are conducted. The concurrent use of the enzymes results in greater amounts of the partially hydrolyzed starch being converted to dextrose than does sequential use except when the steps employed in sequential use are repeated a large number of times.

A number of different types of α-amylase may be used, although it is preferred that saccharifying- or pancreatic-type α-amylase be used. Microorganisms, such as *Bacillus subtilis* var. *amylosaccharíticus* Fukumoto elaborate saccharifying-type α-amylase. Generally, it is also preferred that α-amylase preparations, which are to be used for immobilization, have an S/L value of preferably 50 and most preferably of 100. The S/L value is defined as one thousand times the saccharifying activity measured in saccharifying units (S) per g of α-amylase preparation divided by the dextrinizing activity measured in liquefons per g of preparation.

The following example illustrates the use of glucoamylase immobilized on DEAE-cellulose and α-amylase derived from different sources immobilized on aminoethylcellulose for converting a partially hydrolyzed starch solution to dextrose.

Example: Immobilization of Glucoamylase – 53.0 g of a dry glucoamylase preparation (from *Aspergillus awamori*, free of transglucosylase activity) having a glucoamylase activity of 83.2 GU g^{-1} was incorporated into 3.8 liters of deionized water. The mixture was stirred for 30 minutes and filter aid added. The mixture was filtered, the filter cake washed, the filtrate and washings combined, and the pH of the combined solutions adjusted to 5.5 using 4 N HCl. 13.3 g DEAE-cellulose (Whatman DE 23) was added, the mixture stirred for 60 minutes at ambient temperature and then filtered and the filter cake washed with deionized water. The recovered moist filter cake had a glucoamylase activity of 44 GU g^{-1}. The moist filter cake is referred to as immobilized glucoamylase.

Immobilization of α-Amylase – α-amylase derived from various sources was immobilized by coupling with activated aminoethylcellulose (activated AE). The activated AE was prepared by slurrying 20 g of aminoethylcellulose in 500 ml of a 0.5 M phosphate buffer at pH 7, stirring for 20 minutes at ambient temperature and then maintaining the mixture for 7 hours without stirring. The mixture was filtered, the filter cake washed with deionized water and suspended for 12 hours in 500 ml of 0.5 M phosphate buffer at pH 7. 140 ml of a glutaraldehyde solution (50%) was added to the slurry, the slurry stirred for 90 minutes at ambient temperature, filtered and the filter cake washed with deionized water. 74.9 g of filter cake (75.2% moisture) was recovered.

16.0-g portions of activated AE were added to 50 ml of each of the following four α-amylase solutions:

(1) Solution of *Bacillus subtilis* saccharifying α-amylase (var. *amylosaccharíticus* Fukumoto, twice recrystallized, S/L = 257) containing 0.33 mg protein per ml and having an activity of 134 liquefons per ml;

(2) Solution of *Bacillus subtilis* liquefying α-amylase (Bacterial-type, II-A, 4 x crystallized, S/L = 5) containing 0.45 mg protein per ml and having an activity of 1,166 liquefons per ml;

(3) Solution of *Aspergillus oryzae* fungal α-amylase (3 x crystallized, S/L = 76) containing 0.45 mg protein per ml and having an activity of 322 liquefons per ml; and

(4) Solution of hog pancreatic α-amylase (2 x crystallized, S/L = 227) containing 0.70 mg protein per ml and having an activity of 290 liquefons per ml.

The mixtures were stirred for 2 hours and filtered. The filter cakes were washed with 0.005 M calcium acetate at pH 7 with small portions of 0.5 M NaCl (total

100 ml) and then with 50 ml of 0.05 M calcium acetate solution. The immobilized α-amylases exhibited the following potencies:

α-Amylase	Potency (liquefons g^{-1})
Immobilized saccharifying	12.6
Immobilized liquefying	125.0
Immobilized fungal	7.3
Immobilized pancreatic	41.5

To each of five stirred reactors maintained in a water bath at 50°C was added 138 g of partially hydrolyzed starch solution (pH 5.0, 16.9 DE, 32.5% dry substance) prepared by the procedure described above which had been filtered through a cellulose ester membrane (HAWP 04700, 0.45 μ) and then saturated with toluene. The purpose of the addition of the toluene was to prevent bacterial growth.

0.51 g of immobilized glucoamylase was added to each of the reactors and sufficient immobilized α-amylase was added to four of the reactors to provide a total of 90 liquefons of α-amylase activity per reactor. The reactors were continually stirred and, at various time intervals, samples were taken from the reactors and filtered and the filtrates assayed for percent dextrose. The results of this example are set forth in the table below:

Saccharification Using Immobilized Glucoamylase and Various Types of Immobilized α-Amylase

		Percent Dextrose			
	Average	Hours			
Immobilized Enzyme System	pH*	46	70	106	142
Glucoamylase (control)	5.1	71.5	76.3	79.6	80.9
Glucoamylase and saccharifying α-amylase	5.1	92.3	92.9	93.0	92.7
Glucoamylase and liquefying α-amylase	5.0	89.0	91.5	93.1	93.2
Glucoamylase and fungal α-amylase	5.2	87.2	89.7	91.5	91.9
Glucoamylase and pancreatic α-amylase	5.1	93.1	94.7	94.9	94.2

*During saccharification.

From the table, it is seen that a combination of immobilized glucoamylase and immobilized α-amylase resulted in a more complete conversion of starch to dextrose than when immobilized glucoamylase alone was used. Also, the combination of immobilized enzymes resulted in a more rapid conversion of starch to dextrose. Moreover, in general, the immobilized α-amylase preparations prepared from the soluble α-amylase preparation having a high S/L ratio are more beneficial in the conversion of partially hydrolyzed starch to dextrose.

Immobilized α-1,6-glucosidases may also be used in this process, preferably pullulanase. It is also preferred to immobilize the pullulanase by covalently binding it to an inert carrier.

Third Stage Hydrolysis with Immobilized Glucoamylase

Very high dextrose hydrolysates are produced by *R.E. Hebeda, D.J. Holik, and H.W. Leach; U.S. Patent 4,132,595; January 2, 1979; assigned to CPC International Inc.* using a multistep hydrolysis process. This process comprises the steps of:

(1) Reacting starch with hydrolytic enzymes or acid to produce a low DE starch hydrolysate having a DE from 2 to 20;
(2) Treating the low DE starch hydrolysate with a soluble glucoamylase preparation to produce a starch hydrolysate having a DE less than 85;
(3) Reacting the soluble-glucoamylase-treated starch hydrolysate with an effective amount of enzyme consisting essentially of immobilized form; and
(4) Recovering a dextrose product.

In this process, the high dextrose equivalent starch hydrolysate of less than 85 DE is treated solely with an immobilized glucoamylase preparation (i.e., devoid of free or immobilized active α-amylase) to produce the dextrose. The immobilized glucoamylase preparation is used alone without the addition or combination of other soluble or immobilized α-amylase enzymes, and no further steps or enzymatic treatment are required to produce the dextrose level desired in the product. Dextrose products, containing about 95% dextrose, were produced in column operation using immobilized glucoamylase preparations at 25% solids and 45°C when the starch hydrolysate feed was greater than 30 DE. Dilution of the starch hydrolysate feed to 10% solids gave an effluent containing about 97% dextrose.

Immobilized glucoamylase preparations are well known and can be produced by any of the known methods during which the α-amylase portion of the glucoamylase becomes inactivated and can, therefore, be present as the inactive form of α-amylase.

It is preferred to use the immobilized glucoamylase preparation produced according to U.S. Patent 3,783,101 in which the glucoamylase preparation is covalently coupled to silica.

Example: A low DE starch hydrolysate of 11 DE was treated with soluble glucoamylase at 60°C and pH 4.3 for exactly 16 hours at 25% solids by dry weight basis. Glucoamylase dosage and the DE attained are shown below:

Glucoamylase Dosage Units/100 g Dry Substance	DE Attained
1.4	30
3.1	47
6.4	65
13.4	80

The reaction was stopped by heating at 95°C for 15 minutes. Each substrate was readjusted to 25% solids by dry weight basis. Saccharide distribution for each hydrolysate feed is as shown on the following page.

Feed Composition

Feed DE	Dextrose*	Maltose	Isomaltose	Maltulose	DP-3	DP-4+**
	Percent Dry Basis by Paper Chromatography					
11	0.8	–	3.2***	–	4.7	91.3
30	26.0	5.6	0.1	0	7.9	60.4
47	44.0	8.6	0	0.1	6.3	41.0
65	59.9	10.6	0	0.6	0.7	28.2
80	76.6	3.9	0.5	0.6	0.7	17.7

*Dextrose by difference (i.e., 100% minus sum of percent nondextrose).
**Degree-of-polymerization-of-4 and greater as total percent.
***Total degree-of-polymerization-of-2 components.

A column containing a 100 ml bed of glucoamylase preparation immobilized on porous silica was operated at 1 bed volume per hour at 45°C and pH 4.3 using the prepared feeds at 25% solids and also the 11 DE starch hydrolysate. The procedure used for binding glucoamylase was essentially the same as that presented in U.S. Patent 3,783,101. Composition of the dextrose-containing product obtained at the various DE levels is as follows:

Product Composition

Feed DE	Dextrose**	Maltose	Isomaltose	Maltulose	DP-3	DP-4+***
		Percent Dry Basis by Paper Chromatography*				
11	93.7	0.8	2.4	0.9	0.3	1.9
30	94.1	1.0	2.3	0.9	0.3	1.4
47	94.2	0.8	2.6	0.9	0.4	1.1
65	94.7	0.7	2.6	0.8	0.4	0.8
80	94.6	0.8	2.8	0.8	0.5	0.5

*Normalized.
**Glucose oxidase method.
***Degree-of-polymerization-of-4 and greater as total percent.

SUGAR MIXTURES USING ENZYME MIXTURES

Treatment of Whey with Immobilized Lactase and Isomerase

Whey or an ultrafiltrate of whey is treated by the process developed by *H.H. Weetall and S. Yaverbaum; U.S. Patent 3,852,496; December 3, 1974; assigned to Corning Glass Works* to produce useful sweeteners. The process involves reacting whey with immobilized lactase to hydrolyze lactose to glucose and galactose, removing calcium ions from the whey, adjusting the pH of the whey to above 7 and reacting the whey with immobilized glucose isomerase to isomerize glucose to fructose.

The lactase enzymes were immobilized on zirconia-coated porous 96% silica glass particles having an average pore diameter of 550 A ± 10% (Corning Code MZO-3900). The zirconia-coated porous glass particles were silanized with gamma-aminopropyltriethoxysilane by an aqueous method described in an article by H.H. Weetall and N.B. Havewala, *Biotechnol. Bioeng. Symp.* No. 3, 241 (1972). The enzymes were covalently attached to the alkylamine glass particles, using glutaraldehyde, by the procedure generally described in the above article.

The soluble lactase-Y used in the example was assayed in 20% lactose dissolved in 0.025 M sodium phosphate buffer, pH 6.5, containing 5 x 10^{-4} M manganese and cobalt chlorides, respectively. The pH profiles were also determined for both the soluble and immobilized lactase enzymes. The lactase-Y assay results were unusual. Immobilization caused a pH shift of 3.25 pH units from pH 6.25 to pH 3.0, such that the immobilized enzyme had only 40% maximum activity at 6.25. The difference in the behavior of the two lactases is not fully understood. The following example shows the treatment of a large batch of whey ultrafiltrate for use in ice cream as a sucrose replacement.

Example: Approximately 4 gal of whey ultrafiltrate were continuously passed through a lactase-Y plugged flow-through column containing 22,000 units of activity over a 7-day period. The starting material was approximately 15% solids of which 70% was lactose. The remainder included calcium, lactic acid, riboflavin, and some residual protein. The hydrolyzed sample was collected over 7 days of operation, pooled, filtered through activated charcoal to remove the riboflavin, and neutralized with a 50% solution of NaOH for treatment with the immobilized glucose isomerase. On neutralization, a large white precipitate formed. This was removed and found to consist mostly of calcium phosphate. The remaining clear neutralized hydrolysate was then tested with immobilized glucose isomerase for conversion to fructose.

Glucose isomerase was prepared in a manner similar to that used for the preparation of the immobilized lactase. Glutaraldehyde was used for the coupling procedure. 1 g of immobilized glucose isomerase containing 525 units (1 unit is equivalent to the production of 1 micromol of fructose at 60°C per minute) was added to 50 ml of whey ultrafiltrate previously adjusted to pH 7.5 and buffered to 0.005 M in magnesium ions by adding $MgCl_2$. The test was carried out for 1 hour at 60°C in a shaking water bath. No conversion was observed. It became apparent that the glucose isomerase had been inactivated by the presence of calcium ions found in whey. The calcium ions were removed by ion exchange with a cation exchange resin. The assay was repeated again and with success. The quantity of glucose in the assay solution was decreased by 50%, indicating the maximum isomerization possible.

The whey was continuously passed through a similar column containing the immobilized glucose isomerase. To insure that the final product would be at least equivalent to sucrose on a sweetening basis, an additional 5% by weight glucose was added to the remainder of the 4 gal of hydrolyzed whey ultrafiltrate before isomerization. The isomerization was carried out in a column containing 50 g of the immobilized glucose isomerase described above. The flow rate was maintained at 200 ml/hr. The product was submitted to a dairy for use in preparing an ice cream sample.

The final product, after hydrolysis with lactase-Y, the addition of glucose and magnesium chloride, and after isomerization, was found to have the following constituents by weight of solids: 1% lactose, 45% galactose, 20% fructose, and approximately 25% glucose. The total solids were initially 15% while the final solids content was approximately 10%.

Hydrolysis of Starch to Glucose and Fructose

A process of obtaining high yields of glucose and fructose from liquified starch has been disclosed by *W. Colilla and N.E. Lloyd; U.S. Patent 4,111,750; Sept. 5,*

1978; assigned to Standard Brands Incorporated. This process uses an enzyme system comprising immobilized glucoamylase, immobilized glucose isomerase and immobilized debranching enzyme at a pH and temperature whereby substantially all the starch is converted to glucose and fructose.

The liquefied starch used in this process is preferably obtained by enzyme liquefaction and has a DE greater than 25. At lower DE's, there is the possibility of retrograded particles being present in the liquefied starch which precipitate on the immobilized enzyme thereby decreasing the efficiency of the enzymes. It is preferred that the debranching enzyme used have an action pattern such that it readily hydrolyzes the α-1,6 bond of branched molecules of the lower molecular weight dextrins. An example of such an enzyme is pullulanase, while an example of an enzyme which more readily cleaves branched molecules of higher molecular weight is isoamylase.

Optimum conditions for the catalytic action of glucose isomerase, glucoamylase and pullulanase differ somewhat. In certain instances, immobilization of enzymes will change the optimum pH and thermal stability characteristics. When the three enzymes are used simultaneously, it is preferred that the reaction be carried out under conditions which represent some degree of compromise relative to all three in terms of temperature and pH. While the amounts of immobilized enzymes may vary widely, typically the ratio of activity of isomerase to glucoamylase is at least 2 IGIU per GU and, in the case of the ratio of pullulanase to glucoamylase, it is at least 0.1 IU per GU. Most preferably, the ratio of activity of isomerase to glucoamylase is at least 5 IGIU per GU and, in the case of the ratio of activity of pullulanase to glucoamylase, it is at least 2 IU per GU.

There are a number of advantages associated with the use of the multicomponent enzyme system of the process for converting liquefied starch to a mixture of glucose and fructose. In the preferred embodiment, an immobilized three-enzyme system comprising immobilized glucose isomerase, glucoamylase, and pullulanase is used to simultaneously debranch, saccharify and isomerize liquefied starch. Using this multicomponent enzyme system results in substantially increased rate of conversion of liquefied starch to monosaccharides as well as increasing the overall conversion of the starch to monosaccharides.

Although it is preferred that the glucose isomerase be immobilized on a separate inert carrier from that on which glucoamylase and pullulanase are immobilized, all three enzymes may be immobilized on the same carrier. Alternatively, each enzyme may be immobilized on a separate carrier. The amounts of the various immobilized enzymes used and their activities will vary depending on a number of factors, e.g., the particular immobilized enzyme, reaction conditions, resulting end product, etc.

The conditions of pH and temperature at which the immobilized enzymes are used will also vary but should not be such as would inactivate any of the enzymes or deleteriously affect the reaction products. When all three enzymes are used in a mixed bed, the process may be carried out at a pH of 4.5 to 8 and at 5° to 60°C. The preferred pH and temperature are from 6 to 7 and from 35° to 55°C, respectively.

This process may be carried out in batch or continuous systems including mixed beds, sequential beds, single or multiple columns, batch recycling, differential reactors, fluidized beds, etc. When the three component immobilized enzyme system is used in a sequential manner, the conditions of such use can be changed to fit the optimum conditions of each particular immobilized enzyme. For instance, the pH and the temperature of the substrate may be adjusted either before or after each immobilized enzyme treatment.

In carrying out this process in a sequential manner, there are two treatment sequences possible. The first treatment would comprise contacting the liquefied starch with immobilized glucoamylase, next with immobilized pullulanase, then with immobilized glucose isomerase and then again with immobilized glucoamylase. The second treatment would comprise contacting the liquefied starch with immobilized pullulanase, then with immobilized glucoamylase, next with immobilized glucose isomerase and then again with immobilized glucoamylase.

Example: This example illustrates the use of immobilized glucoamylase, pullulanase and glucose isomerase in a stirred reactor to convert liquefied starch to a mixture comprising essentially fructose and glucose. To a solution containing glucose isomerase at a pH of 6.5 and a temperature of 40°C was added sufficient DEAE-cellulose to adsorb all the isomerase. The mixture was filtered and the filter cake washed extensively with deionized water. Co-immobilization of glucoamylase and pullulanase was carried out by dispersing sufficient DEAE-cellulose (1 g DEAE-cellulose per 200 GU and 200 IU) in a solution at a pH of 7 and ambient temperature containing glucoamylase and pullulanase (about 1 GU/ml and about 1 IU/ml, respectively). The mixture was stirred for 1 hour and filtered and the filter cake washed with 0.01 M maleate buffer (pH 6.7).

A stirred reactor was charged with 200 g of a solution of 30% (w/w) oxalate-treated, calcium-free, 29 DE liquefied starch, pH 6.8, which contained 0.005 mols per liter magnesium sulfate, 0.01 mols per liter sodium bisulfite, and 0.02% (w/v) sodium azide (as preservative). Sufficient amounts of the immobilized enzymes were then added to achieve a level of enzyme activity per g of substrate of 1.30 IU pullulanase, 1.69 GU glucoamylase, and 7.35 IGIU glucose isomerase. The reaction mixture was maintained for a number of hours at pH 6.8 and 45°C under a blanket of nitrogen while being continuously stirred. The carbohydrate composition of the mixture was determined at various intervals. The results are given in the table below.

Conversion of Liquefied Starch by Three-Component Immobilized Enzyme System

Reaction Time (hours)	..... Carbohydrate Composition in Percent*		
	Dextrose	Fructose	Total Monosaccharides
21	46.6	15.3	61.9
45	54.1	26.1	80.2
69	57.4	35.6	93.0
114	54.3	41.6	95.9
136	52.7	43.3	96.0
160	51.4	44.4	95.8
184	51.2	45.0	96.2

*Ash-free carbohydrate basis.

From the data in the above table, it can be seen that substantially complete conversion of a liquefied starch substrate to monosaccharides can be achieved.

Using Mixed Glucosidases Coupled to Same Matrix

It is known to produce maltose using a starch solution or a solution of a starch hydrolysate by subjecting it to α-1,4-glucosidase having β-amylase effect in combination with an α-1,6-glucosidase in such a way that the enzymes are used freely in the solution of starch or the hydrolysate. This method has the drawback that the end product must be separated from residual enzymes and the manufacturing of maltose must take place batchwise. The relatively expensive enzymes are further consumed in the process which entails unnecessarily high manufacturing costs.

In the improved method of *K.B. Martensson; U.S. Patent 3,996,107; Dec. 7, 1976; assigned to AB Stadex, Sweden*, the substrate is brought into contact with a matrix to which both an α-1,6-glucosidase and an α-1,4-glucosidase having β-amylase activity have been coupled. By this coupling, a two-enzyme system has been obtained. The use of this matrix bonded enzyme system eliminates the cleansing step which was formerly needed. The use of the matrix bonded system also permits the continuous manufacturing of a starch conversion product having a high maltose content by bringing the substrate solution into contact with the matrix bonded system in a suitable column. The above method will further permit recovering the enzymes throughout the process and, consequently, they can be reused as long as there is enough activity left.

The matrix materials used should be inert, i.e., resistant to enzymatic and microbial degradation under the conditions used and include cellulose derivatives, acrylic polymers, glass, dextran derivatives, agar derivatives, phenolic resins. Acrylic polymers have been found to be very suitable for matrix material.

In this process, the starting material is a solution of a substrate comprising starch or a partial starch hydrolysate and it is preferable to use as a starting material a starch hydrolysate having a DE of preferably between 5 and 10. A solution of starch hydrolysate in a concentration of between 30 and 35%, having a pH between 5.5 and 6.5, and at a temperature of between 40° and 50°C, is passed through a column containing the matrix bonded two-enzyme system. An example of the production of the matrix bonded enzyme system will be given in Example 1 below and the production of maltose, when the enzyme system is employed, will be found in Example 2.

Example 1: Manufacturing of a Matrix Bonded Two-Enzyme System, β-Amylase/Pullulanase Coupled to an Acrylic Polymer — Dry matrix grains of Bio-Gel CM 100 (a crosslinked copolymer of acrylamide-acrylic acid), 100 mg, were subjected to swelling in a 0.1 M citrate-phosphate buffer, pH 4.0 (20 ml) for 2 hours under venting in a desiccator at room temperature. After excess liquid has been removed on a sintered glass filter, the swollen gel is mixed with a solution of β-amylase protein (1.71 mg) in 0.1 M citrate-phosphate buffer, pH 4.0 (10 ml) in a 20-ml test tube at 4°C. The adsorption of the enzyme on the gel took place under careful shaking for 30 minutes. Then, 150 mg of 1-cyclohexyl-3-(2-morpholinoethyl)-carbodiimide-metho-p-toluenesulfonate (CMDI) were added and the shaking was continued for 18 hours.

The grains of enzyme-gel were then filtered away and washed on a sintered glass filter by means of distilled water (200 ml) and 0.1 M citrate-phosphate buffer, pH 4.2 (50 ml). To a solution of pullulanase protein (9.3 mg) in 0.1 M citrate-phosphate buffer, pH 4.2 (10 ml), in a 20 ml test tube, pullulane (60 mg) was added at 4°C. After 5 minutes, the newly washed β-amylase-gel grains were mixed with the solution. The adsorption process was performed as described above followed by the admixture of a new portion of CMDI (50 mg). The coupling continued under careful shaking for another 18 hours at 4°C. The enzyme-gel grains were then filtered, washed and stirred for 30 minutes in distilled water (200 ml) and for 30 minutes in 0.01 M phosphate buffer, pH 7.0 (600 ml).

The coupling was performed in two consecutive steps owing to the different optimal coupling conditions of the enzymes, β-amylase being coupled in the first step due to its greater stability to high concentrations of CMDI. The following table gives the results of the two couplings. The matrix contained 323 units of β-amylase activity and 49 units of pullulanase activity per g of dry polymer. One unit of β-amylase activity is defined as the quantity of enzyme that liberates 1 μmol of maltose from soluble starch per minute at a pH of 4.8 and at 35°C. One unit of pullulanase activity corresponds to the quantity of enzyme producing 1 μmol of maltotriose per minute at pH 5.0 and at 30°C when pullulane is used as a starting material.

Covalent Coupling of β-Amylase and Pullulanase to an Inert Acrylic Copolymer

	Bonded Protein (mg/g dry protein)		Coupling Yields* (%)		β-Amylase Activity		Pullulanase Activity	
	β-Amylase	Pullulanase	β-Amylase	Pullulanase	Units per mg Bonded Protein	Residual Enzyme Activity (%)	Units per mg Bonded Protein	Residual Enzyme Activity (%)
Bio-Gel CM-100	6.8	35.3	40	38	47.5	22	1.39	32

*Based on the quantity of added enzymes.

Example 2: The Use of a Matrix Bonded Enzyme System for Manufacturing Maltose — A gelatinized substrate solution of an α-amylase hydrolyzed potato starch (DE 7.0) having a dry solid content of 15% in 0.01 M phosphate buffer of pH 6.0 is set at 45°C and is passed through a column comprising a bed of the matrix bonded enzyme system described in Example 1. The column is temperature regulated to 45°C. The bed has a height of 30 cm and a diameter of 2.4 cm and the quantity of matrix amounted to 1.35 g of dry gel. The analysis of the starch hydrolysate rich in maltose, obtained at various flow rates, appears in the table below:

Flow Rate (ml/hr)	Glucose	Maltose	Maltotriose	Other Sugar
	(% by wt)			
13.8	0.3	70.0	15.8	13.9
34.5	0.3	65.9	14.2	19.6
69.0	0.2	60.5	13.4	25.9

IMMOBILIZED ISOMERASE FOR PRODUCTION OF FRUCTOSE

Fructose, also known as levulose, is a ketose monosaccharide which occurs naturally in a large number of fruits. As fructose is sweeter than sucrose and less costly, it finds a large market in the preparation of processed foods and drinks.

Isomerase on Porous Alumina

R.A. Messing; U.S. Patent 3,868,304; February 25, 1975; assigned to Corning Glass Works has converted glucose to fructose by incubating a glucose-containing solution with glucose isomerase adsorbed within the pores of a porous alumina carrier. The carrier must have an average pore diameter between 100 and 1000 A to permit maximum loading of the enzyme. The lower limit of 100 A is determined by the largest dimension of the glucose isomerase molecule which is at least approximately 100 A. It may be necessary to comminute the porous alumina to a desired mesh size which will be compatible with the reactor to be used with the immobilized glucose isomerase. The preferred carrier particle size is between 4 and 200 mesh, United States Standard Sieve. An especially preferred average carrier is between 25 and 80 mesh and carriers having an average particle size within that range were used in the experiments.

After the porous alumina bodies have been ground and sieved to the desired particle size, they are hydrated or preconditioned with a suitable buffer or salt system; hydration at a pH of 7.0 is preferred. The buffer system external to the pores is then removed but the porous alumina bodies are kept wet prior to the adsorption step. Further details of the process are given in the example below.

Example: A crude glucose isomerase of about 444 IGIU/g derived from a Streptomyces organism was used for the adsorption. About 5 g of the glucose isomerase was added to 28.7 ml of 0.1 M magnesium acetate solution in a 50 ml beaker. The slurry was stirred for 25 minutes at room temperature and then filtered through filter paper. The residue was washed with 14.3 ml of 0.1 M magnesium acetate followed by washes with 14 ml of 0.5 M $NaHCO_3$ and 4.3 ml of 0.5 M $NaHCO_3$. The washes were collected directly into the original enzyme filter. The total volume of the enzyme-wash solution (filtrate) was 50 ml.

500 mg of porous alumina bodies were placed in a 50 ml round bottom flask. 10 ml of the glucose isomerase solution was then added to the flask. The flask was attached to a rotary evaporator. Vacuum was applied to the apparatus and the flask was rotated in a bath maintained at 30° to 45°C for 25 minutes. An additional 10 ml of glucose isomerase solution was then added to the flask, and evaporation was continued over the next 35 minutes under the same conditions. This addition and evaporation was carried out over a 4-hour period under the same conditions. The procedure was repeated two more times with separate 10 ml glucose isomerase solutions.

A final 7 ml aliquot of glucose isomerase solution was then added to the flask and evaporation was continued for an additional hour and 10 minutes at 45°C. The flask and contents were removed from the apparatus and placed in a cold room over the weekend. A total of 47 ml of glucose isomerase solution had been added to the alumina. 50 ml of buffer (0.01 M sodium maleate, pH 6.8 to 6.9, containing 0.001 M cobalt chloride, 0.005 M magnesium sulfate) was added to the composite and the sample was extracted over the next hour at room temperature. The extract, volume 50 ml, was saved for assay.

The composite was then washed with 200 ml of water, followed by 10 ml of 0.5 M sodium chloride. The final wash was performed over a fritted glass funnel with 50 ml of water. The composite was then transferred to a 50 ml Erlenmeyer flask and stored in buffer at room temperature. The enzyme extract assay (50 ml) was 39.8 IGIU/ml (enzyme activity recovery in extract was 90%). The composite average assay value was 130 IGIU/500 mg sample (enzyme activity recovery on composite was 6%). The average loading value was 260 IGIU/g.

To show the long term stability and enzymatic half life of the enzyme composite, a sample was prepared for use in a flow-through column through which a glucose-containing solution flowed under conditions encountered in an industrial application. Alumina bodies (175 A average pore size, 25 to 60 mesh) were pretreated as follows. 11 g of porous alumina were transferred to a 100 ml glass stoppered cylinder. 100 ml of 0.05 M magnesium acetate, 0.01 M cobalt acetate, pH 7.5, were added, the cylinder stoppered, the contents gently mixed by inversion, and placed in a 60°C bath.

After 15 minutes of reaction, the cylinder was inverted and the fluid decanted. 100 ml of fresh magnesium-cobalt acetate (described above) was added to the cylinder, inverted, and allowed to stand at room temperature for 2½ hours. A crude glucose isomerase solution from a streptomyces organism and consisting of 590 IGIU/ml in 0.6 M saturated ammonium sulfate was purified for reaction with the alumina as follows. To 40 ml of the glucose isomerase solution an additional 1.4 g of ammonium sulfate was added to precipitate the enzyme and the slurry was stirred at room temperature for 20 minutes and then centrifuged at 16,000 rpm at 2°C for 30 minutes. The supernatant fluid was decanted and discarded.

12 ml of the magnesium-cobalt acetate solution (described above) was added to the precipitate and stirred, and then 3 ml of 0.5 M sodium bicarbonate was added to dissolve the enzyme. The solution was placed in a 60°C water bath for 15 minutes. After removal from the bath, the solution was centrifuged for 15 minutes at 16,000 rpm and 2°C. The clear supernatant enzyme (volume 28 ml) was found to have a pH of 7.5. If no activity was lost during the purification procedure, this solution should have contained 23,600 IGIU.

The magnesium-cobalt acetate solution was then decanted from the porous alumina bodies. The enzyme solution, 28 ml, was added to the porous alumina in the cylinder. The cylinder was stoppered, mixed by inversion, and placed in a 60°C water bath. The enzyme was permitted to react with the porous alumina (11 g) over a 2 hour and 30 minute interval at 60°C with the cylinder inverted every 15 minutes. After removal from the 60°C bath, reaction was continued at room temperature with inversion at 30 minute intervals over the next 2 hours. The reaction was continued without inversion overnight at room temperature. The enzyme solution was decanted (volume 28.5 ml, pH 7.1) and saved for assay.

The composite was washed with 60 ml of distilled water, followed by 40 ml of 0.5 M sodium chloride, and finally by 40 ml of the magnesium-cobalt acetate solution. Three samples (totaling 1 g) were removed from the batch for the assay determinations. The remaining 10 g were transferred to a column, thermostated at 60°C. The column was fed with a solution containing 50% glucose and 0.005 M magnesium sulfate, buffered with sodium sulfite to a pH of 7.7 to 8.0.

During the initial 26 hours, 0.001 M cobalt chloride was incorporated into the feed of the column. After 26 hours, the cobalt was no longer included in the feed and the only activator present was magnesium ions during the remainder of the column life. The initial flow rate of the column was approximately 190 ml per hour. The fructose conversion was maintained between 80 and 85% of theoretical by reducing the flow rate over various intervals of time.

The column was run for approximately 31 days. At this time, due to a shortage of feed, the column dried out completely and was dry for approximately 15 hours. An approximate value for the composite half-life was obtained by calculating the amount of product collected and the conversion of the product in the collecting vessel. The projected half-life for the composite was found to be about 42 days.

An assay of the product prior to use indicated a loading of 381 IGIU/g. This represents an activity recovery of 17.8% of the total IGIUs exposed to the carrier. An assay of the reacted enzyme solution (28.5 ml) indicated 161 IGIU/ml, an activity recovery in the reacted enzyme solution of about 19.4%. The pH of the column had been maintained between 7.4 and 7.8 by appropriate adjustment of the feed. Total column loading was 3,810 IGIU.

Isomerase Sorbed on $MgCO_3$

R.E. Heady and W.A. Jacaway, Jr.; U.S. Patents 3,847,740; November 12, 1974; and 3,941,655; March 2, 1976; both assigned to CPC International, Inc. have also disclosed an immobilized xylose isomerase preparation having improved retained isomerase activity and flow rate characteristics, as compared to former enzyme preparations. It has been found that cell-free, soluble xylose isomerase enzymes, when contacted with basic magnesium carbonate, will become bound or sorbed, to provide an immobilized enzyme which has a high effective isomerase activity.

Other related materials, such as calcium carbonate, do not provide a highly active and stable xylose isomerase. Also, intracellular xylose isomerase does not become sorbed onto basic magnesium carbonate. The immobilized xylose isomerase of this process is characterized as having a very high enzyme efficiency, whereby the sugar contact time necessary to produce a fructose-bearing syrup of at least 45% by weight fructose from a dextrose-containing solution is less than 2 hours.

Due to the lower contact time needed by the improved activity of the bound enzymes, the resulting fructose-bearing syrups have lighter color, lower organic acid, and lower psicose content, that is, generally less than 1%, and quite often less than 0.3% by weight, dry basis, psicose. These advantages are quite significant from the standpoint of capital investment and inventory in commercializing a process for enzymatically preparing fructose-bearing syrups from dextrose-containing solutions.

The stabilized xylose isomerase enzyme preparation is preferably prepared by contacting cell-free xylose isomerase with particulate basic magnesium carbonate. The particulate basic magnesium carbonate may be either in the form of a powder or a granular structure. However, the granular structure is preferred from the standpoint of flow properties when used in a deep bed converter.

The stabilized enzyme preparation can be recovered by conventional means, such as by filtration and the like. Alternatively, the particulate basic magnesium carbonate may be first placed in a converter which is later to be used in the enzymatic isomerization reaction and, secondly, a solution of cell-free xylose isomerase may then be pumped through the column until no more enzyme is sorbed from the solution by the particulate basic magnesium carbonate. Following the preparation of the stabilized enzyme preparation, the converter is ready for use simply by supplying to the converter a dextrose-containing solution.

The xylose isomerase enzyme preparation used is in its soluble form; in other words, the xylose isomerase is freed from the microbial cell where it is formed. The enzyme may be released from its cellular material by any conventional means, and the enzyme in the crude material may be then sorbed on the basic magnesium carbonate. The xylose isomerase becomes bound to the basic magnesium carbonate.

The immobilized enzyme preparation can then be recovered by filtration, centrifugation and the like, and then washed with a buffer. The filtered and washed immobilized enzyme composition can be used in its wet form, as is, or the cake can be dried by conventional drying techniques useful with enzyme preparations, such as by cell, rotary drum or spray-drying or freeze-drying techniques. Superior results are obtained, however, when the enzyme is purified before it is sorbed. The immobilized xylose isomerase preparation of this process is characterized as having at least about 100 units of effective isomerase activity per g of immobilized enzyme preparation, dry basis (that is, 100 units per g effective isomerase activity), preferably at least about 175 units per g. Another striking characteristic of the immobilized and stabilized enzyme preparation is its long half-life. This enzyme preparation is characterized as being capable of producing 45% fructose, DB, at flow rates of greater than 0.5 bed volume per hour with an enzyme half-life in excess of 15 days.

The xylose isomerase enzyme is preferably derived from a microorganism belonging to the genus Streptomyces, and more preferably a microorganism strain that is selected from the group of mutant strains consisting of *S. olivochromogenes*, ATCC No. 21713, *S. olivochromogenes*, ATCC No. 21714, and *S. olivochromogenes*, ATCC No. 21715.

Example: Continuous Enzymatic Conversion of Dextrose to Fructose Using Immobilized Xylose Isomerase – A granular basic magnesium carbonate material having a particle size of –12 to +20 (55.8 g) was placed in a column having a volume of 73.3 ml. A substantially pure, soluble, cell-free xylose isomerase enzyme derived from *S. olivochromogenes* ATCC No. 21713 in a 0.01 M $MgSO_4$ solution, having a 50 units per ml activity, was contacted with granular basic magnesium carbonate to load the column with 5 million units per ft^3 of enzyme.

The immobilized xylose isomerase enzyme had more than 100 units per g of effective activity per g of immobilized enzyme. An aqueous dextrose feed liquor was then supplied to the column at 27°Bé (50% ds). The feed liquor contained $MgCl_2$ (0.005 M with respect to $MgCl_2$). The pH of the feed was adjusted to the range from 8.4 to 8.8 with 4 N NaOH at 58°C.

The column containing the immobilized enzyme preparation was maintained at a temperature of 58°C. The residence time in the column was less than 1 hour. The initial bed volume per hour for the production of 45% fructose, dry basis (BVH_{45} fructose) was about one. The effluent syrup contained about 45% by weight fructose, dry basis, and less than about 0.3% by weight, dry basis, psicose.

The following is an analysis of the pH of the feed syrup and effluent syrup after a prolonged continuous conversion campaign.

Hours of Operation	Feed Syrup pH at 25°C	Effluent pH at 25°C
261	8.60	8.60

The above experiment was repeated, except that the xylose isomerase was bound on an alumina carrier. The same dextrose feed liquor, preadjusted to a pH of 8.3 to 8.4 with 4 N NaOH, was pumped through the column, which was maintained at a temperature of 58°C at a rate to provide 45% by weight fructose, dry basis, in the effluent. The residence time was less than 1 hour. The following is an analysis of the pH of the feed syrup and effluent syrup after a prolonged continuous conversion campaign.

Hours of Operation	Feed Syrup pH at 25°C	Effluent pH at 25°C
16	8.3	8.4
376	8.15	8.25
428	8.4	8.3

In the operation of a column where the residence time may exceed about 4 hours, it is particularly preferred to feed an alkaline material such as sodium hydroxide or $Mg(OH)_2$ solutions or a combination of the two into the column at different elevations. A suitable means for accomplishing this is to provide alkaline inlets at one or more places along the column, so that the pH of the reaction solution is maintained at a pH of at least about 7.5.

In another example of the process, the pH of the enzymatic isomerase solution can be programmed by the use of one or more alkaline inlets along the column. In such a technique, the initial pH of the feed liquor can be 7.5 or greater, preferably from 7.5 to 8.2. By introducing alkaline materials such as sodium hydroxide or $Mg(OH)_2$ or a combination of the two into the reaction solution via inlets along the column, the pH of the reaction solution can be conveniently increased to 9.5. Preferably, the pH will be increased by at least 0.1 from supply liquor to effluent, although more beneficial results in terms of stability of the basic magnesium carrier and quality of the isomerase are realized when the pH is increased by at least 0.2.

Flocculated Isomerase

Microbial cell material having a glucose-isomerizing enzyme has been subjected to flocculation conditions to produce a flocculated aggregate containing the cell material. In the process developed by *M.E. Long; U.S. Patent 4,060,456; Nov. 29, 1977; assigned to R.J. Reynolds Tobacco Company*, glucose syrups are brought into contact with the flocculated aggregates and a portion of the glucose is converted to fructose.

Briefly, this process involves the use of a polyelectrolyte as an aggregate-forming support material for microbial cellular material containing isomerase. The enzyme-containing aggregates are subsequently dried and reduced to a relatively uniform particle size. The material thus obtained is packed into a column or formed into some other suitable reactor bed through which a glucose solution is passed to effect isomerization. Depending on the nature of the feed stock, further treatment of the emerging glucose-fructose solution (e.g., ion exchange resin clean-up) may be carried out if desired.

The polyelectrolytes used in this process are preferably water-soluble polymeric substances containing monomeric units which possess polar or ionizable groups. They are generally classified into three main categories: anionic, cationic and nonionic. Anionic polyelectrolytes usually contain carboxylic, sulfonic or phosphonic acid groups and examples of such materials include polyacrylic acid, polystyrenesulfonic acid, polyvinylphosphonic acid, carboxymethylcellulose, alginic acid and pectic acid. Most cationic polyelectrolytes involve the use of quaternary ammonium, sulfonium or phosphonium groups including the protonated forms of polyamines such as polyethylenimine and polyvinylpyridine. Nonionic polyelectrolytes are exemplified by polyacrylamide and polyvinyl alcohol. Among the specific polyelectrolytes which are effective for this process are Catfloc, Delfloc 40 and Delfloc 763, Dow XD-1923, Natron 86, Primafloc A-10 and Primafloc C-7

The formation of the aggregate comprising the polyelectrolyte material and the enzyme-containing material is preferably effected by combining solutions or suspensions of the materials.

The source of glucose isomerase-containing material used is not critical. A number of microorganisms are known as being capable of elaborating glucose isomerase including species of the genera *Arthrobacter, Lactobacillus, Pasteurella, Aerobacter, Streptomyces* and *Leuconostoc*. Species of *Arthrobacter* and *Streptomyces* are particularly preferred. The microorganism is cultured in an appropriate nutrient medium to induce formation of isomerase. At the conclusion of the fermentation period, the isomerase-containing cell material is recovered and immobilized.

For glucose isomerase-producing microorganisms, the cationic polyelectrolytes are generally most effective and the amount of polyelectrolyte required relative to the wet weight of the enzyme-containing material is preferably in the range of 0.5 to 50% by weight. The temperature of the flocculating medium is maintained between 10° and 40°C and the pH is maintained at 5.0 to 9.5. The polyelectrolyte is preferably added in the form of a 1 to 2% aqueous solution and flocculation of the enzyme-containing material usually begins immediately, being complete within a few minutes.

Following treatment with the polyelectrolyte, the aggregate containing the cell material is removed from the flocculating medium by procedures such as centrifugation or filtration. The collected aggregate is then subjected to an optional extrusion step to convert the material into a shape and size which permit more uniform drying and/or more uniform packing into a bed for glucose isomerization.

The isomerase-containing material produced by the polyelectrolyte treatment is very effective for converting glucose to fructose. This conversion is conveniently carried out continuously by passing a glucose syrup through a suitable bed of the material although a batch-type process may be used if desired. For a continuous process, a suspension of the isomerase-containing material is first prepared by adding the material to a solution containing a magnesium salt and a suitable buffer which will maintain the solution at a pH of 8 to 9. The magnesium salt serves to enhance the isomerase activity and the buffering agent tends to remove any acidic materials which could cause denaturation of the enzyme. It has been found that a solution which is 0.05 M with respect to sodium bicarbonate and 0.01 M with respect to magnesium chloride is satisfactory for preparing a suspension of the isomerase-containing material. Further details for the isomerization process are given in the following example.

Example: Arthrobacter nov. sp. NRRL B-3728 was cultivated to give 400 gal of culture broth having a pH of 5.7. This broth was gently agitated and to it were added 53.3 gal of a 1.5% solution of Primafloc C-7 (previously adjusted to pH 5.0) followed immediately by 53.3 gal of a 1.5% solution of Primafloc A-10. Addition of the flocculating agents was effected over a period of 15 minutes and flocculation of the cells was essentially complete within 20 minutes. Centrifugation yielded 125 lb of wet, flocculated cells having a moisture content of approximately 80%. This wet floc was then extruded through a 30-mesh screen and the extruded floc was subsequently dried at 55° to 60°C by means of a fluidized bed dryer giving 30 lb of material having a moisture content of slightly less than 10%. The dried, flocculated cells retained their isomerase activity which activity was undiminished even after storage for 7 months at room temperature.

The dried, flocculated cell material was milled and sieved to give particles in the 16 to 20 mesh range (U.S. Bureau of Standards). A 3-g portion of the 16 to 20 mesh material was suspended in an aqueous solution that was 0.05 M with respect to sodium bicarbonate and 0.01 M with respect to magnesium chloride. After the particles had swelled to the maximum extent, they were packed into a 1" diameter glass column and the floc particles were washed by passing additional quantities of the sodium bicarbonate-magnesium chloride solution through the packed bed equivalent to approximately three times the volume occupied by the packed bed.

The packed column was heated to 60°C and a 2.0 M dextrose solution containing 0.004 M magnesium chloride (previously adjusted to pH 8.5 with sodium hydroxide) was passed through the column at a constant flow rate of 950 ml per day. The syrup emerging from the column was analyzed daily for fructose content. The degree of glucose to fructose conversion obtained initially was 35.9% and, after 28 days, it had declined to 32.8%.

Isomerase Entrapped in Acryloyl Polymers

I. Chibata, T. Tosa, and T. Sato; U.S. Patent 4,081,327; March 28, 1978; assigned to Tanabe Seiyaku Co., Ltd., Japan have developed an immobilized microorganism which affords high activity of glucose isomerase for a long period of time. In this process, D-fructose can be prepared by polymerizing at least one acryloyl monomer in an aqueous suspension containing a glucose-isomerase-producing microorganism to produce an immobilized glucose-isomerase-producing microorganism, and subjecting the immobilized glucose-isomerase-producing microorganism to enzymatic reaction with D-glucose.

Preferred examples of glucose-isomerase-producing microorganisms which are used in this process include *Streptomyces griseus* IFO 3430, *Streptomyces griseus* IFO 3356, *Streptomyces aureus* IFO 3175, *Streptomyces olivaceus* IFO 3409 and *Bacillus coagulans* IFO 12714.

The polymerization reaction can be carried out using a polymerization initiator and a polymerization accelerator. Potassium persulfate, ammonium persulfate, vitamin B_2 and methylene blue are suitable as polymerization initiators. On the other hand, β-(dimethylamino)propionitrile and N,N,N',N'-tetramethylethylenediamine are used as polymerization accelerators. Suitable amount of the polymerization initiator which is added to the aqueous suspension of the glucose isomerase-producing microorganism is 5 to 50 mg/g of the acryloyl monomer or monomers. Suitable amount of the polymerization accelerator to be added is 10 to 50 mg/g of the acryloyl monomer or monomers. It is preferred to carry out the reaction at 20° to 40°C. The reaction may be completed within 10 to 60 minutes.

The acryloyl monomers suitable for use include acryloylamide, N,N'-lower alkylenebisacryloylamide, bis(acryloylamidomethyl) ether and N,N'-diacryloylethyleneurea. It is suitable to entrap the glucose-isomerase-producing microorganism with a polymer obtained from one or two monomers mentioned above, particularly with a copolymer of acryloylamide and an acryloyl monomer selected from N,N'-lower alkylenebisacryloylamide, bis(acryloylamidomethyl) ether and N,N'-diacryloylethyleneurea, or with a homopolymer of N,N'-lower alkylenebisacryloylamide, bis(acryloylamidomethyl) ether or N,N'-diacryloylethyleneurea.

N,N'-methylenebisacryloylamide and N,N'-propylenebisacryloylamide are preferably used as the N,N'-lower alkylenebisacryloylamide. Suitable amount of N,N'-lower alkylenebisacryloylamide, bis(acryloylamidomethyl) ether or N,N'-diacryloylethyleneurea which is used to copolymerize with acryloylamide is 25 to 80 mg/g of acryloylamide. After the polymerization reaction is completed, the resultant immobilized glucose-isomerase-producing microorganism is granulated by passing it through a sieve to form granules of 3 to 4 mm in diameter.

D-fructose can be prepared from D-glucose by enzymatic reaction of the immobilized glucose-isomerase-producing microorganism. The enzymatic reaction is carried out at 40° to 70°C and at pH 7 to 8.5. The enzymatic activity can be stabilized effectively by adding magnesium and cobaltous ions to the reaction solution. Suitable amount of magnesium ion added to the reaction solution is 3 to 10 mmol/l of the reaction solution. Suitable amount of cobaltous ion added to the reaction solution is 0.5 to 5 mmol/l of the reaction solution.

In the following example, the potency of a microorganism or immobilized microorganism, which afforded 1 mg of D-fructose on the reaction of the microorganism or immobilized microorganism with D-glucose at pH 8.0 to 60°C for an hour, was taken as 1 unit.

Example: An aqueous nutrient medium (pH 7.0) containing the following ingredients is prepared:

	%, w/v
Peptone	1
Yeast extract	0.25
Meat extract	0.5
D-glucose	0.3
D-xylose	0.7
Magnesium sulfate heptahydrate	0.05
Cobaltous chloride hexahydrate	0.024
Sodium chloride	0.5

Streptomyces griseus IFO 3430 is inoculated into 200 ml of the medium. The medium is cultivated at 30°C for 72 hours under shaking. The medium is then centrifuged. The microbial cells collected show the glucose isomerase activity of 30 u/g. 17 g of the microbial cells are suspended in 68 ml of a physiological saline solution. 12.75 g of acryloylamide, 680 mg of N,N'-methylenebisacryloylamide, 7.5 ml of 5% β-(dimethylamino)propionitrile and 7.5 ml of 2.5% potassium persulfate are added to the suspension.

Then, the suspension is allowed to stand at 37°C for 30 minutes. After the reaction is completed, the stiff gel thus obtained is granulated by passing it through a sieve to form granules of 3 mm in diameter. Then, the granules are washed with 1,700 ml of a physiological saline solution. 170 ml of an immobilized preparation of *Streptomyces griseus* IFO 3430 are obtained. Glucose isomerase activity: 25 u/ml.

170 ml of the immobilized preparation of *Streptomyces griseus* IFO 3430 are charged into a 4 x 13.5 cm column. 200 ml of an aqueous 40% D-glucose solution (pH 8.0) containing 5 mM concentration of magnesium ion and 1 mM concentration of cobaltous ion are passed through the column at 60°C at the flow rate of 60 ml/hr. 200 ml of an aqueous 0.04 M sodium tetraborate solution are added to 200 ml of the effluent. Then, the mixture is passed through the column of an ion exchange resin [Dowex 1 x 2 (H^+ type)] which was previously washed with an aqueous 0.02 M sodium tetraborate solution. The effluent is then passed through the column of an ion exchange resin [Dowex 50 x 8 (H^+ type)]. The effluent is concentrated. 13.9 g of a syrup of D-fructose are obtained. The D-fructose content in the syrup is 90%.

ACTIVATION OF IMMOBILIZED ISOMERASE

Use of Cobalt and/or Magnesium Ions

A. Bouniot and M. Guerineau; U.S. Patent 3,990,943; November 9, 1976; assigned to Rhone-Poulenc SA, France have provided a process for isomerizing glucose to levulose which comprises treating an aqueous solution of glucose with glucose isomerase in the presence of an activating cation (magnesium or magnesium and cobalt) which is fixed to a support consisting of a cation exchanger.

The cation exchanger is preferably a resin carrying sulfonic acid groups or carboxylic acid groups. The primary activating ion is the magnesium cation. The resin, initially in the acid form, is preferably saturated beforehand with magnesium hydroxide. When a Mg^{++} and Co^{++} ion-containing resin is to be used, once the resin has been saturated with magnesium hydroxide, it is contacted with a cobalt-ion-containing solution which may contain Co^{++} cations or Co^{++} and Mg^{++} cations.

It appears that the surface of the resin carries $-SO_3Mg(OH)$ groups (in the case of the sulfonic acid resins) or $-COO-Mg(OH)$ groups (in the case of the carboxylic acid resins). The excess magnesium hydroxide is then removed by washing the resin with water saturated with magnesium carbonate to avoid eluting the fixed Mg^{++} ions. The resulting resin is then preferably drained. The enzyme, or the remains of cells of a microorganism containing the enzyme, is then mixed with the treated ion exchanger. The resulting composition may be placed in a column through which the aqueous solution of glucose is passed continuously. The concentration of this solution of glucose preferably does not exceed 60% by weight. The isomerization reaction is carried out at 50° to 75°C. The pH of the solution on issuing from the column is 5.0 to 6.5. The solution of glucose introduced can itself contain activating cations (especially Mg^{++} and Co^{++}) in the amounts usually used for this reaction, namely 2 to 200 ppm of Mg^{++} cations and 1 to 100 ppm of Co^{++} cations. The presence of Mg^{++} ions in the glucose solution limits the elution of the active Mg^{++} ions fixed to the support.

The degree of conversion of glucose in the issuing solution depends on the flow rate of the solution through the column, the limit given by the chemical equilibrium being of the order of 50% for the ratio glucose/(glucose + levulose).

Example 1: 30 ml of a cation exchange resin possessing carboxylic acid groups (Amberlite IRC 50 resin) are activated by treatment with hydrochloric acid, and then washed with distilled water. The resin is then contacted for 1 hour with 60 ml of a 15% solution of milk of magnesia (magnesium hydroxide), after which the excess milk of magnesia is removed by washing the resin copiously with water saturated with magnesium carbonate. The resin obtained is drained and mixed with 565 mg of glucose isomerase (Nagase), corresponding to 1,960 GIU. The mixture is introduced into a vertical glass column of internal diameter 25 mm, which is surrounded by a jacket through which water thermostatically controlled at 60°C flows.

3 ml/hr of an aqueous solution containing 30% by weight of glucose, which also contains 0.26 g of $CoCl_2$/l and 0.10 g of $MgCO_3$/l, are passed through the column filled with treated resin and enzyme. The pH of the solution as it issues from the column is approximately 5.7 and the degree of conversion of glucose to levulose is 43%. After 500 hours of operation, the results are the same. After 1,500 hours of operation under the same conditions, the degree of conversion is still 22%.

When no enzyme was mixed with the resin, only a 1% conversion of glucose to levulose was obtained. The pH of the solution as it issues from the column is 5.5. It is apparent that the process is an enzymatic reaction and not a chemical reaction.

Example 2: Example 1 was repeated except that 30 ml of a cation exchange resin possessing sulfonic acid groups (Allassion CS) activated by treatment with hydrochloric acid were used as the support. From the beginning of the experiment onwards, a degree of conversion of glucose to levulose of 45% is achieved. A 43% conversion is still obtained after 300 hours of operation. Thereafter, the reaction rate decreases more quickly than in Example 1. The pH of the solution as it issues from the reaction column is between 5 and 5.5.

Use of Iron and Magnesium Salts

Almost all of the glucose isomerases found in microbial cells require metal ion which may vary depending upon the origin of isomerase. It has been known that glucose isomerase found in *Streptomyces* requires magnesium ion and that activity and resistance to heat are enhanced in the presence of additional cobaltous ion. In this connection, in commercial practice of fructose production from glucose, magnesium and cobaltous ions have been present in the reaction mixture. However, various heavy metal ions including cobaltous ion are not allowed as additives to foodstuff in many countries and, therefore, the use of such cobaltous ion in the production of fructose should be avoided.

Y. Fujita, A. Matsumoto, H. Ishikawa, T. Hishida, H. Kato, and H. Takamisawa; U.S. Patent 4,008,124; February 15, 1977; assigned to Mitsubishi Chemical Industries Ltd., and Seikagaku Kogyo Co., Ltd., Japan have found that the activity of glucose isomerase is maintained for a long time without using cobaltous ion by addition of a trace amount of water-soluble iron salt together with magnesium salt to a glucose isomerization mixture.

The examples of water-soluble iron salts which may be used in this process include ferrous sulfate, ferric sulfate, ferrous chloride, ferric chloride, ammonium ferric sulfate, ammonium ferrous sulfate and mixtures thereof. The concentration of iron ion to be maintained in the reaction mixture ranges from 0.005 to 5 mmol/l and preferably 0.02 to 0.5 mmol/l.

Examples of water-soluble magnesium salts which may be used together with the iron salt in this process include magnesium sulfate, magnesium chloride, magnesium sulfite, magnesium carbonate and mixtures thereof. The concentration of magnesium salt ranges from 2 to 20 mmol/l.

This process is adaptable for a continuous process using insolubilized glucose isomerase in which an aqueous glucose solution is passed through a column into which glucose isomerase adsorbed on a carrier is packed to effect conversion of glucose into fructose. The carrier suitable for the purpose of insolubilizing glucose isomerase includes: (1) a macroporous anion exchange resin, the matrix of which is styrene-divinyl copolymer and anion exchange group is a quarternary ammonium type; and (2) a cyanogen halide derivative of crosslinked agar in particle form.

The latter is produced by treating agar with a crosslinking agent such as epichlorohydrin under alkaline condition to introduce crosslinkages followed by heating at a high temperature and washing with hot water to remove hot-water-soluble materials, then treating the crosslinked agar with an aqueous solution of cyanogen halide such as cyanogen bromide under alkaline conditions. It has been observed that by using this process the activity of such insolubilized isomerase is maintained for a long period and is satisfactory for use on a commercial basis.

An aqueous glucose solution is conveniently subjected to a continuous process at a pH ranging from 5.5 to 9, a concentration of 20 to 70% by weight and a temperature of 40° to 80°C and such solution is passed through a column at a space velocity of 0.1 to 20 hr^{-1}.

Example 1: Glucose isomerase was extracted by a conventional ultrasonic treatment from Glucose Isomerase Nagase which consisted of cells of strain belonging to *Streptomyces phaeochromogenes*, intracellularly fixed by heat treatment and available from Nagase Sangyo Kabushiki Kaisha, Osaka, Japan. The activity titer of such extract was 200 U/ml. An insolubilized glucose isomerase was prepared by mixing 75 ml of such extract with 15 ml in wet state of a macroporous strong basic anion exchange resin (Dianion HPA, Mitsubishi Chemical Industries, Ltd., a styrene-divinyl benzene copolymer having a quarternary ammonium group) at room temperature for 6 hours to effect adsorption of the isomerase on the anion exchange resin, the degree of adsorption being measured as more than 99%. The insolubilized glucose isomerase thus prepared was packed into a jacketed column with an 8 mm inner diameter.

A 3 M aqueous glucose solution containing 5 mmol/l magnesium sulfate and 1 mmol/l of ferrous sulfate as ion concentrations, respectively, was passed through such column maintained at a temperature of 67°C to effect continuous isomerization, while passing warmed water through into the jacket. The degrees of activity retention after 10 and 20 days are given in the table below.

Examples 2 through 6: Procedures similar to those of Example 1 were repeated excepting that instead of ferrous sulfate various metal salts at given concentrations listed in the table below were used. The degrees of activity retention of the glucose isomerase after 10 and 20 days are also shown in the table from which superior results derived from the addition of iron salts to other salts are proved in respect that prolonged isomerase activity is maintained.

Ex. No.	Compound	Concentration of the Ion (mmol/l)	Percent of Activity Retention 10 Days	Percent of Activity Retention 20 Days
1	Ferrous chloride	1.0	70	45
2	Ferric chloride	0.5	62	37
3	Iron tartarate	1.0	68	40
4	–	–	33	11
5	Cobaltous chloride	1.0	55	25
6	Calcium chloride	1.0	30	13

Magnesium-Iron-Thiol Activator System

It is also the object of the process disclosed by *T.L. Hurst; U.S. Patent 4,113,565; September 12, 1978; assigned to A.E. Staley Manufacturing Company* to provide an acceptable level of enzymatic conversion to fructose without requiring the addition of cobalt ions to a dextrose feed syrup.

Iron ions may be effectively used as a metal activator or co-metal activator in glucose isomerization when used in conjunction with thiol activators. The thiol activators are characterized as being able to cleave disulfide linkages and include reducing reagents such as water-soluble $SO_3^=$ producing metal salts, ascorbic acid, thiocyanates, thioglycolates, etc. The thiol activator-iron ion activating system is especially effective when used in conjunction with at least one other metal ion activator such as magnesium, manganese or cobalt ions. Effective enzymatic glucose isomerization, without using cobalt ions, is achievable by a magnesium-iron-thiol activator-isomerase system.

Water-insoluble or water-soluble glucose isomerases may be used. Since immobilized glucose isomerases are known to be more stable against enzymatic deterioration and generally possess a significantly longer half-life than water-soluble or unbound glucose isomerases, immobilized glucose isomerases are preferably used. Conventional, immobilized glucose isomerases are useful for this purpose. Iron ions are essential to activate the glucose isomerase. Either ferrous ions or ferric ions or mixtures thereof may be used for this purpose. Suitable iron ion sources include the water-soluble, organic and inorganic acid salts of ferric or ferrous compounds. Water-soluble iron salts include ferrous or ferric acetate, ammonium chlorides, ammonium sulfate, ammonium oxalates, hyposulfite, manganese chloride and citrate, sulfate, magnesium citrate, magnesium lactate, magnesium sulfate, sulfite, thiocyanate, tartrate, chloride, mixtures thereof and the like. Iron salts containing magnesium and/or thiol activator ions may be used to perform dual activating functions.

The iron ion requirements for this activity system will depend upon the character and composition of the isomerase, the type of isomerization process used (e.g., batch or continuous), whether or not the isomerase has been pretreated. Advantageously, the amount of iron present in the isomerization reaction in conjunction with the thiol activator should be sufficient to provide a measurable increase in the glucose isomerase activity over that which is achieved when magnesium is used as a sole activator for the isomerase (i.e., without iron activator ions).

Illustrative iron ion concentrations used to activate the glucose isomerase may range from minute amounts (e.g., 1 ppm or 2×10^{-5} M) to 1 M or higher. In batch operations, iron levels in excess of 1.3 M do not appear to adversely affect glucose isomerase activity, but can lead to refining difficulties. It is desirable in a continuous operation to replenish or continuously provide the isomerase's active site with a sufficient amount of iron ions (and other co-metal activators if used) to compensate for iron ion or co-metal activator losses which inherently arise as a result of prolonged usage of the isomerase in a continuous isomerization process.

In addition to the iron ions, the isomerase activator system relies upon a thiol activator. The reason why thiol activators, as a class, will measurably increase the activity of the isomerase-iron complex is not understood, especially since certain analysts have reported the absence of thiol-bearing amino acids in isomerase for which these thiol activators have been found to be effective. Typically, the thiol activator will enhance the isomerase activity of the iron-isomerase complex by at least 5% and more typically by at least 10%.

Thiol activators which cleave disulfide linkages in equilibrium may be illustrated by the following reaction: $RSSR' + A^{=} \rightleftharpoons RS^{-} + R'SA^{-}$, where R and R' in the RSSR' represent organo groups (e.g., including protein molecules) joined together by a disulfide linkage, and $A^{=}$ represents reducing agent. Such reducing agents are representative thiol activators. Chemical reagents which are precursors or form $SO_3^{=}$ in aqueous solutions are particularly effective thiol activators. Such sulfite precursors include sulfur dioxide and water-soluble salts of sulfurous acid.

Illustrative water-soluble salts of sulfurous acid include the alkali metal sulfites (e.g., potassium or sodium sulfites, bisulfites, and pyrosulfites; lithium sulfite, etc.), water-soluble salts of metal activators which form $SO_3^{=}$ anions (e.g., the sulfites, bisulfites, hyposulfites, of ferrous, ferric, cobaltous, nickel, magnesium, etc.) and other water-soluble $SO_3^{=}$ producing salts of cations (e.g., ammonium sulfite, bisulfite, etc.), mixtures thereof and the like. Metal salts known to deactivate or inhibit glucose isomerase activity should be avoided (e.g., zinc, cupric, aluminum salts, etc.). Thiol activators which do not form sulfite ions, but function as reducing agents, include ascorbic acid, isoascorbic acid, water-soluble thiocyanate salts (e.g., thiocyanates of lithium, potassium, sodium, ammonium, cobaltous, magnesium, manganous, ferric, ferrous), cystein, mercaptoethanol, thioglycolate, etc., mixtures thereof and the like.

The thiol activator should be provided to the glucose isomerase in an amount sufficient to increase the activating effect of the iron ion. Although the thiol activator requirements may be provided to the isomerase on a discontinuous or intermittent basis, it is advantageous to supply the isomerase with its thiol activator and iron ion requirements as dextrose feed syrup additives.

For most isomerization reactions, effective isomerase activation will usually be achieved at thiol activator concentrations ranging from 0.0002 to 0.5 M. Somewhat higher thiol activator ranges are advantageously used in batch operations (e.g., about 0.005 to 0.2 M) as opposed to continuous operations which are advantageously operated at a thiol activator concentration ranging from 0.0003 to 0.1 M. Thiol activator concentrations from 0.0004 to 0.05 M have been found to be particularly effective.

Example: An immobilized, dry isomerase preparation (41.3 g) obtained from *Bacillus coagulans*, identified as Novo SP 113 A, was pretreated by uniformly admixing and slurrying the dry isomerase initially with 75 ml glucose syrup (50% dry solids of which 94% was glucose) followed by the admixing of a 5 ml solution containing 1 M sodium sulfite (1.3% of the enzyme dry weight) and 2.1 g ferrous sulfate (Melanterite-a, 0.1 M Fe^{++}) and 5 ml aqueous solution of 1 M Mg^{++} ion (magnesium sulfate).

The slurry was mixed for 30 minutes at 23°C and charged to an isomerization reactor (3-necked flask) containing 1 liter of a 55°C glucose syrup (at 94% glucose and 50% dsb) which contained 0.005 M Mg^{++} ion (magnesium sulfate), 0.005 M sodium sulfite and 20 ppm Fe^{++} ion (0.00036 M ferrous sulfate). The contents of the isomerization reaction were gently stirred and maintained for 6.5 hours at 55°C and pH 7.0. Thereafter, fresh glucose syrup which contained 0.005 M Na_2SO_3, 0.005 M Mg^{++} ion and 0.00036 M ferrous ion at a rate of 1 liter/day was continuously charged to the reactor with a corresponding rate withdrawal of isomerized syrup therefrom. The reactor was continuously run for 13 days.

The total amount of fructose produced was determined by high pressure liquid chromatography for each day, allowing calculation of the isomerase activity (in International Glucose Isomerase Units per gram) for each 24-hour period.

Fructose productivity (in grams) for each of the 13 days was: 1 (day)–180 g and 106 IGIU of isomerase; 2–223 g and 110 IGIU/g of isomerase; 3–231 g and 105 IGIU/g of isomerase; 4–226 g and 96 IGIU/g of isomerase; 5–202 g and 84 IGIU/g of isomerase; 6–209 g and 85 IGIU/g of isomerase; 7–215 g and 85 IGIU/g of isomerase; 8–206 g and 69 IGIU/g of isomerase; 9–200 g and 63 IGIU/g of isomerase; 10–195 g and 56 IGIU/g of isomerase; 11–201 g and 63 IGIU/g of isomerase; 12–173 g and 45 IGIU/g of isomerase; and 13th day–175 g and 48 IGIU/g of isomerase.

The calculated glucose isomerase units per gram isomerase for each day were plotted on semilogarithmic graph paper (IGIU on Y axis and days on X axis) and the isomerase half-life was determined by drawing a line through the above recorded points and recording the points of time at which the glucose isomerase had one-half of its initial activity. Initial isomerase activity was 142 IGIU/g isomerase with an isomerase half-life of about 8 days.

For comparative purposes, this example was repeated eliminating the ferrous ion and sodium sulfite from the pretreatment solution and the isomerization reaction (i.e., relying upon Mg^{++} as the isomerase activator). Initial glucose isomerase activity for the Mg^{++} activated system without the cooperative effect was 105 IGIU/g glucose isomerase and a half-life of 4.7 days which illustrates the Fe^{++} and thiol activators enhanced isomerase activity by 135% and half-life by 170%.

In another disclosure by *T.L. Hurst; U.S. Patent 4,026,764; May 31, 1977; assigned to A.E. Staley Manufacturing Company*, the emphasis is on using an aqueous solution of the activators with a dry isomerase. Enzyme activity can be increased up to 20% by pretreating the isomerase with such solutions.

As a general rule, the metal ion activators for isomerases have a valence of two and are most generally characterized as having an atomic number of 28 or less. The Period IIa metal ions (e.g., magnesium) and metal ions of atomic numbers 22 to 27 inclusive (particularly manganous, ferrous and cobalt) and mixtures thereof are most commonly used and reported as isomerase metal ion activators in an isomerization process. These metal ion activators form a metal complex with the isomerase and enable it to isomerize dextrose to fructose. Divalent metals having an atomic number 29 or greater (e.g., copper and zinc) generally deactivate the isomerases.

Another class of pretreating solutes having a favorable effect in restoring a dry isomerase in a more active state is the isomerizable monosaccharides. The term isomerizable monosaccharides refers to those monosaccharides which the isomerase can isomerize. Accordingly, the isomerizable monosaccharides used in the pretreating solution will depend upon the isomerization characteristics of the isomerases to be pretreated therewith. As a general rule, most glucose isomerases are capable of isomerizing certain pentoses (e.g., xylose and/or xylulose) as well as certain hexoses (e.g., dextrose and/or fructose). These isomerizable monosaccharides may be utilized in the pretreating solution.

Another class of pretreatment reagents having a beneficial effect upon restoring previously dried isomerases to a higher degree of activity are thiol activators. The thiol activators, as a class, chemically react with isomerase disulfide linkages to provide thiol-containing reaction products. Why these disulfide reducing agents increase isomerase activity is not fully understood. It is evident, however, that dry isomerases contain disulfide linkages in a form which adversely affects their enzymatic activity. The thiol activators which cleave isomerase disulfide linkages in equilibrium are particularly effective.

Example: An immobilized, dry isomerase preparation (1.062 g) obtained from a *Bacillus coagulans*, immobilized in accordance with Patent No. 2,345,185, West Germany and known as Novo SB 113 A was pretreated by uniformly admixing the dry isomerase with a pretreating solution consisting of 20 ml glucose syrup (58.2%) dry solids, 14.6 g ds of which 94% was glucose) at 0.001 M sodium sulfite and a 3 ml aqueous solution of 1 M Mg^{++} (magnesium sulfate) and 0.15 ml of a 1 M Co^{++} aqueous solution (as cobaltous nitrate) for 45 minutes at 23°C. An isomerization reactor was charged with 300 ml glucose syrup having an identical composition as used in the pretreatment step, heated to 60°C, at pH 6.8 and the pretreated isomerase and pretreatment solution were added thereto. The isomerization medium was then maintained at pH 6.8 and 60°C. An identical dry weight of untreated dry isomerase was directly admixed into an isomerization medium of an identical composition and permitted to isomerize the glucose syrup under the same isomerization conditions. The tabulated results of this example are as follows:

Isomerization Time (hr)	Pretreated Isomerase	Untreated Isomerase
	.... (g fructose/100 g dry solids).	
6	7.8	6.5
35	28.0	24.8
70	37.0	34.0
94	39.6	37.1

Control of Co, Ca and Mg Ions in Continuous Process

Virtually all known glucose isomerases are cobalt activated, a factor which adds significantly to syrup conversion costs because any cobalt added to incoming glucose syrup must be removed from the product glucose-fructose by relatively high cost ion exchange techniques.

It has been discovered by *P.B.R. Poulsen and L.E. Zittan; U.S. Patent 4,025,389; May 24, 1977; assigned to Novo Industri A/S, Denmark* that the biochemical needs of the immobilized *B. coagulans* glucose isomerase enzymes for cobalt may remain completely satisfied without any cobalt supplementation during the course of isomerization. Actually it has been found that by omitting Co^{++} from the gluose syrup, the productivity of the isomerization process and the stability of the enzyme are not impaired and in some cases improved.

It has also been discovered that the biochemical needs of glucose isomerase enzyme for magnesium are much lower than had been believed. As a matter of fact, the addition of Mg^{++} may be avoided altogether, if the concentration of Ca^{++} is low. The activation needs of the glucose isomerase for Mg^{++} is fully satisfied at pH 7.8 or higher by less than 10^{-3} M of Mg^{++}. However, calcium ions in the syrup are inhibitory, perhaps more so than had been appreciated. Apparently the Mg^{++} in the syrup acts mostly to counter Ca^{++}inhibition. A low calcium content in the syrup allows reduction in the magnesium content. The Mg^{++}/Ca^{++} ratio in the syrup should exceed 5:1 on a molar basis, preferably the ratio should exceed 10:1, but not 500:1, if $Mg^{++} > 10^{-3}$M.

Conduct of an enzymatic glucose isomerization reaction at above pH 8.0 is, however, contrary to usual practices of isomerizing at pH below pH 8.0 in order to reduce color formation. Removal of color from the product syrup, e.g., by treatment with activated carbon, adds to the processing expense. Minimizing color formation is highly desirable. Unfortunately, the rate of color formation in the glucose syrup increases with increasing pH.

Investigating of the color forming propensity revealed that the rate of color formation at all pH levels is some function of time. In short, the higher color formation rate at pH 8.0^{+} may be countered by carrying out the isomerization more quickly. Limiting the total holding or contact time of the glucose syrup in the enzymatic conversion reactor to below 3.5 hours produces a product syrup of acceptable color. If contact time is limited to less than 1 hour, color cleanup treatment of the product syrup might become unnecessary. Further details of the process are shown in the following examples.

Example 1: Isomerization in Presence and Absence of Co^{++} – All isomerizations were performed as continuous packed bed plug flow reactions. The *B. coagulans* enzyme was prepared according to Example 2, U.S. Patent 3,980,521. The particle size varied between 150 and 2,800μ. This enzyme preparation was presoaked in 40 wt % glucose syrup at room temperature for 1 hour. This presoaked glucose isomerase was packed in different water-jacketed columns. A current of glucose syrup having the concentration, pH and other characterizing data, according to the following table, was sent upward through the enzyme material. The flow rate was adjusted to give an output syrup conversion of 45%. The linear flow rate always exceeds 10 cm/hr (film diffusion limit). The Ca^{++} concentration in this example was below 2.5×10^{-5} M.

Run	Temp (°C)	Glucose Conc (% by wt)	pH in Feed	Co^{++} Added (mol/liter)	Mg^{++} Added (mol/liter)
1	60	40	8.5	0	8×10^{-3}
2	60	40	8.5	3.5×10^{-4}	8×10^{-3}
3	65	40	8.5	0	8×10^{-3}
4	65	40	8.5	3.5×10^{-4}	8×10^{-3}
5	65	40	7.6	0	8×10^{-3}
6	65	40	7.6	3.5×10^{-4}	8×10^{-3}

Different sizes of columns were used, i.e., columns of sizes (1.5 x 35 cm) and (5.8 x 45 cm), whereby the first dimension refers to the diameter and the second to the height of the column. For the sake of brevity, the columns are hereafter referred to as 60 ml and 1 liter columns, respectively. 9 g of enzyme was packed in the 60 ml column and 225 g of enzyme in the 1 liter column. All isomerizations were continued for 450 hours. Thereafter, the residual activity was measured, and the productivity per 100 IGIC/g and the contact time per 100 IGIC/g were calculated. The results are shown in the following table.

Performance	Run 1	2	3	4	5	6
60 ml Column						
Productivity,*	340	315	420	375	395	410
Residual activity, %	65	60	52	42	28	41
Contact time, min**	90-140	90-150	75-145	75-180	75-270	75-185
1 liter Column						
Productivity,*	350	320	440	370	400	415
Residual activity, %	69	59	60	45	44	46
Contact time, min**	90-130	90-155	75-125	75-170	75-170	75-170

*Per 100 IGIC per gram (IGIC = immobilized glucose isomerase column unit. One IGIC is the amount of enzyme which converts glucose to fructose at an initial rate of 1μ/mol/min at 40 wt % glucose, pH 8.5, 65°C, 4×10^{-3} M Mg^{++}, no Ca^{++}, in a 2.5 x 40 cm continuous packed bed column.)

**Per 100 IGIC per gram. First figure, early in run when activity is high; second figure, end of run when activity is low.

Example 2: Isomerization in Presence and Absence of Mg^{++} – The isomerizations were performed as described in Example 1. The following parameters were kept constant during the experimental series: glucose concentration, 40% w/w; inlet pH, 8.5; Co^{++} addition, none; Ca^{++} concentration, less than 2.5 x 10^{-5} M; temperature, 65°C; and magnesium addition varied from 0 to 8 x 10^{-3} M. After 450 hours of isomerization, the experiments were interrupted. The residual activity was determined, and productivity per 100 IGIC/g and the contact time per 100 IGIC/g were calculated. The results are listed in the following table.

	Mg^{++} Addition, mol/l				
Performance	**0**	**4 x 10^{-4}**	**8 x 10^{-4}**	**4 x 10^{-3}**	**8 x 10^{-3}**
60 ml Column					
Productivity,*	400	419	424	415	420
Residual activity, %	50	50	50	48	52
Contact time, min*	75-150	75-150	75-150	75-155	75-145
1 liter Column					
Productivity,*	410	435	435	450	440
Residual activity, %	58	60	62	62	60
Contact time, min*	75-130	75-125	75-125	72-125	78-125

*Per 100 IGIC/g.

It appears from the table that the addition of Mg^{++} does not affect the productivity per 100 IGIC/g or the residual activity.

PROCESS IMPROVEMENTS USING IMMOBILIZED ISOMERASE

Programmed Temperature for Isomerization

R.E. Heady and W.A. Jacaway, Jr.; U.S. Patent 3,847,741; November 12, 1974; assigned to CPC International Inc. have disclosed an improved conversion of dextrose to fructose using two temperature stages, preferably using a fixed enzyme. Generally, the process involves conducting the isomerization at an initial, fairly constant temperature of at least 50°C, followed by increasing the operating temperature, either in several increments or in a single step, to a value that is 5°C or more higher than the initial operating temperature, to an operating temperature not exceeding 80°C.

In one preferred mode, the process involves subjecting a stream of a dextrose solution to an immobilized isomerization enzyme preparation at a pH of 7.5 to 8.5, and at a temperature level of at least 50°C but that is at least 10°C below the temperature at which rapid inactivation of the enzyme occurs, during an initial isomerization phase.

For enzyme preparations derived from microorganisms of the *Streptomyces* genus, ordinarily the temperature above which rapid inactivation of the enzyme preparation can be observed is about 70°C, so that the preferred operating temperature during the initial phase of isomerization is 60°C or lower, that is, preferably 50° to 60°C. At temperatures in the range from 50° to 60°C, some enzyme inactivation occurs, but the rate of inactivation is low relative to the rate of inactivation at 70°C. After some material loss of enzyme activity is observed after operating in the range from 50° to 60°C or so, the temperature is raised at least 5°C, either gradually, in increments of 2° to 3°C or less, as needed to maintain the desired ketose level in the product, or alternatively, the temperature may be raised at least 5°C in a single step. Preferably, however, the increase in temperature for the second phase of operation is at least 10°C, preferably through gradual or small incremental increases in operating temperature. Among the advantages of this process are better enzyme stability, more desirable carbohydrate content of the isomerized product, improved efficiency as compared to batch conversions, and operating economy.

Example: Continuous Isomerization Using a Fixed Bed of Microbial Cells, Single Step Temperature Adjustment – For this example, a jacketed column was used. The jacket was connected to a source of hot water for controlling column temperature during isomerization. A strain of a microorganism of the *Streptomyces* genus was grown under submerged, aerobic conditions on a medium containing xylose to produce intracellular isomerase.

After fermentation, magnesium hydroxide was added to the fermenter broth in the ratio of 2 parts by weight of magnesium hydroxide for each 1 part by weight of the cell mass in the broth. The slurry thus obtained was filtered, and the filter cake was then dried in an open pan at room temperature. The activity of the dry enzyme preparation obtained was 330 units per gram. The dry enzyme preparation was dispersed in a 50% w/v solution of dextrose.

The slurry was then placed in the jacketed column, and as the slurry was added to the column, small glass beads, approximately 3 mm in diameter, were added simultaneously. The glass beads served as a support and also prevented the enzyme preparation from packing and thus plugging up the column. In this manner, approximately 750 units of the enzyme were charged to the column. A dextrose syrup at 50% w/v concentration was adjusted to a pH of 7.0 to 7.5 by adding magnesium hydroxide. The syrup was sparged with nitrogen, and was then fed to the top of the column under a nitrogen atmosphere.

The isomerized product was collected in aliquots of 15 ml in test tubes. The test tubes each contained 5 ml of 0.5 N perchloric acid to inactivate any soluble isomerase that might be present in the product. The temperature of the column was maintained at 60°C during an initial phase of operation. The flow rate of dextrose solution through the column was maintained at a substantially uniform rate, and the ketose content of the effluent was 40 to 50% on a dry solids basis.

The isomerization was conducted in this manner for 9 days before a substantial decrease in enzyme activity became apparent as evidenced by a dropping off in the ketose value observed in the effluent. During that initial phase of operation, the average ketose content of the effluent was 38.7%. At the end of this initial phase of operation, the temperature of the column was increased (from the initial level of 60°C) to 70°C in a single step.

Isomerization was then continued, at the increased temperature, for an additional period of 24 hours. The ketose content of the product averaged out at about 49%, for the second phase of operation. The results are summarized in Table 1 shown on the following page.

Table 1: Continuous Isomerization, Single Step Temperature Adjustment

Days	Temp, °C	Throughput Bed Volumes per Hour	Output of 42% db Fructose Product (lb ds/ft³ of bed) Expected at 60°C	Obtained
1	60	1.27	951	951
2	60	1.27	951	951
3	60	1.27	951	951
4	60	1.27	951	951
5	60	1.25	936	936
6	60	1.17	876	876
7	60	1.10	823	823
8	60	1.03	771	771
9	70	1.65	734	1,235
10	70	1.65	681	1,235

When the isomerization was repeated with an enzyme preparation prepared from a mixture of diatomaceous earth with the microbial cells, closely comparable results were obtained. When a similar reaction was done using small step temperature adjustment, the results obtained are shown in Table 2.

Table 2: Fixed Bed Isomerization with Small Step Temperature Increases

Days	Temperature, °C		Output of 42% db Fructose Product (lb ds/ft³ of bed) Expected at 60°C	Obtained
1	60		787	787
2	60		787	787
3	60		787	787
4	60		787	787
5	60		622	622
6	60		516	516
7	62		418	590
8	62		343	516
9	62		286	516
10	64		233	509
11	64		187	509
12	66		158	473
13	68		127	674
14	70		106	622
		Total	6,144	8,695

Continuous Fluid Flow Isomerization

A continuous process for enzymatically isomerizing glucose to fructose has been provided by *K.N. Thompson, R.A. Johnson, and N.E. Lloyd; U.S. Patent 3,909,354; September 30, 1975; assigned to Standard Brands Incorporated*. This process involves continuously introducing a glucose-containing solution having a viscosity of 0.5 to 100 cp, a pH of 6 to 9, a temperature of 20° to 80°C and containing from 5 to 80% glucose by weight into a zone containing particles of glucose isomerase bound to anion exchange cellulose or to a synthetic anion exchange resin whereby the particles of the bound glucose isomerase are maintained in suspension and up to 54% of the glucose is converted to fructose. The color of the converted solution is increased by less than 2 color units and there is no substantial production of psicose. The converted solution is withdrawn from the zone at a rate substantially equivalent to the rate the glucose-containing solution is introduced into the zone, the particles of the bound glucose isomerase being characterized as having a glucose isomerase activity of at least 3 International Glucose Isomerase Units (IGIU) per cubic centimeter when packed in a bed and a stability value of at least 50 hours.

The stability value of the bound glucose isomerase used in this process should be at least 300 hours and preferably about 400 hours. Forming the bound glucose isomerase may be accomplished in any convenient manner so long as the bound glucose isomerase has the stability set forth above. For instance, glucose isomerase may be bound to DEAE-cellulose (diethylaminoethylcellulose) or like material and excellent results will be obtained. Of course, to effectuate binding, the glucose isomerase must be removed from the cells and there must be no interfering substances present during binding. The binding may be accomplished in an aqueous medium or in a sugar solution, e.g., corn syrup.

Also, the glucose isomerase may be bound on inert carriers either along with cellular material or in a relatively pure state. Various polymeric materials may be suitable for this purpose but, of course, the porosity of such materials must be such that allows the glucose to contact the glucose isomerase. The following example further illustrates the use of glucose isomerase bound to a porous, synthetic anion exchange resin in a fluidized bed.

Example: Streptomyces sp. ATCC 21175 was grown under submerged aerobic conditions and sufficient $CoCl_2 \cdot 6H_2O$ and a cationic surfactant (Variquat 50 MC) was added to obtain in the resulting broth 240 ppm and 1,600 ppm, respectively. The broth was continuously stirred for 5 hours while maintained at 40°C and pH 6.5. The broth was then cooled to 15°C and filtered. The filtrate had a glucose isomerase activity of 11.4 IGIU/ml. To 4 liters of the filtrate was added 12 g of cationic cellulose (Whatman DE-23) to adsorb nonisomerase proteinaceous materials. The cationic cellulose was removed by filtration. The filtrate had a glucose isomerase activity of 11.2 IGIU/ml.

830 g of moist Amberlite IRA-938 anion exchange resin was added to 3 liters of 1.5 N NaOH. The resin was stirred at 60°C for 15 minutes, filtered, rinsed with demineralized water, restirred in 4 liters of 2 N HCl at 60°C for 15 minutes, filtered, rinsed with demineralized water, stirred in demineralized water at a pH of 7 and filtered. This resin was added to 4 liters of the previously described glucose isomerase filtrate having a glucose isomerase activity of 11.2 IGIU/ml and maintained therein for 1 hour at 60°C, filtered and washed. The filtrate and wash water were collected and the glucose isomerase activity determined.

Based upon the difference between the glucose isomerase activity of this filtrate and wash water and the glucose isomerase activity of the filtrate to which the resin was added, it was calculated that the resin adsorbed 46.5 IGIU/g. Based upon this calculated value, the resin had a glucose isomerase activity of 29.9 IGIU/cm^3 when packed in a bed and an estimated stability value of 600 hours. A refined glucose-containing solution containing 90.5% glucose, 2.0% fructose and 52.1% dry substance was diluted with water to obtain solutions containing about 10, 20, 30 and 40% dry substance. To each solution was added 0.005 mols of $MgSO_3$/l and the pH of the solutions was adjusted to 7.4

90 g of resin containing the adsorbed glucose isomerase was stirred in 500 ml of the solution containing 10% dry substance, heated to 40°C and stirred under vacuum to remove entrapped air. This slurry was poured into a jacketed column having a diameter of 1" and a height of 24" wherein a constant temperature of 65°C was maintained. The upper and lower ends of the column were equipped with screens to retain the resin in the column.

The previously prepared solutions were pumped through a heated coil to adjust their temperature to 65°C and then pumped into the bottom of the column wherein the bound glucose isomerase was fluidized. Set forth in the table below are the flow rates of the solutions pumped through the columns, the percent of glucose converted to fructose and the psicose content of the converted solutions.

Fluidized Bed Isomerization

Percent Dry Substance	Solution Flow Rate (ml/min)	Percent Glucose Converted to Fructose	Percent Psicose
10.0	47.4	14.1	<1
20.2	28.5	12.6	<1
30.2	13.0	17.3	<1
40.4	7.4	16.8	<1

Binding of Fresh Isomerase to Carrier During Continuous Operation

The continuous isomerization of dextrose to levulose is efficiently conducted in a fixed bed reactor by the process developed by *M. Tamura, S. Ushiro, and S. Hasagawa; U.S. Patent 3,960,663; June 1, 1976; assigned to CPC International Inc.* The immobilized dextrose isomerase bed is periodically recharged by adding fresh isomerase to the feed liquor. The carrier, on which the isomerase is immobilized, binds the fresh isomerase so that enzyme activity is maintained in the reactor at an effective level, without interrupting operations.

Preferred microorganisms for producing isomerase for use in this process are the members of the *Streptomyces* genus. Particularly preferred species among this genus are *S. venezuelae* ATCC 21113 and *S. olivochromogenes* ATCC 21114. The most preferred microorganisms are mutant strains of *Streptomyces olivochromogenes*, especially *S. olivochromogenes* ATCC Nos. 21713, 21714, 21715 and their equivalents. These microorganisms form appreciable quantities of isomerase when cultivated in nutrient media free of xylose and xylose-supplying material and free of added cobalt. One unit of enzyme activity is defined as the amount of the enzyme activity which forms one micromol of levulose in 1 minute under isomerization conditions.

The macroreticular (MR) or porous-type strongly basic anion exchange resins used in the process include Amberlite IRA-904 and IRA-938 (Rohm & Haas Co.) as MR-type strongly basic anion exchange resins and Dianion PA-308 and PA-304 (Mitsubishi Chemical Co.) as porous-type strongly basic ion exchange resins. The resin or polymer particles are used in the form of granules or beads and usually in the range of 16 to 100 mesh particle size (U.S. Standard Screen Series) and more preferably in the form of 20 to 50 mesh beads. For convenience of use, the beads are placed or arranged in a column.

Macroreticular resins are characterized by the presence, throughout the polymeric matrix, of a network of extra-cellular microchannels or pores. While these microchannels are very small, they are large in comparison with the pores in conventional homogeneous crosslinked gells. Macroreticular resins, suitable for use in this process, may have specific surface areas of up to 2,000 m^2/g or more.

The isomerase is used in the form of solutions: 0.01 to 0.1 M Tris-HCl or phosphate buffer solutions; salt solution such as $(NH_4)_2SO_4$, $MgCl_2$ or KNO_3 solution, all adjusted to pH 7 to 8; or simply in water or in dextrose solutions; all having concentrations ranging from 3 to 50 units per milliliter.

The ion exchange resin is packed in a column of the proper size and then equilibrated by passing the same solution as used for dissolving the isomerase through the column at a flow rate between 1 and 3 SV for 5 to 10 hours (SV is an abbreviation for substrate velocity and refers to the flow rate in bed volumes per hour; one bed volume is the volume of substrate per hour that is equivalent to the column volume that is taken up by the resin in the column). Then, an amount of the dextrose isomerase solution corresponding to 10 to 100 units (preferably 50 units) of the enzyme per g of wet resin is passed through the resin column at a flow rate between 1 and 3 SV. After all of the enzyme solution has passed through the column, water is passed through the column to wash away unadsorbed isomerase.

If a continuous isomerizing reaction is carried out at about pH 8.0, using the adsorbed isomerase, the equilibrium rate of isomerization (52% levulose) is maintained for 15 days. After that, however, the rate lowers gradually dropping to one-half of the initial value (26%) in 22 days. The time in which the rate of isomerization of an immobilized isomerase drops to one-half of the initial value is defined as its half-life.

The method of reactivating columns, whose isomerizing activity has begun to drop, is as follows. First, a dextrose isomerase is dissolved in the same glucose solution as that which is used for isomerization. Then, an amount of the resulting enzyme solution, containing 25 to 50 units of isomerase per g of wet resin, is passed through the columns at the same flow rate as that at which the glucose solution to be isomerized is passed. As it passes through the column, more enzyme is adsorbed at the same time that the glucose is being isomerized.

In this manner, the resin columns regain their original rate of isomerization (52%). Moreover, the immobilized isomerase, which has been reactivated, shows the same half-life as the original one (17 to 22 days). If this procedure is repeated, it is possible to carry out the isomerization without repacking columns as long as the ion exchange resins last. Further, the levulose content in the effluent from resin columns can be held constant at a given isomerization rate up to about 52% (the normal equilibrium value) in the following way. When the isomerization rate is found to drop, as can be detected by measurement with a polarimeter, the resin column is immediately reactivated in the above manner, or, alternatively, the flow rate is gradually reduced to keep the isomerization rate constant, and then the column is reactivated at a proper time.

Example 1: 50 g of moist Amberlite IRA-904, an MR strongly basic anion exchange resin, was packed in a column (2.5 x 20 cm). With the resin equilibrated with a 0.01 M $MgCl_2$ solution adjusted to pH 8.0, 3,000 ml of a solution of crude dextrose isomerase in the same solution (containing 5,000 units of isomerase) was passed through the column at a flow rate of SV 3. After all the isomerase solution had been passed through the column, a 60% dextrose solution containing 0.01 M $MgCl_2$ and adjusted to pH 8.0 was passed through the column at a flow rate of SV 1 at 70°C.

The rate of isomerization was measured with a polarimeter and expressed as the percentage of levulose of the solid substance in the effluent from the column. The rate remained 52% (equilibrium value) for the first 12 days. After that, however, it lowered gradually and dropped to 26% on the 17th day. At this time, 1,500 ml of a solution of isomerase in a 60% dextrose solution containing 0.01 M $MgCl_2$ and adjusted to pH 8.0 (corresponding to 2,500 units of isomerase) was passed through the column at a flow rate of SV 1. Then, the isomerizing reaction was continued on by passing the same dextrose solution (without isomerase) through the column under the same conditions as those used for the isomerizing reaction.

As the result of this procedure, the resin column regained its original rate of isomerization of 52% and retained this value for further 12 days. In the succeeding 5 days, however, the rate of isomerization dropped to 26%. At this time, the resin column was again reactivated in the same manner. It was again returned completely to its initial activity.

Example 2: 50 g of moist Dianion PA-308, a porous strongly basic anion exchange resin, was packed in a column (2.5 x 2.0 cm). After the resin had been equilibrated with a 0.01 M $MgCl_2$ solution adjusted to pH 8.0, 200 ml of a solution of partially purified dextrose isomerase in the same solution (containing 5,000 units of isomerase) was passed through the column at a flow rate of SV 1.

Then, a 60% dextrose solution containing 0.01 M $MgCl_2$ and adjusted to pH 8.0 was passed through the column at a flow rate of SV 3 with the column kept at 70°C. The rate of isomerization was measured continuously with a polarimeter. It remained at 52% for the initial 3 days and began to drop gradually thereafter. At that time, the flow rate of the dextrose solution was lowered so that the rate of isomerization could be held at 52%. This procedure was made possible by reducing the flow rate of the dextrose solution by SV 0.25 every day.

10 days after the time when the reduction of the flow rate was started (when the flow rate dropped down to SV 0.5), 100 ml of a solution of isomerase in a 60% dextrose solution containing 0.01 M $MgCl_2$ and adjusted to pH 8.0 (corresponding to 2,500 units of glucose isomerase) was passed through the column at a flow rate of SV 0.5. After that, the isomerizing reaction was continued by passing the dextrose solution (without isomerase) through the column at the initial flow rate, namely, at SV 3. By repeating this procedure, it was possible to continue the isomerization at a rate of isomerization of 52% as long as the ion exchange resin lasted. It was unnecessary to repack the column during the period.

Renewal of Isomerase on Anion Exchange Resins

Y. Fujita, A. Matsumoto, I. Miyachi, N. Imai, I. Kawakami, T. Hishida, and A. Kamata; U.S. Patent 4,113,568; September 12, 1978; assigned to Mitsubishi Chemical Industries, Ltd. and Seikagaku Kogyo Co., Ltd., Japan have developed means of treating exhausted immobilized glucose isomerase to remove adsorbed materials. The regenerated resin can then adsorb fresh glucose isomerase to recover activity. The support onto which glucose isomerase is adsorbed include a wide variety of synthetic anion exchange resins. Among them, the preferred resin is a macroporous strong basic tertiary or quaternary ammonium-type, the matrix of which is a styrene-divinyl benzene copolymer, for example, Amberlite IRA 900, Amberlite IRA 93 and Amberlite IRA 904 which are available from Rohm & Haas Co. and Dianion HPA 11 available from Mitsubishi Chemical Industries, Ltd.

The regeneration of the supporter resin is by a treatment of the insolubilized glucose isomerase with a combination of an aqueous mineral acid solution with an aqueous alkali solution, an aqueous electrolytic salt solution, an aqueous mineral acid solution or a mixture thereof. More particularly, the regeneration is carried out by passing the following one or more aqueous solutions through the column packed with the deactivated insolubilized glucose isomerase:

(1) An aqueous mixture of a mineral acid and an electrolytic salt;
(2) An aqueous mineral acid and an aqueous alkali solution, or vice versa;
(3) An aqueous mineral acid and an aqueous electrolytic salt solution, or vice versa;
(4) An aqueous mineral acid and an aqueous mixture of an alkaline material and a electrolytic salt, or vice versa;
(5) An aqueous mineral acid and an aqueous mixture of electrolytic salt and mineral acid;
(6) Instead of the aqueous mineral acid in cases 2 to 4, an aqueous mixture of a mineral acid and an electrolytic salt is used; and
(7) In addition to cases (1) to (4), an aqueous mineral acid is used in third step.

The supporter anion exchange resin as a salt-type, thus regenerated, can adsorb fresh isomerase to form a renewed insolubilized glucose isomerase which is used for further isomerization.

Details of this process are given in the examples which used materials prepared as follows. Glucose isomerase Nagase, which is produced from the strain belonging to *Streptomyces phacochromoges* and is available from Nagase Sangyo Kabushiki Kaisha, was suspended in water and, after addition of a small amount of egg white lysozyme, the suspension was stirred to effect extraction. The extracted solution had an activity of glucose isomerase of 240 units per milliliter. (This was designated as glucose isomerase extract A.)

A sulfate-type macroporous strong basic anion exchange resin Dianion HPA 11, which is a styrene-divinyl benzene crosslinked copolymer having quaternary ammonium ion exchange group, was suspended in an aqueous glucose isomerase extract in an amount 10 times the volume of the swollen resin and stirred at 50°C for 6 hours to obtain an insolubilized glucose isomerase.

100 ml of the insolubilized glucose isomerase was packed in a jacketed column of 20 mm internal diameter through which an aqueous solution containing glucose, magnesium sulfate and iron sulfate at a concentration of 3 M, 5 mM and 0.1 mM, respectively, and having a pH of 7.5 was passed downwardly at a temperature of 60°C and at such a space velocity that the average degree of isomerization was 45% for 30 days to convert glucose into fructose.

Examples 1 and 2: Glucose isomerization reaction was carried out with an insolubilized glucose isomerase which had been prepared by adsorbing glucose isomerase extract A on Dianion HPA 11 and had a degree of adsorption of glucose isomerase of 100% and an activity of 2,400 units per milliliter resin. After completion of the isomerization, 1 liter of desalted water was passed downwardly through the column to wash the insolubilized isomerase and 600 ml of 1 N sulfuric acid and 21 ml of desalted water were in turn passed downwardly through the column. From the supporter resin, 50% of adsorbed protein and 97% of adsorbed iron ion were desorbed.

The glucose isomerase extract A was adsorbed onto the supporter thus regenerated, the degree of adsorption of glucose isomerase being 100% and the activity being 2,400 units per milliliter resin. Then, the glucose isomerization was carried out by the process described above.

After repeating 4 cycles of adsorption-isomerization-desorption, the degree of adsorption was observed to be 51% and the activity was 1,224 units per milliliter resin, which values show inferior performance of the insolubilized isomerase. This insolubilized isomerase was used for carrying out the glucose isomerization. Then, two portions of the insolubilized isomerase of 20 ml were packed in two columns through which the aqueous solutions in predetermined amount given in the table below were passed downwardly to regenerate the supporter resin. Glucose isomerase extract A was adsorbed on each of the regenerated supporter resins and the activity and the degree of adsorption thereof were measured. The results are given in the following table.

	Example 1		Example 2	
Regenerating agent				
	1 N H_2SO_4	120 ml	1 N H_2SO_4	120 ml
	H_2O	120 ml	H_2O	120 ml
	1 N NaOH	120 ml	aqueous mixture of 0.1 N NaOH and 0.5 M Na_2SO_4	120 ml
	H_2O	120 ml	H_2O	120 ml
	1 N H_2SO_4	120 ml	1 N H_2SO_4	120 ml
	H_2O	400 ml	H_2O	400 ml
Activity after re-adsorption (U/ml resin)	1,560		2,208	
Degree of bonded isomerase, %	65		92	
Degree of activity retention, %	85		88	

Note: H_2O used as desalted water.

From the table, it is clear that, according to this process, the ability of a supporter resin to adsorb glucose isomerase is recovered entirely.

OTHER USES OF IMMOBILIZED ENZYMES

CHEMICAL PROCESSES

Production of Cephalothin by Enzymatic Acylation

This process developed by *T. Fujii and Y. Shibuya; U.S. Patent 3,853,705; December 10, 1974; assigned to Toyo Jozo KK, Japan* relates to the production of cephalothin, that is, 7-(2-thienyl acetamido)-cephalosporanic acid, by enzymatic acylation of 7-amino cephalosporanic acid (designated 7-ACA). More particularly this process relates to the enzymatic process for production of cephalothin in which 7-ACA is reacted with thienyl acetic acid or a derivative thereof in the presence of the acylating enzyme preparation derived from a microorganism strain, which produced an amino group acylating enzyme for 7-ACA, belonging to genus *Bacillus* in an aqueous medium.

It was found that when the enzyme preparation derived from *Bacillus megaterium* B-400 NRRL B-5385 is used, the production of bad odors and objectionable wastewater is avoided. The acylation enzyme-producing microorganism used in this process can be prepared by aerobic cultivation with a nutrient medium containing organic or inorganic nitrogen sources such as peptone meat extract, cornsteep liquor, yeast extract, dry yeast, nitrate, ammonium salt or the like, a carbon source such as molasses, glucose, starch hydrolysate or the like and inorganic salts, and if desired other suitable growth-stimulating substances, at 25° to 37°C, for 12 to 60 hours. For industrial production, submerged aerobic cultivation is generally used.

The acylating enzyme for the amino group in 7-ACA may generally be present as an exoenzyme. As for the enzyme preparations microbial cultures, culture filtrate or enzyme prepared therefrom may be used in the enzyme reaction. Furthermore, culture filtrate treated by chemical or physical procedures, for example, refined enzymes obtained by known separation and refining procedures such as salting-out, fractionation precipitation, dialysis, adsorption chromatography, ion-exchange chromatography or gel-filtration of culture filtrate; and the solid phase enzyme preparations or the insolubilized enzymes prepared from adsorption of the acylation enzyme on an inert carrier can be used. The carriers used are selected with regard to their properties; they should not inactivate the acylating enzyme activity; the adsorbed enzyme must not be removed by washing; the carrier must be inert for each substrate and it must not adsorb the cephalothin. Examples of the carriers which are advantageously used for adsorbing an aqueous enzyme solution include active alumina, diatomaceous earth, active clay, calcium phosphate, hydroxyapatite and the like.

It is recommended first to adjust the pH to a stable pH for the acylating enzyme when adsorbing the enzyme on a carrier. The adsorption operation can be carried out batchwise or in a column, the latter being preferable for a continuous enzyme reaction. The amount of carrier used varies with the volume of the culture filtrate, the enzyme potency and the adsorption ratio of the carrier. Generally, in the case of a batch type adsorption, the amount of carrier is 5 to 15 w/v for culture filtrates. In the case of a column type operation, a column packed with carrier is wetted with water or buffer solution having the optimum pH for the enzyme, passing through the cultured broth or enzyme solution and thereafter washing the column with water or buffer solution in order to obtain the column type solid phase enzyme preparation.

The carrier containing adsorbed acylating enzyme, i.e., a solid phase enzyme preparation, may become denatured or lose its activity if it becomes too dry. Therefore, it should be kept moist. The 7-ACA is then reacted with thienyl acetic acid in the presence of a solid phase enzyme preparation. Since the concentration of substrate may vary mainly according to the enzyme potency or rate of flow through the column, it should preferably be determined so that there is no unreacted 7-ACA or thienyl acetic acid in an eluate. The acylating reaction should be carried out at the optimum pH and temperature for the enzyme activity. The reaction time in the case of a column type operation can be adjusted by changing the effluent volume of the substrate.

Usually, the reaction can be carried out by passage through a column of a solid phase enzyme preparation and continuous operation can easily be carried out by continuously adding the substrates. If, however, the substrates are found in the effluent, the effluent rate should be reduced or the effluent should be recycled through the column. These processes can be systematized as an automatic controlled plant system. When the activity of the enzyme decreases or is lost, the operation should, of course, be terminated. The cephalothin produced is separated by known isolation procedures.

Example: 600 liters of medium (pH 7.0) containing 1% glucose, 1% peptone, 1% meat extract and 0.5% NaCl was inoculated with 30 liters of seed culture of *Bacillus megaterium* B-400 NRRL B-5385 and cultured submerged with aeration for 72 hours at 26°C. After fermentation, 580 liters of culture filtrate were obtained by centrifugation. The filtrate was adjusted to pH 7.5 and 6 kg of celite was then added and stirred for 40 minutes. The enzyme adsorbed carrier was packed into a column (diameter 12 cm) with outer jacket. The column was washed with 0.1 M phosphate buffer pH 7.5 and the mixed solution containing 1 kg of 7-ACA (10 mg/ml), 15 liters of 2-thienyl acetic acid methyl ester and 100 liters of 0.1 M phosphate buffer (pH 7.5) was passed through with a specific velocity (sv) of 0.5. The eluate was twice washed with 20 liters of ethyl acetate and the celite column was also washed with 15 liters of ethyl acetate.

The ethyl acetate was discarded and aqueous layer was adjusted to pH 1 to 2 by addition of 6 N HCl solution. The aqueous layer was again extracted twice with 20 liters of ethyl acetate and dried in vacuo to obtain 1.86 kg of crude cephalothin (purity 80%): overall yield 96%. The crude product was recrystallized to obtain 1.34 kg of the crystals (purity 98%); total yield 85%.

6-Aminopenicillanic Acid Preparation Using Penicillin Acylase

It has been found by *M.A. Cawthorne; U.S. Patent 3,887,432; June 3, 1975; assigned to Beecham Group Limited, England* that penicillin deacylase enzymes can be linked to water-soluble polymeric materials to produce water-soluble enzyme complexes and that such complexes retain the deacylase activity of the free enzyme.

Moreover, it has also been found that, when such an enzyme complex is used in the production of 6-aminopenicillanic acid (6-APA), more concentrated solutions can be used than with the water-insoluble complexes and this leads to the economic advantages of greater throughput or of using a chemical plant of smaller size. Furthermore, it has been found that the water-soluble enzyme complex can be separated from the aqueous solution of 6-APA produced by the use of ultrafiltration techniques.

One type of polymer that has been found to be particularly suitable is water-soluble copolymers of maleic anhydride or acrylic anhydride with vinyl methyl ether, ethylene, styrene, and/or vinyl acetate. One particularly suitable range of water-soluble copolymers is the methyl vinyl ether/maleic anhydride copolymers (particularly Gantrez AN 149, a copolymer of molecular weight about 750,000 and Gantrez AN 169, a copolymer of molecular weight about 1,125,000). Other very suitable water-soluble polymer substrates are polysaccharides. Thus there may be used such a polysaccharide as dextran, dextrin or cellulose, provided that such is water-soluble and leads to a water-soluble enzyme complex after it has been coupled to the penicillin deacylase enzyme.

It has further been found suitable to use polymers that are modified saccharides or oligosaccharides, such as sucrose, dextrose or lactose. An especially useful polymer of this type is Ficoll, which is a sucrose-epichlorohydrin copolymer. A Ficoll polymer with a molecular weight of about 400,000 has been found to be particularly suitable. The water-soluble enzyme complexes for use may be prepared by any of the known methods for linking enzymes to polymers including the use of linking agents such as cyanogen halides, s-triazines, azides, organic cyanates and diorganocarbodiimides. The process is illustrated by the following examples in which the activity is expressed in μmol of 6-aminopenicillanic acid produced from benzylpenicillin at pH 7.8 and 37°C/min/g or ml of preparation.

Example 1: 1 g of maleic anhydride/methyl vinyl ether copolymer of molecular weight about 750,000 (Gantrez AN 149) was stirred into 100 ml of 0.2 M phosphate buffer at pH 7.4 at 4°C. After 3 minutes, 5 ml of a solution of partially purified penicillin deacylase enzyme of activity 63.4 μmol/min/ml was added and the whole stirred for 16 hours at 4°C. The resulting viscous solution was subjected to ultrafiltration through an XM-300 membrane (Amicon Corp.), which has a molecular weight cut-off at 300,000. The retentate was washed in the ultrafiltration cell and finally freeze dried. The resulting enzyme complex was water-soluble and had an activity of 220 μmol/min/g when used to prepare 6-aminopenicillanic acid from benzylpenicillin.

Example 2: Ficoll (5 g) which is a sucrose-epichlorohydrin copolymer of MW 400,000, was dissolved in water (165 ml) and the pH adjusted to 10.0 with 2 N NaOH. Cyanogen bromide (400 mg) was added to the Ficoll solution and the solution stirred magnetically, the temperature being maintained at 20°C. The pH was maintained at 11.0 by the addition of 2 N NaOH. 20 minutes after the addition of the cyanogen bromide the solution was adjusted to pH 8.5 with 2 N HCl and mixed with 170 ml of an aqueous solution of penicillin deacylase (560 mg protein) having an activity such that it will produce 6.9 μmol of 6-APA per mg of protein per minute at pH 7.8 and 37°C.

The solution was agitated overnight at 4°C and was then concentrated by ultrafiltration through an XM-300 Amicon membrane. This membrane has a nominal MW cut-off of 300,000. The retentate in the ultrafiltration cell, which contained the Ficoll-penicillin deacylase complex, was washed. The volume of the retentate was then reduced to 44 ml and the product was stored at 4°C. The retentate had a similar viscosity to the native enzyme. The Ficoll penicillin deacylase enzyme complex retained 86.2% of the activity of the native deacylase enzyme.

Example 3: Reusability Study—The solution of Ficoll-penicillin deacylase complex (4 ml = 44.4 mg protein), which had been prepared according to the method in Example 2, was added to a solution of 6.3 g potassium benzylpenicillin in water and stirred at 37°C. The pH value of the reaction mixture was kept constant at 7.8 by the continuous addition of 0.1 N NaOH using a titrimeter. The reaction was complete after 4 hours although it was continued for 6 hours.

The uptake of 0.1 N NaOH indicated that the yield of 6-APA was 96.2% of the theoretical. The enzyme was separated from the reactants of small MW by ultrafiltration through a membrane having a MW cut-off of 50,000. After washing the enzyme in the ultrafiltration cell with distilled water, the enzyme-polymer was again subjected to ultrafiltration until the volume was reduced to 150 ml. The retentate was reused for the conversion of a further batch of 6.3 g potassium benzylpenicillin, which was converted to 6-APA with an efficiency of 96.5%. The whole cycle was repeated for a further three times, the yields of 6-APA being 92.0, 96.0 and 91.0% of the theoretical value.

Immobilized L-Histidine Ammonialyase for Producing Urocanic Acid

A process for producing urocanic acid using an immobilized enzyme has been developed by *I. Chibata, T. Tosa, T. Sato and K. Yamamoto; U.S. Patent 3,898,127; August 5, 1975; assigned to Tanabe Seiyaku Company, Ltd., Japan.*

In this process urocanic acid or a mixture of urocanic acid and D-histidine can be prepared by the steps of polymerizing at least one acrylic monomer in an aqueous suspension containing an L-histidine ammonialyase-producing microorganism, and subjecting the resultant immobilized L-histidine ammonialyase-producing microorganism to enzymatic reaction with L-histidine, DL-histidine or an acid addition salt thereof.

The polymerization for immobilizing the enzyme can be carried out in the presence of a polymerization initiator and a polymerization accelerator. Potassium persulfate, ammonium persulfate, vitamin B_2 and methylene blue are suitable as the polymerization initiators. On the other hand, β-(dimethylamino)-propionitrile and N,N,N',N'-tetramethylethylenediamine are used as polymerization accelerators.

It is preferred to carry out the reaction at 5° to 80°C, especially at 10° to 50°C. The reaction may be completed within 10 to 60 minutes. In some cases, in order to carry out the subsequent enzymatic reaction advantageously, it may be preferred to heat the L-histidine ammonialyase-producing microorganism at 60° to 80°C for 30 minutes prior to the immobilization reaction, or to heat the immobilized L-histidine ammonialyase-producing microorganism at 60° to 80°C for about 30 minutes. The acrylic monomers which are suitable for use in the process include acrylamide, N,N'-lower alkylene-bis-acrylamide and bis(acrylamidomethyl) ether.

Urocanic acid can be prepared by enzymatic reaction of the resultant immobilized microorganism with L-histidine or an organic or inorganic acid addition salt thereof. Alternatively, urocanic acid and D-histidine can be prepared by using DL-histidine or an organic acid addition salt thereof instead of L-histidine.

Suitable examples of the organic or inorganic acid addition salt of L- or DL-histidine include hydrochloride, sulfate, nitrate, acetate, etc. It is preferred to carry out the enzymatic reaction at 0° to 60°C, especially at 37°C. The enzymatic reaction can be accelerated by carrying it out in the presence of a surfactant. For example, when an aqueous 0.25 M L-histidine solution is reacted with 1 g of an immobilized L-histidine ammonialyase-producing microorganism at 37°C for 1 hour in the presence or absence of cetyltrimethyl ammonium bromide, urocanic acid is produced as shown in the table. The amount of urocanic acid produced is given in micrograms.

The concentration of a substrate employed is not critical. For example, L- or DL-histidine is dissolved in water at any concentration. The immobilized microorganism is suspended in the solution of L- or DL-histidine, and the suspension is stirred. After the reaction is completed, the mixture is filtered or centrifuged to recover the immobilized microorganism for subsequent use.

Immobilized Microorganisms	Cetyltrimethyl Ammonium Bromide Added	
	None	0.05% w/v
Achromobacter aquatilis OUT 8003	847	938
Achromobacter liquidum IAM 1667	946	1,054
Agrobacterium radiobacter IAM 1526	148	479
Flavobacterium flavescens IFO 3085	575	745
Sarcina lutea IAM 1099	26	194

Urocanic acid or a mixture of urocanic acid and D-histidine is recovered from the filtrate or supernatant solution. The optimum reaction condition for complete conversion of L-histidine or DL-histidine, respectively, to urocanic acid or a mixture of urocanic acid and D-histidine can be readily obtained by adjusting the reaction time. Alternatively, the enzymatic reaction of the process can be performed by a column method.

Example: An aqueous nutrient medium (pH 7.0) containing the following ingredients is prepared:

	%, w/v
Glucose	1
Dipotassium phosphate	0.2
Monopotassium phosphate	0.05
Ammonium chloride	0.1
Magnesium sulfate septahydrate	0.02
Yeast extract	0.1
L-histidine hydrochloride	0.02

Achromobacter liquidum IAM 1667 is inoculated into 200 ml of the medium. The medium is cultivated at 30°C for 24 hours with shaking. Then, the medium is centrifuged. The microbial cells thus collected are suspended in 12 ml of a physiological saline solution, and the suspension is heated at 70°C for 30 minutes. 2.25 g of acrylamide, 0.12 g of N,N'-methylene-bis-acrylamide, 1.5 ml of 5% β-(dimethylamino)-propionitrile and 1.5 ml of 2.5% potassium persulfate are added to the suspension. Then, the suspension is allowed to stand at 25°C for 10 minutes. The insoluble product is ground and washed with a physiological saline solution. 25 g of an immobilized preparation of *Achromobacter liquidum* IAM 1667 are obtained.

25 g of the immobilized preparation of *Achromobacter liquidum* IAM 1667 are charged into a 1.6 cm x 25.5 cm column, and 500 ml of an aqueous 0.25 M L-histidine hydrochloride solution (pH 9.0) are passed through the column at 37°C at the flow rate of 6 ml/hr. 500 ml of the effluent is adjusted to pH 4.7 with concentrated sulfuric acid. Then, the effluent is allowed to stand at 5°C overnight. The crystalline precipitate is collected by filtration, washed with ice water, and dried. 21.5 g of urocanic acid dihydrate are obtained. MP 225°C.

Hydrolysis of Prostaglandin Esters with Lipases

The prostaglandins (PG) are derivatives of prostanoic acid, 7⟨[2-(3-octyl)cyclopent-1α-yl]⟩heptanoic acid. These compounds have useful pharmacological effects on the reproductive, cardiovascular, respiratory and circulatory organs. Several total chemical syntheses have been worked out for the preparation of the prostaglandins which were originally isolated from natural products. In some of these methods, a critical step is the preparation of the free acid from the PG-ester without damaging or isomerizing the end product.

The hydrolysis step disclosed by *I. Stadler, G. Kovacs, Z. Meszaros, J. Radoczi, V. Simonidesz, C. Szantay, I. Szekely and C. Szathmary; U.S. Patent 3,875,003; April 1, 1975; assigned to Chinoin Pharmaceutical and Chemical Works Ltd., Hungary* can be carried out with a lipase of microbic or animal origin, with carboxylesterhydrolase, acetylcholinesterase or acetylesterase. Homogeneous, commercial enzyme products are preferably used. With the use of these, the PG-methylesters can be quantitatively decomposed in an aqueous suspension.

The process is preferably carried out as follows. The substance is hydrolyzed under pH control using a pH-stat with automatic burette at room temperature under nitrogen or argon atmosphere with an enzyme having a carboxylesterhydrolase effect on the esters. The hydrolysis is followed by thin layer chromatography or automatic alkali addition.

The end of the hydrolysis is indicated by the termination of the alkali consumption. The deviation from the theoretical alkali consumption considering the pH values of the prostaglandins indicates the purity of the substance under given reaction conditions as well. The process can be used for the determination of the purity of the active ingredients too. After the hydrolysis is ended, the reaction mixture is extracted with a water-immiscible solvent to remove the contaminant. The aqueous phase is acidified with an acid to pH 3.5 to 3.0 and the PG free acid thus obtained is shaken in a water-immiscible organic solvent. The water is removed from the organic phase, the mixture is evaporated and the residue is recrystallized from a suitable solvent.

The enzyme can also be applied on a suitable carrier, while the process becomes continuous by regenerating the enzyme. Some possibilities for the preparation of the solid enzyme products are (a) crosslinking by using diazotized polyaminostyrene, (b) the enzyme is bound to CM-cellulose, or CM-Sephadex, while transforming the cellulose-derivative into acid azide, or (c) the enzyme is enclosed in polyacrylamide gel capsules. The following examples gives details of the hydrolysis of the PGE_1 ester to form PGE_1 having the following formula

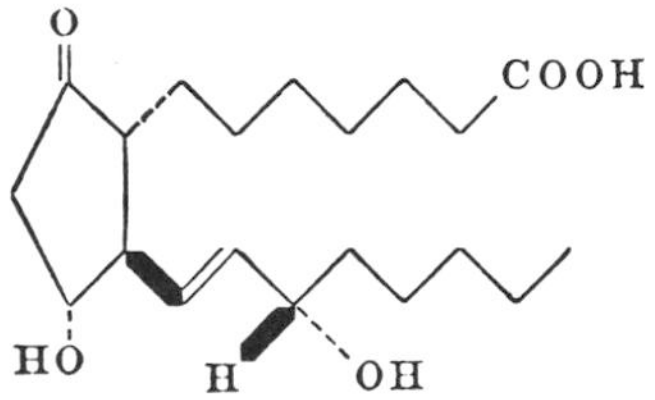

Example: The following two solutions are prepared. (a) 20 mg (54×10^{-5} mol) of natural l-PGE_1-methylester (obtained from natural l-PGE_1 by the reaction with diazomethane) are dissolved in 0.4 ml of 95% ethanol and 15 ml of distilled water are added. The emulsion is adjusted to pH 7.4 and nitrogen is bubbled through the mixture for 15 minutes to remove the absorbed carbon dioxide. (b) 20 mg of Nagase Saiken Lipase A (having a specific activity of 30 to 35 Desnuelle units) are dissolved in 5 ml of distilled water, and the pH is adjusted to 7.4. The mixture is stirred at room temperature for 30 minutes. The hydrolysis is carried out with 0.01 N sodium hydroxide at pH 7.4 placing the solutions (a) and (b) into a pH-stat system in nitrogen current at $25 \pm 1°C$.

The end of the reaction is indicated by the termination of alkali consumption, or by approximation of the theoretical alkali consumption. After the end of automatic alkali addition, thin layer chromatography was carried out and no traces of ester could be detected. That indicates the end of the hydrolysis. Alkali consumption: 5.3 ml of 0.01 N sodium hydroxide. Considering the pK value of PGE_1 acid (calculated consumption is 5.4 ml) the attained efficiency of the hydrolysis is 98%. Reaction time: 30 minutes.

After the reaction has taken place, the mixture is cooled to 0° to 5°C, the pH is adjusted to 3.5 by adding 0.1 N sulfuric acid and the mixture is shaken out with ethyl acetate several times. The water is removed from the ethyl acetate phase, which is then evaporated at a reduced pressure and the precipitated crystals are recrystallized from the mixture of ethyl acetate and hexane. The crystals are dried. MP 114° to 115°C, 18.5 mg of PGE_1 acid are obtained: yield 92.5%.

Conversion of Glucose to Gluconic Acid

Glucose oxidase (GOD) when used for oxidation of glucose to gluconic acid in the dissolved state, is not recovered from the reaction. Theoretically, these difficulties can be overcome by using GOD bound to an insoluble carrier. However, it has been found that, although the carrier-bound GOD can be easily separated, it is inactivated in a very short period of time. Starting from the assumption that the hydrogen peroxide formed is at least partially responsible for this inactivation, investigations were made to overcome this difficulty by the addition of the enzyme catalase to the reaction batch.

However, the experiments carried out showed that even in the presence of catalase, the rapid inactivation of GOD was not avoided. The attempt was, therefore, made to use a mixture of carrier-bound GOD with carrier-bound catalase. In order to achieve the most intimate possible contact and thus an immediate removal of the formed hydrogen peroxide, each of the enzymes was fixed on finely divided carrier material and these finely divided carriers then mixed together. However, this also did not overcome the above difficulties so that it was assumed that the formation of hydrogen peroxide was not of decisive importance for the ascertained rapid inactivation of the GOD.

H.U. Bergmeyer and D. Jaworek; U.S. Patent 3,935,071; January 27, 1976; assigned to Boehringer Mannheim GmbH, Germany have now found that the above problem can be overcome by using glucose oxidase and catalase bound together on a suitable carrier in immediate proximity with one another. In other words, a molecule of GOD can be regarded as being present bound to the carrier next to a molecule of catalase. It is to be assumed that, in this way, a kind of cyclic reaction takes place at the molecular level in which the hydrogen peroxide formed by the GOD is again decomposed by the catalase present in the immediate proximity before the hydrogen peroxide can act upon another GOD molecule.

Example 1: 50 mg of a commercially available GOD preparation with a specific activity of 220 U/mg were dissolved in 1 ml 0.5 M triethanolamine/hydrochloric acid buffer (pH 8.0), mixed at 30°C with 0.25 ml acrylic acid 2,3-epoxypropylester, while passing in nitrogen, and stirred for about 30 minutes at this temperature. Subsequently, the reaction mixture was cooled to 10°C.

Thereafter, 3.0 g acrylamide, 0.1 g N,N'-methylene-bis-acrylamide and 300 mg of an enzyme preparation with a GOD activity of 20 U/mg and a catalase activity of 260 U/mg were successively dissolved in 18 ml distilled water and introduced into the reaction vessel. The polymerization reaction was then initiated by the addition of 50 mg benzoyl peroxide and 0.5 ml N,N-dimethylaminopropionitrile solution. When the polymerization was finished (after about 1 hour), the reaction mixture was left to stand for 12 to 18 hours at 4°C. The polymer formed was then forced through a sieve of mesh size 4 mm, washed and dried. The specific activity of the carrier-bound enzyme thus obtained was: GOD 281 U/mg, catalase 185 U/mg (manometric determination of the liberated oxygen). In the following, the preparation thus obtained is referred to as GOD-catalase A.

Example 2: Batchwise Carrying Out of the Process—20 g (0.11 mol) fructose and 1.1 g (0.0055 mol) glucose hydrate were dissolved in water to give 200 ml of solution. The solution was placed in a thermostatically controlled reaction vessel and the temperature was adjusted to 35°C. The reaction vessel was provided with a pH electrode which was connected, via a pH-stat, with a burette filled with 0.2 N aqueous sodium hydroxide solution. The pH value was maintained at 5.5 by automatically adding an appropriate amount of aqueous sodium hydroxide solution to neutralize the gluconic acid formed in the course of the reaction.

Furthermore, the reaction vessel contained an oxygen inlet tube provided with a frit, through which 10 liters oxygen/hour were introduced. The fructose-glucose solution was now mixed with 250 mg of a carrier-bound GOD with a specific activity of 300 U/g. At intervals of 50 minutes, the glucose content of the solution was determined. After 100 minutes, 20% of the glucose initially present had been converted into gluconic acid.

In a further experiment, the above described process was repeated but with the addition of 2 mg of a catalase with a specific activity of 40,000 U/mg to the solution. After 100 minutes, the glucose conversion was 30% and after 200 minutes was 40%. In the case of still longer reaction times, the GOD was deactivated and could not be reactivated. In another experiment, instead of using carrier-bound GOD and dissolved catalase, there were added 250 mg of the GOD-catalase preparation A of Example 1. After 100 minutes, the conversion of glucose to gluconic acid was 50% and after 200 minutes was 70%. Inactivation of the enzyme did not occur and in the case of continued reaction, practically the whole of glucose was converted into gluconic acid.

Resolution of DL-Lysine Esters Using Immobilized Protease

A method has been developed by *H. Hirohara, S. Nabeshima and T. Nagase; U.S. Patent 4,108,723; August 22, 1978; assigned to Sumitomo Chemical Company, Limited, Japan* for optical resolution of DL-lysine alkyl ester or their acid-addition salts by contacting with a nonspecific immobilized protease.

It has been found that a nonspecific protease produced by bacteria of the genus *Streptomyces* exhibits a powerful esterase activity to a lysine alkyl ester having an unacylated nitrogen at the α position and is effective for resolution. Heretofore, proteases were thought to have no specificity on optical isomerism with respect to amino acid esters having no acylated nitrogen atom in the amino group. For these reasons, examples of optical resolution by use of proteases of microorganic origin are not very numerous. However, nonspecific proteases produced by bacteria of the genus *Streptomyces* are found to exhibit a very high activity to the L-form lysine ester as compared to the activity to the D-form isomer.

Furthermore, the present enzymatic reaction proceeds very efficiently in a slightly acidic pH range and substantially no nonenzymatic spontaneous hydrolysis proceeds. In the hydrolysis of lysine alkyl ester with the nonspecific protease produced by bacteria of the genus *Streptomyces,* the optimum pH is 5.7.

The process comprises contacting an aqueous solution of the DL-lysine alkyl ester or acid-addition salts thereof with a nonspecific protease having ability to resolve the DL-lysine alkyl ester or acid-addition salts thereof produced by culturing a microorganism of *Streptomyces sp.,* under a weakly acidic condition at a temperature lower than 60°C, and recovering L-lysine or acid-addition salts thus produced.

The substrate, DL-lysine alkyl ester or acid addition salts thereof, is prepared by any known method. The number of carbon atoms in the alkyl moiety of the lysine ester is preferably 1 to 6, e.g., methyl, ethyl, n-propyl, isopropyl, n-butyl, secondary butyl and tertiary butyl groups. Methyl and ethyl are most preferable.

Any of the bacteria of genus *Streptomyces* that produce nonspecific proteases can be effectively used insofar as it is able to produce an enzymatic component which exhibits activity to the lysine alkyl ester. Particularly, *Streptomyces griseus, S. fradiae, S. erythreus, S. rimosus* and *S. flavovirens* are advantageously used since they produce highly active lysine alkyl ester hydrolases.

With regard to the immobilization of the nonspecific protease, any of the known methods can be used. The requirement to be met is that the nonspecific protease immobilized on the insoluble carrier should retain an activity relative to the asymmetric hydrolysis of the DL-lysine alkyl ester or acid-addition salt thereof. Owing to the immobilization of the enzyme, the enzyme which has heretofore been used in a homogeneous reaction system in a solution and, as a natural consequence, discarded at the end of each batchwise use can be used repeatedly or continuously.

Thus, the immobilization of the enzyme brings about outstanding economic advantages such as the reduced amount of enzyme used, reduction in the amount of secondarily produced waste and simplification of the process involved in the purification of the product. Further details of the process are shown in the following examples.

Example 1: In 40 ml of water, 1 g of Sephalose 4B (Farmacia) was immersed in conjunction with 1 g of cyanogen bromide at pH 11 and 15°C for 15 minutes to activate the Sephalose 4B. Then, the mixture was held in contact with the enzyme pronase at pH 7.4 at room temperature for 2 hours, to afford an immobilized enzyme containing 23 mg of the enzyme. The activity of this immobilized enzyme to DL-lysine methyl ester dihydrochloride at 40°C was 2.5 μmols/mg/min, which corresponds to 40% of that of the enzyme in the form of a solution. This immobilized enzyme was packed in a jacketed column. With the column maintained at 30°C, a 10% solution of DL-lysine methyl ester dihydrochloride in a phosphate buffer solution at pH 6.85 was fed upflow through the lowest portion of the column at space velocity of 8.6 hr^{-1}.

By the quantitative analysis of the effluent from the column head for its ester content by the hydroxamic acid process, the conversion relative to the DL-form was found to be 40.5% (81% relative to the L-form). The effluent was mixed with a volume of alcohol several times as large to cause precipitation of salts which were removed by filtration. The filtrate was concentrated and was again mixed with a volume of alcohol about 15 times as large to produce a precipitate (1). Specific optical rotation, $[\alpha]_{546}$ 20°C, of the precipitate was found to be +15.6 (c=2, 6N HCl). This precipitate (1) was purified by recrystallization from water-alcohol to produce crystals (2) having specific optical rotation, $[\alpha]_{546}$ 20°C, of +19.7. When the filtrate remaining after the separation of the precipitate (1) was evaporated to expel methanol therefrom, the residue was found to have specific optical rotation, $[\alpha]_{546}$ 20°C of -6.7.

Example 2: In methanol, 2 g of a macroporous weakly acidic cation-exchange resin of acrylic acid-divinylbenzene copolymer (having an exchange group of –COOH and a particle size of 24 mesh) was esterified with thionyl chloride and converted into the hydrazide with hydrazine hydrate. In 1 N hydrochloric acid, this resin was allowed to react with 2.1 g of sodium nitrite at 2°C for 1 hour to produce the azide. An aqueous solution of the enzyme (containing 155 mg of the enzyme) prepared in advance from *Streptomyces fradiae,* was mixed with the azidized resin and agitated at pH 7.5 and 4°C for 4 hours to have the enzyme immobilized on the resin.

The immobilized enzyme was washed with a 0.2 M buffer solution and a 5 M NaCl solution and the amount of enzyme immobilized was found to be 47 mg/g of the carrier. A test for dependency of the enzyme activity upon pH revealed that the optimum pH value had shifted to a range of 6.5 to 6.7. In 20 ml of a 0.1 M potassium chloride solution, 4.12 g of DL-lysine ethyl ester dihydrochloride was dissolved. The immobilized enzyme was added to the solution and the mixture was agitated to 35°C with the pH kept at 6.6 by addition of a 1 N NaOH solution. After 40 minutes of the agitation, the immobilized enzyme was separated by filtration and the remaining solution was concentrated by evaporation to 5 ml.

To the concentrate was added 95 ml of ethanol. The precipitate which occurred was separated by filtration, washed and dried to give 1.67 g of L-lysine dihydrochloride, which was found to have specific optical rotation, $[\alpha]_{546}$ 20°C, of +19.1 (c=2, 6N HCl). The 95% ethanol solution of the filtrate was evaporated and the residue (slightly hygroscopic) was found by a test to have specific optical rotation, $[\alpha]_{546}$ 20°C, of -4.5 (c=2, 6N HCl).

PHARMACOLOGICAL USES

Inactivation of Aerosol Borne Pathogens

D.J. Kirwan and J.L. Gainer; U.S. Patent 3,849,254; Nov. 19, 1974; assigned to The University of Virginia have discovered that enzymatic reactions can be effected in an aerosol form in the presence of an immobilized enzyme. It had been believed that such reactions required a continuous liquid phase medium, and preferably an aqueous substrate. It has now been discovered that enzymatic reactions between aerosol substrate and immobilized enzyme reactor will proceed at a significantly faster rate than the same reaction in a continuous liquid medium.

A wide variety of carriers can be used for immobilizing the enzymes. For instance, suitable carriers include siliceous material such as glass, Fiberglas, ceramic, colloidal silica (such as Cab-O-Sil), wollastonite, dried silica gel, and bentonite: nonsiliceous metal oxides such as nickel oxide, aluminum oxide, natural and synthetic polymeric materials such as polyacrylamide, cellulose, polyacrylic acid, polyaminostyrene, polysaccharides, polymethylacrylic acid, collagen, polygalacturonic acid, polyaspartic acid, L-alanine and L-glutamine copolymers, maleic anhydride and ethylene copolymers, polyamide, ion exchange resins. Also carbon and platinum are suitable carriers.

A wide variety of enzymes can be immobilized onto the substrates, using any of the known immobilizing techniques. The enzymes can be used either as a single variety or combinations of multiple varieties for specialized purposes.

For instance, the enzymes suitably immobilized for use in this process include a wide variety of enzymes which may be classified under three general headings: hydrolytic enzymes, redox enzymes, and transferase enzymes. The first group, hydrolytic enzymes, include proteolytic enzymes which hydrolyze proteins, cabohydrases which hydrolyze carbohydrates, esterases which hydrolyze esters, lysozyme, nucleases which hydrolyze nucleic acid, and amidases which hydrolyze amides.

The second group are redox enzymes that catalyze oxidation or reduction reactions. These include glucose oxidase, catalase, peroxidase, lipoxidase, and cytochrome reductase. In the third group are transferase enzymes such as glutamic-oxaloacetic transaminase, transmethylase, phosphopyruvic transphosphorylase.

In effecting a reaction wherein the substrate is in an aerosol form the aerosol containing the enzymatically treatable substrate is passed over the immobilized enzyme. The immobilized enzyme may be in the form of a mesh screen, a fibrous pad, a membrane or sheet, honeycomb, particles or porous monolith. For instance, a tube may be filled with large particles and the aerosol forced through the tube whereby the particles of the aerosol are caused to pass over and around the immobilized enzyme particles thereby assuming intimate contact between the substrate and the immobilized enzymes. Suitably placed fans and/or baffles can be set into place to assure uniform or enhanced contact of the substrate with the immobilized enzymes.

The enzymatically active substrates which are catalytically acted upon by the enzymes fall into one of two categories: (1) pathogens which are inactivatable by the enzyme contact; and (2) reactants which can be enzymatically catalyzed. The first category, pathogens, includes the deleterious and harmful bacteria and viruses. The viruses include any RNA (ribonucleic acid) and DNA (deoxyribonucleic acid) viruses. To destroy a DNA containing virus, a deoxyribonuclease enzyme could be used in the immobilized form, and to destroy an RNA containing virus, a ribonuclease enzyme could be used. Specific viruses which are contemplated as inactivatable include adenovirus, poxvirus, reovirus, picornavirus, enterovirus, picornavirus rhinovirus, vaccinia, herpesvirus and hepatitis.

Many viruses which have been shown to be inactivatable by contact of the virus with an immobilized enzyme in aqueous solution, have now been found to be inactivatable when the substrate is in the form of an aerosol. For instance, the enzyme ribonuclease has been found to inactivate London and Hong Kong A2 viruses in aerosol form, whereas the literature indicates that many influenza viruses are not inactivatable in aqueous solutions by ribonuclease.

Inactivatable bacteria include *Staphylococcus, Streptococcus, Pneumococcus, E. coli, Pseudomonas* or the like. Both gram positive and gram negative bacteria have been shown to be inactivatable by the techniques of this process.

The pathogens are usually carried in aqueous droplets, sometimes containing other components such as mucus or the like. The droplet size is preferably 0.01 to 1.5 μ. The aerosol can be produced naturally such as is produced by human sneezing, coughing or even breathing, but may also be produced synthetically by a conventional atomizer.

Example: Influenza virus is transmitted through the air in drops or droplet nuclei in the 1 μ size range. A reactor was prepared to handle virus containing infectious droplets directly rather than having to scrub the air and treat the resulting solution. A solution containing virus was atomized to micron size with an air stream and the resulting aerosol passed through a reactor.

The reactor was formed by attaching enzymes to ceramic monoliths which are currently manufactured as catalyst supports in automotive exhaust systems. They have straight-through pores approximately 1 mm on a side and a geometrical surface to volume ratio of about 30 cm^2/cm^3. Ribonuclease, trypsin and lysozyme were attached to these materials using a γ-aminopropyltriethoxysilane agent followed by glutaraldehyde fixation of the enzyme to the silane agent. Enzyme loadings were about 1.5 mg/g of ceramic based on disappearance of enzyme from solution.

The system was then tested for its effectiveness for influenza (Type A2 Hong Kong) disinfection. The results are presented in the following table. The infectious virus concentration in the effluent from an active enzyme reactor to that from a blank reactor (heat denatured enzyme) treated in an identical fashion were compared. Differences between the two values were attributable to the effect of the active enzyme. RNase A at a 1 second residence time was found to destroy over 99% of the infectious particles entering the reactor. Sequential reactors, were also used, the first half having lysozyme or trypsin, while the second half contains immobilized RNase A, and the results are shown below.

Results of Room Temperature Aerosol Phase Studies Influenza Type A2 Hong Kong

Enzyme System	Air Residence Time (sec)	Percent Virus Removed Control*	Reactor
		.... %	
Trypsin (borine pancreatic)	1.0	79	79
Lysozyme	0.4	97	97
RNase A	0.4	74	97.2
RNase A	1.0	78	99.3
Trypsin/RNase A	1.0	78.6	97.9
Lysozyme/RNase A	1.0	78.6	97.9

*Reactor containing heat denatured enzyme.

Insoluble Urease for Removing Urea from Dialysate

Dialysis with an artificial kidney is common treatment for kidney failure or renal disease. The procedures using such artificial kidneys provide for the conducting of blood outside the body and across a semipermeable membrane system with a saline solution passing on the other side of the membrane. The saline solution used is known as dialysate. The undesirable waste products within the blood are caused to pass through the membrane into the dialysate by dialysis.

Once the dialysate solution has picked up the waste products of the blood, either the dialysate solution must be discarded or the concentration of the waste products must be reduced prior to the solution being recirculated through the artificial kidney. The dialysate in itself is relatively inexpensive. However, between

50 and 100 gallons of the dialysate solution will be passed through the artificial kidney for each treatment which substantially increases the cost. Therefore, it has been desirable to use only a small volume of the dialysate by reducing the concentration of the waste products and recirculating it back through the artificial kidney. The most difficult waste material to remove from the dialysate is urea.

Urease, sometimes supported on diatomaceous earth, has been used for this purpose. However diatomaceous earth does not make the enzyme completely insoluble in the dialysate. An immobilized urease has been developed by *L.B. Marantz and M.G. Giorgianni; U.S. Patent 3,989,622; Nov. 2, 1976; assigned to CCI Life Systems, Inc.* for use with an artificial kidney having a recirculating dialysate.

Urease is adsorbed in an aqueous environment on a material selected from aluminum oxide and magnesium silicate so that when contacted by liquid containing urea, the urea is converted into ammonium carbonate while the urease remains adsorbed on the material. The column used in this process is shown in the following Figure 13.1.

Figure 13.1: Cross-Sectional Vertical Section of Column

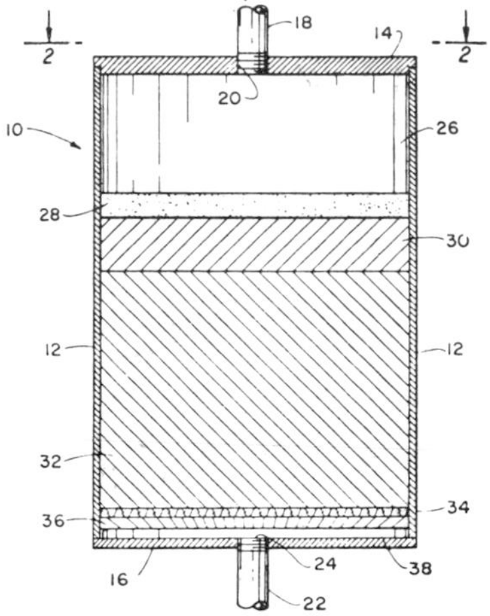

Source: U.S. Patent 3,989,622

The column **10** is constructed in the shape of a container having a circular side wall **12** closed at one end by wall **14** and closed at the opposite end by wall **16**. The dialysate enters the column from passage **18** which connects with opening **20** within the end wall **14**. The dialysate is to exit from the column through passage **22** which connects with opening **24** located in the end wall **16**. The open space **26** adjacent the end **14** receives the dialysate entering from **18** and serves to disperse the solution over the area of the column. The open space terminates into a first layer **28** of material which initially comprises a dry powder consisting of a mixture of finely divided urease and a urease retaining material of magnesium silicate or of aluminum oxide.

Each one of these materials could be used singularly or in combination, however, normally a single material of the group would be used and urease added. In use, the dialysate stream enters from **18** through **20** into **26** and through **28**. As the urease and urease retaining material are wetted, the urease is then adsorbed on the urease retaining material and becomes insoluble to water. The ability of the urease to hydrolyze urea is not impaired. Located immediately below the first layer **28** is the second layer **30** comprised of one or a mixture of the urease retaining materials of layer **28** not intermixed with the urease. In the event that any granules of urease are carried beyond layer **28**, this urease will be brought in contact with layer **30**, and be adsorbed upon these particles. If the second layer **30** is not employed this urease would then pass on through the column.

In operation, neither urease nor urea is found in the solution which has passed through layers **28** and **30**. The insolubility of the urease on aluminum oxide or magnesium silicate is independent of temperature between 0° and 50°C. Below 0°C the solution freezes, and above 50°C the urease becomes inactive. The insolubility of urease is also independent of the state of the urease as it is added to the layer material, or the rate of introduction to the layer of material.

In an artificial kidney system, it is also desirable to remove the ammonium carbonate which results from the conversion of urea by the urease. This may be accomplished by the use of zirconium phosphate ion exchange in fine particle form, shown as a third layer **32** located adjacent the second layer **30**. A filter **34** is located between the zirconium phosphate layer **32** and the flow director device **36**, to retain the powdered material. The function of the flow director device is to direct the dialysate flow to outlet opening **24** within the bottom wall **16**. A space **38** exists between the flow director and the outlet opening to permit the flow of dialysate solution to the passage **22**.

To obtain an indication of the degree of effectiveness of the urease retaining materials in this process, an aqueous solution of urease was passed through a number of fixed weight samples of alumina (aluminum oxide) and magnesium silicate of suitable mesh sizes and the excess urease was washed out. The results of these tests showed that the calcined and activated alumina such as produced by Aluminum Company of America (Alcoa), Reynolds Metals, or Kaiser hold more urease in active form/unit weight than similar quantities of tabular or hydrated alumina manufactured by one or more of the same companies.

The test showed that the more tabular and hydrated alumina used at the same mesh size, the greater the conversion of urea into ammonium carbonate. Furthermore, tests with different mesh ranges of tabular and hydrated alumina showed that the finer the mesh size of the alumina, the greater the conversion of urea into ammonium carbonate. In some mesh sizes, five times the amount of tabular or hydrated alumina were required to obtain the same percentage conversion of urea into ammonium carbonate as obtained from a unit amount of activated or calcined alumina.

Tabular and hydrated alumina can be ground very fine so that the conversion of urea to ammonium carbonate would be very high but the pressure drop characteristics of the material for a given liquid flow rate would not be as satisfactory for the shown embodiment as that for calcined or activated alumina. A typical mesh size of calcined alumina for use in this process falls within the range of

100 to 325 American Standard mesh. Alcoa utilizes the following designation for alumina: A Grade for calcined; C Grade for hydrated; and T Grade for tabular. Reynolds Aluminum uses RC numbers for calcined and RA numbers for activated alumina and Kaiser Aluminum uses KC numbers for calcined.

Fibrinolytic Compositions for Dispersal of Blood Clots

Enzyme systems with fibrinolytic and/or plasminogen activation (peptidolytic) activity, but with low caseinolytic activity have now been developed by *J. Green and M.A. Cawthorne; U.S. Patent 4,055,635; Oct. 25, 1977; assigned to Beecham Group Limited, England*. It was discovered that the formation of water-soluble complexes of proteolytic enzymes linked to polymeric substances can lead to the desired properties.

Accordingly, the process provides a fibrinolytic pharmaceutical composition in unit dosage form which comprises a water-soluble complex of a proteolytic enzyme linked covalently to a polymeric substance. The proteolytic enzyme used is preferably trypsin, due to its correct peptidase specificity, but the use of similar enzymes, for example, chromotrypsin, bromelain, and papain is also possible from the point of direct fibrinolytic activity. There may also be used complexes prepared from proteolytic enzymes commercially available, for example, pronase, a broad spectrum proteolytic enzyme, isolated from *Streptomyces griseus;* and brinase, which is isolated from *Aspergillus oryzae.* Both of these have fibrinolytic and plasminogen activation activity.

The polymeric substance is preferably one that is water-soluble and digestable by in vivo processes. Thus it is preferred to use polysaccharides such as dextran, dextrins, cellulose, or starch. Dextrans are preferred, especially those having a MW of 10,000 to 500,000, for example, those designated as T40 and T70. Some polysaccharides may be modified by reaction with, e.g., epichlorhydrin or it may be carboxymethyl-, hydroxyethyl- or aminoethyl-modified cellulose or a partially degraded starch.

It has further been found particularly suitable to use polymers that are modified saccharides or oligosaccharides, such as sucrose, dextrose or lactose. An especially useful polymer of this type is one that is commercially available under the trade name Ficoll, which is a sucrose epichlorhydrin copolymer. A Ficoll polymer with a MW of about 400,000 has been found to be suitable. It is also possible to use synthetic water-soluble polymers, such as polymers of polyvinyl alcohol and copolymers of maleic or acrylic anhydrides with ethylene, styrene, methyl vinyl ether, divinyl ether or vinyl acetate. One suitable range of water-soluble copolymers are the methyl vinyl ether/maleic anhydride copolymers sold under the trade name Gantrez AN, particularly Gantrez AN 119 (a copolymer of MW about 250,000).

The water-soluble enzyme complexes for use may be prepared by any of the known methods for linking enzymes to polymers provided that the resulting complex can be regarded as water-soluble. The enzyme is linked to the polymer in an amount such that the weight ratio of polymer:enzyme is preferably 2:1 to 50:1. For use, the fibrinolytic compositions of the process are to be prepared in a form adapted for injection, either in the form as sold or more usually after dissolution in an appropriate solvent.

Thus these compositions are preferably presented in sealed ampoules containing a predetermined amount of sterile freeze-dried enzyme complex.

Example: Preparation of Water-Soluble Enzyme Complex of Trypsin on Gantrez— Gantrez AN-119 (a methyl vinyl ether-maleic anhydride copolymer having a MW of 250,000) (10 g) was added to 300 ml of 0.2 M phosphate pH 7.4 buffer and the mixture was stirred for 5 minutes to produce a suspension. A solution of trypsin (1 g) in phosphate buffer (40 ml) was added and the mixture was mechanically stirred for 24 hours at 4°C. The resulting solution after considerable dilution with deionized water was purified by repeated ultrafiltration through a membrane that had a MW cut-off of 300,000. By this means free unreacted trypsin, which passed through the membrane was separated from the adduct, which was retained in the ultrafiltration cell.

This process also served to remove the phosphate buffer from the adduct and this was further accomplished by dialyzing the adduct against deionized water. The adduct was finally freeze-dried to yield 8.9 g of a white fluffy product. The protein content of the adduct was determined by the method of Lowry et al., using trypsin as standard, and was found to be 47 μg/mg. The peptidolytic activity of the adduct was measured by its esterolytic ability to release N-benzoyl arginine from N-benzoyl arginine ethyl ester. Using this assay each mg of the adduct had an activity equivalent to 125 μg of trypsin, therefore the adduct had a greater esterolytic activity and therefore a greater peptidolytic activity than free trypsin.

The caseinolytic activity of trypsin and the adduct were measured by the method of Remmert & Cohen, *J. Biol. Chem,* 181, 431-448, (1949). Each mg of adduct had an activity equivalent to 10 μg of enzyme and therefore the caseinolytic activity of the adduct was reduced relative to the activity of free trypsin.

The fibrinolytic activities of trypsin and the adduct were measured by their ability to lyse plasma clots. Human plasma containing a trace of I^{125}-fibrinogen was clotted with thrombin. The clots were then incubated in a solution of the enzyme or adduct and the release of I^{125} products from the clot into the medium was measured as shown in the table below. The adduct was relatively much more active than free trypsin; each mg of adduct being the equivalent of about 300 μg trypsin.

Comparison of Trypsin-Gantrez vs Trypsin

	Trypsin Equivalent (μg/mg)	Ratio
Protein content	47	1
Esterolytic cleavage of N-benzoyl arginine ethyl ester	125	2.5
Fibrinolysis of plasma clot	300	7.0
Caseinolytic test	10	0.2

Thus it will be noted that the water-soluble enzyme complex had fibrinolytic activity that was 7 times that of the free enzyme and an esterolytic activity 2.5 times that of the free enzyme, but a caseinolytic activity that was only one-fifth that of the free enzyme.

Fixed Enzymes for Hydrolyzing Polyribonucleotides

It is known to fractionate polyribonucleotides by action of enzymes, to obtain various polynucleotides and oligonucleotides. These enzymatic actions comprise essentially two stages, of which the first consists of subjecting the polyribonucleotides used to enzymatic hydrolysis by using a ribonuclease which cleaves the polymeric chain at the 5'-phosphate junction, giving rise to a 3'-phosphate group, and it is followed by a second stage, which is a dephosphorylation by an alkaline phosphatase. The enzymes successively used in such operations are in solution, so that it is necessary, at the end of the operations to separate the products obtained from the enzymes used, which are often very resistant so that these processes are difficult to apply and the yield of oligonucleotide obtained is very low.

It is the object of the process developed by *M-N. Thang and A. Guissani; U.S. Patent 4,039,382; August 2, 1977; assigned to Choay, SA, France* to provide matrices which permit the fractionation of polyribonucleotides into copolymers or into oligonucleotides, in a single operation, with excellent yields.

The polyribonucleotides used are selected from poly$(A_mU_l)_n$, poly$(A_{m'}C_l)_{n'}$, poly$(A_{m''}G_l)_{n''}$ when it is sought to obtain polymers A_nU_{OH}, A_nC_{OH}, and A_nG_{OH}; and they are selected from homopolynucleotides poly U, poly C, poly A and poly G or their copolymers, when it is sought to obtain oligonucleotides. U, C, A and G have their usual meaning, i.e., uridine, cytosine, adenine and guanosine, respectively. The process uses a matrix carrying simultaneously a nuclease and an alkaline phosphatase.

The insoluble support on which the enzymes are bound is selected from among nondenaturing supports effecting the irreversible physical adsorption of the enzymes, such as supports of glass or quartz beads, highly crosslinked gels notably of the agarose or, in particular, of the cellulose type. The amounts of active enzymes fixed on the support are between 0.001 and 0.15% of alkaline phosphatase, and between 0.01 and 0.05% of nuclease by weight with respect to the weight of the support.

To prepare these matrices carrying several different enzymes, the solid support previously activated by cyanogen bromide, for example, is placed in contact, in a suitable buffer solution, with a nuclease and an alkaline phosphatase, for a time and at a temperature which are interdependent, after which any unbound enzymes are removed by washing by means of a buffer solution identical with that used for the binding, and the free activated groups of the support are neutralized by means of a free amino organic base such as lysine, arginine, ethanolamine, aniline and the like.

Example: Production of Supported Alkaline Phosphatase and Ribonuclease on Cellulose—The cellulose is washed on sintered glass with distilled water, then water and a fresh 2% solution of cyanogen bromide (BrCN) are added so as to obtain a final concentration of 1.5% of BrCN, namely 50 mg of BrCN/g of dry cellulose. It is stirred at 4°C while maintaining the pH at 11 by gradual addition of 4 N NaOH (the operation lasting about 10 minutes), then it is filtered on sintered glass, with washing 3 times with a total of 10 volume of sodium bicarbonate. For immediate use, it is rinsed an additional 5 to 6 times with sodium bicarbonate.

For later use, the washings are done with water mixtures richer and richer in acetone, then the paste obtained is dried in a desiccator. Just before use, it is washed carefully with sodium bicarbonate in order to remove all traces of acetone. For binding of the enzymes, a solution containing 2.5 ml of activated cellulose (representing about 0.4 g of dry cellulose), 350 μg of alkaline phosphatase, representing 100 units and 40 μg of beef pancreatic ribonuclease, representing 80 units, in a 0.02 M Tris buffer at pH 8.3, is stirred gently for 16 hours at 4°C.

The unbound enzymes are removed by carrying out 5 successive washings followed by filtrations, in 5 times 5 ml of 0.02 M Tris buffer, pH 8.3, then it is resuspended in the same buffer. The amount of enzymes bound is 92.5 units out of 100 of alkaline phosphatase and 53 units of 80 of ribonuclease, namely 92.5% and 66% respectively, binding yield. The remaining free groups on the cellulose are then saturated, by addition of 0.1 M aniline in 0.1 M Tris buffer, pH 8.3, followed by stirring for 16 hours at 4°C. The unbound aniline is then removed by carrying out 5 successive washings followed by filtrations, in 5 times 5 ml of 0.1 M Tris buffer, pH 8.3.

The enzymes bound on the cellulose are then resuspended in 0.02 M Tris buffer, pH 8.3, and then the activity of the bound enzymes is measured by titration of the suspension and of the filtrate; there are thus found: alkaline phosphatase, 40 units in suspension and 25 units in the filtrate, namely a final yield of 43% of active bound enzyme; beef pancreas ribonuclease, 25 units in the suspension and 16 units in the filtrate, namely a final yield of 47% active bound enzyme.

Such immobilized enzymes may then be used for hydrolyzing polyribonucleotides of the formula $(A_m U_l)_n$, $(A_{m'},C_l)_{n'}$, or $(A_{m''}G_l)_{n''}$ which have been prepared by known methods from ADP, UDP, CDP and GDP in the presence of polynucleotide-phosphorylase. The polymer of formula $(A_m U_l)_n$, $(A_{m\text{-}207}C_l)_n$, or $(A_{m''}G_l)_{n''}$ is then subjected to the action of the bifunctional matrix which degrades it into a polymer of the formula A_nU-P, A_nC-P or A_nG-P under the action of a ribonuclease, and causes the dephosphorylation of the polymer into $A_n U_{OH}$ $A_n C_{OH}$, or $A_n G_{OH}$ by the action of alkaline phosphatase. It should be noted that the alkaline phosphatase only exerts its action on the terminal phosphate group, so that the presence of an excess of phosphatase is without importance, contrary to what occurs with ribonuclease whose degradation effect is capable of continuing to the monomer, that is to say beyond the desired degradations, if it is present in excess.

FOOD AND FEED USES

Malt Diastase Impregnated Laminates for Controlling Spoilage of Silage

An improved fermentation control-containing laminate for retarding the spoilage of silage and like materials on storage has been disclosed by *J.G. Forest and E. J. Czarnetzky; U.S. Patent 3,871,949; March 18, 1975.* This laminate comprises a polyvinylidene chloride layer and a layer of a Kraft paper impregnated with sodium sulfate and a malt diastase.

The polyvinylidene chloride layer of the above laminate is a polymer of vinylidene chloride and any of the conventionally known film-forming polyvinylidene chlorides or other nontoxic, water-insoluble, oxygen and carbon dioxide impervious resins of suitable molecular weights can be used. One suitable form of the polyvinylidene chloride useful in the laminate of this process is Saran. The second layer of this laminate comprises a Kraft paper layer having absorptive properies. Almost any type of Kraft paper or other strong and inexpensive paper can be used, the only condition being that it be absorptive to the active ingredients impregnated therein, i.e., the sodium sulfate and malt diastase.

The laminate can be prepared according to known techniques, as for example, by extruding a layer of polyvinylidene chloride onto a previously prepared Kraft paper, thereby laminating the polyvinylidene chloride layer to the Kraft paper layer. The laminate can range in thickness from a diaphanous film to a laminate which is several mm in thickness. The particular thickness is not critical and will be dependent upon the end use to which the laminate product is subjected. Generally the thickness of the laminate will range from 25 to 200 mils in thickness, with the polyvinylidene chloride layer ranging from 0.25 to 100 mils in thickness, preferably from 1.0 to 20 mils in thickness, and the Kraft paper layer ranging from 1.5 to 150 mils in thickness, preferably from 2.5 to 50 mils in thickness.

The active ingredients, i.e., the sodium sulfate and the malt diastase, can be applied to the Kraft paper layer in any conventional manner. Generally the amount of sodium sulfate per square foot of a 50 lb weight paper will range from 0.05 to 2.0 pounds per square foot. The amount of the malt diastase will generally range from 1 part to 5 parts of malt diastase to each 50 parts of sodium sulfate.

Example: A laminate was prepared from a polyvinylidene chloride layer (1½ mils thick) and a forty pound Kraft paper layer (4½ mils thick). A 20% solution of sodium sulfate containing malt diastase at a level of 1 part by weight to each 30 parts of sodium sulfate was applied to the Kraft paper side of the laminate. Absorption by the paper of the solution is such that the Kraft paper contained 6% by weight sodium sulfate, based on the weight of the paper. This laminate was then cut into sheets approximately 5" x 5" and inserted into quart containers such that the laminate formed a top-to-bottom divider. These containers were then filled with milo grain (having a moisture content of 18%) or with a high moisture content corn (having a 30% moisture content).

In another run, quart containers were filled with milo grain (having the moisture content of 18%) and high moisture content corn (having a 30% moisture content) each saturated with the sodium sulfate/malt diastase composition in the manner described in U.S. Patent 3,184,314. A control of the laminate was prepared in exactly the same manner with the exception that the Kraft paper layer was not impregnated with the sodium sulfate and malt diastase and was inserted into containers and subsequently filled with milo grain or high moisture content corn. Microaerophilic conditions were established for each of the samples and the containers were set at 79°F. After 5 days the containers containing the laminate of the process and the containers containing the milo grain and high moisture content corn saturated with the sodium sulfate/malt diastase composition were examined with results shown on the following page.

Sample	Results
Milo with laminate impregnated with sodium sulfate and malt diastase	Mold had grown all over the sides of the container as well as over the top, except for a distance of about ¼ to ½ inch next to the Kraft paper side of the laminate, both as observed from the outside of the container and as observed after emptying the container.
High moisture corn with laminate impregnated with sodium sulfate and malt diastase	The same results as were obtained above with the milo were observed, but the mold was much more luxuriant. However, mold was not observed on the Kraft paper side of the laminate similar to the above results.
Milo impregnated with sodium sulfate and malt diastase per se	No mold growth was observed.
High moisture corn impregnated with sodium sulfate and malt diastase	No mold growth was observed.
Control (laminate without sodium sulfate and malt diastase)	Molds of several types grew luxuriantly. The milo sprouted after ten days. Sprouting occurred in all cases with the milo. The high moisture corn did not sprout.

Immobilized Sulfhydryl Oxidase to Remove Cooked Flavor from Milk

There are processes known for producing ultrahigh temperature (UHT) sterilized milk having a prolonged shelf-life without the necessity of refrigeration. While UHT processing of milk has a slightly different meaning in the U.S. and Canada from that in Europe each involves treatment of milk at temperatures of at least 190°F for varying periods of time or at about 300°F for around 4 seconds. Milk when heat treated at such high temperatures (or even lower for more prolonged periods of time), develops an undesirable cooked flavor thus rendering the milk unmarketable, especially in the United States. Much research has been carried out to determine the nature of the cooked flavor and provide a means for its elimination.

It has been speculated by some workers that the cooked flavor of heated milk is due to the liberation of sulfhydryl groups in the milk. However, others have expressed doubts as to the significance of free sulfhydryl groups in milk.

In the process developed by *H.E. Swaisgood; U.S. Patent 4,053,644; Oct. 11, 1977; assigned to Research Triangle Institute*, this objectional cooked flavor is removed by contacting the heat treated fluid milk with an immobilized sulfhydryl oxidase enzyme.

The term cooked flavor refers to the chalky flat or insipid taste attendant fluid milk treated at temperatures generally in excess of 155°F. Such milk may be unpleasant both to taste and smell resembling boiled cabbage. While such a cabbagy smell may disappear within several days after treatment of the milk, the flat, chalky cooked flavor which eventually produces a cardboardy cooked flavor remains.

The immobilized sulfhydryl oxidase enzyme used may be prepared according to the procedure described in U.S. Patent 4,087,328, pp 6-8. The sulfhydryl oxidase enzyme obtained from raw whole milk and used according to the process catalyzes the conversion of sulfhydryls to disulfides. Apparently the oxidation of sulfhydryl groups in heated milk through contact with immobilized sulfhydryl oxidase eliminates the cooked flavor. Use of the immobilized sulfhydryl oxidase catalyst has the advantage that the enzyme is obtained from whole raw milk and is thus a natural constituent. Furthermore, use of the enzyme in its immobilized form means that there is no additive to the milk being treated.

Example 1: A biocatalytic reactor containing immobilized sulfhydryl oxidase is formed by filling a 6 mm diameter glass column with 1.4 cc of sulfhydryl oxidase enzyme immobilized by attachment to glass beads having a pore size of about 2000A. The reactor is a glass column which contains the immobilized enzyme and is adapted to receive heat treated milk (inlet) which flows through the column and bed of immobilized enzyme with appropriate means for subsequent discharge of the milk after contacting the immobilized enzyme. The reactor can be placed in such a manner to provide a continuous flow process by directly coupling to a UHT processing unit.

Example 2: Bovine milk which is subjected to a temperature of 300°F for about 4 seconds has a noticeable cooked flavor and is passed through the biocatalytic reactor described in Example 1 at a temperature of 30° to 35°C and a flow of about 40 ml/hr. The unpleasant odor and cooked flavor present in the milk prior to treatment is eliminated. The percentage oxidation of sulfhydryl groups in heat treated milk and the stability of the reactor described above is illustrated by the data in the following table.

Stability of a Sulfhydryl Oxidase Reactor Operating with UHT Sterilized Skim Milk*

Day of Operation	Duration of Operation (hr)	Oxidation (%)
Fresh	5	45
5	3	34
10	5	31

*A 1.4 cc reactor was operated at a flow rate of 40 ml/hr at 35°C.

It is preferred to use glass beads of a pore size greater than 700A to immobilize the enzyme for use in treating milk. Smaller sizes yield unfavorable results, and optimum activity is achieved with a pore size of about 2000A. One should avoid contacting the milk and immobilized enzyme at high temperatures since high temperatures may cause the enzyme to lose activity. Contact temperatures of 30° to 35°C may be used without significant loss of enzyme activity. As the activity of the immobilized enzyme drops through normal use, reactivation may be achieved through contacting the immobilized enzyme with a fresh aqueous solution of ferrous sulfate.

The consequences of contacting fluid heat treated milk with immobilized sulfhydryl oxidase enzyme may be more far reaching than simply removal of the cooked flavor as such treatment also appears to prevent development of other undesirable flavors and destabilization of milk proteins. Treatment of high temperature treated milk provides a means whereby aseptic and/or sterile fluid milk having a taste and flavor comparable to ordinary pasteurized homogenized milk may be obtained.

Continuous Production of Beer

M. Moll, G. Durand and H. Blachere; U.S. Patent 4,009,286; Feb. 22, 1977; assigned to Groupement d'Interet Economique, France have developed a continuous process for producing beer under sterile conditions. The process uses the replacement of fresh hops with a hop extract that can be sterilized plus the use of supported enzymes in two stages of the process, the fermentation tower and the treatment tower. Review of this process is limited to the use of the immobilized enzymes in these two steps.

The fermentation tower used is a fermenter of the homogeneous or heterogeneous type where yeast is in liquid medium or immobilized on a support which is inert with respect to the fermentation. This support may be formed either of diatoms, mixed with yeast, or polyvinylchloride in granules mixed with yeast or other feed plastics and yeast. In the case of using an inert support, there is used a fixing circuit which is traversed by the suspension of yeast which, in connective contact with the support is retained by physical-chemical bonds. As a result of repetition of passages of the yeast there takes place, not a filtration, but an adsorption on the support permitting a relatively large heterogeneous particle size of the order of several mm.

In this way a clogging of the support by sedimentation is avoided. In order to effect this circulation, a conduit is provided which makes it possible to fix the yeast by recycling on the adsorbent support. Means are also provided in this tower for the sterile introduction of additional yeast as needed and sterile sampling of the wort.

Within the fermentation tower a certain proliferation of the yeast cells is favored by controlling the initial composition of the wort coming from a sterilization chamber and the flow of the pump feeding the fermentation tower. Under these conditions, the beer obtained has a composition in volatile products which is very close to a conventional beer, as shown by the following table.

	Beer, mg/l	
Volatile Products	Continuous	Conventional
Ethyl acetate	22.4	22-33
Isobutyl acetate	0.1	0.08-0.10
Propanol	42.4	5-45
Isobutanol	34.8	25-40
Isoamyl acetate	0.8	2.0-3.0
Isoamyl alcohol	87.3	75-100
Ethyl caproate	0.2	0.07-0.2
Ethyl caprylate	1.2	0.2-1.0
Caproic acid + phenylethyl acetate	3.2	3-10
β-phenylethanol	18.2	20-40
Caprylic acid	3.3	2-4
Capric acid	0.7	0.3-1.0
Total volatile products Σ	214.6	–
Diacetyl	traces	0.05-0.15
2,3-pentanedione	0	0

It is well-known, as a matter of fact, that a large part of the volatile substances present in a beer come from the metabolism of the amino acids utilized by the yeast cells in order to multiply. If there were not this multiplication of cells, the final product emerging from the fermentation tower would be different.

The treatment tower is provided with a support of natural or synthetic organic polymers, brick, silica, glass, sand, compounds having a base of silica, in particular halogenated silicas and amino silicas, previously activated clay materials mixed with a protease and the beer passes into this treatment tower on which proteases are fixed, such as papain, pepsin, chymotrypsin, ficin, bromelin and others. The liquid flows through the treatment tower and the proteases are retained by physical-chemical bonds. The treatment tower is for the purpose of reducing the amount of diacetyl, decreasing the quantity of sulfur compounds, and improving the organoleptic quality of the product.

ENZYME-CONTAINING DETERGENTS

Enzymes Encapsulated in Detergent Beads

M.H. Win, W.A. De Salvo and E.J. Kenney; U.S. Patents 3,858,854; Jan. 7, 1975; and 4,016,040; April 5, 1977; latter assigned to Colgate-Palmolive Co. have prepared a granular free-flowing nondusting product of high enzyme content for use in detergents by spraying a blend of molten nonionic detergent and enzyme concentrate into cool air to form tiny spherical beads.

Preferably the nonionic detergent is a waxy water-soluble material having a MP of at least 50°C and not above 60°C, which contains a hydrophilic polyethylene oxide chain attached to a hydrophobic radical. Suitable types of materials are those which are ethylene oxide adducts of long chain alkanols (e.g., alkanols of 12 to 20 carbons) or long chain alkyl phenols (e.g., phenols having alkyl side chains of 8 to 18 carbons).

Other nonionic detergents are ethylene oxide adducts of long chain alkyl thiophenols; ethylene oxide adducts of monoesters of hexahydric alcohols and inner ethers thereof such as sorbitan monolaurate, sorbitol monooleate and mannitan monopalmitate, ethylene oxide adducts of polypropylene glycol; and ethylene oxide adducts of partial esters (e.g., monoesters) of fatty acids and glycerol; and ethylene oxide adducts of esters or amides of long chain fatty acids.

The enzyme mixed with the molten nonionic detergent may be an extremely fine, often substantially, impalpable powder. In a typical powdered enzyme preparation the particle diameter is mainly below 0.15 mm, generally above 0.01 mm, e.g., about 0.1 mm; for example, as much as 75% or more of the material may pass through a 100 mesh sieve.

Commercial enzyme preparations are usually mixtures of an organic enzyme concentrate with a solid diluent salt, such as calcium sulfate, sodium chloride, sodium sulfate and other inert materials (e.g., clay). The process however, finds its greatest utility when the organic enzyme concentrate (as produced, and substantially free of diluent salt) is blended directly with the substantially anhydrous molten nonionic detergent and the heated blend is sprayed into a cool atmosphere to produce the tiny spherical beads.

Examples of proteolytic enzymes which may be used in this process include pepsin, trypsin, chymotrypsin, papain, bromelin, collagenase, keratinase, carboxylase, aminopeptidase, elastase, subtilisin and aspergillopeptidase A and B. Preferred enzymes are subtilisin enzymes manufactured and cultivated from special strains of spore-forming bacteria, particularly *Bacillus subtilis.*

There are indications that the beads have a skin of highly crystalline nonionic detergent encapsulating a mixture of nonionic detergent, enzyme concentrate and carrier salt. Thus when a sharp scalpel blade was pressed diagonally against the top of a bead (under a dissecting microscope, using reflected light) the blade appeared to pass easily through the outer portion of the bead and to be deflected by a core (whose thickness is greater than that of the outer portion), giving a cut skin fragment which (when viewed with a transmitted polarized light) was seen to have a highly crystalline structure.

Example: 19 parts of a proteolytic enzyme concentrate (in the form of a dark brown salt-containing fine powder whose enzyme content is 4 AU/g) are mixed with 81 parts of a molten nonionic detergent having a 48°C MP (Plurafac A-38) to form a free-flowing liquid slurry, which is then sprayed, at a temperature about 5°C above the MP of the nonionic detergent, and under pressure, through a single fluid nozzle (of standard type, having a small outlet orifice and having, just upstream of the orifice, a stationary four-vaned core which is arranged to impart a swirling motion to the liquid). During spraying most of the hot slurry is continuously recirculated from the nozzle to the heated storage vessel from which it is pumped continuously to the nozzle.

The spray emerges continuously from the nozzle into a circular tower about 8' in diameter and about 40' high, to the bottom of which there is supplied a continuous stream of air at a temperature of about 13°C, so that the cool air flows upward into contact with the sprayed droplets, cooling and solidifying them within seconds after they leave the nozzle. The solid beads are collected at the base of the tower. The resulting free-flowing tan beads have the following screen analysis:

Screen Mesh	Opening, mm	% Remaining on Screen
10	2.0	0
20	0.84	trace
40	0.42	0.7
60	0.25	33.2
80	0.177	39.8
100	0.149	14.9
200	0.074	11.2
270	0.053	0
325	0.044	0
Pan	0.044	0

The dust content of the beads is about 18 ppm. To make a laundry detergent for use in the automatic machine washing of clothes, one part of the beads is blended with 99 parts of spray dried hollow white granules of heavy duty built detergent composition having the following approximate screen analysis:

Screen size	10	20	40	60	80	100	-100
% retained	0.2	2.7	29.4	40.4	13.1	6.0	8.2

The enzyme-containing beads are not noticeable, to the naked eye, in the resulting mixture. The spray dried granules of the heavy duty built detergent composition have the following approximate overall composition: 10% sodium linear tridecylbenzenesulfonate; 2% of the ethoxylation product made from ethylene oxide and primary alkanols of C_{14} to C_{15} chain length, the ethoxylation product containing 11 mols of oxyethylene/mol of alkanol; 2% of sodium soap of a mixture of 3 parts of tallow fatty acids and 1 part of coconut oil fatty acids; about 8.5% of total moisture; 34% of phosphate solids; 7% of sodium silicate solids ($Na_2O:SiO_2$ mol ratio 1:2.35); 0.5% of sodium carboxymethylcellulose; 0.2% of water-soluble polyvinyl alcohol; and the balance sodium sulfate together with small amounts of fluorescent brighteners.

The granules of built detergent composition are prepared by spray drying a heated aqueous slurry containing the ingredients described and having a solids content of about 60% (i.e., the slurry has a total moisture content of about 40%). This aqueous slurry is prepared by vigorous agitation in a crutcher and is at a temperature of about 60°C; in making the aqueous slurry the phosphate (supplied as a powder of anhydrous pentasodium tripolyphosphate) is added last, just before spraying. Then the aqueous slurry is sprayed into a spray tower to which heated air, at a temperature well above the boiling point of water, is fed to evaporate off the water, in conventional manner.

Enzymes Combined with Aminated Polysaccharides

Stabilized enzyme-containing detergents have been developed by *F.L. Diehl, E. Zeffren and E.J. Milbrada; U.S. Patents 3,944,470; March 16, 1976;* and *4,011,169; March 8, 1977; both assigned to The Procter & Gamble Company.* These stabilized enzymes comprise the enzyme and an aminated polysaccharide having 0.10 to 2% by weight of nitrogen in its elemental composition, the weight ratio of the polymer to enzyme being 500:1 to 1:1.

These stabilized enzymes are obtained by codissolving in an essentially aqueous medium an enzymatic ingredient in combination with the aminated polysaccharides. The detergent compositions contemplated comprise (1) 5 to 99.9% by weight of an organic surface-active agent selected from the group consisting of anionic, nonionic, zwitterionic and ampholytic detergents and mixtures thereof; and (2) 95 to 0.1% by weight of a mixture comprising (a) an enzyme suitable for use in detergent compositions; and (b) an aminated polysaccharide having 0.01 to 2% by weight of nitrogen in its elemental composition, the weight ratio of enzyme to aminated polysaccharide being from 1:500 to 1:1.

The enzymes suitable for the detergent compositions include all those which degrade or alter or facilitate the degradation or alteration of soil and stains encountered in cleansing situations so as to either remove more easily the soil or stain from the fabric or object being laundered or make the soil or stain more removable in a subsequent cleansing step. Both degradation and alteration improve soil removability. Well-known and preferred examples of these enzymes are proteases, lipases and amylases. Extensive lists of specific enzymes which may be included in this detergent composition are given in the complete patent.

The operable aminated polysaccharides contain 0.01 to 2% nitrogen in their elemental composition, and are prepared by reacting a polysaccharide starting material with an aminating agent such as an amine, preferably a tertiary amine, or a quaternary ammonium compound.

These N-containing substituents preferably impart a cationic charge to the aminated polysaccharide, when they are maintained at a pH which is equal to or below their pK_a. The aminated polysaccharide component can be made using known condensation techniques. Preparation of aminated polysaccharides used in this process is as follows: 1.485 g epichlorohydrin was mixed with 0.945 g trimethylamine in 20 ml water, stirred for 8 hours at room temperature, and stored at room temperature overnight. Volatile components were removed under reduced pressure.

30 g of methylcellulose and 0.225 g sodium hydroxide in 600 ml water were added and stirred for 20 hours at 40°C. The reaction mixture was then stored at ambient conditions and a pH of 7 for 48 hours. The aminated polysaccharide was recovered by freeze drying. The yield was 29.2 g, i.e., more than 90%. Kjeldahl analysis showed 0.02% nitrogen (AP1). Using the same procedure, but with 5.94 g epichlorohydrin, 3.78 g trimethylamine and 0.9 g sodium hydroxide, 39 g of product were obtained having a nitrogen content of 0.17% (AP2).

The detergents preferably used in these compositions are the alkali metal alkyl benzene sulfonates, in which the alkyl group contains 9 to 20 carbons in straight chain or branched chain configuration. Especially valuable are straight chain alkyl benzene sulfonates in which the average of the alkyl groups is about 11.8 carbons, commonly abbreviated as $C_{11.8}$ LAS. The following examples clearly show that a significant stabilization and enhancement of enzymatic activity in an aqueous solution occurs from the use of enzymes in conjunction with the aminated polysaccharides.

			----Relative Enzymatic Activity at 100°F----				
Example	**Additive**	**Enzyme**	**t = O**	**1 Week**	**2 Weeks**	**4 Weeks**	**6 Weeks**
	None	Peroxidase (EC 1.1.11.17)	1	0.70	0.64	0.60	-
1	AP1	Peroxidase (EC 1.1.11.17)	2.40	1.84	1.74	-	-
2	AP2	Peroxidase (EC 1.1.11.17)	2.47	1.85	1.71	-	-
3	Cato* (saturated solution; <1%)	Peroxidase (EC 1.1.11.17)	1.28	-	1.18	1.26	-
4	JR-IL** 1%	Peroxidase (EC 1.1.11.17)	1.07	0.97	0.81	0.75	-
			-----Relative Esterase Activity at 100°F-----				
	None	Alcalase (EC 3.4.4.16)	1	-	-	-	0.13
5	AP2 (0.5%)	Alcalase (EC 3.4.4.16)	1.46	-	-	-	0.41
			-------- Relative Protease Activity --------				
	None	Alcalase (EC 3.4.4.16)	1	-	-	-	-
6	AP2 (0.5%)	Alcalase (EC 3.4.4.16)	3.3	-	-	-	-

*Cationic potato starch. **Cationic cellulose.

Note: Peroxidase experiments done in phosphate buffer pH 6.0-7.0. Alcalase experiments done in phosphate trishydroxymethylaminomethane HCl at pH 7.8.

Enzymes in Film Carriers for Automatic Dishwasher Detergents

The disclosure by *D.L. Richardson and F.J. Mueller; U.S. Patent 4,115,292; Sept. 19, 1978; assigned to The Procter & Gamble Company* relates to compositions which permit the convenient, safe and efficient incorporation of enzymes into detergent products. More particularly, this process relates to dispersing an enzyme into a water-soluble resin and forming the resin, e.g., by casting or extruding, into a sheet. The sheet is then dried, if necessary, and, if required, cut into "ribbons" for incorporation into the detergent product. The ribbons of enzyme-dispersed resin may be mixed with a detergent composition in granular, viscous liquid, paste or gel form. The resulting mixture may be used directly in the washing process, particularly in an automatic dishwasher, or it may be incorporated within a water-soluble packet, for easy and convenient dispensing.

Resins suitable for use with the enzymes are water-soluble, are film-formers, and, preferably, are organic. The water-soluble resin should have proper characteristics, such as strength and pliability, to permit machine handling. Preferred water-soluble resins include polyvinyl alcohol, cellulose ethers, polyethylene oxide, starch, polyvinylpyrrolidone, polyacrylamide, polyvinyl methyl ether-maleic anhydride, polymaleic anhydride, styrene-maleic anhydride, hydroxyethylcellulose, methylcellulose, polyethylene glycols, carboxymethylcellulose, polyacrylic acid salts, alginates, acrylamide copolymers, guar gum, casein, ethylene-maleic anhydride resin series, polyethyleneimine, ethyl hydroxyethylcellulose, ethyl methylcellulose, hydroxyethyl methylcellulose. Lower molecular weight water-soluble, polyvinyl alcohol film-forming resins are preferred.

The weight percent of water-soluble, film-forming resin in the final articles of the process is from 15 to 75%, preferably 20 to 50%. The second essential component of this process is the enzyme which is dispersed in the water-soluble, film-forming resin. This component comprises an effective amount of a single enzyme, such as a proteolytic or amylolytic enzyme, or an enzyme mixture, suitable for use in a cleaning, particularly an automatic dishwashing, composition. Such enzymes are disclosed in U.S. Patent 3,627,688, McCarty et al., Dec. 14, 1971. It is preferred that the enzyme component constitute from 10 to 60% by weight, more particularly from 15 to 50% by weight of the articles.

Where the articles are incorporated into a detergent composition for use in automatic dishwashers, a preferred enzyme component is an effective amount of an enzyme mixture which comprises a proteolytic enzyme having a proteolytic activity of 80 to 100% of maximum activity when measured at pH 12 using the Anson Hemoglobin method carried out in the presence of urea, and an amylolytic enzyme. Particularly preferred proteolytic enzymes are the strain numbers C372, (NCIB 10 317); C303, (NCIB 10 147); C367, (NCIB 10 313); and C370, (NCIB 10 315). Another preferred enzyme is that cultivated from the microorganism of *Bacillus firmus* strain NRS 783, (NRRL B 10017).

Particularly preferred proteolytic enzymes are those cultivated from strains NCIB 10147 and NRRL B 10017 and their mixtures. Preferred, commercially available proteolytic enzymes for use in the compositions are available under the tradenames SP-72 (ESPERASE) and SP-88, produced by Novo Industri. The selected proteolytic enzyme is combined with an amylolytic enzyme, derived from bacteria or fungi. Preferred amylolytic enzymes are those which exhibit an amylolytic activity of greater than 50% of maximum activity when measured at pH 8 by the SKB method at 37°C.

Particularly preferred amylolytic enzymes are those cultivated from the strains of *Bacillus licheniformis* NCIB 8061; NCIB 8059; ATCC 6334; ATCC 6598; ATCC 11945; ATCC 8480; and ATCC 9945A. A particularly preferred, commercially available amylolytic enzyme, is produced and distributed under the tradename SP-95 (Termamyl), by Novo Industri.

The proteolytic and amylolytic enzymes described above, are combined in a ratio of 4:1 to 1:4 by weight, and the preferred enzyme mixture is present in the detergent composition in an amount such that the detergent composition has an amylolytic activity of at least 150 Kilo Novo units/kg, preferably at least 300 Kilo Novo units/kg, and a proteolytic activity of at least 6.0 Anson units/kg. The enzyme-containing articles of the process may also contain other additives, such as plasticizers, solids, and suds modifiers.

The detergent compositions which may be used in conjunction with the articles of the process contain water-soluble anionic, nonionic, ampholytic or zwitterionic surface-active agents, or mixtures of these surfactants. Where the detergent compositions are for use in automatic dishwashers, it is preferred that they contain a nonionic surface-active agent, particularly an alkoxylated nonionic surface-active agent, wherein the alkoxy moiety is ethylene oxide, propylene oxide, or mixtures thereof. The surface-active component should comprise at least 0.5% by weight of the detergent composition.

However, by choosing an appropriate nonionic surfactant system, along with small quantities of materials such as solubilizers, thickeners, and the like, stable paste-form compositions containing up to 55% of the nonionic surfactant system may be prepared. Preferred detergent compositions contain 3 to 30% of the nonionic surfactant.

Example 1: 100 g of Monsanto Gelvatol 20-30 polyvinyl alcohol (MW 10,000 and percent hydrolysis 85.5 to 88.7) was dissolved in 233.3 g of distilled water. A Waring blender was used to wet out the polyvinyl alcohol. The mechanical heat of mixing raised the temperature of the mixture to 130°F. 100 g of Novo SP-72 Tergitol (enzyme) slurry (1:9 approximate weight proportions of enzyme powder, including small amounts of sodium sulfate and sodium chloride, to surfactant, 8 KNPU/g proteolytic activity) was then added into the Waring blender jar as agitation was being maintained. The resulting mixture was allowed to stand for about 20 minutes allowing for deaeration, then poured onto a flat Lucite plate, and spread to a 0.080 inch (80 mils) thickness using a wire-wound rod.

After air drying for 24 hours, the film was separated from the Lucite, turned over, and allowed to air dry for an additional 4 hours. The weight of the mix which was actually delivered onto the Lucite sheet, and the weight of the ultimately dry film was then determined. The dried film weighed 189.24 g and had the approximate composition of: polyvinyl alcohol water-soluble resin, 48.7 wt %; Novo SP-72 Tergitol (enzyme) slurry, 48.7 wt %; and water, 2.6 wt %. The resulting enzyme-containing film was cut into segments with an area of 1 square inch and had a thickness of 80 mils.

Results substantially comparable to those of Example 1 can also be obtained when the film is cut into ribbons, tapes, wedges, or wafers having one dimension not less than about 3 mm, 5 mm, 10 mm, 15 mm, and 20 mm in size, and being about 1,000 mils, 750 mils, 500 mils, 100 mils, 50 mils, 15 mils and 5 mils thick.

Excellent performance is obtained when polyvinyl alcohol water-soluble, film-forming resin is replaced by cellulose ethers, polyethylene oxide, starch, polyvinylpyrrolidone, polyacrylamide, polyvinylmethylether-maleic anhydride, polymaleic anhydride, and polystyrene-maleic anhydride in comparable weight proportions. Excellent performance is also obtained when the enzyme dispersed in the film-forming resin is an SP-88-Termamyl enzyme mixture (1:1 proportion of protease to amylase, 8 KNPU/g proteolytic activity and 8 KNU/g amylolytic activity) in comparable weight percentages to Example 1.

When the plasticizers such as glycols, glycerol, sorbitol, triethanolamine and urea are incorporated into the above films from about 10 to about 30% by weight, improved handling and elongation characteristics of the above film are obtained. Addition of solid extenders, such as granular starch, urea, and sodium sulfate in amounts about 30% by weight permits the production of a less expensive film which exhibits excellent physical and handling characteristics and is more quickly soluble.

Example 2: A detergent composition for use in automatic dishwashers is formulated in the following manner. The composition of Example 1 is cut into articles about 0.75" square and about 0.2 g of the articles are combined with about 24 g of granular detergent composition having the following formulation:

Component	% by Wt
Ethylene oxide/propylene oxide condensate of trimethylol propane (Pluradot HA-433 Wyandotte)	10.00
Monostearyl acid phosphate (contained in Pluradot HA-433 Wyandotte)	0.30
Sodium cumene sulfonate	10.00
Sodium carbonate	20.00
Sodium bicarbonate	10.00
Silicate solids ratio: $SiO_2:Na_2O$ = 2.0	20.00
$(NaPO_3)_{21}$	2.00
Sodium sulfate	Balance to 100%

COMPANY INDEX

INVENTOR INDEX

U.S. PATENT NUMBER INDEX

NOTICE

ANTIOXIDANTS 1979
RECENT DEVELOPMENTS

by William Ranney

Chemical Technology Review No. 127

Antioxidants prolong the useful life of countless products including plastics, elastomers, resins, lubricants and foodstuffs.

Once oxidation of an organic substrate begins, a chain reaction of autocatalytic oxidation ensues, ultimately causing gross changes in properties. An antioxidant will inhibit the initial oxidation or suppress the subsequent chain reaction. When two antioxidants with diverse actions are combined, the overall synergistic effect can be more than merely additive.

Over the years the antioxidant field has progressed from simple phenols to those of much more complex and sophisticated structures having a high specificity.

The 256 processes in *Antioxidants* detail manufacturing technology and compositions applicable to many fields. The partial table of contents below gives chapter headings and **examples of some** subtitles. The number of processes per topic is in parentheses.

ISBN 0-8155-0747-X **372 pages**

FOOD PROCESSING ENZYMES 1979
RECENT DEVELOPMENTS

by Nicholas D. Pintauro

Food Technology Review No. 52

This book presents developments in enzyme technology according to pertinent food categories, as listed below. Subject matter is organized to demonstrate means to solve a manufacturing problem, to improve storage, or to provide greater convenience in recipe preparation.

Enzyme technology is still in its infancy, but chemical elucidation of protein structure and binding mechanisms of enzymes with substrates are beginning to unfold as researchers probe for answers to problems of enzyme reactions. The 185 processes described in this book attest to the progress already made in various sectors of the food industry. Chapter headings and **examples of some** important subtitles are given in the condensed table of contents below. Numerals in parentheses specify the number of processes per topic.

ISBN 0-8155-0748-8 **420 pages**

EMULSIFIERS AND EMULSIFYING TECHNIQUES 1979

by Jeanne C. Johnson

Chemical Technology Review No. 125

Emulsifiers have become indispensable in the manufacture of all kinds of better-quality products running the gamut from thixotropic emulsions for fighting forest fires to emulsifiers for flavoring oils. This book comprises about 250 processes dating from January 1973 on preparation and applications of emulsifiers and emulsifying techniques. While a large segment covers food processing, a considerable number deal with the petroleum, petrochemical and polymer industries, with roughly a third concerned with chemical, cosmetic, pharmaceutical, ink, coating and other industries.

Food technologists will find interesting new ideas in this book; it is also of value to petroleum engineers for flooding emulsions in oil recovery, to specialty chemical researchers for compounding new formulations, and to cosmeticians.

Most emulsions described conform to the definition of an emulsion as a two-phase system of two incompletely miscible liquids, one dispersed in the other. A few processes, however, apply to micelle formation or solubilization.

Following is a condensed table of contents, with chapter headings and some subtitles. Numbers in parentheses indicate the number of processes per topic.

ISBN 0-8155-0740-2 **448 pages**

VITAMINS SYNTHESIS, PRODUCTION AND USE 1978

Advances Since 1970

by Charles S. Sodano

Chemical Technology Review No. 119

Vitamins are complex organic substances that occur naturally in plant and animal tissues. Vitamins are essential for proper functioning of the metabolic processes, and their lack produces deficiency diseases, which in the Middle Ages, were largely responsible for the short average life span of around 35 years. Scurvy, produced by lack of dietary vitamin C, is the classic example of a deficiency disease and is most often reported in past centuries. Pernicious anemia, much feared as late as the 1940s, responds dramatically to oral and injectable doses of vitamin B_{12}.

While eating diversified foods is an acceptable form of insurance against vitamin deficiencies, it is not easy to eat wisely and at the same time cheaply. Also, many individuals will not consume foods that are distasteful to them. This is where vitamin supplements and vitamin-enriched foods [often by government decree] fill a real need. With the introduction of many slimming diets, a new era of vitamin consciousness and demand for vitamin supplements seems to have arisen.

This book is designed for the manufacturing chemist and reviews the newer syntheses and practical formulations, essential for the production and consumption of most vitamins. A partial and condensed table of contents follows here. Chapter headings and examples of subtitles are given.

ISBN 0-8155-0728-3 **305 pages**